Dedicated to
Sherry and Linda

Nuclear Receptors and Genetic Disease

Edited by

Thomas P. Burris

Gene Regulation
Lilly Research Laboratories
Lilly Corporate Center
Indianapolis, Indiana, USA

and

Edward R. B. McCabe

Department of Pediatrics
UCLA School of Medicine
Los Angeles
California, USA

ACADEMIC PRESS

A Harcourt Science and Technology Company

San Diego San Francisco New York Boston
London Sydney Tokyo

Academic Press
A Harcourt Science and Technology Company
Harcourt Place, 32 Jamestown Road, London NW1 7BY, UK
http://www.academicpress.com

Academic Press
A Harcourt Science and Technology Company
525 B Street, Suite 1900, San Diego, California 92101-4495, USA
http://www.academicpress.com

ISBN 0-12-146160-2

Library of Congress Catalog Number: 00-105498

A catalogue record for this book is available from the British Library

Typeset by Paston PrePress Ltd, Beccles, UK
Printed and bound in Great Britain by MPG Books Ltd, Cornwall, UK

00 01 02 03 04 05 MP 9 8 7 6 5 4 3 2 1

Contents

Plate section appears between pages 52 and 53

List of Contributors

Alan Adams Department of Basic Chemistry, Merck Research Laboratories, RY 80N-C31, Rahway, NJ 07065-0900, USA

George F. Allan Department of Reproductive Therapeutics, R.W. Johnson Pharmaceutical, Research Institute, 1000 Rte 202 South, Raritan, NJ 08886, USA

Aria Baniahmad Genetic Institute, University of Giessen, Heinrich Buff Ring 58–62, Giessen, 35392, Germany

Albert O. Brinkman Department of Endocrinology and Reproduction, Erasmus University Rotterdam, The Netherlands

Alex J. Brown Department of Pharmacological/Physiological Science, St. Louis University School of Medicine, St. Louis, MO 63104-1004, USA

Thomas P. Burris Gene Regulation, Lilly Research Laboratories, Lilly Corporate Center, DC-0434, Indianapolis, IN 46285, USA

George P. Chrousos Section on Pediatric Endocrinology, Developmental Endocrinology Branch, NICHD, National Institutes of Health, Bldg. 10, Rm. 10N262, 10 Center Drive MSC 1862, Bethesda, MD 20892, USA

Arthur C.-K. Chung Department of Cell Biology, Baylor College of Medicine, One Baylor Plaza, Houston, TX 77030, USA

Austin J. Cooney Department of Cell Biology, Baylor College of Medicine, One Baylor Plaza, Houston, TX 77030, USA

Uwe Dressel Genetic Institute, University of Giessen, Heinrich Buff Ring 58–62, Giessen, 35392, Germany

Alex Elbrecht Department of Molecular Endocrinology, Merck Research Laboratories, PO Box 2000, RY 80N-C31, Rahway, NJ 07065-0900, USA

Guido Jenster Department of Urology, Erasmus University Rotterdam, PO Box 1738, The Netherlands

Tomoshige Kino Section on Pediatric Endocrinology, Developmental Endocrinology Branch, NICHD, National Institutes of Health, Bldg. 10, Rm. 10N262, 10 Center Drive MSC 1862, Bethesda, MD 20892, USA

Dennis M. Kraichely Department of Pharmacological/Physiological Science, St. Louis University School of Medicine, St. Louis, MO 63104-1004, USA

David M. Lonard Department of Cell Biology, Baylor College of Medicine, One Baylor Plaza, Houston, TX 77030, USA

Paul N. MacDonald Department of Pharmacological/Physiological Science, St. Louis University School of Medicine, 1402 S Grand Blvd, St. Louis, MO 63104-1004, USA

Edward R. B. McCabe Department of Pediatrics, UCLA School of Medicine, 10833 LeConte Ave, Los Angeles, CA 90095-1752, USA

David E. Moller Department of Molecular Endocrinology, Merck Research Laboratories, RY 80N-C31, Rahway, NJ 07065-0900, USA

Zafar Nawaz Department of Cell Biology, Baylor College of Medicine, One Baylor Plaza, Houston, TX 77030, USA

Shawn D. Seidel Environmental Toxicology Program, University of California, Riverside, CA 9252-0314, USA

Frances M. Sladek   Environmental Toxicology Program, University of California, 5419 Boyce Hall, Riverside, CA 9252-0314, USA

Jan Trapman   Department of Pathology, Erasmus University Rotterdam, The Netherlands

Eric Vilain   Departments of Human Genetics and Pediatrics, UCLA School of Medicine, Los Angeles, CA 90095-1752, USA

Alessandro Vottero   Section on Pediatric Endocrinology, Developmental Endocrinology Branch, NICHD, National Institutes of Health, Bethesda, MD 20892, USA

Preface

The history of the nuclear hormone receptor superfamily is the history of biomedical research in the twentieth century. Early in that century the technologies were those of chemistry, and the lipophilic steroid hormones were isolated. The actions of these hormones were studied using the physiological tools available at the time. Subsequently, investigators applied the methods of cell biology and biochemistry to localize the hormones to the nucleus and to identify the highly specific nuclear receptors within hormone-responsive tissues. Observations indicated that the transit of the receptor-bound hormone from the cytoplasm to the nucleus was followed by binding of the receptor complex to DNA, and investigations began to include the approaches of molecular biology. Ensuing research showed that these hormones and their nuclear protein complexes influenced the transcription of specific messenger RNAs and their protein products. The tools of molecular genetics were used to clone and characterize hormone-responsive genes, permitting identification of hormone response elements within the promoters of these genes. Additional members of the nuclear hormone receptor superfamily have been identified by methodologies developed in the course of the Human Genome Project. The three-dimensional organization of individual hormone–receptor complexes has been solved by structural biologists. Most recently, interactions between proteins in the complex are being elucidated by the nascent field of proteomics.

The investigations describing the normal roles of the individual members of the nuclear hormone receptor superfamily represent beautiful examples of systems biology research. Systems biology is a concept that emerged toward the end of the twentieth century and represents a reconsideration of physiology in the context of modern technologies. This concept may also be referred to as functional genomics. Whatever it is called, the purpose is to understand the functional integration of molecular processes within individual cells, whole organs and complete individuals. Implicit in such investigations is the acknowledgement that such systems are not static, but change during development and in response to external influences. Therefore, such research must consider the developmental biology and communication mechanisms of the system.

The history of the nuclear receptors is replete from its beginning with the concepts of the role of these molecules in an integrated system within cells and the organism, including the developmental biology of these molecules and their roles in communication. A principal theme in this field has been the importance of hormone–receptor complexes in the normal development of all aspects in the organism and in the functional integration of the neuroendocrine axis. The latter represents an organized mechanism for receiving and transducing information to permit the timely response of the organism to external influences such as nonspecific stress, mediated, for example, by adreno-cortical hormones. In addition, this axis is involved in the dynamic processes of sex determination, sexual development, ovulation and pregnancy. The developmental influence of these hormone receptors extends well beyond the confines of the hypo-thalamic–pituitary–adrenal/gonadal axis to include the execution of the basic body plan including limb formation, the integration of carbohydrate, fat and energy metabolism, and the regulation of calcium utilization. Fundamental to the execution, regulation and integration of what may appear superficially to be disparate processes is the essential unifying characteristic of communication, and the steroid hormones and their receptors are critical molecular components of essential biological communications systems.

A catch phrase in biomedical research toward the end of the twentieth century was the importance of translating information 'from the bench and the bedside'. This book examines the state of our understanding regarding not only the normal biology of selected nuclear receptors, but also the diseases associated with the genetic disruptions of these normal activities. We have become well aware of the importance of astute clinical observation in expanding our knowledge base beyond the limits that might be otherwise imposed by anticipation of the effects of genes and their mutations, and their normal and mutant gene products. Experiments of nature, spontaneous in humans and spontaneous or induced in model organisms, are described in each of the chapters in this volume. Although the molecular explanation(s) for each of the observed phenotypes may not be rigorously proven at this time, the phenotypes give us novel directions for future investigation.

Additional fundamental concepts in modern biology and genetics are the relationships between structure and function, and between genotype and phenotype. Exquisite structural and functional investigations of the nuclear hormone receptors have dissected the molecular basis for these interactions within individual proteins, establishing basic principles that may be generalized to additional superfamily members based on similarities in protein sequence. These investigations, however, have also demonstrated unique features within specific proteins. Regarding the relationships between genotype and phenotype, while some appear obvious, upon more detailed investigation, others are not explained by our current level of knowledge and may be more complex than anticipated. For example, identical mutations do not always produce identical phenotypes among individuals in different families or even within the same family. We know the importance of modifying genes in model organisms and are beginning to elucidate these modifiers in humans. Studies of the nuclear receptors are consistent with another emerging theme in human genetics: even diseases inherited in a simple mendelian fashion have phenotypes that are, in fact, complex traits.

Much work remains to be done. We have known for some time of the dimeric nature of most family members, but we are still examining how the distinct components communicate with one another. We are still completing the inventory of additional proteins involved in the functional complexes and their networking regulatory cascades. Identification of these additional gene products will provide better insight into the modifiers influencing phenotypes. Details of structure–function relationships are deficient for many of the nuclear receptors. Ligands for the orphan receptors are, by definition, unknown, and whether they are ligand-responsive or true orphans will require additional investigation. Knowledge about the extent of the influence of nuclear hormone receptors on genes involved in cell-specific regulatory networks is also incomplete. This area will benefit from the rapidly emerging chip technology that will eventually permit examination of a receptor's influence on the complete catalog of the transcripts within an individual cell, organ or developing organism.

This volume is not intended to be the final discussion of the role of nuclear receptors in health and disease. It is a discussion of our knowledge at this time. As such, it is intended to stimulate future research, while providing the current state of the evidence base for clinical practice involving nuclear hormone receptors.

T.P. Burris
E.R.B. McCabe

Chapter 1

The Nuclear Receptor Superfamily

Thomas P. Burris

INTRODUCTION AND HISTORICAL PERSPECTIVE

Identification of steroid hormones

Nuclear receptors regulate a myriad of biologic events in a variety of organisms ranging from the nematode to the human. These transcription factors, many of which function as receptors for lipophilic hormones, control differentiation, development, homeostasis and behavior by directly regulating the expression of select target genes. The field of nuclear receptor biology has its origin in the characterization of a subclass of these receptors known as the steroid receptors.

In the early part of the twentieth century, significant effort was expended attempting to isolate lipophilic hormones based on their physiologic activity. In a landmark paper, Allen and Doisy (1923) described the biochemical course of their initial purification of a lipid-soluble hormone with estrogenic activity. Between 1929 and 1930, several groups isolated a pure crystalline estrogenic hormone (Butenandt, 1929; D'Amour and Gustavson, 1930; Dingenmanse *et al.*, 1930; Doisy *et al.*, 1930). Independently, both Doisy and Butenandt proposed the correct chemical formula, $C_{18}H_{22}O_2$, for this hormone. This steroid, dubbed 'folliculine' by Butenandt, was later renamed estrone. Corner identified a second lipophilic ovarian hormone extracted from the corpus luteum with a distinct series of physiologic actions (Corner and Allen, 1929). Within 5 years of Corner's description of this hormone, which he termed 'progestin', four laboratories were able to crystallize the active hormone, progesterone, and determine its steroidal structure (Allen and Wintersteiner, 1934; Butenandt and Westphal, 1934; Butenandt *et al.*, 1934; Hartmann and Wettstein, 1934; Slotta *et al.*, 1934a,b; Wintersteiner and Allen, 1934). Male sex steroid hormones, such as androsterone (Butenandt, 1931) and testosterone (David *et al.*, 1935), were also isolated about the same time. It was soon demonstrated that the synthetically derived androgen steroids had biological activity identical to those purified from biologic tissue (Butenandt and Harrisch, 1935; Ruzicka and Wettstein, 1935). Steroid hormones of adrenal origin, corticosteroids, were also identified during this era (Reichstein and Shoppee, 1943; Simpson *et al.*, 1953; Simpson and Tait, 1955).

Nuclear Receptors and Genetic Disease
ISBN 0-12-146160-2

Before the identification of the steroid hormones, another hydrophobic hormone, thyroxine, derived from extracts of the thyroid gland with the ability to induce metamorphosis in amphibians (Gudernatsch, 1912) was crystallized and determined to be derived from iodinated tyrosine residues (Kendall, 1915). The role of lipophilic vitamin A derivatives in development and differentiation had also been well characterized (Sporn *et al.*, 1976); however, the similarity of the mechanism of action of all of these diverse hormones would not be fully appreciated until the 1980s when their receptors were cloned, identifying all the receptors as belonging to the nuclear hormone receptor superfamily.

Discovery of steroid receptors

Although it had been postulated that the distribution of hormone 'receptors' may be responsible for the selective actions of various steroids in target tissues, direct characterization of the receptors would await development of radiolabeled hormones. Glasscock and Hoekstra (1959) and Jensen and Jacobson (1960, 1962), using tritiated estrogens, found that these hormones were retained selectively by estrogen-responsive tissues, such as the uterus, leading to the concept of estrogen receptors within target tissues. Using autoradiography, the vast majority of signal was localized within the nucleus; thus 'estrogen receptors' appeared to be contained within the nucleus in contrast to other hormone receptors, which appeared to be associated with the plasma membrane (Jensen *et al.*, 1967). Biochemical studies indicated that most of the estrogen receptor (ER) was associated with the nuclear–myofibrillar fraction of rat uterus and was most likely a protein (Noteboom and Gorski, 1965). The nuclear localization of ER, along with studies indicating that the receptor was specifically associated with chromatin (King *et al.*, 1966; Maurer and Chalkley, 1967; Shyamala and Gorski, 1967; Teng and Hamilton, 1968; Mainwaring and Peterken, 1971; Spelsberg *et al.*, 1971) was consistent with the previous suggestion that estrogen action may be associated with the control of RNA synthesis (Mueller *et al.*, 1958).

Utilizing sucrose gradients, Toft and Gorski (1966) were the first to isolate a steroid hormone receptor. These investigators purified a protein from the soluble fraction of rat uteri that specifically bound to estrogenic steroids, but not nonestrogenic steroids. Consistent with receptor theory, the 'binding entity' was not detected in tissues that were unresponsive to the actions of estrogens. Further studies suggested that the ER exists in cells in two forms: a cytoplasmic 9.5S protein and a nuclear 5S protein (De Sombre *et al.*, 1967; Shyamala and Gorski, 1967). These observations led to the two-step theory of the mechanism of action of the ER (Jensen *et al.*, 1967; Shyamala and Gorski, 1967; Gorski *et al.*, 1968; Jensen *et al* 1968) (Fig. 1.1). This theory suggested that estrogens enter the cell and first interact with the large 9.5S 'untransformed' cytoplasmic ER. This interaction would initiate a conformational change leading to migration of the complex to the nucleus. This 'transformation' results in the appearance of the 5S nuclear ER. It was also suggested that the 9.5S receptor is composed of subunits that may separate upon entry of the complex into the nucleus. The two-step theory proved to be a general mechanistic pattern for receptors of other classes of steroids (Jensen and De Sombre, 1972).

Mechanism of action of steroid receptors

The proposed entry of the steroid receptor complex into the nucleus along with data suggesting association of the receptor with chromatin led investigators to assess the

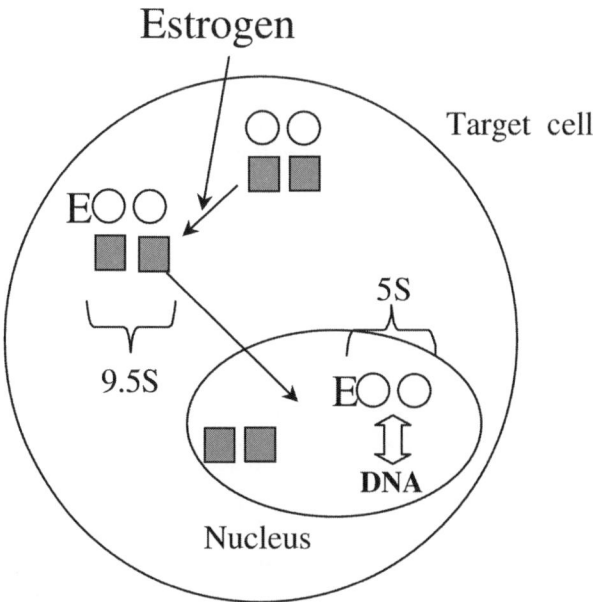

Fig. 1.1 The two-step theory of estrogen action. This theory suggested that estrogens enter the cell and first interact with the large 9.5S 'untransformed' cytoplasmic estrogen receptor (ER). This interaction would initiate a conformational change leading to migration of the complex to the nucleus. The 'transformation' results in the appearance of the 5S nuclear ER. It was also suggested that the 9.5S receptor is composed of subunits that may separate upon entry of the complex into the nucleus. The two-step theory proved to be a general mechanistic pattern for receptors of other steroids.

ability of several steroid receptors to bind directly to DNA. Initial studies indicated that several steroid receptors have significant affinity for DNA; however, it appeared that there was relatively nonselective affinity for double-stranded DNA (King and Gordon, 1972; Musliner and Chader, 1972; Toft, 1972; Andre and Rochefort, 1973; Beato *et al.*, 1973; Yamamoto and Alberts, 1974). Interestingly, a steroid-resistant mouse lymphoma cell line provided insight into the significance of the DNA-binding activity of steroid receptors. Both the resistant and parental steroid-responsive cell line retained the ability to bind to the glucocorticoid ligand, dexamethasone; however, glucocorticoid receptor (GR) isolated from the resistant cell line exhibited significantly reduced affinity for DNA. These data suggested that DNA-binding activity was essential for biologic activity of the steroid receptor and that the steroid and DNA binding activities of the receptor were separable (Gehring and Tomkins, 1974; Yamamoto *et al.*, 1974). Although the significance of the DNA-binding activity was apparent, the nonspecific affinity for the receptors for double-stranded DNA was perplexing. Yamamoto and Alberts (1975) suggested, based on an analogy to prokaryotic DNA-binding proteins, that steroid receptors act by binding to a small number of high-affinity binding sites within the genome that directly mediate hormone responsiveness; however, the receptors also maintain relative low affinity for nonspecific DNA, which masks the detection of the high-affinity sites.

Specific sites for action of steroids within the genome was suggested by the work of

Ashburner (1971, 1972, 1973, 1974), revealing that the insect steroid hormone, ecdysone, selectively induced six major *Drosophila* polytene chromosome band early puffs within minutes of treatment. This suggested that the steroid induces transcription of a limited number of targeted genes. The first demonstration of steroid-induced RNA synthesis was made by Mueller *et al.* (1958); however, the form of RNA that was specifically stimulated was not determined until some time later. Clearly, steroid hormones act to increase RNA synthesis (Greenman *et al.*, 1965; Wicks *et al.*, 1965; Means and Hamilton, 1966; O'Malley *et al.*, 1968) by stimulating RNA polymerase activity (Gorski, 1964; Glasser *et al.*, 1972; Mohla *et al.*, 1972). The stimulatory effects of steroids such as estrogens on RNA synthesis are essential for the actions of this steroid since most of the cellular changes induced by estrogens are blocked by pretreatment with an RNA synthesis inhibitor such as actinomycin D (O'Malley *et al.*, 1968; DeAngelo and Gorski, 1970). The observation that steroid hormone-induced increases in the synthesis of specific proteins in target tissues are preceded by increased RNA synthesis (Kenney *et al.*, 1965; DeAngelo and Gorski, 1970) led to the proposal that steroid hormones may specifically increase transcription of a few targeted messenger RNAs (mRNAs). Indeed, it was soon demonstrated that transcription of specific mRNAs was directly stimulated by estrogens and progestins (Comstock *et al.*, 1972; Means *et al.*, 1972; Rosenfeld *et al.*, 1972; Chan *et al.*, 1973) and was the rate-limiting step in steroid-dependent induction of protein synthesis. These effects of steroids on the transcription of target genes were promptly confirmed utilizing a model for GR action involving the hormone-dependent induction of expression of murine mammary tumor virus (MMTV) RNA (Parks *et al.*, 1974; Ringold *et al.*, 1975, 1977; Scolnick *et al.*, 1976; Young *et al.*, 1977).

The identification and cloning of several steroid-regulated genes led to direct investigation of the mechanisms of their regulation. The elements required for steroid regulation are contained within the target genes themselves since transcription of several cloned genes including MMTV, α_{2u}-globulin, growth hormone, lysozyme and ovalbumin, can be induced by steroid hormone after transfer of the genes to cells expressing the cognate receptors (Buetti and Diggelmann *et al.*, 1981; Hynes *et al.*, 1981; Kurtz, 1981; Ucker *et al.*, 1981; Fasel *et al.*, 1982; Robins *et al.*, 1982; Dean *et al.*, 1983, 1984; Lai *et al.*, 1983; Renkawitz *et al.*, 1984). The 5' flanking sequences of the target genes have been shown to be essential for hormone responsiveness and by simply transferring the 5' flanking sequence of a target gene to a heterologous gene it can be rendered hormone responsive. Furthermore, deletion of regions within the 5' flanking sequences of the chimeric constructs allows for localization of the regions responsible for hormone responsiveness (Huang *et al.*, 1981; Lee *et al.*, 1981; Robins *et al.*, 1982; Buetti and Diggelmann, 1983; Chandler *et al.*, 1983; Dean *et al.*, 1983; Hynes *et al.*, 1983; Majors and Varmus, 1983; Karin *et al.*, 1984a; Ostrowski *et al.*, 1984; Renkawitz *et al.*, 1984; Slater *et al.*, 1985). The response elements also conferred hormone responsiveness to a heterologous gene in a position and orientation independent manner consistent with the element functioning as a hormone-dependent enhancer (Chandler *et al.*, 1983; Payvar *et al.*, 1983; Ostrowski *et al.*, 1984). These regions within the 5' flanking sequences of the regulated genes that accord hormone responsiveness were also demonstrated to be high-affinity binding sites for steroid hormone receptors (Govindan *et al.*, 1982; Mulvihill *et al.*, 1982; Pfahl, 1982; Compton *et al.*, 1983; Payvar *et al.*, 1983; Scheidereit *et al.*, 1983; Yamamoto *et al.*, 1983; Jost *et al.*, 1984; Karin *et al.*, 1984b; Renkawitz *et al.*, 1984; Scheidereit and Beato, 1984).

Characterization of the receptor binding sites within the promoters of several steroid-regulated genes led to the identification of consensus hormone response elements

(HREs; sequence-specific DNA elements to which the steroid receptor binds to mediate the hormone-responsive expression of the gene) for various steroid receptors including the GR (Scheidereit *et al.*, 1983; Karin *et al.*, 1984b; Scheidereit and Beato, 1984), progesterone receptor (PR) (Mulvihill *et al.*, 1982; von der Ahe *et al.*, 1985) and ER (Jost *et al.*, 1984, 1985; Walker *et al.*, 1984; Klein-Hitpass *et al.*, 1986). Interestingly, the PR, GR and androgen receptor (AR) share identical HREs, with the hormone specificity dictated by the pattern of tissue-specific expression of the various receptors (von der Ahe *et al.*, 1985; Strahle *et al.*, 1987, 1989; Ham *et al.*, 1988). Several HREs from various hormone-regulated genes are shown in Fig. 1.2. The dyad symmetry of the HREs shown in Fig. 1.2 suggests that steroid receptors bind to DNA as dimers.

EREs

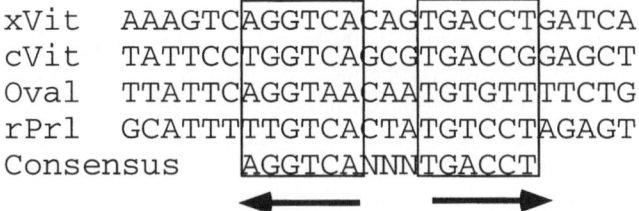

```
xVit      AAAGTCAGGTCACAGTGACCTGATCA
cVit      TATTCCTGGTCAGCGTGACCGGAGCT
Oval      TTATTCAGGTAACAATGTGTTTTCTG
rPrl      GCATTTTTGTCACTATGTCCTAGAGT
Consensus      AGGTCANNNTGACCT
```

GREs
PREs

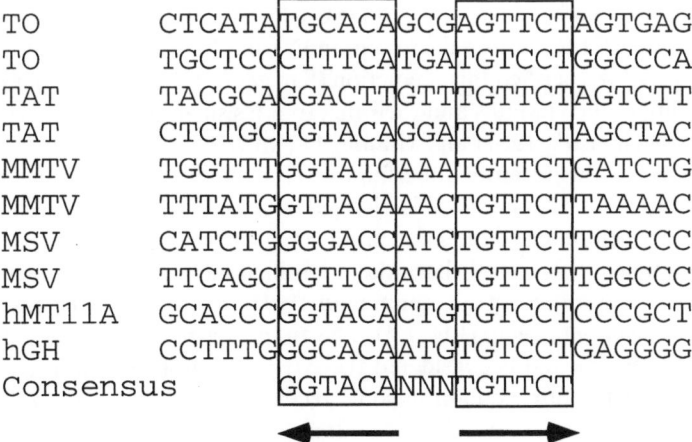

```
TO        CTCATATGCACAGCGAGTTCTAGTGAG
TO        TGCTCCCTTTCATGATGTCCTGGCCCA
TAT       TACGCAGGACTTGTTTGTTCTAGTCTT
TAT       CTCTGCTGTACAGGATGTTCTAGCTAC
MMTV      TGGTTTGGTATCAAATGTTCTGATCTG
MMTV      TTTATGGTTACAAACTGTTCTTAAAAC
MSV       CATCTGGGGACCATCTGTTCTTGGCCC
MSV       TTCAGCTGTTCCATCTGTTCTTGGCCC
hMT11A    GCACCCGGTACACTGTGTCCTCCCGCT
hGH       CCTTTGGGCACAATGTGTCCTGAGGGG
Consensus      GGTACANNNTGTTCT
```

Fig. 1.2 Examples of several hormone response elements (HREs) identified in steroid receptor target genes. The dyad symmetry of the HREs suggested that steroid receptors bind to DNA as dimers. ERE, estrogen response element; GRE, glucocorticoid response element; hGH, human growth hormone; hMT11A, human metallothionein; MMTV, murine mammary tumor virus; MSV, murine sarcoma virus; Oval, chicken ovalbumin; PRE, progesterone response element; rPrl, rat prolactin; TAT, tyrosine aminotransferase; TO, tyrosine oxidase; cVit, chicken vitellogenin, xVit, *Xenopus* vitellogenin.

NUCLEAR RECEPTOR STRUCTURE

The domain structure of nuclear receptors was suggested by experiments in which purified GR was subjected to limited proteolysis yielding fragments of the protein that retained individual functional properties of the native receptor. Three discernible domains of the GR can be identified after digestion that contain either the DNA-binding activity, steroid-binding activity or an 'immunogenic' domain (Wrange and Gustafsson, 1978; Carlstedt-Duke *et al.*, 1982). Cloning of the complementary DNAs (cDNAs) of several nuclear receptors would demonstrate that this discrete domain structure is a hallmark of the superfamily.

Characterization of nuclear receptor cDNAs

Molecular cloning of the GR in the mid 1980s provided the first glimpse into the genetic organization of a nuclear receptor (Hollenberg *et al.*, 1985). Based on the predicted amino acid sequence, data concerning the domain nature of the receptor, and homology to the v-*erbA* oncogene, Weinberger *et al.* (1985) predicted the localizations of the DNA-binding, steroid-binding and immunogenic domains within the GR. Receptors for all of the steroid receptors were soon cloned, including the ER (Walter *et al.*, 1985; Green *et al.*, 1986; Greene *et al.*, 1986; Krust *et al.*, 1986), the PR (Conneely *et al.*, 1986, 1987; Loosfelt *et al.*, 1986; Gronemeyer *et al.*, 1987; Misrahi *et al.*, 1987), the mineralocorticoid receptor (MR) (Arriza *et al.*, 1987; Patel *et al.*, 1989) and the AR (Chang *et al.*, 1988; Govindan *et al.*, 1988; Lubahn *et al.*, 1988; Trapman *et al.*, 1988; Tilley *et al.*, 1989).

Significant homology between the GR, ER and the v-*erbA* oncogene led to the prediction that the c-*erbA* gene may encode a receptor for a steroid or related molecule (Weinberger *et al.*, 1985; Green *et al.*, 1986). Cloning of the cDNA for c-*erbA* and functional analysis of the expressed protein, indicating that it encodes the receptor for thyroid hormone (TR), verified this prediction (Sap *et al.*, 1986; Weinberger *et al.*, 1986). Receptors for lipophilic vitamins and their metabolites such as retinoic acid (Giguere *et al.*, 1987; Petkovich *et al.*, 1987) and vitamin D (McDonnell *et al.*, 1987; Baker *et al.*, 1988) also belong to this receptor superfamily. Thus, lipophilic hormones with diverse chemical structures, such as the steroids, thyroid hormones and some vitamins, function through receptors belonging to the same family (Fig. 1.3).

Sequence alignment of some of the earlier identified members of the nuclear receptor superfamily led to the definition of six subregions (regions A through F) based on degree of homology (Krust *et al.*, 1986) (Fig. 1.4). The amino-terminus of the receptors is most divergent of the regions and is considered the A/B domain; it functions as a hormone-independent transactivation domain in some receptors. The size of the A/B domain is quite variable and ranges from several hundred amino acids in length for the steroid receptors to only a few amino acids in some of the nonsteroid nuclear receptors.

The C region is the most well conserved region; it is rich in cysteines and basic amino acid residues. The positions of the cysteines are absolutely conserved in nearly all members of the nuclear receptor superfamily (Fig. 1.5). Exceptions to this conservation include a limited number of nuclear receptors that lack a classic DNA-binding domain, such as dosage sensitive sex reversal AHC critical region on chromosome X, gene 1 (DAX-1) and short heterodimer partner (SHP). The cysteine-rich organization of region C bears resemblance to the zinc finger region of the *Xenopus laevis* transcription factor IIIA (TFIIIA) (Ginsberg *et al.*, 1984; Miller *et al.*, 1985), suggesting that this region of

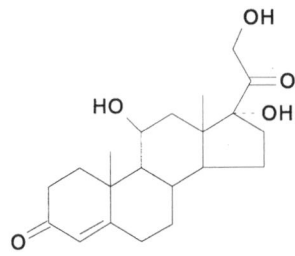

21
20
11 12 18 17
1 19 11 13 C D 16
2 9 14 15
10 A B 8
3 5 7
4 5 7

Steroid nucleus

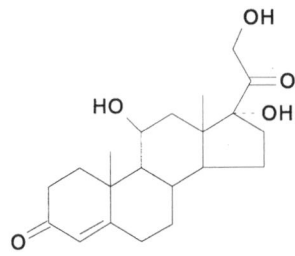

Cortisol
Glucocorticoid receptor (GR)

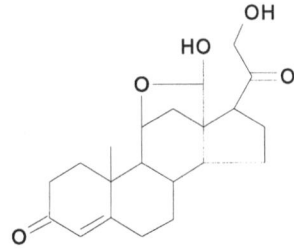

Aldosterone
Mineralocorticoid receptor (MR)

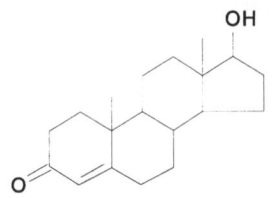

Testosterone
Androgen receptor (AR)

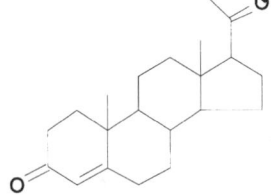

Progesterone
Progesterone receptor (PR)

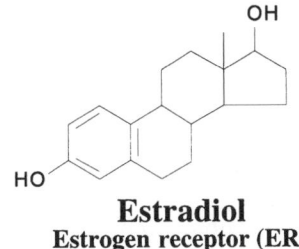

Estradiol
Estrogen receptor (ER)

(a)

Fig. 1.3 Chemical structures of various ligands for members of the nuclear hormone receptor superfamily. (a) Type I nuclear receptor ligands. *Continued.*

3,5,3′-L-triiodothyronine
Thyroid hormone receptor (TR)

1,25-dihydroxyvitamin D$_3$
Vitamin D receptor (VDR)

9-*cis* retinoic acid
Retinoid X receptor (RXR)

All-*trans* retinoic acid
(b)   Retinoic acid receptor (RAR)

15-deoxy-$\Delta^{12,14}$-prostaglandin J$_2$
Peroxisome proliferator-activated receptor γ (PPARγ)

Fig. 1.3 *continued.*   (b) Type II nuclear receptor ligands.

the nuclear receptors may also form a zinc finger-like DNA-binding domain (Weinberger *et al.*, 1985).

The D region, often called the 'hinge domain' because of its localization between the DNA-binding domain and the ligand-binding domain, is a relatively short domain with a low degree of conservation among receptors. It has functions in both transcriptional silencing and modulation of DNA-binding activity. The carboxy-terminal E domain is the second most conserved region among nuclear receptors and contains many overlapping functional regions including ligand binding, transactivation, dimerization and nuclear localization. The F region is present in only a limited number of nuclear receptors and has an unclear function.

Three regions within the receptors (I, II and III) have been identified based solely on

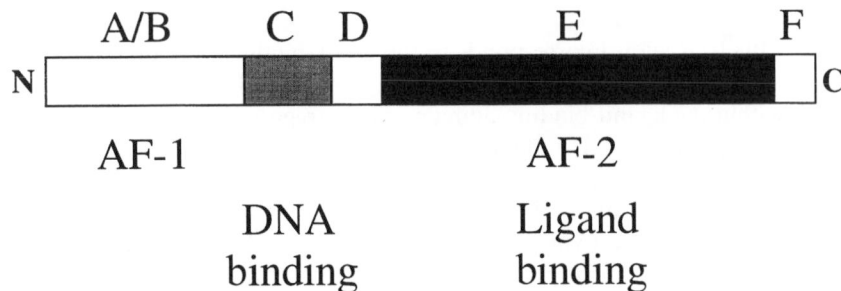

Fig. 1.4 Conserved domain structure of nuclear hormone receptor superfamily members. The amino-terminus of the receptors is most divergent of the regions and is considered the A/B domain, which functions as a hormone-independent transactivation domain (activation function-1 (AF-1)) in some receptors. The C region is the most well conserved region; it is rich in cysteines and basic amino acid residues, and serves as the DNA-binding domain. The D region, often called the 'hinge domain' because of its localization between the DNA-binding domain and the ligand-binding domain, is a relatively short domain with a low degree of conservation among receptors; it has functions both in transcriptional silencing and modulation of DNA-binding activity. The carboxy-terminal E domain is the second most conserved region among nuclear receptors and contains many overlapping functional regions including ligand binding, transactivation, dimerization and nuclear localization. The F region is present in only a limited number of nuclear receptors and has an unclear function.

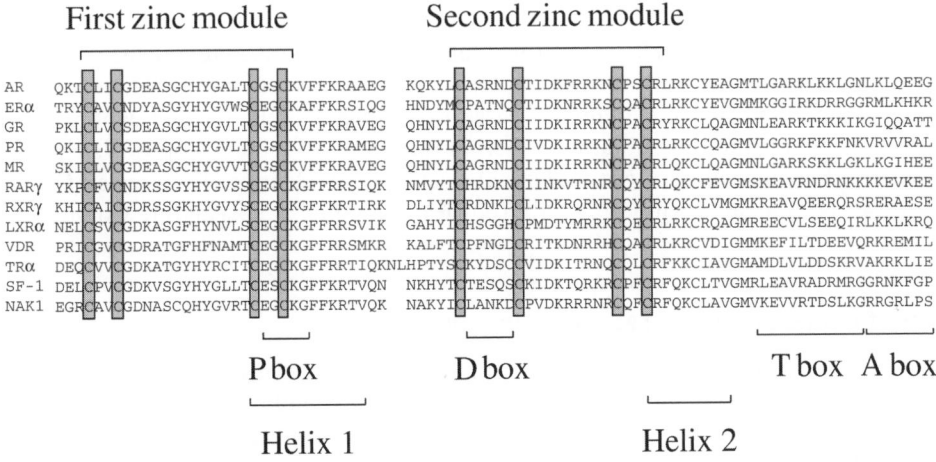

Fig. 1.5 Alignment of region C and D reveals several elements of the DNA-binding domain. The absolute conservation of the cysteine residues is illustrated along with the relative positions of the zinc modules. DNA-binding subdomains such as the proximal box (P box) and distal box (D box), involved in response element recognition and half-site spacing respectively, are also indicated.

maximal amino-acid sequence similarity upon alignment of members of the superfamily (Wang *et al.*, 1989). Region I corresponds to the DNA-binding domain (region C) and thus has the highest degree of conservation among regions I, II and III. Regions II and III are both within the ligand-binding domain (LBD) (region E) and are less conserved among superfamily members than region I.

The high degree of sequence homology within regions of the nuclear receptors such as the DNA-binding domain allowed for identification of novel members of the super-family using low-stringency hybridization or polymerase chain reaction (PCR) with degenerate oligonucleotide primers. Using the cross-hybridization strategy, a new class of nuclear receptors were identified, EAR (v-*erbA* related receptor) and ERR (estrogen related receptor), utilizing the u-*erbA* or estrogen receptor cDNAs as probes (Giguere *et al.*, 1988; Miyajima *et al.*, 1988). These 'orphan receptors', so called because they lack an identified ligand, now compose the largest segment of the nuclear receptor superfamily. Orphan receptors have also been identified with techniques other than sequence homology, including biochemical methods (e.g. chicken ovalbumin upstream promoter transcription factor (COUP-TF); Wang *et al.*, 1988), positional cloning (e.g. DAX-1; Burris *et al.*, 1996) and interaction cloning (e.g. receptor-interacting protein (RIP) 14, RIP-15 and SHP; Seol *et al.*, 1995, 1996). More recently, ligands for several of the orphan receptors have been identified including 9-*cis* retinoic acid (retinoid X receptor (RXR); Heyman *et al.*, 1992; Levin *et al.*, 1992), 15-deoxy-Δ12,14-prostaglandin J2 (peroxisome proliferator activated receptor γ (PPARγ); Forman *et al.*, 1995; Kliewer *et al.*, 1995; Yu *et al.*, 1995), oxysterols (liver X receptor (LXR); Janowski *et al.*, 1996, 1999; Forman *et al.*, 1997; Lehmann *et al.*, 1997), bile acids (farnesoid X receptor (FXR); Makishima *et al.*, 1999; Parks *et al.*, 1999; Wang *et al.*, 1999) and androstanes (constitutive androstane receptor β (CARβ); Forman *et al.*, 1998). Various ligands for members of the nuclear receptor superfamily are illustrated in Fig. 1.3b and demonstrate the diversity of structures that interact with this class of receptors.

Nuclear receptor nomenclature

The immense number of nuclear receptors that have been identified to date has required that a unified nomenclature system be employed for standardization. This system divides the superfamily into seven subfamilies based on sequence similarity and evolutionary relatedness, and within the subfamilies there is further division into groups (Table 1.1). In general, receptors within a group share at least 80–90% identity within region C, and at least 40–60% identity within region E (Nuclear Receptor Nomenclature Committee, 1999).

Current model of nuclear receptor function

The studies described above, along with studies that are outlined later in this chapter focusing on the action of cloned receptors, led to the model of nuclear hormone action that is illustrated in Fig. 1.6. Based on function, the superfamily can be divided into two groups. Type I or group A receptors include the receptors for the classic steroid hormones. These receptors reside primarily in the cytoplasm associated with heat shock proteins (Joab *et al.*, 1984; Catelli *et al.*, 1985; Sanchez *et al.*, 1985; Denis *et al.*, 1988) and are unable to bind to DNA. In this form, the receptors are considered to be in the 'untransformed' 8S–10S complex. Steroid hormones diffuse freely across the plasma membrane of the target cell and bind directly to the receptor causing 'transformation' of

Table 1.1 Nuclear receptor nomenclature

Group	Designation	Common name
Subfamily 1		
A	NR1A1	TRα, c-erbA-1, THRA
	NR1A2	TRβ, c-erbA-1, THRB
B	NR1B1	RARα
	NR1B2	RARβ, HAP
	NR1B3	RARγ, RARD
C	NR1C1	PPARα
	NR1C2	PPARβ, NUC-1, PPARδ, FFAR
	NR1C3	PPARγ
D	NR1D1	REVERBα, EAR-1, EAR-1A
	NR1D2	REVERBβ, EAR-1β, BD73, RVR, HZF2
	NR1D3	E75
E	NR1E1	E78, DR-78
F	NR1F1	RORα, RZRα
	NR1F2	RORβ, RZRβ
	NR1F3	RORγ, TOR
	NR1F4	HR-3, DHR-3, MHR-3, GHR-3, CNR-3, CHR-3
G	NR1G1	CNR-14
H	NR1H1	ECR
	NR1H2	UR, OR-1, NER-1, RIP-15, LXRβ
	NR1H3	RLD-1, LXR, LXRα
	NR1H4	FXR, RIP-14, HRR-1
I	NR1I1	VDR
	NR1I2	ONR-1, PXR, SXR, BXR
	NR1I3	MB-67, CAR-1, CARα
	NR1I4	CAR-2, CARβ
J	NR1J1	DHR-96
K	NR1K1	NHR-1
Subfamily 2		
A	NR2A1	HNF-4
	NR2A2	HNF-4G
	NR2A3	HNF-4B
	NR2A4	DHNF-4, HNF-4D
B	NR2B1	RXRA
	NR2B2	RXRB, H-2RIIBP, RCoR-1
	NR2B3	RXRG
	NR2B4	USP, Ultraspiracle, 2C1, CF-1
C	NR2C1	TR2, TR2-11
	NR2C2	TR4, TAK-1
D	NR2D1	DHR78
E	NR2E1	TLL, TLX, XTLL
	NR2E2	TLL, Tailless
F	NR2F1	COUP-TFI, COUP-TFA, EAR-3, SVP-44
	NR2F2	COUP-TFII, COUP-TFB, ARP-1, SVP-40
	NR2F3	SVP, COUP-TF
	NR2F4	COUP-TFIII, COUP-TFG
	NR2F5	SVP-46
	NR2F6	EAR-2

continued overleaf

Table 1.1 *Continued*

Group	Designation	Common name
Subfamily 3		
A	NR3A1	ERα
	NR3A2	ERβ
B	NR3B1	ERR1, ERRα
	NR3B2	ERR2, ERRβ
C	NR3C1	GR
	NR3C2	MR
	NR3C3	PR
	NR3C4	AR
Subfamily 4		
A	NR4A1	NGFI-B, TRS, N10, NUR-77, NAK-1
	NR4A2	NURR-1, NOT, RNR-1, HZF-3, TINOR
	NR4A3	NOR-1, MINOR
	NR4A4	DHR-38, NGFI-B, CNR-8, C48D5
Subfamily 5		
A	NR5A1	SF1, ELP, FTZ-F1, AD4BP
	NR5A2	LRH-1, xFF1rA, xFF1rB, FFLR, PHR, FTF
	NR5A3	FTZ-F1
Subfamily 6		
A	NR6A1	GCNF, RTR
Subfamily 0		
A	NR0A1	KNI, Knirps
	NR0A2	KNRL, Knirps related
	NR0A3	EGON, Embryonic gonad, EAGLE
	NR0A4	ODR7
	NR0A5	Trithorax
B	NR0B1	DAX-1, AHCH
	NR0B2	SHP

MR, mineralocorticoid receptor; GR, glucocorticoid receptor; PR, progesterone receptor; AR, androgen receptor; ER, estrogen receptor; TR, thyroid hormone receptor; RAR, retinoic acid receptor; VDR, vitamin D receptor; RXR, retinoid X receptor; LXR, liver X receptor; GCNF, germ cell nuclear factor; PPAR, peroxisome proliferator-activated receptor; FXR, farnisoid X receptor; CAR, constitutive androstane receptor; ERR, estrogen related receptor; PXR, pregnane X receptor; HNF, hepatocyte nuclear factor; COUP-TF, chicken ovalbumin upstream promoter transcription factor; SVP, seven-up; NGFIB, nerve growth factor induced gene B; SF1, steroidogenic factor 1; FTZ-F1, fushi tarazu factor 1; DAX1, dosage sensitive sex reversal adrenal hypoplasia congenita critical region chromosome X gene 1; SHP, short heterodimer partner; ECR, ecdysone receptor; UR; ubiquitous receptor; RIP, RXR interacting protein; SXR, steroid and xenobiotic receptor; BXR; benzoate X receptor; HAP, hepatoma; AD4BP, adrenal 4 binding protein; RTR; retinoid receptor-related testis-associated receptor; TR2, testicular receptor 2; TR4, testicular receptor 4; ARP, apolipoprotein AI regulatory protein 1; ear, v-erb A related receptor; FTF, fetoprotein transcription factor; RVR, Rev-erbA-related receptor; ROR, retinoid-related orphan receptor; RZR, retinoid Z receptor.

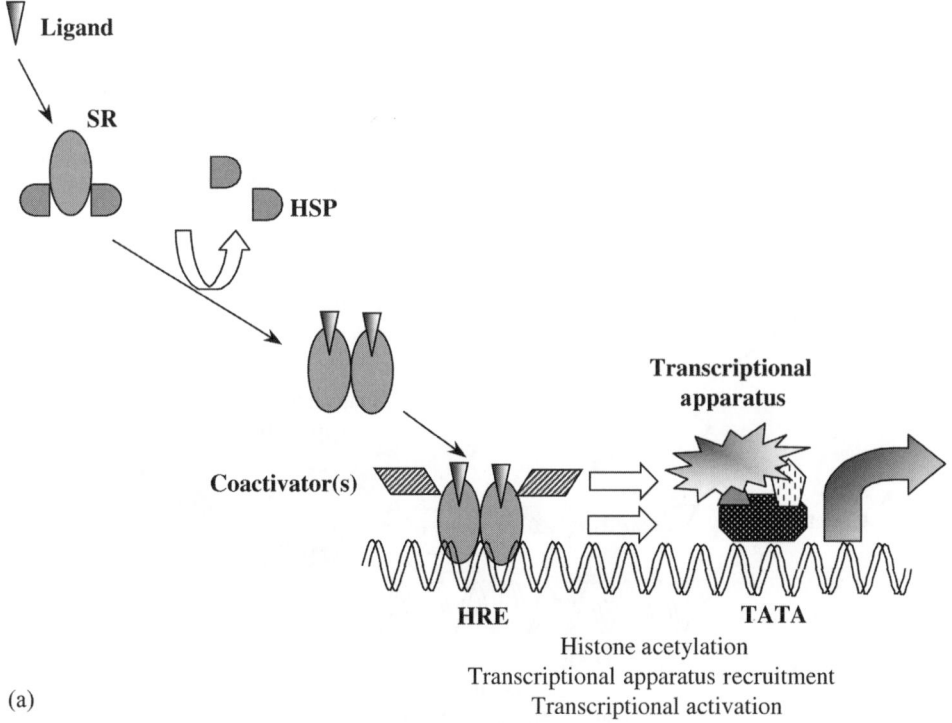

(a)

Fig. 1.6 General models of nuclear receptor function. (a) Mechanism of action of a type I nuclear receptor. The steroid receptor (SR) is associated with heat shock proteins and is not bound to DNA in the absence of ligand. Upon binding ligand, the receptor dissociates from the heat shock proteins (HSPs), homodimerizes and binds to specific hormone response elements (HREs) in the promoter regions of target genes. Once localized to the promoter, the receptor mediates transcriptional activation of the target gene by interaction with general transcription factors and by recruitment of transcriptional coactivators. *Continued.*

the receptor to the 4S complex, mediated by the dissociation of the heat shock proteins. This complex is now sufficient to allow for nuclear translocation and association of the receptor with DNA. The receptor binds as a homodimer to HREs in the promoters of target genes and increases the rate of transcription of these genes resulting in increased expression of the proteins and alteration of cellular function.

Type II or group B receptors, which serve as receptors for the nonsteroid hormones of the superfamily, are localized to the nucleus in both the absence and the presence of hormone. They do not associate with heat shock proteins, are constitutively associated with chromatin, and at least some of the receptors in this group actively inhibit basal transcription (silencing) of their respective target genes in the absence of hormone. Like type I receptors, they bind to their respective HREs as dimers; however, these receptors function as heterodimers with the RXR.

The vast majority of the identified orphan receptors also appear to function as type II receptors. However, as illustrated in Fig. 1.7, at least two other categories of receptors have been defined based on their method of dimerization (or lack thereof) and

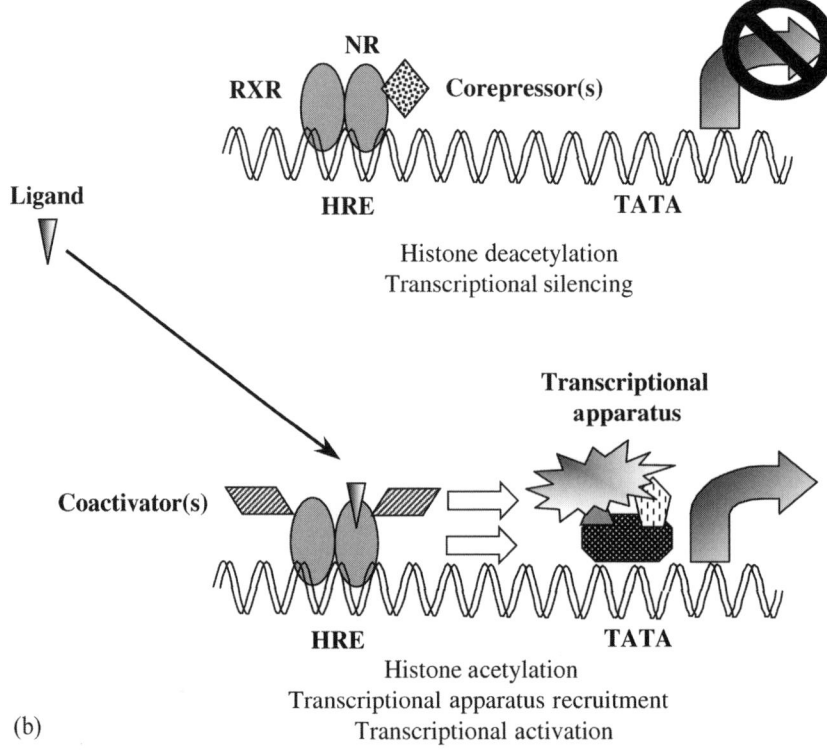

Fig. 1.6 *continued*   (b) Mechanism of action of a type II nuclear receptor. The nuclear receptor (NR) is constitutively bound to its DNA hormone response element (HRE), and the majority of the type II receptors exist as a heterodimer with retinoid X receptor (RXR), although there are some type II receptors that function as monomers (e.g. steroidogenic factor 1 (SF-1) and nerve growth factor induced gene B (NGFI-B)) or homodimers (e.g. hepatocyte nuclear factor-4 (HNF-4) and germ cell nuclear factor (GCNF)). Many of the type II receptors actively silence target gene transcription in the absence of ligand by recruiting transcriptional corepressors. Upon ligand binding, the corepressors are released and the receptor activates transcription in a manner similar to that of type I receptors.

mechanism of interaction with DNA. Several orphan receptors have been identified that prefer to function as homodimers, but unlike type I receptors still have significant similarities to type II receptors in terms of nuclear localization, P box sequence (see DNA Binding below) and HRE. Yet another group of orphans prefers to bind to DNA as a monomer and interact with a HRE 'half-site', which is a fragment of the typical HRE that is recognized by most nuclear receptors. Although the manner in which members of these receptor groups dimerize and recognize their DNA response elements differ, they do share similar mechanisms to activate transcription of target genes (see Transactivation below).

Type I

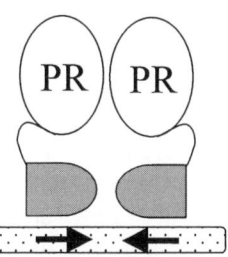

Homodimer

Progesterone receptor
Glucocorticoid receptor
Estrogen receptor
Mineralocorticoid receptor
Androgen receptor

Type II

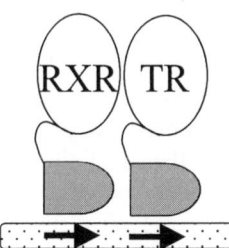

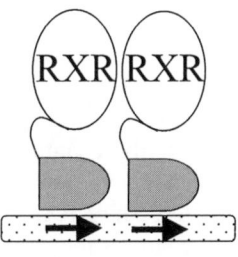

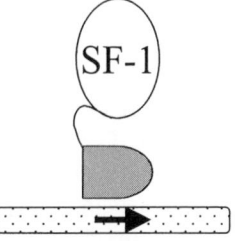

RXR heterodimer

Thyroid hormone receptor
Retinoic acid receptor
Vitamin D receptor
PPAR
LXR

Homodimer

RXR
GCNF
COUP-TF
HNF-4

Monomer

SF-1
NGFI-B

Fig. 1.7 Classification of the types of nuclear receptors within the superfamily based on dimerization properties and the class of response element recognized. Type I receptors include the steroid receptors that homodimerize and typically recognize response elements composed of inverted repeats of half-sites. Type II receptors typically recognize response elements organized into direct repeats. Although the majority of type II receptors form heterodimers with retinoid X receptor (RXR), homodimer and monomers may be formed with a limited number of receptors of this type. COUP-TF, chicken ovalbumin upstream promoter transcription factor; GCNF, germ cell nuclear factor; HNF, hepatocyte nuclear factor-4; LXR, liver X receptor; NGFI-B, nerve growth factor induced gene B; PPAR, peroxisome proliferator activated receptor; SF, steroidogenic factor; TR, thyroid hormone receptor.

NUCLEAR RECEPTOR FUNCTION

DNA binding

Response elements

Nuclear receptors bind to DNA by recognizing a hexameric nucleotide sequence known as a core recognition motif or 'half-site'. The sequence, arrangement and spacing of the half-sites define the nature and responsiveness of an HRE to various nuclear receptors. Four paradigms of half-site orientations have been demonstrated to constitute HREs for nuclear receptors (Fig. 1.8). The majority of nuclear receptors bind to DNA as dimers with each of the receptors occupying one of the half-sites. The arrangement of the half-sites within the HREs allows the homodimers or heterodimers to be arranged in a manner that permits efficient protein–DNA and protein–protein interactions. The half-sites may be arranged as inverted repeats (IRs; palindromes), direct repeats (DRs) or everted repeats (ERs; inverted palindromes). The variable nucleotide spacing between the half-site is also designated for a particular HRE (e.g. the progesterone and glucocorticoid HRE is an IR3 – inverted repeat (palindrome) with a three-nucleotide spacer between the half-sites). Nuclear receptors that bind as monomers recognize a single half-site; however, this class of HRE is typically defined as an extended half-site as additional specific 5′ nucleotides are required for efficient binding of this class of nuclear receptor.

The nucleotide sequences of the half-sites have been defined through identification and analysis of both natural and synthetic HREs. Although there is considerable variability in these half-site sequences, which appear to be able to regulate the relative affinity of the nuclear receptors for the half-site, two categories of consensus half-site sequences have been identified. Type I receptors (steroid receptors) preferentially recognize AGAACA as their half-site sequence and typically bind as homodimers to an IR3 element. Type II receptors preferentially recognize AGGTCA as their half-site and typically bind as heterodimers (with RXR as a partner) to DR elements with variable nucleotide spacing between the half-sites. The ER does not behave as a typical steroid receptor in this classification since it recognizes the AGGTCA half-site, but still binds to this half-site in an IR not a DR mode. Monomeric nuclear receptors, such as steroidogenic factor (SF) 1 and nerve growth factor induced gene B (NGFI-B), also recognize the AGGTCA half-site but require additional specific nucleotides upstream of this half-site for efficient binding (extended half-site). For instance, the consensus SF-1

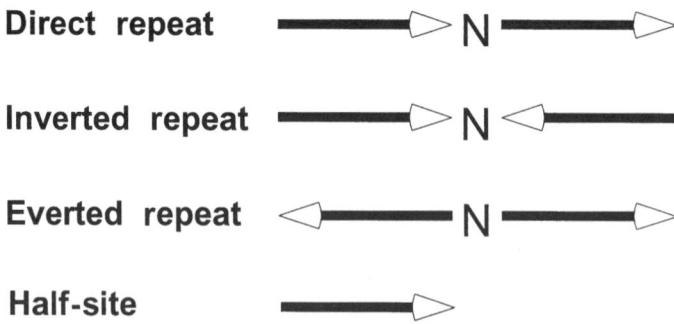

Fig. 1.8 Various orientations of half-sites lead to different types of response elements. Direct repeat (DR), inverted repeat (IR) and everted repeat (ER) orientations are illustrated along with a single half-site.

HRE is an extended half-site of the sequence *CCA*AGGTCA while NGFI-B binds to the consensus *AAA*GGTCA.

With the preponderance of nuclear receptors belonging to the type II class of receptors, functioning as obligate heterodimers with RXR and recognizing the identical half-site consensus sequence, the mechanism by which a particular receptor activates transcription of a specific target gene in response to its cognate ligand is unclear. Analysis of HREs for type II receptors has revealed that they prefer a direct repeat orientation of the half-sites, and that the number of nucleotide spacers between the half-sites dictates nuclear receptor specificity of the HRE (Naar *et al.*, 1991; Umesono *et al.*, 1991).

The studies that defined the 3-4-5 rule demonstrated that the specificity of a type II receptor for an HRE is not defined by the precise sequence of the half-sites, but by the number of nucleotide spacers between those half-sites. Thus, a direct repeat of AGGTCA with a five-base-pair spacer (DR5) defines a retinoic acid response element, a DR4 element confers responsiveness to thyroid hormone, and a DR3 is a vitamin D response element. These DR elements have been identified in hormone-responsive genes, although imperfect repeats may be utilized. Retinoic acid response elements of the DR5 class were identified in the retinoic acid receptor type (βRAR) promoter (de The *et al.*, 1990; Sucov *et al.*, 1990); vitamin D_3 response elements of the DR3 class were identified in the osteoponin and osteocalcin promoters (Demay *et al.*, 1990; Noda, *et al.*, 1990; Ozono *et al.*, 1990); and thyroid hormone response elements of the DR4 class were identified in the myosin heavy chain, malic enzyme and growth hormone promoters (Koenig *et al.*, 1987; Izumo and Mahdavi, 1988; Petty *et al.*, 1990).

The 3-4-5 rule can be further expanded to include other spacers including a DR1, which is a response element for RXR and PPAR found in the promoters of the cellular retinol-binding protein type II gene and the acyl-coenzyme A oxidase genes respectively (Mangelsdorf *et al.*, 1991; Kliewer *et al.*, 1992c). In addition, several orphans bind to various DR elements, adding complexity to the 3-4-5 rule (Fig. 1.9).

Everted repeat arrangements of the half-sites have also been reported to be response elements for some type II receptors. Two natural everted repeat elements have been identified which serve as response elements for either thyroid hormones or retinoic acid. The chicken lysozyme gene promoter contains an ER5 element which serves as a silencer for the thyroid hormone receptor (Baniahmad *et al.*, 1990), while

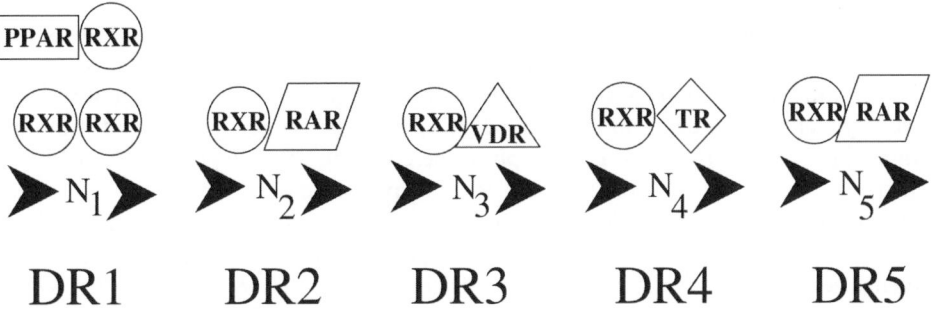

Fig. 1.9 The '1 to 5' rule indicating half-site spacing preference of various type II receptor heterodimers on direct repeat (DR) response elements. PPAR, peroxisome proliferator activated receptor; RXR, retinoid X receptor; RAR, retinoic acid receptor; VDR, vitamin D receptor; TR, thyroid hormone receptor.

an ER8 element confers retinoic acid responsiveness to the γF-crystallin gene (Tini et al., 1993).

DNA-binding domain function

Functional domains of the nuclear receptors have been exhaustively investigated using a variety of molecular biological techniques including deletion analysis, site-directed mutagenesis, and transfer of specific regions of the receptors to heterologous functional proteins. Various mutagenesis techniques allowed for localization of the DNA-binding activity of the nuclear receptors to the 66-amino-acid long, highly conserved, C region containing the two putative zinc fingers. Insertion or deletion of amino acids within the C region or site-directed mutagenesis of the conserved cysteine residues resulted in receptors with reduced ability to bind to DNA and thus activate transcription (Green et al., 1986; Kumar et al., 1986, 1987; Danielsen et al., 1987; Godowski et al., 1987; Hollenberg et al., 1987; Rusconi and Yamamoto, 1987; Rusconi et al., 1987; Hollenberg and Evans, 1988). Although the C region is absolutely required for the DNA-binding activity of the receptors, this region alone cannot bind to DNA. Carboxy-terminal flanking sequences that extend into the D region are required for recognition and binding to the HRE (Mader et al., 1993a). The proposal that the C region contains two Zn^{2+} fingers or modules due to the homology to other Zn^{2+} finger-containing transcription factors was supported by studies indicating that the ER or a fragment of GR containing the DNA-binding domain lost the ability to bind to DNA when metal ions where chelated, and regained this activity upon replacement of Zn^{2+} ions (Sabbah et al., 1987; Freedman et al., 1988). Biophysical studies indicated that the DNA-binding domain reversibly ligates two Zn^{2+} ions, consistent with arrangement of two pairs of cysteine residues coordinating the ions in a tetrahedral arrangement (Freedman et al., 1988). An alignment of the DNA-binding domain of several nuclear receptors illustrating the absolute conservation of the cysteine residues responsible for the Zn^{2+} coordination is shown in Fig. 1.5.

The DNA-binding domain is transferable and was first demonstrated by inserting the C region of one nuclear receptor (GR) to replace the C region of another receptor (ER), conferring the DNA-binding specificity of the GR to the ER chimera while retaining responsiveness to estrogens, and vice versa (Green et al., 1986; Kumar and Chambon, 1988). This 'finger swap' was utilized to characterize the hormone-dependent transactivation properties of the retinoic acid receptor (RAR) without knowledge of the sequence of the retinoic acid response element. By replacing the C domain of RAR with that of either ER (Petkovich et al., 1987) or GR (Giguere et al., 1987), the ability of retinoic acid to activate the receptor was characterized.

The regions within the DNA-binding domain responsible for determination of HRE specificity were identified by more detailed 'swapping' experiments in which the amino-terminal zinc finger GR was used to replace the corresponding finger in ER, yielding a chimeric receptor that activated glucocorticoid response element containing reporters in an estrogen-dependent fashion. Exchange of the carboxy-terminal zinc finger did not affect HRE specificity, indicating that the amino-terminal zinc finger defines the sequence selectivity of DNA binding for nuclear receptors (Green et al., 1988). Green et al. (1988) also proposed that, analogous to helix-turn-helix DNA-binding motifs, the first zinc finger of the nuclear receptor recognizes a half-site of the HRE by lying in the major groove of DNA, and the second finger plays a role in stabilization of the complex through dimer interactions.

The precise amino acid residues responsible for HRE selectivity are contained within a

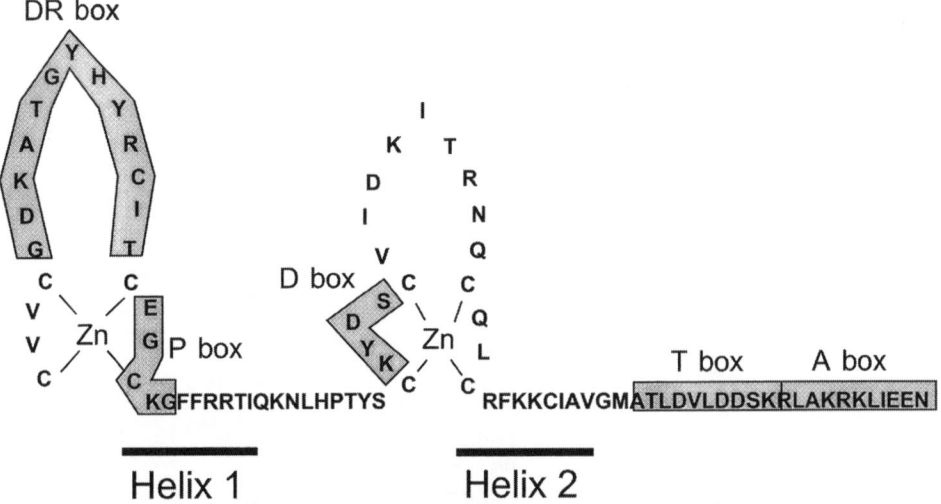

Fig. 1.10   Schematic representation of the structural motifs within the DNA-binding domain of nuclear hormone receptors. Localization of the zinc modules as well as the DR, P, D, T and A boxes is indicated.

cluster of conserved residues at the carboxy-terminal base of the first zinc finger, denoted as the P (proximal) box (Figs 1.5 and 1.10). Alteration of only three amino acids within the P box of the GR with the corresponding residues from the ER transforms the HRE specificity to that of ER, and vice versa (Danielsen *et al.*, 1989; Mader *et al.*, 1989; Umesono and Evans, 1989). Analysis of the amino acid sequences within the P box of nuclear receptors demonstrates that the receptors fall into several groups, with identical residues within the box yielding receptors with similar DNA half-site preferences. Several groups of receptors have been defined based on P box sequences (Cooney and Tsai, 1994) (Table 1.2). Steroid receptors such as GR, MR, PR and AR have the P box sequence GSCKV and prefer to bind to the half-site sequence AGAACA. The ER has a unique P box sequence relative to the other steroid receptors, EGCKA, and binds to an AGGTCA consensus half-site. The majority of nuclear receptors have the P box sequence, EGCKG, and also prefer the AGGTCA half-site. Although more rare, P box sequences such as EGCKS, DGCKG and ESCKG have been identified in various receptors. Like receptors containing the P box sequence EGCKG, these receptors also prefer the AGGTCA half-site and illustrate the uniqueness of the steroid receptors (P box sequence GSCKV) as a very small subset of the nuclear receptor superfamily that prefers the AGAACA half-site.

Amino acid sequences within the amino-terminal base of the second zinc finger in a region defined as the D (distal) box were shown to play an essential role in discrimination of the spacing between the half-sites of an HRE (Umesono and Evans, 1989) (Fig. 1.10 and Table 1.2). Replacement of the five residues comprising the D box (AGRND) of the GR with the homologous region of the TR (KYEGK) alters the specificity of the resulting chimeric receptor from an inverted repeat sequence with three-base-pair spacing to one with no spacing between the half-sites. This region allows for discrimination of half-site spacing by serving as a dimerization interface, consistent with reports that the DNA-binding domain itself contributes to dimerization (Kumar and Chambon, 1988; Tsai *et al.*, 1988). There is considerably more divergence in the D box relative to

Table 1.2 Amino acid sequences of the P and D boxes in several members of the nuclear hormone receptor superfamily

Receptor	P box	D box
AGAACA half-site binding subclass		
MR	GSCKV	AGRND
GR	GSCKV	AGRND
PR	GSCKV	AGRND
AR	GSCKV	ASRND
AGGTCA half-site binding subclass		
ERα	EGCKA	PATNQ
ERβ	EGCKA	PATNQ
TRα	EGCKG	KYDSC
TRβ	EGCKG	KYEGK
RARα	EGCKG	HRDKN
RARβ	EGCKG	HRDKN
VDR	EGCKG	PFNGD
RXRα	EGCKG	RDNKD
RXRγ	EGCKG	RDNKD
LXRα	EGCKG	HSGGH
GCNF	EGCKG	SRDKN

MR, mineralocorticoid receptor; GR, glucocorticoid receptor; PR, progesterone receptor; AR, androgen receptor; ER, estrogen receptor; TR, thyroid hormone receptor; RAR, retinoic acid receptor; VDR, vitamin D receptor; RXR, retinoid X receptor; LXR, liver X receptor; GCNF, germ cell nuclear factor.

the P box, which is consistent with the wide variety of spacer sequences that can be recognized by nuclear receptors that prefer identical half-sites.

Two additional subdomains within the DNA-binding domain localized on the carboxy-terminal side of the zinc fingers also function in DNA recognition for nuclear receptors, as is particularly apparent within the subclass of receptors that bind to DNA as monomers. The A box is a region composed of six amino acids that is key for recognition of the 5′ region of the extended half-site for receptors that bind to DNA as monomers. The amino acid sequence of the NGFI-B A box is essential for recognition of the 5′ *AA* nucleotide sequence portion of the *AA*AGGTCA NGFI-B response element (Wilson *et al.*, 1991, 1992). Similarly, the A box of SF-1 or fushi tarazu-factor 1 (FTZ-F1) facilitates binding to the 5′ *CCA* nucleotides in the *CCA*AGGTCA SF-1 response element (Ueda *et al.*, 1992; Wilson *et al.*, 1993). A box residues may also function in nuclear receptors that act as dimers. Regions carboxy-terminal to the zinc fingers extending into region D are required for efficient DNA-binding activity and, similar to monomeric receptors, nucleotides outside the defined half-site may also play a role in the specificity of binding of the dimeric receptors (Mader *et al.*, 1993a,b).

The T box is situated between the second zinc finger and the A box, and is crucial for the DNA-binding activity of nuclear receptors (Fig. 1.10). The T box was originally identified as a region important for receptor dimerization since deletion of the amino-terminal portion of the T box of RXR results in a receptor that is deficient in its ability to homodimerize (Wilson *et al.*, 1992). Complete deletion of the T box eliminates DNA binding altogether, suggesting that this region is also necessary for protein–DNA interactions. Additional functional and structural studies have confirmed that the T box is important for both protein–protein dimer interactions and protein–DNA interactions (Lee *et al.*, 1993; Predki *et al.*, 1994) and, unlike the DNA-binding domains of steroid receptors that utilize the D box for homodimerization to inverted repeats, homodimers and heterodimers of RXR also employ the T box on direct repeats, consistent with the unique orientation requirements for binding to this class of HRE (Wilson *et al.*, 1992).

The DR box, composed of the extended portion of the first zinc finger, functions as a dimer interface for receptors binding to direct repeat elements (Fig. 1.10). Assessment of the mechanism of heterodimerization between RXR and TR/RAR showed that the DR box of TR or RAR interacts directly with the D box of RXR on direct repeat elements (Perlmann *et al.*, 1993). Thus, RXR binds to the 5′ half-site, exposing the D box for interaction with the DR box of either RAR or TR, which occupies the 3′ half-site (Kurokawa *et al.*, 1993; Perlmann *et al.*, 1993). Thus, a model for the various protein–protein interactions within the DNA-binding domains of nuclear receptor dimers on different classes of HRE can be devised. Steroid receptors bind to inverted repeats and the dimers interact through D box–D box association. Direct repeats require that the receptor occupying the 5′ half-site orient itself to expose its D box to the heterodimeric partner, while the 3′ receptor is in the reverse orientation exposing the DR box to the partner, allowing for a D box–DR box interaction. The DR box itself specifies the spacing between the half-sites of the direct repeat with the TR's DR box requiring four-base-pair spacing and RAR's DR box preferring a five-base-pair spacing (Perlmann *et al.*, 1993). Everted repeats would require opposite polarity of the D box–D box orientation and thus most likely would require a DR box–DR box interaction.

Physical structure of the DNA-binding domain

The three-dimensional structure of the DNA-binding domain of several nuclear receptors, including the GR (Hard *et al.*, 1990; Luisi *et al.*, 1991), the ER (Schwabe *et al.*, 1990, 1993), the RAR (Knegtel *et al.*, 1993), the RXR (Lee *et al.*, 1993), an RXR/TR heterodimer (Rastinejad *et al.*, 1995), RevErb (Zhao *et al.*, 1998) and monomeric NGFI-B (Meinke and Sigler, 1999), have been determined using both nuclear magnetic resonance (NMR) and X-ray crystallography techniques. The physical structure of the DNA-binding domains is very well conserved among the nuclear receptors (Fig. 1.11 see color plate section). The DNA-binding domain is composed of a globular structure organized from two amphipathic α helices arranged perpendicular to each other. Consistent with functional studies, two zinc ions are coordinated within the structure. Each zinc finger contains a single α helix, with the helix initiating at the second pair of coordinating cysteines within each respective finger. The zinc fingers of the nuclear receptors are quite distinct from those classically defined in TFIIIA. As described above, the two nuclear receptor zinc fingers form a globular structure in which each unit is closely associated and functionally dependent on the other. In contrast, TFIIIA zinc fingers form independent structural units that can retain autonomous function (Miller *et al.*, 1985; Pavletich and Pabo, 1991). Thus, this structural unit within the nuclear receptor DNA-binding domain is not truly a zinc finger, but has been defined as a class II zinc-binding motif (Harrison, 1991).

Structural studies of the DNA-binding domain have confirmed many of the conclusions from the functional studies of the domain. The amino-terminal α helix of the DNA-binding domain that contains the P box interacts directly with the response element half-site (Fig. 1.12 see color plate section). This α helix binds within the major groove of the DNA at nearly a right angle to the DNA axis and makes base-specific contacts with the nucleotides within the half-site. As predicted, the nonconserved P box residues of the ER and PR, which were determined to be critical for discrimination between their respective response elements, are involved in base-specific interactions. Unexpectedly, several conserved residues also play a role in recognition of the two distinct half-sites. Although the identity of these residues is conserved between the receptors, their function is not: they appear to make discrete base-pair contacts and thus have different roles in the two receptors. The homodimer interface between the two DNA-binding domains is composed of residues previously defined as D box (Luisi et al., 1991; Schwabe et al., 1993). The second zinc finger module also makes phosphate backbone contacts with the DNA.

The NMR structure of RXR identified a third α helix carboxy-terminal to the other helices which corresponds to the T box that was functionally identified in the RXR and NGFI-B (Wilson et al., 1992; Lee et al., 1993; Predki et al., 1994). Based on this structure, it was predicted that the third α helix would make minor groove contacts with the DNA and also be in a position to make direct contacts with the D box of a dimer partner on tandem repeats (Lee et al., 1993). In contrast to the type I receptors that interact with a symmetric response element in a 'head-to-head' arrangement with a fixed D box–D box dimer interface, type II receptors must utilize a novel dimerization interface to facilitate a 'head-to-tail' arrangement on the direct repeat response elements preferred by this type of receptor. Thus, a T box–D box interface such as that proposed for the RXR may facilitate the head-to-tail arrangement. The crystal structure of the RXR/TR heterodimer illustrates the importance of the T box helix in the head-to-tail formation (Rastinejad et al., 1995) (Fig. 1.13 see color plate section). In this structure, the DNA-binding domains are complexed with a DR4 response element and, although the half-sites are identical, the RXR occupies the 5′ half-site while the TR occupies the 3′ half-site. The T box of the TR forms a loop interrupted by a 3_{10}-type helical turn that, along with regions within the first zinc finger, serves as a dimerization interface with RXR via several arginine residues in the second zinc finger. The amino acid residues forming the dimer interface are reasonably conserved between receptor isoforms, but generally not conserved among other members of the nuclear receptor superfamily. Thus, the dimerization interface for various heterodimers may be unique, providing a partial explanation for the variability in the half-site spacing in response elements for type II receptors.

The A box may also contribute to half-site spacing selectivity. The A box of TR forms a long α helix which makes extensive contacts with the minor groove of the DNA and would produce unfavorable contacts on half-site spacing less than four (Rastinejad et al., 1995). Thus, the A box of other nuclear receptors may provide alternate conformations allowing for other half-site spacers. The structure of the RevErb homodimer complexed to a DR2 HRE is consistent with this model. The A box or 'grip box' of RevErb makes significant minor groove contacts with DNA, extending five base pairs preceding the 5′ half-site and six base pairs preceding the 3′ half-site (Zhao et al., 1998). Sequence comparison of the carboxy-terminal extension of the DNA-binding domain which includes the A box followed by modeling suggests that this region acts as a 'ruler' regulating the spacing between the half-sites of the HRE (Zhao et al., 1998). As predicted, the A box is essential for recognition of the 5′ extension of the core half-site by nuclear receptors that bind to DNA as monomers. This is exemplified in the structure of NGFI-B complexed with its respective response element where, in a manner similar to RevErb, the residues of the A box make direct contact with the minor groove in upstream region of the core half-site (Meinke and Sigler, 1999).

Ligand binding

Ligand binding induces a dramatic effect on nuclear receptors, transforming them from either inactive receptors or transcriptional silencers to transcriptional activators. In some cases, the transformation induced by ligand results in transcriptional repression of certain target genes with unique HREs. The hydrophobic ligands for this class of receptor typically pass through the cellular membrane by passive diffusion and bind to their cognate receptors with high affinity (dissociation constants ranging from 10^{-11} to 10^{-9} nmol L^{-1}). Some of the more recently identified ligands for orphan nuclear receptors have affinities that reach into the micromolar range.

The LBD within nuclear receptors is localized to the carboxy-terminal region of the receptor within region E. Expression of region E, alone, is sufficient to reconstitute ligand-binding activity, and deletions, insertions or point mutations within region E diminish or eliminate the ability of the receptor to bind to ligand (Danielsen *et al.*, 1986; Giguere *et al.*, 1986; Kumar *et al.*, 1986, 1987; Gronemeyer *et al.*, 1987; Hollenberg *et al.*, 1987; Rusconi and Yamamoto, 1987; Dobson *et al.*, 1989). Within the LBD regions responsible for dimerization, nuclear localization, and transcriptional activation or silencing have been delineated.

Heat shock protein interactions

The 'two-step' model of steroid receptor action proposed that ligand induced a conformational change leading to activation of the receptor. For type I receptors, ligand binding results in dissociation of the receptor from heat shock proteins such as hsp90, hsp70 and hsp56 (Smith and Toft, 1993). Type II receptors do not exhibit any affinity for these proteins. It was postulated that, for type I receptors, ligand-mediated dissociation of the heat shock proteins from the receptor (transformation to the 4S complex) may be sufficient to allow the receptors to bind to DNA and transactivate target genes (Groyer *et al.*, 1987; Picard *et al.*, 1988; Pratt *et al.*, 1988). Thus, removal of the heat shock proteins by biochemical means independent of ligand binding resulting in the 4S complex should yield receptors that are competent for DNA binding and transactivation, which was initially found to be the case in electrophoretic mobility shift assays and *in vitro* transcription systems (Tsai *et al.*, 1988, 1990; Klein-Hitpass *et al.*, 1989, 1990; Elliston *et al.*, 1990; Rodriguez *et al.*, 1990).

Taking advantage of the ability of steroid receptors to function in yeast, Picard *et al.* (1990) examined the function of the ER and GR in a strain with a 20-fold reduction in the expression of hsp90. Reduced expression of hsp90 did not result in hormone-independent activation of the receptors, suggesting that dissociation of hsp90 is not sufficient for receptor activation; however, an inhibitory role of other heat shock proteins could not be ruled out. Subsequent studies illustrated that purified steroid receptors, free from heat shock proteins, require ligand binding for activation, indicating that simple dissociation of the heat shock proteins is not sufficient for receptor activation (Bagchi *et al.*, 1990, 1991; Kalff *et al.*, 1990; Elliston *et al.*, 1992).

Ligand-induced conformation change

Several lines of evidence suggest that ligand-bound receptor maintains a conformation distinct from that of the aporeceptor. A variety of biophysical methods has been utilized to illustrate conformational changes in steroid receptors induced by ligand (Jasper *et al.*, 1985; Pavlik *et al.*, 1985; Attardi and Happe, 1986; Faye *et al.*, 1986) and ligand-induced alterations in the immunoreactivity of various nuclear receptors have been demonstrated, which are also consistent with the notion of a ligand-induced conformational shift (Tate *et al.*, 1984; Martin *et al.*, 1988; Weigel *et al.*, 1992). Also conforming with this model are observations that the mobility of nuclear receptor–DNA complexes is altered in electrophoretic mobility shift assays upon addition of ligand (Kumar and Chambon, 1988; El-Ashry *et al.*, 1989; Fawell *et al.*, 1990a; Sabbah *et al.*, 1991). Additionally, several studies have indicated that the act of ligand binding to a nuclear receptor decreases surface hydrophobicity, which is likely due to structural rearrangement of the macromolecule (Hansen and Gorski, 1985, 1986; Fritsch *et al.*, 1992).

Protease digestion assays have provided convincing evidence that ligands induce a direct conformational change within nuclear receptors. Utilizing PR synthesized *in vitro*, it was demonstrated that a progestin binding to the receptor 'protected' a large (approximately 30 kDa) fragment of the receptor from digestion with proteases (Allan *et al.*, 1992a). This finding is consistent with the supposition that, if ligand induces a novel conformation within the receptor, the availability of sites for protease digestion will be altered. This conformational change occurs before the dissociation of heat shock proteins (Allan *et al.*, 1992b) as the protease-resistant band can still be detected even in the presence of molybdate, which stabilizes the 8–10S form of the receptor (Nishigori and Toft, 1980). The protease-resistant fragment was determined to correspond to the LBD or region E (Allan *et al.*, 1992b) and a number of additional studies has demonstrated that induction of the protease-resistant fragment by ligand can be shown in virtually any nuclear receptor examined, including the ER, GR, AR, TR, vitamin D receptor (VDR) and RAR, as well as PPARα/γ, RXR and LXRα (Fritsch *et al.*, 1992; Beekman *et al.*, 1993; Leng *et al.*, 1993; Zeng *et al.*, 1994; Berger *et al.*, 1996; Janowski *et al.*, 1996; Nayeri *et al.*, 1996; Dowell *et al.*, 1997; Benkoussa *et al.*, 1998; Quack *et al.*, 1998).

Nuclear localization

Early studies indicated that steroid receptors reside in the cytoplasm until activation by ligand-inducing translocation into the nucleus. However, there is still considerable disagreement on the cellular localization of various type I receptors. Although type II receptors are generally recognized as being invariably localized within the nucleus, the localization of type I receptors in the absence of ligand is controversial, most likely due to the variability in localization of these receptors under differing experimental conditions (Papamichail *et al.*, 1980; Antakly and Eisen, 1984; Gasc *et al.*, 1984; King and Greene, 1984; Perrot-Applanat *et al.*, 1985; Welshons *et al.*, 1985; Picard and Yamamoto, 1987; Gasc *et al.*, 1989; Press *et al.*, 1989; Husmann *et al.*, 1990; Sar *et al.*, 1990; Martins *et al.*, 1991; Simental *et al.*, 1991; Brink *et al.*, 1992). More recent reports utilizing green fluorescent protein to tag nuclear receptors in order to follow localization within living cells indicate that localization in the absence of ligand may be a function of the individual type I nuclear receptor in question (Htun *et al.*, 1996, 1999; Georget *et al.*, 1997; Ogawa and Umesono, 1998; Lim *et al.*, 1999).

Putative nuclear localization signals have been identified in regions spanning the carboxy-terminal portion of the DNA-binding domain, hinge region and amino-terminal portion of the LBD. This region has significant homology to the nuclear localization signal of the simian virus 40 (SV 40) large T antigen, and deletion analysis in several receptors suggests that this signal is important for nuclear localization of nuclear receptors (Picard *et al.*, 1988; Guiochon-Mantel *et al.*, 1989, 1991; Ylikomi *et al.*, 1992). The proximity of the nuclear localization signal to the LBD may be significant in the ligand dependency of nuclear localization of some type I receptors.

Dimerization and DNA binding
Ligand also plays an important role in inducing homodimerization of type I receptors (Fawell *et al.*, 1990a,b; DeMarzo *et al.*, 1991; Allan *et al.*, 1992a,b). After the ligand-induced conformational change and heat shock protein dissociation, the receptors appear to undergo dimerization which precedes DNA binding (Kumar and Chambon, 1988; Tsai *et al.*, 1988; Fawell *et al.*, 1990a,b; Drouin *et al.*, 1992). In contrast, type II nuclear receptors such as TR and RAR were shown to require heterodimerization with nuclear 'auxiliary' proteins in order for high-affinity DNA binding to take place (Glass *et al.*, 1990; Beebe *et al.*, 1991; Darling *et al.*, 1991; Yen *et al.*, 1991). The hetero-dimerization partner was subsequently determined to be another member of the nuclear hormone receptor superfamily, RXR, which acts as a heterodimerization partner with most type II receptors (Berrodin *et al.*, 1992; Bugge *et al.*, 1992; Durand *et al.*, 1992; Kliewer *et al.*, 1992a,b,c; Marks *et al.*, 1992; Yao *et al.*, 1992; Zhang *et al.*, 1992; Mader *et al.*, 1993b). Also in contrast to the type I receptors, the DNA-binding activity of type II receptors is believed to be constitutive and not ligand dependent (Tsai and O'Malley, 1994).

Heterodimerization with RXR among type II nuclear receptors also introduces the possibility of 'cross-talk' between various hormone-signaling systems. Thus, it may be possible for a nuclear receptor heterodimer to respond to the cognate hormone of either of the partners. Although RXR heterodimerization is required for efficient DNA binding and function of some type II receptors, it appears to be a silent partner when complexed with certain receptors. Thus, receptors such as TR or RAR are considered to be nonpermissive as RXR is a silent partner in heterodimers containing these receptors, but receptors such as PPARs and LXR are permissive receptors because heterodimers of these receptors with RXR are responsive to ligands of either partner (Leblanc and Stunnenberg, 1995; Willy *et al.*, 1995; Direnzo *et al.*, 1997; Mukherjee *et al.*, 1997, 1998; Schulman *et al.*, 1998; Westin *et al.*, 1998).

Phosphorylation
Nuclear receptors have been demonstrated to be targets for various kinases, including mitogen-activated protein (MAP) kinases, cyclin-dependent kinases, glycogen synthase kinase, casein kinase and cyclic adenosine monophosphate (cAMP)-dependent protein kinase (Auricchio, 1989; Orti *et al.*, 1992; Weigel, 1994; Shao and Lazar, 1999). Thus, 'cross-talk' between kinase cascades and nuclear receptor hormone signaling is a significant possibility. Several nuclear receptors undergo substantial enhancement of phosphorylation upon binding to their cognate ligand (Pike and Sleator, 1985; Washburn *et al.*, 1991; Beck *et al.*, 1992; Rochette-Egly *et al.*, 1992) and activators of protein kinases have been shown to increase the ability of ER, GR and PR to activate reporter genes, suggesting that phosphorylation increases the transactivation activity of these receptors (Denner *et al.*, 1990; Somers and DeFranco, 1992; Aronica and Katzenellenbogen, 1993; Cho and Katzenellenbogen, 1993). In contrast, phosphoryla-tion of PPARγ at a MAP kinase consensus site in the A/B domain decreases its ability to transactivate and induce adipogenesis (Hu *et al.*, 1996; Adams *et al.*, 1997).

Besides transcriptional activation, phosphorylation has also been shown to modulate a variety of activities of various nuclear receptors, including DNA binding, dimerization and ligand-binding affinity (Weigel, 1994; Shao and Lazar, 1999). Direct phosphorylation of some nuclear receptors may bypass the need for ligand in the activation of the receptor, leading to a ligand-independent activation pathway (Power *et al.*, 1991). The existence of a ligand-independent activation pathway raises the possibility that at least some orphan nuclear receptors that have been identified to date may utilize this method of activation exclusively and may not have a natural hormone activator.

Physical structure of the ligand-binding domain

Crystal structures of the LBD of several nuclear receptors have been deduced, including RXRα, TRα, RARγ, ERα, PR, PPARγ and PPARδ (Bourguet *et al.*, 1995; Renaud *et al.*, 1995; Wagner *et al.*, 1995; Brzozowski *et al.*, 1997; Nolte *et al.*, 1998; Shiau *et al.*, 1998; Tanenbaum *et al.*, 1998; Uppenberg *et al.*, 1998; Williams and Sigler, 1998; Xu *et al.*, 1999). All share a similar structure composed of 12 α helices (H1–H12) organized into an antiparallel helical sandwich consisting of three layers (Figs 1.14 and 1.15 see color plate section). Minor variations in the helical organization between various nuclear receptors have been noted, but the overall helical sandwich organization remains constant. For example, PR contains only 10 helices because of a lack of H2 and the contiguity of H10 and H11 (Williams and Sigler, 1998). Similarly, RXRα lacks H2, but PPARγ contains an additional helix, H2' (Bourguet *et al.*, 1995; Nolte *et al.*, 1998).

Examination of the structures of holo-receptors illustrates the significant conformational changes within the LBD induced by ligand. Ligand binding generates notable rearrangements of the helices, yielding a more compact structure. H12, which contains the activation function (AF) 2 domain (see Transactivation below), undergoes a large conformational shift upon ligand binding, whereupon it rotates from an exposed position directed away from the LBD in the apo-receptor to a position along the surface of the LBD where it serves to seal the ligand-binding cavity of the receptor. The ligand is generally buried within the structure, with H12 providing the 'lid' of the cavity.

With the exception of the PPARs, no routes of entry for the ligand into the ligand-binding cavity can be identified, which has led to the 'mouse-trap' model for ligand binding (Moras and Gronemeyer, 1998). In this model, the flexibility of H12 allows for entry of a ligand into the ligand-binding cavity created by the displacement of the H12 'lid'. Once the ligand has entered the cavity lined with hydrophobic amino acid residues, contacts are made inducing the conformational shift which also includes closing the H12 lid. As many ligands make direct contact with residues within H12, the positioning of the helix along the surface of the LBD is stabilized. The positioning and stabilization of H12 is necessary to create a surface on the LBD for recruitment of coactivator proteins, which are required for transcriptional activation (Fig. 1.14 see color plate section) (Nolte *et al.*, 1998; Xu *et al.*, 1998). Interestingly, receptor antagonists appear to occupy the ligand-binding cavity, but do not allow for positioning of H12 that is compatible with coactivator recruitment (Brzozowski *et al.*, 1997; Shiau *et al.*, 1998; Oberfield *et al.*, 1999).

Of the nuclear receptor LBDs that have been crystallized as dimers, most utilize a conserved surface for the dimerization interface. H10 provides the major dimerization surface for RXRα, ERα and PPARγ, with H9, H7 and the connecting loop between H7 and H8 also contributing to the interface (Bourguet *et al.*, 1995; Brzozowski *et al.*, 1997; Nolte *et al.*, 1998; Shiau *et al.*, 1998; Tanenbaum *et al.*, 1998) (Fig. 1.15 see color plate section). PR, however, utilizes a unique dimerization interface consisting primarily of

H11 and H12 (Tanenbaum *et al.*, 1998; Williams and Sigler, 1998). All the dimerization interfaces that have been described are those of homodimers, including PPARγ which is not believed to function as a homodimer. The heterodimerization interfaces between a type II receptor and RXR have not yet been characterized.

Transcriptional regulation

Nuclear receptors exert their effects on cellular function by regulating the rate of transcription of target genes containing HREs. The classic effect of a nuclear receptor, once activated, is to increase the rate of transcription of target genes via interaction with the *cis*-acting DNA response element – a mechanism called transactivation. It is also clear that a subclass of nuclear receptors can decrease transcription of their respective target genes below the basal rate via interactions with *cis*-acting DNA elements – a mechanism called silencing. Characterization of the transcriptional regulatory properties has been facilitated by the 'cotransfection' assay in which vectors directing the expression of a nuclear receptor are cotransfected into a cell line normally lacking the receptor along with a reporter vector containing HRE(s) upstream of an easily detectable 'reporter' gene such as chloramphenicol acetyltransferase (CAT) or luciferase (Danielsen *et al.*, 1986; Giguere *et al.*, 1986; Miesfeld *et al.*, 1986, 1987; Conneely *et al.*, 1987; Godowski *et al.*, 1987; Gronemeyer *et al.*, 1987; Kumar *et al.*, 1987). This assay, along with later developed 'cell-free' transcription systems, has proven to be essential for identification of the regions of the nuclear receptor responsible for the transcriptional regulatory properties, as well as the mechanism by which the receptors interact with the cellular machinery to facilitate alterations in the rate of transcription of mRNA.

Transactivation

Activation functions. Functional analysis of the regions responsible for transactivation in type I receptors using site-directed, insertional and deletion mutants coupled to the cotransfection assay demonstrated that there are two distinct transactivation domains, one in the amino-terminus within the A/B region (activation function (AF-1)) and another in the carboxy-terminus within region E (AF-2) (Giguere *et al.*, 1986; Danielsen *et al.*, 1987; Godowski *et al.*, 1987; Gronemeyer *et al.*, 1987; Hollenberg *et al.*, 1987; Kumar *et al.*, 1987; Hollenberg and Evans, 1988; Lees *et al.*, 1989; Tora *et al.*, 1989). Construction of chimeric proteins consisting of either the AF-1 or AF-2 transactivation domain fused to a heterologous DNA-binding domain such as one derived from the yeast Gal4 protein or the bacterial LexA repressor illustrates that these domains can function independently since either is able to mediate transactivation of a Gal4- or LexA-responsive reporter gene (Godowski *et al.*, 1988; Hollenberg and Evans, 1988; Webster *et al.*, 1988; Tora *et al.*, 1989).

Not only are the two transactivation domains functionally independent, but they also differ in terms of their responsiveness to hormone. AF-1 transactivates in a consititutive, hormone-independent manner; however, the relative level of transactivation varies according to the cellular and promoter context. For example, truncation of ER to remove AF-2 yields a receptor with constitutive transactivation activity ranging from 3 to 70% of the wild-type receptor depending on the cell line utilized for the cotransfection (Kumar *et al.*, 1987; Bocquel *et al.*, 1989; Lees *et al.*, 1989). Interestingly, the activity of AF-1 is also promoter dependent, as illustrated by the differential activity of a mutant ER with AF-1 removed. Removal of AF-1 had no effect on the activity of ER at a reporter driven by the estrogen-responsive vitellogenin promoter, but the activity at a reporter driven by the pS2 promoter was reduced by 20% (Kumar *et al.*, 1987). The AF-1 domain of the GR has also been shown to be required for the full transactivation activity of this receptor (Danielsen *et al.*, 1987; Hollenberg *et al.*, 1987; Miesfeld *et al.*, 1987).

In contrast to AF-1, AF-2 acts as a hormone-inducible transactivation domain (Miesfeld *et al.*, 1987; Hollenberg and Evans, 1988; Webster *et al.*, 1988; Lees *et al.*, 1989; Tora *et al.*, 1989a). The colocalization of the LBD and the hormone-inducible AF-2 suggested that the ligand-induced conformational change reveals the activation function, a prediction that was confirmed upon the solution of the co-crystal structure of a nuclear receptor with a coactivator peptide (see below). The transactivation activity of both AF-1 and AF-2 is transferable to a heterologous DNA-binding domain with the retention of the respective hormone responsiveness or lack thereof (Godowski *et al.*, 1988; Hollenberg and Evans, 1988; Webster *et al.*, 1988).

As described above, AF-1 was originally identified in type I receptors that contain large A/B regions ranging into hundreds of amino acids in size. However, type II receptors typically have short or sometimes nonexistent A/B regions, bringing into question whether this subclass of receptor would retain AF-1 activity. It appears that the existence of AF-1 within the type II receptor subclass may be receptor specific. For example, weak AF-1 activity has been described for the PPARγ, but is nonexistent in the VDR (Adams *et al.*, 1997; Sone *et al.*, 1991). Since type II receptors interact with DNA in a constitutive manner, the presence of a strong AF-1 may adversely affect the hormone dependency of this subclass of receptor. Type I receptors, however, interact with DNA only in the presence of ligand, thus limiting AF-1 activity due to cellular localization.

Within the ligand-dependent AF-2 domain lies a short stretch of amino acids at the extreme carboxy-terminus of the LBD that acts as a constitutively active transactivation domain when fused to a heterologous DNA-binding domain (Danielian *et al.*, 1992; Saatcioglu *et al.*, 1993; Durand *et al.*, 1994; Baniahmad *et al.*, 1995). This short region is well conserved among nuclear receptor superfamily members and has a sequence consistent with an α-helical structure, with the core of the helix composed of a central acidic residue, typically a glutamic acid, flanked by hydrophobic residues (Fig. 1.16).

MR	S	H	A	L	K	V	F	P	A	M	L	V	**E**	I	I	S	D	Q	L	P
AR	H	M	V	S	V	D	F	P	E	M	M	A	**E**	I	I	S	V	Q	V	P
GR	K	T	M	S	I	E	F	P	E	M	L	A	**E**	I	I	T	N	Q	I	P
ER	C	K	V	V	V	P	L	Y	D	L	L	L	**E**	M	L	D	A	H	R	L
PR	R	A	L	S	V	E	F	P	E	M	M	S	**E**	V	I	A	A	Q	L	P
TRβ	E	C	P	T	E	L	L	P	P	L	I	L	**E**	V	F	E	D			
RXRα	L	I	G	D	T	P	I	D	T	F	L	M	**E**	M	L	E	A	P	H	Q
RARα	M	E	I	P	G	S	M	P	P	L	I	Q	**E**	M	L	E	N	S	E	G

Fig. 1.16 Alignment of the activation function-2 (AF-2) helix region of several nuclear receptors. The highly conserved glutamate residue is in bold and the flanking hydrophobic regions are boxed. AR, androgen receptor; ER, estrogen receptor; GR, glucocorticoid receptor; MR, mineralocorticoid receptor; PR, progesterone receptor; RAR, retinoic acid receptor; RXR, retinoid X receptor; TR, thyroid hormone receptor.

Not only does the AF-2 helix function as a constitutively active autonomous trans-activation domain, but when this region is deleted or mutated in the context of the entire LBD the LBD loses the ability to transactivate while retaining the ability to bind ligand, indicating that this region is responsible for the transactivation activity within the LBD (AF-2). Crystal structures of several nuclear receptors indicate that the AF-2 helix is helix 12, the helix that undergoes a significant conformational change in response to ligand binding. In the unliganded state, the AF-2 helix projects away from the globular core of the LBD. Upon ligand binding, the AF-2 helix folds back on to the globular surface of the LBD forming a cleft that is able to recruit and dock coactivators necessary for transcriptional activation.

Mechanism of transactivation. Conceivably, nuclear receptors may utilize several mechanisms to increase the rate of transcription of target genes. First, nuclear receptors may directly interact with and recruit general transcription factors that are components of the preinitiation complex (PIC) to increase the rate of initiation of transcription. Second, the receptors may act by altering chromatin structure so as to render the DNA of the target gene more accessible to various transcription factors and to RNA polymerase II (PolII). Third, the receptor may interact with other cellular components that act to bridge interactions with members of the PIC and/or proteins that alter chromatin structure.

Initiation of transcription of PolII-transcribed genes requires the stepwise assembly of general transcription factors on the promoter region of a gene, typically on a TATA box, followed by recruitment of PolII itself. Nuclear receptors are able to activate transcription from minimal promoters, suggesting that there may be direct interactions between a given nuclear receptor and a component(s) of the general transcriptional machinery. Using a cell-free steroid receptor-dependent transcription system, several nuclear receptors have been shown to stabilize the PIC, thus facilitating the initiation of transcription (Elliston et al., 1990; Klein-Hitpass et al., 1990; Tsai et al., 1990). Interestingly, the first demonstration that a member of the nuclear receptor superfamily interacted directly with a general transcription factor was made by analysis of a protein, S300-II, which copurified with COUP-TF (Tsai et al., 1987). Sequence analysis of S300-II indicated that this protein is actually the general transcription factor TFIIB (Ing et al., 1992). Subsequently, several nuclear receptors were shown to interact with TFIIB in a similar fashion (Ing et al., 1992; Baniahmad et al., 1993; Blanco et al., 1995; MacDonald et al., 1995). Direct interaction between various nuclear receptors and other components of the general transcriptional machinery has also been described, including interactions with TFIIH, TFIIF and the TATA box-binding protein (TBP) (Sadovsky et al., 1995; Schulman et al., 1995; McEwan and Gustafsson, 1997; Rochette-Egly et al., 1997) as well as TBP-associated factors (TAFs) such as TAFII30 and TAFII110 (Jacq et al., 1994; Schwerk et al., 1995), indicating that multiple interactions between nuclear receptors and general transcription factors may underlie the ability of the receptors to stabilize the PIC.

The existence of limiting accessory proteins or coactivators required for nuclear receptor transactivation activity was suggested by the observation that the activity of the PR could be downregulated by coexpression and activation of ERs in the same cell (Meyer et al., 1989). This transcriptional interference effect, also known as squelching, was proposed to be caused by competition of the two receptors for a limiting intermediary factor(s) within the cell that is required for transactivation. Additional studies indicated that nuclear receptors contain multiple classes of transactivation domains that utilize distinct coactivators (Tasset et al., 1990).

Using biochemical methods, Halachmi *et al.* (1994) identified putative intermediary factors for the ER by characterizing proteins that interact with ER's LBD only in the presence of an agonist. These investigators indentified two ER-associated proteins (ERAP-140 and ERAP-160) that interacted with the ER only in the presence of agonist. Interestingly, ER mutants that are transcriptionally defective are unable to recruit ERAPs, and ER antagonists disrupt the interaction of ERAPs with the receptor. A similar protein, glucocorticoid receptor interacting protein 170 kDa (GRIP 170), was discovered with biochemical methods using GR to identify proteins that interact in a ligand-dependent manner (Eggert *et al.*, 1995). Another approach yielded the sequence of this class of proteins, which appear to serve as general coactivators for members of the nuclear receptor superfamily. The cDNA for steroid receptor coactivator-1 (SRC1) was identified using the LBD of PR as bait in the yeast two-hybrid screen (Onate *et al.*, 1995). SRC-1 was found to interact with the LBD of PR as well as several other nuclear receptors, including ER, GR, TR and RXR, in a ligand-dependent manner. Consistent with the expected behavior of a coactivator for AF-2, antagonists preclude the interaction of SRC-1 with the LBD. Also, if SRC-1 is the limiting (or one of several limiting) coactivator that is responsible for the transcriptional interference effect noted between the ER and the PR when coexpressed, overexpression of SRC-1 should relieve the interference, and, as illustrated by Onate *et al.* (1995), SRC-1 performs as expected for an authetic coactivator in this paradigm.

It was quickly noted that SRC-1 is but one member of a family of related SRC-1-like proteins, and work from several laboratories identified three SRC-1-like coactivators: (1) SRC-1/NCoA-1 (nuclear receptor coactivator 1), (2) SRC-2/TIF-2 (transcriptional intermediate factor-2)/NCoA-2 and (3) SRC-3/p/CIP (p300/CBP/cointegrator protein)/ RAC-3 (receptor-associated coactivator 3)/ACTR (activator of retinoid receptors)/ AIB-1 (amplification in breast cancer 1)/thyroid hormone receptor activator molecule 1 (TRAM-1) (Hong *et al.*, 1996, 1997; Voegel *et al.*, 1996; Anzick *et al.*, 1997; Chen *et al.*, 1997; Li *et al.*, 1997a; Takeshita *et al.*, 1997; Torchia *et al.*, 1997; Suen *et al.*, 1998). The redundancy within this family of coactivators is the likely cause of the limited hormone resistance phenotype detected in SRC-1 null mice (Xu *et al.*, 1998). SRC family members share significant sequence homology and have a similar domain organization. SRCs typically contain multiple transactivation and nuclear receptor interaction domains (Fig. 1.17). Additionally, a basic helix-loop-helix (bHLH) and two Per-ARNT-Sim (PAS; regions with similarity to the peroid-aryl hydrocarbon receptor-single minded domains) domains are present in the amino-terminal region of these proteins. Although the presence of bHLH and PAS regions are conserved within the SRC family, their role is currently unknown.

Another region of particular interest is within the carboxy-terminal region of these coactivators. Both SRC-1 and the SRC-3 have histone acetyltransferase (HAT) activity associated with this area of the coactivator (Chen *et al.*, 1997; Spencer *et al.*, 1997) and, although SRC-2 has not been characterized as having HAT activity itself, all SRCs appear to have the ability to recruit other proteins (directly or indirectly), such as cyclic adenosine monophosphate response element-binding (CREB) protein (CBP)/p300 and p300/CBP-associated factor (PCAF), both of which have intrinsic HAT activity (Bannister and Kouzorides, 1996; Ogryzko *et al.*, 1996; Voegel *et al.*, 1998). Additionally, CBP can interact directly with nuclear receptors themselves, allowing for formation of a ternary nuclear receptor–CBP–SRC complex (Chakravarti *et al.*, 1996; Hanstein *et al.*, 1996; Kamei *et al.*, 1996; Smith *et al.*, 1996; Yao *et al.*, 1996; Fronsdal *et al.*, 1998; Zhou *et al.*, 1998).

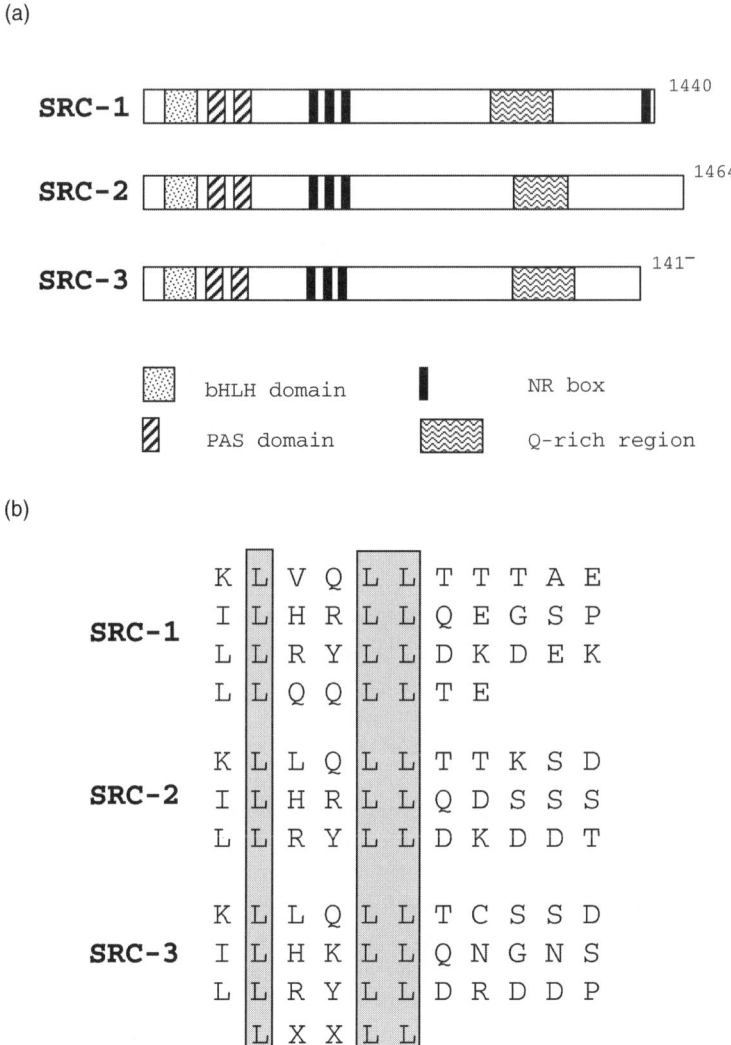

Fig. 1.17 The steroid receptor coactivator (SRC) family of coactivators. (a) Alignment of SRC-1, 2 and 3 illustrating the relative positions of the basic helix-loop-helix (bHLH) and Per-ARNT-Sim (PAS) domains as well as the nuclear receptor (NR) boxes and Q-rich regions. (b) Alignment of the amino acid sequences of the NR boxes from SRC-1, 2 and 3. The LXXLL (X is any amino acid) motif is clearly conserved in these boxes.

The number of coactivators with HAT activity recruited by ligand-bound nuclear receptors suggests that acetylation of the histones is an important mechanism involved in transactivation. Acetylation of the histones by receptor-recruited factors allows for opening of the repressed local chromatin structure leading to a derepressed state that is transcriptionally permissive and may facilitate additional steps leading to PIC formation. Alteration of chromatin structure is not the only mechanism by which SRCs can facilitate transactivation, as it has recently been demonstrated that SRC-1 can also coactivate PR-dependent transactivation from nonchromatin templates (Liu *et al.*, 1999).

The nuclear receptor interaction domains of the SRC family are of particular interest as they play an essential role in recognition of ligand-bound receptor. Analysis of the interaction domains of these coactivators allowed for the identification of a conserved motif that was named an NR (nuclear receptor) box (Heery *et al.*, 1997; Torchia *et al.*, 1997). The NR box is a short motif consisting of the amino acid sequence LXXLL (L = leucine; X = any amino acid), which mediates interaction of the large coactivator protein with ligand-bound receptor. Several NR boxes are repeated within the receptor-interacting domains of SRC family members (Fig. 1.17). NR boxes have also been localized to other classes of nuclear receptor coactivators and are described below and in Chapter 11. Structural analysis of the NR box indicates that this short stretch of amino acids forms an amphipathic α helix. Functionally, based on co-crystal structures of a NR box complexed with ligand-bound nuclear receptors, the NR box recognizes a cleft on the surface of the LBD formed, in part, by the AF-2 helix of the receptor folding back upon the globular LBD core structure in response to ligand binding (Fig. 1.14 see color plate section).

A novel multicomponent coactivator complex distinct from the SRC family, known as the TRAP or DRIP complex, was identified independently by two laboratories, both utilizing biochemical methods. Fondell *et al.* (1996) utilized HeLa cells stably transfected with epitope-tagged TR followed by immunoprecipitation to identify a series of nine polypeptides (TR-associated proteins; TRAPs) that interacted with the TR in a ligand-dependent manner. The TRAP complex was able to enhance TR-mediated transactivation *in vitro* in a chromatin-free system. At least one of the TRAPs, TRAP-220, interacts directly with nuclear receptors in a ligand-dependent manner (Yuan *et al.*, 1998). A similar method was used to identify the DRIP complex. The LBD of VDR was utilized as an affinity matrix to identify 10 proteins that interact with VDR (VDR-interacting proteins; DRIPs) in a ligand-dependent manner (Rachez *et al.*, 1998). DRIPs interact with several nuclear receptors including TR and, like TRAPs, they are able to enhance the transcriptional activity of a nuclear receptor *in vitro* in a chromatin-free system.

Comparison of the DRIP and TRAP complexes indicated that the two were apparently identical. The DRIP/TRAP complex also shares components with the suppressor of RNA polymerase B (SRB) and mediator (MED)-containing cofactor complex (SMCC), a complex that was originally identified on the basis of homology to components of the yeast SRB/mediator complex (Gu *et al.*, 1999; Ito *et al.*, 1999; Naar *et al.*, 1999; Rachez *et al.*, 1999). The DRIP/TRAP/SMCC complex also appears to act as a coactivator for transcription factors other than nuclear receptors such as p53, VP16, SREBP-1a and SP1 (Ito *et al.*, 1999; Naar *et al.*, 1999). Thus, the DRIP/TRAP/SMCC complex may be a general coactivator for many types of transcriptional activators, although unique components of the complex are targeted by various activators, for example TRAP-220 is the target for nuclear receptors while TRAP-80 is the target for VP16 and p53 (Yuan *et al.*, 1998; Ito *et al.*, 1999).

Like the SRC family of coactivators, several protein components of the DRIP/TRAP/SMCC complex contain NR boxes (Yuan *et al.*, 1998; Ito *et al.*, 1999; Zhang and Fondell, 1999); thus the mechanism of ligand-dependent interaction of these coactivators with the LBD of a nuclear receptor is likely to be analogous to those described above for the SRC NR boxes. The NR boxes of these two classes of coactivators may compete for the identical cleft formed on the surface of the ligand-bound receptor; however, the distinct roles or need for the two types of coactivator are not yet clear. Although these coactivators function as mediators of AF-2 transactivation, it is possible that components of the DRIP/TRAP/SMCC complex may also serve as targets for

AF-1. SRC-1 has been shown to play a role in transactivation by AF-1 (Onate *et al.*, 1998) and coactivators with a degree of AF-1 specificity have been identified (Lanz *et al.*, 1999). Additional putative nuclear receptor coactivators have been described, but are beyond the scope of this work and are reviewed in detail elsewhere (Chapter 11; Chen and Li *et al.*, 1998; McKenna *et al.*, 1999).

Transcriptional silencing

Transcriptional silencing refers to the ability of a limited number of type II receptors, in the unliganded state, to actively inhibit basal transcription of their respective target genes. Silencing is distinct from other mechanisms of repression that may function through competition of transcription factors for DNA response elements or dimerization partners. The ability of a nuclear receptor to silence transcription was initially described for the TR and occurred only in the unliganded state (Damm *et al.*, 1989; Graupner *et al.*, 1989; Sap *et al.*, 1989; Baniahmad *et al.*, 1990). The related v-*erbA* oncogene product also silenced transcription, but in a constitutive manner owing to mutations within its coding sequence.

Utilizing chimeric receptors fused to a heterologous DNA-binding domain, Baniahmad *et al.* (1992) localized the region of the receptor responsible for the silencing activity to regions D and E. Gal4 DNA-binding domain fusions with regions D and E of TR, RAR and v-*erbA* were all found to silence reporter gene transcription. Deletion analysis indicated that regions in both region D and the carboxy-terminus of region E were essential for retention of the silencing activity.

The possible mechanisms underlying the ability of nuclear receptors to silence transcription are very similar to those proposed for transactivation; however, they would tend to interfere with PIC formation and/or induce a closed state within the chromatin rather than promote PIC formation and opening of the chromatin. Sequestration of the general transcription factor TFIIB by unliganded TR has been shown to be one possible mechanism of silencing (Baniahmad *et al.*, 1993), and indeed in a cell-free transcription system unliganded TR is able to inhibit PIC formation directly (Fondell *et al.*, 1993). In addition, in a manner analogous to the function of coactivators in transactivation, silencing requires corepressors (Casanova *et al.*, 1994; Baniahmad *et al.*, 1995). Transcriptional interference studies have demonstrated that limiting intermediary factors, corepressors, required for the silencing activity of TR could be squelched by coexpression of the carboxy-terminus of v-*erbA*, unliganded TR, or a fraction of TR consisting of region D and the amino-terminus of region E (Baniahmad *et al.*, 1995). Like coactivators, corepressors interact with receptors in a ligand-dependent manner; however, the ligand dependency is reversed. Corepressors bind to silencing nuclear receptors in the unliganded state and are released upon ligand binding to the receptor.

Both biochemical and interaction cloning techniques were used to identify two related nuclear receptor corepressors: nuclear receptor corepressor (N-CoR) and silencing mediator for RAR and TR (SMRT) (Chen and Evans, 1995; Horlein *et al.*, 1995). Both SMRT and N-CoR interact with TR and RAR only in the absence of ligand, and expression of either of these corepressors relieves transcriptional interference. Each of the corepressors also interacts with other nuclear receptors that actively silence transcription such as COUP-TF, Rev-erbA-related receptor (RVR), RevErb and v-*erbA* (Downes *et al.*, 1996; Shibata *et al.*, 1997; Zamir *et al.*, 1996). Type I receptors, such as ER and PR, have been shown to interact with corepressors even though they do not silence transcription; however, the interaction occurs only when the receptors are bound to an antagonist, suggesting that corepressors may be important mediators of the pharmacology of type I receptors (Jackson *et al.*, 1997; Smith *et al.*, 1997).

SMRT and N-CoR are large, characterized as 168- and 270-kDa proteins respectively. Independently, partial cDNAs of both N-CoR and SMRT have been identified using interaction cloning methods as RIP-13 and T3 receptor-associating cofactor-1 (TRAC-1) (Soel *et al.*, 1995; Sande and Privalsky, 1996). SMRT and N-CoR are highly related, sharing 41% identity over the shorter SMRT sequence, which was initially characterized as lacking an approximately 1000-amino-acid insertion within the amino-terminus of N-CoR (Chen *et al.*, 1996). Recently, an extended isoform of SMRT with an amino-terminal extention homologous to the 1000-amino-acid region in N-CoR was described (Ordentlich *et al.*, 1999; Park *et al.*, 1999). Throughout the large sequences of N-CoR and the extended isoform of SMRT, 44% identity is retained.

Contained within the corepressors are multiple nuclear receptor interaction domains and autonomous repression domains, the latter of which is transferable to a heterologous DNA-binding domain with retention of the silencing activity (Chen and Evans, 1995; Horlein *et al.*, 1995; Chen *et al.*, 1996; Li *et al.*, 1997b; Nagy *et al.*, 1997).

The corepressors interact directly with nuclear receptors within a section of region D, a domain that has been termed the CoR box (Horlein *et al.*, 1995). This identical region was shown to squelch TR-mediated silencing, and mutations within the CoR box interfere with corepressor binding and silencing activity (Baniahmad *et al.*, 1995; Horlein *et al.*, 1995; Nawaz *et al.*, 1995). Interaction of corepressors with nuclear receptors appears to be more complex than originally described based on the discovery of CoRNR boxes described below (Hu and Lazar, 1999; Nagy *et al.*, 2000; Perissi *et al.*, 2000). Interestingly, the ligand-dependent conformational change in the AF-2 helix, which is essential for recruitment of coactivators, is also essential for release of corepressors (Baniahmad *et al.*, 1995; Li *et al.*, 1997b). Receptors lacking the AF-2 helix or with mutations in this region become constitutive silencers due to the inability to displace corepressors in response to ligand binding.

The corepressor nuclear receptor interaction domains, like coactivators, contain short a-helical motifs that mediate recognition of the state of the nuclear receptor with respect to ligand binding (Hu and Lazar, 1999; Nagy *et al.*, 2000; Perissi *et al.*, 2000). In contrast to the NR box of the coactivators that recognize agonist-bound LBD, the CoRNR box of the corepressors recognize the unliganded LBD. Replacement of the NR box of a coactivator with the CoRNR box of a corepressor is able to reverse the ligand-dependency of a nuclear receptor (Hu and Lazar, 1999).

The corepressors mediate silencing, at least in part, by modulation of chromatin structure. Although neither SMRT nor N-CoR contains intrinsic histone deacetylase activity, both recruit a complex containing mSin3 and histone deacetylases that are sufficient to condense chromatin (Alland *et al.*, 1997; Heinzel *et al.*, 1997; Nagy *et al.*, 1997). Thus, in contrast to coactivation which is associated with histone acetylase activity, corepression is associated with histone deacetylation, which leads to condensation of chromatin and decreased accessibility of general transcription factors to the target gene.

NUCLEAR RECEPTORS AND HUMAN DISEASE

With the vast number of identified members of the nuclear hormone receptor superfamily and their roles in regulation of a multitude of physiologic functions in the organism, it is not surprising that aberrant function of these receptors would lead to pathologic states. In the remaining 10 chapters of this book, several members of the nuclear receptor superfamily are described in greater detail in terms of their normal physiologic function and the pathologic states associated with their dysfunction.

REFERENCES

Adams, M., Reginato, M.J., Shao, D., Lazar, M.A. and Chatterjee, V.K. (1997) Transcriptional activation by peroxisome proliferator-activated receptor gamma is inhibited by phosphorylation at a consensus mitogen-activated protein kinase site. *J. Biol. Chem.* **272**: 5128–5132.

Allan, G.F., Tsai, S.Y., Tsai, M.J. and O'Malley, B.W. (1992a) Ligand-dependent conformational changes in the progesterone receptor are necessary for events that follow DNA binding. *Proc. Natl. Acad. Sci. USA* **89**: 11 750–11 756.

Allan, G.F., Leng, X., Tsai, S.Y. *et al.* (1992b) Hormone and antihormone induce distinct conformational changes which are central to steroid receptor activation. *J. Biol. Chem.* **267**: 19 513–19 520.

Alland, L., Muhle, R., Hou, H., Jr. *et al.* (1997) Role for N-CoR and histone deacetylase in Sin3-mediated transcriptional repression. *Nature* **387**: 49–55.

Allen, E. and Doisy, E.A. (1923) An ovarian hormone. Preliminary report on its localization, extraction and partial purification and action in test animals. *JAMA* **81**: 819–821.

Allen, W.M. and Wintersteiner, O. (1934) Crystalline progestin. *Science* **80**: 190–191.

Andre, J. and Rochefort, H. (1973) Specific effect of estrogens on an interaction between the uterine estradiol receptor and DNA. *FEBS Lett* **29**: 135–140.

Antakly, T. and Eisen, H.J. (1984) Immunocytochemical localization of glucocorticoid receptor in target cells. *Endocrinology* **115**: 1984–1989.

Anzick, S.L., Kononen, J., Walker, R.L. *et al.* (1997) AIB1, a steroid receptor coactivator amplified in breast and ovarian cancer. *Science* **277**: 965–968.

Aronica, S.M. and Katzenellenbogen, B.S. (1993) Stimulation of estrogen-receptor-mediated transcription and alteration in the phosphorylation state of the rate uterine estrogen receptor by estrogen cyclic adenosine monophosphate and insulin-like growth factor-1. *Mol. Endocrinol.* **7**: 743–752.

Arriza, J.L., Weinberger, C., Cerelli, G. *et al.* (1987) Cloning of human mineralocorticoid receptor complementary DNA: structural and functional kinship with the glucocorticoid receptor. *Science* **237**: 268–275.

Ashburner, M. (1971) Induction of puffs in polytene chromosomes of *in vitro* cultured salivary glands of *Drosophila melanogaster* by ecdysone and ecdysone analogs. *Nature New Biol.* **230**: 222–224.

Ashburner, M. (1972) Ecdysone induction of puffing in polytene chromosomes of *Drosophila melanogaster*. Effects of inhibitors of RNA synthesis. *Exp. Cell Res.* **71**: 433–440.

Ashburner, M. (1973) Sequential gene activation by ecdysone in polytene chromosomes of *Drosophila melanogaster*. I. Dependence upon ecdysone concentration. *Dev. Biol.* **35**: 47–61.

Ashburner, M. (1974) Sequential gene activation by ecdysone in polytene chromosomes of *Drosophila melanogaster*. II. Effects of inhibitors of protein synthesis. *Dev. Biol.* **39**: 141–157.

Attardi, B. and Happe, K.H. (1986) Comparison of the physiochemical properties of uterine nuclear estrogen receptors bound to estradiol or 4-hydroxytamoxifen. *Endocrinology* **119**: 904–915.

Auricchio, F. (1989) Phosphorylation of steroid receptors. *J. Steroid Biochem.* **32**: 613–622.

Bagchi, M.K., Tsai, S.Y., Tsai, M.J. and O'Malley, B.W. (1990) Identification of a functional intermediate in receptor activation in progesterone-dependent cell-free transcription. *Nature* **345**: 547–550.

Bagchi, M.K., Tsai, S.Y., Tsai, M.J. and O'Malley, B.W. (1991) Progesterone enhances target gene transcription by receptor free of heat shock proteins hsp90, hsp56 and hsp70. *Mol. Cell. Biol.* **11**: 4998–5004.

Baker, A.R., McDonnell, D.P., Hughes, M. *et al.* (1988) Cloning and expression of full-length cDNA encoding human vitamin D receptor. *Proc. Natl. Acad. Sci. USA* **85**: 3294–3298.

Baniahmad, A., Steiner, C., Kohne, A.C. and Renkawitz, R. (1990) Modular structure of a chicken lysozyme silencer: involvement of an unusual thyroid hormone receptor binding site. *Cell* **61**: 505–514.

Baniahmad, A., Kohne, A.C. and Renkawitz, R. (1992) A transferable silencing domain is present in the thyroid hormone receptor, in the v-*erbA* oncogene product and in the retinoic acid receptor. *EMBO J.* **11**: 1015–1023.

Baniahmad, A., Ha, I., Reinberg, D., Tsai, S.Y., Tsai, M.J. and O'Malley, B.W. (1993) Interaction of human thyroid hormone receptor β with transcription factor TFIIB may mediate target gene derepression and activation by thyroid hormone. *Proc. Natl. Acad. Sci. USA* **90**: 8832–8836.

Baniahmad, A., Leng, X., Burris, T.P., Tsai, S.Y., Tsai, M.J. and O'Malley, B.W. (1995) The τ4 activation domain of the thyroid hormone receptor is required for release of a putative corepressor(s) necessary for transcriptional silencing. *Mol. Cell. Biol.* **15**: 76–86.

Bannister, A.J. and Kouzorides, T. (1996) The CBP coactivator is a histone acetyltransferase. *Nature* **384**: 641–643.

Beato, M., Kalimi, M., Konstam, M. and Feigelson, P. (1973) Interaction of glucocorticoids with rat liver nuclei. II. Studies on the nature of the cytosol transfer factor and the nuclear acceptor site. *Biochemistry* **12**: 3372–3379.

Beck, C.A., Weigel, N.L. and Edwards, D.P. (1992) Effects of hormone and cellular modulators of protein phosphorylation on transcriptional activity, DNA binding, and phosphorylation of human progesterone receptors. *Mol. Endocrinol.* **6**: 607–620.

Beebe, J.S., Darling, D.S. and Chin, W.W. (1991) 3,5,3'-Triiodothyronine receptor auxiliary protein (TRAP) enhances receptor binding by interactions within the thyroid hormone response element. *Mol. Endocrinol.* **5**: 85–93.

Beekman, J.M., Allan, G.F., Tsai, S.Y., Tsai, M.J. and O'Malley, B.W. (1993) Transcriptional activation by the estrogen receptor requires a conformational change in the ligand binding domain. *Mol. Endocrinol.* **7**: 1266–1274.

Benkoussa, M., Nomine, B., Mouchon, A. *et al.* (1998) Limited proteolysis for assaying ligand binding affinities of nuclear receptors. *Recept. Signal Transduction* **7**: 257–267.

Berger, J., Bailey, P., Biswas, C. *et al.* (1996) Thiazolidinedione produces a conformational change in peroxisome proliferator-activated receptor-γ: binding and activation correlate with antidiabetic actions in *db/db* mice. *Endocrinology* **137**: 4189–4195.

Berrodin, T.J., Marks, M.S., Ozato, K., Linney, E. and Lazar, M.A. (1992) Heterodimerization among thyroid hormone receptor, retinoic acid receptor, retinoid X receptor, chicken ovalbumin upstream promoter transcription factor, and an endogenous liver protein. *Mol. Endocrinol.* **6**: 1468–1478.

Blanco, J.C., Wang, I.M., Tsai, S.Y. *et al.* (1995) Transciption factor TFIIB and vitamin D receptor cooperatively activate ligand-dependent transcription. *Proc. Natl. Acad. Sci. USA* **92**: 1535–1539.

Bocquel, M.T., Kumar, V., Stricker, C., Chambon, P. and Gronemeyer, H. (1989) The contribution of the N- and C-terminal regions of steroid receptors to activation of transcription if both receptor and cell-specific. *Nucleic Acids Res.* **17**: 2581–2595.

Bourguet, W., Ruff, M., Chambon, P., Gronemeyer, H. and Moras, D. (1995) Crystal structure of the ligand-binding domain of the human nuclear receptor RXR-α. *Nature* **375**: 377–382.

Brink, M., Humbel, B.M., de Kloet, E.R. and van Driel, R. (1992) The unliganded glucocorticoid receptor is localized in the nucleus, not in the cytoplasm. *Endocrinology* **130**: 3575–3581.

Brzozowski, A.M., Pike, A.C.W., Dauter, Z. *et al.* (1997) Molecular basis of agonism and antagonism in the oestrogen receptor. *Nature* **389**: 753–758.

Buetti, E. and Diggelmann, H. (1981) Cloned mouse mammary tumor virus DNA is biologically active in transfected mouse cells and its expression is stimulated by glucocorticoid hormones. *Cell* **23**: 335–345.

Buetti, E. and Diggelmann, H. (1983) Glucocorticoid regulation of mouse mammary tumor virus: identification of short essential DNA region. *EMBO J.* **2**: 1423–1429.

Bugge, T.H., Pohl, J., Lonnoy, O. and Stunnenberg, H.G. (1992) RXRα, a promiscuous partner of retinoic acid and thyroid hormone receptors. *EMBO J.* **11**: 1409–1418.

Burris, T.P., Guo, W. and McCabe, E.R.B. (1996) The gene responsible for adrenal hypoplasia congenita, DAX-1, encodes a nuclear hormone receptor that defines a new class within the superfamily. *Recent Prog. Horm. Res.* **51**: 241–260.

Butenandt, A.F.J. (1929) Untersuchungen uber das weibliche sexualhormone. Darstellung und eigenshaften des kristallisierten 'progynons'. *Dtsch. Med. Wochenschr.* **55**: 2171–2173.

Butenandt, A. (1931) Uber die chemishce untersuchung der sexualhormone. *Ztschr. F. angwe. Chem.* **44**: 905.

Butenandt, A. and Harrisch, G. (1935) Ukk testosteron-unwandlung des dehydroandrostedions in androstendiol und testosteron; en weg sur. *Z. Physiol. Chem.* **237**: 89.

Butenandt, A.F.J. and Westphal, U. (1934) Zur isolierung und characterisierung des corpusluteum-hormons. *Ber. Dtsch. Chem. Ges.* **67**: 1440–1442.

Butenandt, A., Westphal, U. and Cobler, H. (1934) Uber einen abbau des stigmasterins zu corpus-luteum-wirksamen stoffen; ein beitrag zur konstitution des corpus-luteum-hormons (vorlauf mitteil). *Ber. Dtsch. Chem. Ges.* **67**: 1611–1616.

Carlstedt-Duke, J., Okret, S., Wrange, O. and Gustafsson, J.A. (1982) Immunochemical analysis of the glucocorticoid receptor: identification of a third domain separate from the steroid-binding and DNA-binding domains. *Proc. Natl. Acad. Sci. USA* **79**: 4260–4264.

Casanova, J., Helmer, E., Selmi-Ruby, S. *et al.* (1994) Functional evidence for ligand-dependent dissociation of thyroid hormone and retinoic acid receptors from an inhibitory cellular factor. *Mol. Cell. Biol.* **14**: 5756–5765.

Catelli, M.G., Binart, N., Jung-Testas, I. *et al.* (1985) The common 90-kD protein component of non-transformed '8S' steroid receptors is a heat-shock protein. *EMBO J.* **4**: 3131–3135.

Chakravarti, D., LaMorte, V.J., Nelson, M.C. *et al.* (1996) Role of CBP/p300 in nuclear receptor signalling. *Nature* **383**: 99–103.

Chan, L., Means, A.R. and O'Malley, B.W. (1973) Rates of induction of specific translatable messenger RNAs for ovalbumin and avidin by steroid hormones. *Proc. Natl. Acad. Sci. USA* **70**: 1870–1874.

Chandler, V.L., Maler, B.A. and Yamamoto, K.R. (1983) DNA sequences bound specifically by glucocorticoid receptor *in vitro* render a heterologous promoter hormone responsive *in vivo. Cell* **33**: 489–499.

Chang, C., Kokintis, J. and Liao, S. (1988) Molecular cloning of human and rat complementary DNA encoding androgen receptors. *Science* **240**: 324–326.

Chen, H., Lin, R.J., Schiltz, R.L. *et al.* (1997) Nuclear receptor coactivator ACTR is a novel histone acetyltransferase and forms a multimeric activation complex with P/CAF and CBP/p300. *Cell* **90**: 569–580.

Chen, J.D. and Evans, R.M. (1995) A transcriptional corepressor that interacts with nuclear hormone receptors. *Nature* **377**: 454–457.

Chen, J.D. and Li, H. (1998) Coactivation and corepression in transcriptional regulation by steroid/nuclear hormone receptors. *Crit. Rev. Euk. Gene Exp.* **8**: 169–190.

Chen, J.D., Umesono, K. and Evans, R.M. (1996) SMRT isoforms mediate repression and antirepression of nuclear receptor heterodimers. *Proc. Natl. Acad. Sci. USA* **93**: 7567–7571.

Cho, H. and Katzenellenbogen, B.S. (1993) Synergistic activation of estrogen receptor-mediated transcription by estradiol and protein kinase activators. *Mol. Endocrinol.* **7**: 441–452.

Compton, J.G., Schrader, W.T. and O'Malley, B.W. (1983) DNA sequence preference of the progesterone receptor. *Proc. Natl. Acad. Sci. USA* **80**: 16–20.

Comstock, J.P., Rosenfeld, G.C., O'Malley, B.W. and Means, A.R. (1972) Estrogen-induced changes in translation, and specific messenger RNA levels during oviduct differentiation. *Proc. Natl. Acad. Sci. USA* **69**: 2377–2380.

Conneely, O.M., Sullivan, W.P., Toft, D.O. *et al.* (1986) Molecular cloning of the chicken progesterone receptor. *Science* **233**: 767–770.

Conneely, O.M., Dobson, A.D.W., Tsai, M.J. *et al.* (1987) Sequence and expression of a functional chicken progesterone receptor. *Mol. Endocrinol.* **1**: 517–525.

Cooney, A.J. and Tsai, S.Y. (1994) Mechanism of steroid hormone regulation of gene transcription. In: M.J. Tsai and B.W. O'Malley (eds) *Nuclear Receptor–DNA Interactions*, pp. 25–59. R.G. Landes, Austin, Texas.

Corner, G.W. and Allen, W.M. (1929) Normal growth and implantation of embryos after very early ablation of the ovaries, under the influence of extracts of the corpus luteum. *Am. J. Phys.* **88**: 340–346.

Damm, K., Thompson, C.C. and Evans, R.M. (1989) Protein encoded by v-*erbA* functions as a thyroid hormone receptor antagonist. *Nature* **339**: 593–597.

D'Amour, F.E. and Gustavson, R.G. (1930) Preparation and assay of crystalline female sex hormone. *J. Pharmacol. Exp. Ther.* **40**: 485–488.

Danielian, P.S., White, R., Lees, J.A. and Parker, M.G. (1992) Identification of a conserved region required for hormone dependent transcriptional activation by steroid hormone receptors. *EMBO J.* **11**: 1025–1033.

Danielsen, M., Northrop, J.P. and Ringold, G.M. (1986) The mouse glucocorticoid receptor: mapping of functional domains by cloning, sequencing and expression of wild-type and mutant receptor proteins. *EMBO J.* **5**: 2513–2522.

Danielsen, M., Northrop, J.P., Jonkiaas, J. and Ringold, G.M. (1987) Domains of the glucocorticoid receptor involved in specific and nonspecific deoxyribonucleic acid binding, hormone activation, and transcriptional enhancement. *Mol. Endocrinol.* **1**: 816–822.

Danielsen, M., Hinck, L. and Ringold, G.M. (1989) Two amino acids within the knuckle of the first zinc finger specify DNA response element activation by the glucocorticoid receptor. *Cell* **57**: 1131–1138.

Darling, D.S., Beebe, J.S., Burnside, J., Winslow, E.R. and Chin, W.W. (1991) 3,5,3'-Triiodothyronine (T3) receptor-auxiliary protein (TRAP) binds DNA and forms heterodimers with the T3 receptor. *Mol. Endocrinol.* **5**: 73–84.

David, K., Dingemanse, E., Freud, J. and Laqueur, E. (1935) Uber krystallinisches mannliches hormon aus hoden (testosteron), wirksamer als aus harn order aus cholesterin bereitetes androsterone. *Ztschr. f. physiol. Chem.* **233**: 281.

Dean, D.C., Knoll, B.J., Riser, M.E. and O'Malley, B.W. (1983) A 5'-flanking sequence essential for progesterone regulation of an ovalbumin fusion gene. *Nature* **305**: 551–554.

Dean, D.C., Gope, R., Knoll, B.J., Riser, M.E. and O'Malley, B.W. (1984) A similar 5'-flanking region is required for estrogen and progesterone induction of ovalbumin gene expression. *J. Biol. Chem.* **359**: 9967–9970.

DeAngelo, A.B. and Gorski, J. (1970) Role of RNA synthesis in the estrogen induction of a specific uterine protein. *Proc. Natl. Acad. Sci. USA* **66**: 693–700.

DeMarzo, A.M., Beck, C.A., Onate, S.A. and Edwards, D.P. (1991) Dimerization of mammalian progesterone receptors occurs in the absence of DNA and is related to the release of the 90-kDa heat shock protein. *Proc. Natl. Acad. Sci. USA* **88**: 72–76.

Demay, M.B., Gerardi, J.M., DeLuca, H.F. and Kronenburg, H.M. (1990) DNA sequences in the rat osteocalcin gene that bind the 1,25-dihydroxyvitamin D3 receptor and confer responsiveness to 1,25-dihydroxyvitamin D3. *Proc. Natl. Acad. Sci. USA* **87**: 369–373.

Denis, M., Poellinger, L., Wikstom, A.C. and Gustafsson, J.A. (1988) Requirement of hormone for thermal conversion of the glucocorticoid receptor to a DNA-binding state. *Nature* **333**: 686–688.

Denner, L.A., Weigel, N.L., Maxwell, B.L., Schrader, W.T. and O'Malley, B.W. (1990) Regulation of progesterone receptor-mediated transcription by phosphorylation. *Science* **250**: 1740–1743.

De Sombre, E.R., Hurst, D., Kawashima, T., Jungblut, P.W. and Jensen, E.V. (1967) *Fed. Proc.* **26**: 536.

de The, H., Vivanco-Ruiz, M.M., Tiollais, P., Stunnenberg, H. and Dejean, A. (1990) Identification of a retinoic acid responsive element in the retinoic acid receptor β gene. *Nature* **343**: 177–180.

Dingenmanse, E., DeJonge, S.E., Kober, S. *et al.* (1930) Uber kristallinisches Menformen. *Dtsch. Wochenschr.* **56**: 301–304.

Direnzo, J., Soderstrom, M., Kurokawa, R. *et al.* (1997) Peroxisome proliferator-activated receptors and retinoic acid receptors differentially control the interactions of retinoid X receptor heterodimers with ligands, coactivators, and corepressors. *Mol. Cell. Biol.* **17**: 2166–2176.

Dobson, A.D.W., Conneely, O.M., Beattie, W. *et al.* (1989) Mutational analysis of the chicken progesterone receptor. *J. Biol. Chem.* **264**: 4207–4211.

Doisy, E.A., Veler, C.D. and Thayer, S. (1930) The preparation of the crystalline ovarian hormone from the urine of pregnant women. *J. Biol. Chem.* **86**: 499–509.

Dowell, P., Peterson, V.J., Zabriskie, T.M. and Leid, M. (1997) Ligand-induced peroxisome proliferator-activated receptor α conformational change. *J. Biol. Chem.* **272**: 2013–2020.

Downes, M., Burke, L.J., Bailey, P.J. and Muscat, G.E. (1996) Two receptor interaction domains in the corepressor, N-CoR/RIP13 are required for an efficient interaction with Rev-erbA alpha and RVR: physical association is dependent on the E region of orphan receptors. *Nucleic Acids Res.* **24**: 4379–4386.

Drouin, J., Sun, Y.L., Tremblay, S. *et al.* (1992) Homodimer formation is rate-limiting for high affinity DNA binding by glucocorticoid receptor. *Mol. Endocrinol.* **6**: 1299–1309.

Durand, B., Sauders, M., Leroy, P., Leid, M. and Chambon, P. (1992) All-trans and 9-*cis* retinoic acid induction of CRABPII is mediated by RAR-RXR heterodimers bound to DR1 and DR2 repeated motifs. *Cell* **71**: 73–85.

Durand, B., Saunders, M., Gaudon, C., Roy, B., Losson, R. and Chambon, P. (1994) Activation function 2 (AF-2) of retinoic acid receptor and 9-*cis* retinoic acid receptor: presence of a conserved autonomous constitutive activating domain and influence on the nature of the response element on AF-2 activity. *EMBO J.* **13**: 5370–5382.

Eggert, M., Mows, C.C., Tripier, D. *et al.* (1995) A fraction enriched in a novel glucocorticoid receptor interacting protein stimulates receptor dependent transcription *in vitro*. *J. Biol. Chem.* **270**: 30 755–30 759.

El-Ashry, D., Onate, S.A., Nordeen, S.K. and Edwards, D.P. (1989) Human progesterone receptor complexed with the antagonist RU 486 binds to hormone response elements in a structurally altered form. *Mol. Endocrinol.* **3**: 1545–1558.

Elliston, J.F., Fawell, S.E., Klein-Hitpass, L. *et al.* (1990) Mechanism of estrogen receptor-dependent transcription in a cell free system. *Mol. Cell. Biol.* **10**: 6607–6612.

Elliston, J.F., Beekman, J.M., Tsai, S.Y., O'Malley, B.W. and Tsai, M.J. (1992) Hormone activation of baculovirus expressed progesterone receptors. *J. Biol. Chem.* **267**: 5193–5198.

Fasel, N., Pearson, K., Buetti, E. and Diggelmann, H. (1982) The region of mouse mammary tumor virus DNA containing the long terminal repeat includes a long coding sequence and signals for hormone regulated transcription. *EMBO J.* **1**: 3–7.

Fawell, S.E., White, R., Hoare, S., Sydenham, M., Page, M. and Parker, M.G. (1990a) Inhibition of estrogen receptor-DNA binding by the 'pure' antiestrogen ICI 164 384 appears to be mediated by impaired receptor dimerization. *Proc. Natl. Acad. Sci. USA* **87**: 6883–6887.

Fawell, S.E., Lees, J.A., White, R. and Parker, M.G. (1990b) Characterization and colocalization of steroid binding and dimerization activities in the mouse estrogen receptor. *Cell* **60**: 953–962.

Faye, J.C., Fargin, A. and Bayard, F. (1986) Different interaction of estradiol and antiestrogens with the estrogen receptor of rat uterus. *Mol. Cell. Endocrinol.* **47**: 119–124.

Fondell, J.D., Roy, A.L. and Roeder, R.G. (1993) Unliganded thyroid hormone receptor inhibits formation of a functional preinitiation complex: implications for active repression. *Genes Dev.* **7**: 1400–1410.

Fondell, J.D., Ge, H. and Roeder, R.G. (1996) Ligand induction of a transcriptionally active thyroid hormone receptor coactivator complex. *Proc. Natl. Acad. Sci. USA* **93**: 8329–8333.

Forman, B.M., Tontonoz, P., Chen, J., Brun, R.P., Speigelman, B.M. and Evans, R.M. (1995) 15-Deoxy-Δ12,14-prostaglandin J2 is a ligand for the adipocyte determination factor PPARγ. *Cell* **83**: 803–812.

Forman, B.M., Ruan, B., Chen, J., Schroepfer, G.J. and Evans, R.M. (1997) The orphan nuclear receptor LXRα is positively and negatively regulated by distinct products of mevalonate metabolism. *Proc. Natl. Acad. Sci. USA* **94**: 10 588–10 593.

Forman, B.M., Tzameli, I., Choi, H.S. *et al.* (1998) Androstane metabolites bind to and deactivate the nuclear receptor CAR-β. *Nature* **395**: 612–615.

Freedman, L.P., Luisi, B.F., Korszun, Z.R., Basavappa, R., Sigler, P.B. and Yamamoto, K.R. (1988) The function and structure of the metal coordination sites within the glucocorticoid receptor DNA-binding domain. *Nature* **334**: 543–546.

Fritsch, M., Leary, C.M., Furlow, D.J. *et al.* (1992) A ligand-induced conformational change in the estrogen receptor is localized in the steroid binding domain. *Biochemistry* **31**: 5303–5311.

Fronsdal, K., Engedal, N., Slagsvold, T. and Saatcioglu, F. (1998) CREB binding protein is a coactivator for the androgen receptor and mediates crosstalk with AP-1. *J. Biol. Chem.* **273**: 31 853–31 859.

Gasc, J.M., Renoir, J.M., Radanyi, C., Joab, I., Tuohimaa, P. and Baulieu, E.E. (1984) Progesterone receptor in the chick oviduct: an immunohistochemical study with antibodies to distinct receptor components. *J. Cell Biol.* **99**: 1193–1201.

Gasc, J.M., Delahaye, F. and Baulieu, E.E. (1989) Compared intracellular localization of the glucocorticosteroid and progesterone receptors: an immunocytochemical study. *Exp. Cell Res.* **181**: 492–504.

Gehring, U. and Tomkins, G.M. (1974) New mechanism for steroid unresponsiveness. Loss of nuclear binding activity of a steroid hormone receptor. *Cell* **3**: 301–306.

Georget, V., Lobaccaro, J.M., Terouanne, B., Mangeat, P., Nicolas, J.C. and Sultan, C. (1997) Trafficking of the androgen receptor in living cells with fused green fluorescent protein-androgen receptor. *Mol. Cell. Endocrinol.* **129**: 17–26.

Giguere, V., Hollenberg, S.M., Rosenfeld, M.G. and Evans, R.M. (1986) Functional domains of the human glucocorticoid receptor. *Cell* **46**: 645–652.

Giguere, V., Ong, E.S., Segui, P. and Evans, R.M. (1987) Identification of a receptor for the morphogen retinoic acid. *Nature* **330**: 624–629.

Giguere, V., Yang, N., Segui, P. and Evans, R.M. (1988) Identification of a new class of steroid hormone receptors. *Nature* **331**: 91–94.

Ginsberg, A.M., King, B.A. and Roeder, R.G. (1984) Xenopus 5S gene transcription factor, TFIIIA: characterization of a cDNA clone and measurement of RNA levels throughout development. *Cell* **39**: 479–489.

Glass, C.K., Devary, O.V. and Rosenfeld, M.G. (1990) Multiple cell type-specific proteins differentially regulate target sequence recognition by the α retinoic acid receptor. *Cell* **63**: 729–738.

Glasscock, R.F. and Hoekstra, W.G. (1959) Selective accumulation of tritium-labeled hexosterol by the reproductive organs of immature female goats and sheep. *Biochem. J.* **72**: 673–682.

Glasser, S.R., Chytil, F. and Spelsberg, T.C. (1972) Early effects of oestradiol-17β on the chromatin and activity of the deoxyribonucleic acid-dependent ribonucleic acid polymerases (I and II) of the rat uterus. *Biochem. J.* **130**: 947–957.

Godowski, P.J., Rusconi, S., Miesfeld, R. and Yamamoto, K.R. (1987) Glucocorticoid receptor mutants that are constitutive activators of transcriptional enhancement. *Nature* **325**: 365–368.

Godowski, P.J., Picard, D. and Yamamoto, K.R. (1988) Signal transduction and transcriptional regulation by glucocorticoid-receptor-LexA fusion proteins. *Science* **241**: 812–816.

Gorski, J. (1964) Early estrogen effects on the activity of uterine ribonucleic acid polymerase. *J. Biol. Chem.* **239**: 889–892.

Gorski, J., Toft, D., Shyamala, G., Smith, D. and Notides, A. (1968) Hormone receptors: studies on the interaction of estrogen with uterus. *Recent Progr. Horm. Res.* **24**: 45.

Govindan, M.V., Spiess, E. and Majors, J. (1982) Purified glucocorticoid receptor–hormone complex from rat liver cytosol binds specifically to cloned mouse mammary tumor virus long terminal repeats *in vitro*. *Proc. Natl. Acad. Sci. USA* **79**: 5157–5161.

Govindan, M.V., Burelle, M., Cantin, C. *et al.* (1988) Cloning of the human androgen receptor cDNA. *Prog. Cancer Res. Ther.* **35**: 49–54.

Graupner, G., Wills, K.N., Tzukerman, M., Zhang, X.K. and Pfahl, M. (1989) Dual regulatory role for thyroid-hormone receptors allows control of retinoic-acid receptor activity. *Nature* **340**: 653–656.

Green, S. and Chambon, P. (1987) Oestradiol induction of a glucocorticoid-responsive gene by a chimaeric receptor. *Nature* **325**: 75–78.

Green, S., Walter, P., Kumar, V. *et al.* (1986) Human oestrogen receptor cDNA: sequence expression, and homology to v-*erb-A*. *Nature* **320**: 134–139.

Green, S., Kumar, V., Theulaz, I., Wahli, W. and Chambon, P. (1988) The N-terminal DNA-binding 'zinc finger' of the estrogen and glucocorticoid receptors determine target gene specificity. *EMBO J.* **7**: 3037–3044.

Greene, G.L., Gilna, P., Waterfield, M., Baker, A., Hort, Y. and Shine, J. (1986) Sequence and expression of human estrogen receptor complementary DNA. *Science* **231**: 1150–1154.

Greenman, D.L., Wicks, W.D. and Kenney, F.T. (1965) Stimulation of ribonucleic acid synthesis by steroid hormones. II. High molecular weight components. *J. Biol. Chem.* **240**: 4420–4426.

Gronemeyer, H., Turcotte, B., Quirin-Stricker, C. *et al.* (1987) The chicken progesterone receptor: sequence, expression and functional analysis. *EMBO J.* **6**: 3985–3994.

Groyer, A., Schweizer-Groyer, G., Cadepond, F., Mariller, M. and Baulieu, E.E. (1987) Antiglucocorticoid effects suggest why steroid hormone is required for receptors to bind DNA *in vivo* but not *in vitro*. *Nature* **328**: 624–626.

Gu, W., Malik, S., Ito., M. *et al.* (1999) A novel human SRB/MED-containing cofactor complex, SMCC, involved in transcriptional regulation. *Mol. Cell* **3**: 97–108.

Gudernatsch, J.F. (1912) Wilhelm Roux' Arch. *Entwicklungsmech. Organ.* **35**: 457.

Guiochon-Mantel., A., Lossfelt, H., Lescop, P. *et al.* (1989) Mechanisms of nuclear localization of the progesterone receptor: evidence for interaction between monomers. *Cell* **57**: 1147–1154.

Guiochon-Mantel, A., Lescop, P., Christin-Maitre, S. *et al.* (1991) Nucleocytoplasmic shuttling of the progesterone receptor. *EMBO J.* **10**: 3851–3859.

Halachmi, S., Marden, E., Martin, G., MacKay, H., Abbondanza, C. and Brown, M. (1994) Estrogen receptor-associated proteins: possible mediators of hormone-induced transcription. *Science* **264**: 1455–1458.

Ham, J., Thomson, A., Needham, M., Webb, P. and Parker, M. (1988) Characterization of response elements for androgens, glucocorticoids and progestins in mouse mammary tumor virus. *Nucleic Acids Res.* **16**: 5263–5276.

Hansen, J.C. and Gorski, J. (1985) Conformational and electrostatic properties of unoccupied and liganded estrogen receptors determined by aqueous two-phase partitioning. *Biochemistry* **24**: 6078–6085.

Hansen, J.C. and Gorski, J. (1986) Conformational transitions of the estrogen receptor monomer. Effects of estrogens, antiestrogen, and temperature. *J. Biol. Chem.* **261**: 13 990–13 996.

Hanstein, B., Eckner, R., DiRenzo, J. *et al.* (1996) P300 is a component of an estrogen receptor coactivator complex. *Proc. Natl. Acad. Sci. USA* **93**: 11 540–11 545.

Hard, T., Kellenbach, E., Boelens, R. *et al.* (1990) Solution structure of the glucocorticoid receptor DNA-binding domain. *Science* **249**: 157–160.

Harrison, S.C. (1991) A structural taxonomy of DNA-binding domains. *Nature* **353**: 715–719.

Hartmann, M. and Wettstein, A. (1934) Ein krystallisiertes hormon aus corpus luteum. *Helv. Chim. Acta* **17**: 878–882.

Heery, D.M., Kalkhoven, E., Hoare, S. and Parker, M.G. (1997) A signature motif in transcriptional coactivators mediates binding to nuclear receptors. *Nature* **387**: 733–736.

Heinzel, T., Lavinsky, R.M., Mullen, T.M. *et al.* (1997) A complex containing N-CoR, mSin3 and histone deacetylase mediates transcriptional repression. *Nature* **387**: 43–48.

Heyman, R.A., Mangelsdorf, D.J., Dyck, J.A. *et al.* (1992) 9-*Cis* retinoic acid is a high affinity ligand for the retinoid X receptor. *Cell* **68**: 397–406.

Hollenberg, S.M. and Evans, R.M. (1988) Multiple and cooperative trans-activation domains of the human glucocorticoid receptor. *Cell* **55**: 899–906.

Hollenberg, S.M., Weinberger, C., Ong. E.S. *et al.* (1985) Primary structure and expression of a functional human glucocorticoid receptor cDNA. *Nature* **318**: 635–641.

Hollenberg, S.M., Giguere, V., Sequi, P. and Evans, R.M. (1987) Colocalization of DNA-binding and transcriptional activation functions in the human glucocorticoid receptor. *Cell* **49**: 39–46.

Hong, H., Kohli, K., Trivedi, A., Johnson, D.L. and Stallcup, M.R. (1996) GRIP1, a novel mouse protein that serves as a transcriptional coactivator in yeast for the hormone binding domains of steroid receptors. *Proc. Natl. Acad. Sci. USA* **93**: 4948–4952.

Hong, H., Kohli, K., Garabedian, M.J. and Stallcup, M.R. (1997) GRIP1, a transcriptional coactivator for the AF2 transactivation domain of steroid, thyroid, retinoid and vitamin D receptors. *Mol. Cell. Biol.* **17**: 2735–2744.

Horlein, A.J., Naar, A.M., Heinzel, T. *et al.* (1995) Ligand-independent repression by the thyroid hormone receptor mediated by a nuclear receptor corepressor. *Nature* **377**: 397–404.

Htun, H., Barsony, J., Renyi, J., Gould, D.L. and Hager, G.L. (1996) Visualization of glucocorticoid receptor translocation and intranuclear organization in living cells with a green fluorescent protein chimera. *Proc. Natl. Acad. Sci. USA* **93**: 4845–4850.

Htun, H., Holth, L.T., Walker, D., Davie, J.R. and Hager, G.L. (1999) Direct visualization of the human estrogen receptor α reveals a role for ligand in the nuclear distribution of the receptor. *Mol. Cell. Biol.* **10**: 471–486.

Hu, E., Kim, J.B., Sarraf, P. and Spiegelman, B.M. (1996) Inhibition of adipogenesis through MAP kinase-mediated phosphorylation of PPARγ. *Science* **274**: 2100–2103.

Hu, X. and Lazar, M.A. (1999) The CoRNR motif controls the recruitment of corepressors by nuclear hormone receptors. *Nature* **402**: 93–96.

Huang, A.L., Ostrowski, M.C., Berard, D. and Hager, G.L. (1981) Glucocorticoid regulation of the Ha-MuSV p21 gene conferred by sequences from mouse mammary tumor virus. *Cell* **27**: 245–255.

Husmann, D.A., Wilson, C.M., McPhaul, M.J., Tilley, W.D. and Wilson, J.D. (1990) Antipeptide antibodies to two distinct regions of the androgen receptor localize the receptor protein to the nuclei of target cells in the rat and human prostate. *Endocrinology* **126**: 2359–2368.

Hynes, N.E., Kennedy, N., Rahmsdorf, U. and Groner, B. (1981) Hormone-responsive expression of an endogenous proviral gene of mouse mammary tumor virus after molecular cloning and gene transfer into cultured cells. *Proc. Natl. Acad. Sci. USA* **78**: 2038–2042.

Hynes, N., van Ooyen, A.J.J., Kennedy, N., Herrlich, P., Ponta, H. and Groner, B. (1983) Subfragments of the large terminal repeat cause glucocorticoid responsive expression of mouse mammary tumor virus and of an adjacent gene. *Proc. Natl. Acad. Sci. USA* **80**: 3637–3641.

Ing, N.H., Beekman, J.M., Tsai, S.Y., Tsai, M.J. and O'Malley, B.W. (1992) Members of the steroid hormone receptor superfamily interact with TFIIB (S300II). *J. Biol. Chem.* **267**: 17 617–17 623.

Ito, M., Yuan, C.X., Malik, S. *et al.* (1999) Identity between TRAP and SMCC complexes indicates novel pathways for the function of nuclear receptors and diverse mammalian activators. *Mol. Cell* **3**: 361–370.

Izumo, S. and Mahdavi, V. (1988) Thyroid hormone receptor alpha isoforms generated by alternative splicing differentially activate myosin HC gene transcription. *Nature* **334**: 539–542.

Jackson, T.A., Richer, J.K., Bain, D.L., Takimoto, G.S., Tung, L. and Horwitz, K.B. (1997) The partial agonist activity of antagonist-occupied steroid receptors is controlled by a novel hinge domain-binding coactivator L7/SPA and the corepressors N-CoR and SMRT. *Mol. Endocrinol.* **11**: 693–705.

Jacq, X., Brou, C., Lutz, Y., Davidson, I., Chambon, P. and Tora, L. (1994) Human TAFII30 is present in a distinct TFIID complex and is required for transcriptional activation by the estrogen receptor. *Cell* **79**: 107–117.

Janowski, B.A., Willy, P.J., Devi, T.R., Falck, J.R. and Mangelsdorf, D.J. (1996) An oxysterol signalling pathway mediated by the nuclear receptor LXRα. *Nature* **383**: 728–731.

Janowski, B.A., Grogan, M.J., Jones, S.A. *et al.* (1999) Structural requirements of ligands for the oxysterol liver X receptors LXRα and LXRβ. *Proc. Natl. Acad. Sci. USA* **96**: 266–271.

Jasper, T.W., Ruh, M.F. and Ruh, T.S. (1985) Estrogen and antiestrogen binding to rat uterine and pituitary estrogen receptor: evidence for at least two physiochemical forms of the estrogen receptor. *J. Steroid Biochem.* **23**: 537–545.

Jensen, E.V. and De Sombre, E.R. (1972) Mechanism of action of the female sex hormones. *Ann. Rev. Biochem.* **41**: 203–230.

Jensen, E.V. and Jacobson, H.I. (1960) Fate of steroid estrogens in target tissues. In: G. Pincus and E.P. Vollmer (eds) *Biological Activities of Steroids in Relation to Cancer*, pp. 161–178. Academic Press, New York.

Jensen, E.V. and Jacobson, H.I. (1962) Basic guides to the mechanism of estrogen action. *Recent Prog. Horm. Res.* **18**: 387–414.

Jensen, E.V., Jacobson, H.I., Flesher, J.W. *et al.* (1966) Estrogen receptors in target tissues. In: G. Pincus, T. Nakao and J.F. Tait (eds) *Steroid Dynamics*, pp. 133–156. Academic Press, New York.

Jensen, E.V., De Sombre, E. and Jungblut, P.W. (1967) In: *Proc. 2nd Int. Congr. Hormonal Steroids* (eds L. Martini, F. Franschini and M. Motto), Interaction of estrogens with receptor sites *in vivo* and *in vitro*. **132**: 492–500.

Jensen, E.V., Suzuki, T., Kawashima, T., Stumpf, W.E., Jungblut, P.W. and De Sombre, E.R. (1968) A two-step mechanism for interaction of oestradiol with rat uterus. *Proc. Natl. Acad. Sci. USA* **59**: 632.

Joab, I., Radanyi, C, Renoir, M. *et al.* (1984) Common non-hormone binding component in non-transformed chick oviduct receptors of four steroid hormones. *Nature* **308**: 850–853.

Jost, J.P., Seldran, M. and Geiser, M. (1984) Preferential binding of estrogen–receptor complex to a region containing of estrogen-dependent hypomethylation site preceding the chicken vitellogenin II gene. *Proc. Natl. Acad. Sci. USA* **81**: 429–433.

Jost, J.P., Geisser, M. and Seldran, M. (1985) Specific modulation of the transcription of cloned avian vitellogenin II gene by estradiol–receptor complex *in vitro*. *Proc. Natl. Acad. Sci. USA* **82**: 988–991.

Kalff, M., Gross, B. and Beato, M. (1990) Progesterone receptor stimulates transcription of mouse mammary tumour virus in a cell-free system. *Nature* **344**: 360–362.

Kamei, Y., Xu, L., Heinzel, T. *et al.* (1996) A CBP cointegrator complex mediates transcriptional activation and AP-1 inhibition by nuclear receptors. *Cell* **85**: 403–414.

Karin, M., Haslinger, A., Holtgreve, H., Cathala, G., Slater, E. and Baxter, J.D. (1984a) Activation of a heterologous promoter in response to dexamethasone and cadmium by metallothionein gene 5'-flanking DNA. *Cell* **36**: 371–379.

Karin, M., Haslinger, A., Holtgreve, H. *et al.* (1984b) Characterization of DNA sequences through which cadmium and glucocorticoid hormones induce human metallothionein-IIA gene. *Nature* **308**: 513–519.

Kendall, E.C. (1915) The isolation in crystalline form of the compound containing iodine which occurs in the thyroid: its chemical nature and physiological activity. *JAMA* **64**: 2042.

Kenney, F.T., Wicks, W.D. and Green, D.L. (1965) Hydrocortisone stimulation of RNA synthesis in induction of hepatic enzymes. *J. Cell. Comp. Physiol.* **66**: 125–136.

King, R.J.B. and Gordon, J. (1972) Involvement of DNA in the acceptor mechanism for uterine estradiol receptor. *Nature New Biol.* **240**: 185–187.

King, R.J.B., Gordon, J., Cowan, D.M. and Inman, D.R. (1966) The intranuclear localization of [6,7-^{3}H]-oestradiol-17β in dimethylbenzathracene-induced rat mammary adenocarcinoma and other tissues. *J. Endocrinol.* **36**: 139–150.

King, W.J. and Greene, G.L. (1984) Monoclonal antibodies localize oestrogen receptor in the nuclei of target cells. *Nature* **307**: 745–747.

Klein-Hitpass, L., Schorpp, M., Wagner, U. and Ryffel, G.U. (1986) An estrogen-responsive element derived from 5' flanking region of the xenopus vitellogenin A2 gene functions in transfected human cells. *Cell* **46**: 1053–1061.

Klein-Hitpass, L., Tsai, S.Y., Greene, G.L., Clark, J.H., Tsai, M.J. and O'Malley, B.W. (1989) Specific binding of estrogen receptor to the estrogen response element. *Mol. Cell. Biol.* **9**: 43–49.

Klein-Hitpass, L., Tsai, S.Y., Weigel, N.L. *et al.* (1990) The progesterone receptor stimulates cell-free transcription by enhancing the formation of a stable preinitiation complex. *Cell* **60**: 247–257.

Kliewer, S.A., Umesono, K., Mangelsdorf, D.J. and Evans, R.M. (1992a) Retinoid X receptor interacts with nuclear receptors in retinoic acid, thyroid hormone, and vitamin D3 signaling. *Nature* **355**: 446–449.

Kliewer, S.A., Umesono, K., Heyman, R.A., Mangelsdorf, D.J., Dyck, J.A. and Evans, R.M. (1992b) Retinoid X receptor–COUP-TF interactions modulate retinoic acid signaling. *Proc. Natl. Acad. Sci. USA* **89**: 1448–1458.

Kliewer, S.A., Umesono, K., Noonan, D.J., Heyman, R.A. and Evans, R.M. (1992c) Convergence of 9-*cis* retinoic acid and peroxisome proliferator signaling pathways through heterodimer formation of their receptors. *Nature* **358**: 771–774.

Kliewer, S.A., Lenhard, J.M., Willson, T.M., Patel, I., Morris, D.C. and Lehmann, J. (1995) A prostaglandin J2 metabolite binds peroxisome proliferator-activated receptor γ and promotes adipocyte differentiation. *Cell* **83**: 813–819.

Knegtel, R.M., Katahira, M., Schilthuis, J.G. *et al.* (1993) The solution structure of the human retinoic acid receptor-beta DNA-binding domain. *Biomol. NMR* **3**: 1–17.

Koenig, R.J., Brent, G.A., Warne, R.L., Larsen, P.R. and Moore, D.D. (1987) Thyroid hormone receptor binds to a site in the rat growth hormone promoter required for induction by thyroid hormone. *Proc. Natl. Acad. Sci. USA* **84**: 5670–5674.

Krust, A., Green, S., Argos, P. *et al.* (1986) The chicken oestrogen receptor sequence: homology with v-*erbA* and the human oestrogen and glucocorticoid receptors. *EMBO J.* **5**: 891–897.

Kumar, V. and Chambon, P. (1988) The estrogen receptor binds tightly to its responsive element as a ligand-induced homodimer. *Cell* **55**: 145–156.

Kumar, V., Green, S., Staub, A. and Chambon, P. (1986) Localization of the estradiol-binding and putative DNA-binding domains of the human estrogen receptor. *EMBO J.* **5**: 2231–2236.

Kumar, V., Green, S., Stack, G., Berry, M., Jin, J.R. and Chambon, P. (1987) Functional domains of the human estrogen receptor. *Cell* **51**: 941–951.

Kurokawa, K., Yu, V.C., Naar, A. *et al.* (1993) Differential orientations of the DNA-binding domain and carboxy-terminal dimerization interface regulate binding site selection by nuclear receptor heterodimers. *Genes Dev.* **7**: 1423–1435.

Kurtz, D.T. (1981) Hormone inducibility of rat alpha 2u globulin genes in transfected mouse cells. *Nature* **291**: 629–631.

Lai, E.C., Riser, M.E. and O'Malley, B.W. (1983) Regulated expression of the chicken ovalbumin gene in a human estrogen-responsive cell line. *J. Biol Chem.* **258**: 12 693–12 701.

Lanz, R.B., McKenna, N.J., Onate, S.A. *et al.* (1999) A steroid receptor coactivator, SRA, functions as an RNA and is present in an SRC-1 complex. *Cell* **97**: 17–27.

Leblanc, B.P. and Stunnenberg, H.G. (1995) 9-*Cis* retinoic acid signaling: changing partners causes some excitement. *Genes Dev.* **9**: 1811–1816.

Lee, F., Mulligan, R., Berg, P. and Ringold, G. (1981) Glucocorticoids regulate expression of dihydrofolate reductase cDNA in mouse mammary tumour virus chimaeric plasmids. *Nature* **294**: 228–232.

Lee, M.S., Kliewer, S.A., Provencal, J., Wright, P.E. and Evans, R.M. (1993) Structure of the retinoid X receptor α DNA-binding domain: a helix required for homodimeric DNA binding. *Science* **260**: 1117–1121.

Lees, J.A., Fawell, S.E. and Parker, M.G. (1989) Identification of two transactivation domains in the mouse oestrogen receptor. *Nucleic Acids Res.* **17**: 5477–5488.

Lehmann, J.M., Kliewer, S.A., Moore, L.B. *et al.* (1997) Activation of the nuclear receptor LXR by oxysterols defines a new hormone response pathway. *J. Biol. Chem.* **272**: 3137–3140.

Leng, X., Tsai, S.Y., O'Malley, B.W. and Tsai, M.J. (1993) Ligand-dependent conformational changes in thyroid hormone and retinoic acid receptors are potentially enhanced by heterodimerization with retinoic X receptor. *J. Steroid Biochem. Mol. Biol.* **46**: 643–661.

Levin, A.A., Sturzenbecker, L.J., Kazmer, S. *et al.* (1992) 9-*Cis* retinoic acid stereoisomer binds and activates the nuclear receptor RXRα. *Nature* **355**: 359–361.

Li, H., Gomes, P.J. and Chen, J.D. (1997a) RAC3, a steroid/nuclear receptor-associated coactivator that is related to SRC-1 and TIF-2. *Proc. Natl. Acad. Sci. USA* **94**: 8479–8484.

Li, H., Leo, C. and Chen, J.D. (1997b) Characterization of receptor interaction and transcriptional repression by the corepressor SMRT. *Mol. Endocrinol.* **11**: 2025–2037.

Lim, C.S., Baumann, C.T., Htun, H. *et al.* (1999) Differential localization and activity of the A- and B-forms of the human progesterone receptor using green fluorescent protein chimeras. *Mol. Endocrinol.* **13**: 366–375.

Liu, Z., Wong, J., Tsai, S.Y., Tsai, M.J. and O'Malley, B.W. (1999) Steroid receptor coactivator-1 (SRC-1) enhances ligand-dependent and receptor-dependent cell-free transcription of chromatin. *Proc. Natl. Acad. Sci. USA* **96**: 9485–9490.

Loosfelt, H., Atger, M., Misrahi, M. *et al.* (1986) Cloning and sequence analysis of rabbit progesterone-receptor complementary DNA. *Proc. Natl. Acad. Sci. USA* **83**: 9045–9049.

Lubahn, D.B., Joesph, D.R., Sullivan, P.M., Willard, H.F., French, F.S. and Wilson, E.M. (1988) Cloning of human androgen receptor complementary DNA and localization to the X chromosome. *Science* **240**: 327–330.

Luisi, B.F., Xu, W.X., Otwinowski, Z., Freedman, L.P., Yamamoto, K.R. and Sigler, P.B.

(1991) Crystallographic analysis of the interaction of the glucocorticoid receptor with DNA. *Nature* **352**: 497–505.

MacDonald, P.N., Sherman, D.R., Dowd, D.R., Jefcoat, S.C. and DeLisle, R.K. (1995) The vitamin D receptor interacts with general transcription factor IIB. *J. Biol. Chem.* **270**: 4748–4752.

Mader, S., Kumar, V., de Verneuil, H. and Chambon, P. (1989) Three amino acids of the oestrogen receptor are essential to its ability to distinguish an oestrogen from a glucocorticoid-response element. *Nature* **338**: 271–274.

Mader, S., Chambon, P. and White, J.H. (1993a) Defining a minimal estrogen receptor DNA-binding domain. *Nucleic Acids Res.* **21**: 1125–1132.

Mader, S., Leroy, P., Chen, J.Y. and Chambon, P. (1993b) Multiple parameters control the selectivity of nuclear receptors for their response elements selectivity and promiscuity in response element recognition by retinoic acid receptors and retinoid X receptors. *J. Biol. Chem.* **268**: 591–600.

Mainwaring, W.I.P. and Peterken, B.A. (1971) A reconstituted cell-free system for the specific transfer of steroid–receptor complexes into nuclear chromatin isolated from rat ventral prostate gland. *Biochem. J.* **125**: 285–295.

Majors, J. and Varmus, H.E. (1983) A small region of the mouse mammary tumor virus long terminal repeat confers glucocorticoid hormone regulation on a linked heterologous gene. *Proc. Natl. Acad. Sci. USA* **80**: 5866–5870.

Makishima, M., Okamoto, A.Y., Repa, J.J. *et al.* (1999) Identification of a nuclear receptor for bile acids. *Science* **284**: 1362–1365.

Mangelsdorf, D.J., Umensono, K., Kliewer, S.A., Borgmeyer, U., Ong, E.S. and Evans, R.M. (1991) A direct repeat in the cellular retinol-binding protein type II gene confers differential regulation by RXR and RAR. *Cell* **66**: 555–561.

Marks, M.S., Hallenbeck, P.L., Nagata, T. *et al.* (1992) H-2RIIBP (RXRβ) heterodimerization provides a mechanism for combinatorial diversity in the regulation of retinoic acid and thyroid hormone responsive genes. *EMBO J.* **11**: 1419–1435.

Martin, P.M., Berthois, Y. and Jensen, E.V. (1988) Binding of antiestrogens exposes an occult antigenic determinate in the human estrogen receptor. *Proc. Natl. Acad. Sci. USA* **85**: 2533–2537.

Martins, V.R., Pratt, W.B., Terracio, L., Hirst, M.A., Ringold, G.M. and Housley, P.R. (1991) Demonstration by confocal microscopy that unliganded overexpressed glucocorticoid receptors are distributed in a nonrandom manner throughout all planes of the nucleus. *Mol. Endocrinol.* **5**: 217–225.

Maurer, H.R. and Chalkley G.R. (1967) Some properties of a nuclear binding site of estradiol. *J. Mol. Biol.* **27**: 431–441.

McDonnell, D.P., Mangelsdorf, D.J., Pike, J.W., Haussler, M.R. and O'Malley, B.W. (1987) Molecular cloning of complementary DNA encoding the avian receptor for vitamin D. *Science* **234**: 1214–1217.

McEwan, I. and Gustafsson, J.A. (1997) Interaction of the human androgen receptor transactivation function with the general transcription factor TFIIF. *Proc. Natl. Acad. Sci. USA* **94**: 8485–8490.

McKenna, N.J., Lanz, R.B. and O'Malley, B.W. (1999) Nuclear receptor coactivators: cellular and molecular biology. *Endocr. Rev.* **20**: 321–344.

Means, A.R. and Hamilton, T.H. (1966) Early estrogen action: concomitant stimulations within two minutes of nuclear RNA synthesis and uptake of RNA precursor by the uterus. *Proc. Natl. Acad. Sci. USA* **56**: 1594–1598.

Means, A.R., Comstock, J.P., Rosenfeld, G.C. and O'Malley, B.W. (1972) Ovalbumin messenger RNA of chick oviduct: partial characterization, estrogen dependence, and translation *in vitro*. *Proc. Natl. Acad. Sci. USA* **69**: 1146–1150.

Meinke, G. and Sigler, P.B. (1999) DNA-binding mechanism of the monomeric orphan nuclear receptor NGFI-B. *Nature Struct. Biol.* **6**: 471–477.

Meyer, M.E., Gronemeyer, H., Turcotte, B., Bocquel, M.T., Tasset, D. and Chambon, P. (1989) Steroid hormone receptors compete for factors that mediate their enhancer function. *Cell* **57**: 433–442.

Miesfeld, R., Rusconi, S., Godowski, P.J. *et al.* (1986) Genetic complementation of a glucocorticoid receptor deficiency by expression of a cloned receptor cDNA. *Cell* **46**: 389–399.

Miesfeld, R., Godowski, P.J., Maler, B.A. and Yamamoto, K.R. (1987) Glucocorticoid receptor mutants that define a small region sufficient for enhancer activation. *Science* **236**: 423–427.

Miller, J., McLachlan, A.D. and Klug, A. (1985) Repetitive zinc-binding domains in the protein transcription factor IIA from *Xenopus* oocytes. *EMBO J.* **4**: 1609–1614.

Misrahi, M., Atger, M., d'Auriol, L. *et al.* (1987) Complete amino acid sequence of the human progesterone receptor deduced from cloned cDNA. *Biochem. Biophys. Res. Commun.* **143**: 740–748.

Miyajima, N., Kadowaki, Y., Fukushige, S. *et al.* (1988) Identification of two novel members of erbA superfamily by molecular cloning: the gene products of the two are highly related to each other. *Nucleic Acids Res.* **16**: 11 057–11 074.

Mohla, S., De Sombre, E.R. and Jensen, E.V. (1972) Tissue-specific stimulation of RNA synthesis by transformed estradiol–receptor complex. *Biochem. Biophys. Res. Commun.* **46**: 661–667.

Moras, D. and Gronemeyer, H. (1998) The nuclear receptor ligand-binding domain: structure and function. *Curr. Opin. Cell Biol.* **10**: 384–391.

Mueller, G.C., Herranen, A.M. and Jervell, K.F. (1958) Studies on the mechanism of action of estrogens. *Recent Progr. Horm. Res.* **14**: 95.

Mukherjee, R., Davies, P.J.A., Crombie, D.L. *et al.* (1997) Sensitization of diabetic and obese mice to insulin by retinoid X receptor agonists. *Nature* **386**: 407–410.

Mukherjee, R., Strasser, J., Jow, L., Hoener, P., Paterniti, J.R. and Heyman, R.A. (1998) RXR agonists activate PPARa-inducible genes, lower triglycerides, and raise HDL levels *in vivo*. *Arterioscler. Thromb. Vasc. Biol.* **18**: 272–276.

Mulvihill, E.R., LePennec, J.P. and Chambon, P. (1982) Chicken oviduct progesterone receptor: location of specific regions of high-affinity binding in cloned DNA fragments of hormone-responsive genes. *Cell* **24**: 621–632.

Musliner, T.A. and Chader, G.J. (1972) Estradiol-receptors of the rat uterus. Interaction of the cytoplasmic estrogen-receptor with DNA *in vitro*. *Biochim. Biophys. Acta* **262**: 256–263.

Naar, A.M., Boutin, J.M., Lipkin, S.M. *et al.* (1991) The orientation and spacing of core DNA-binding motifs dictate selective transcriptional responses to three nuclear receptors. *Cell* **65**: 1267–1279.

Naar, A.M., Beaurang, P.A., Zhou, S., Abraham, S., Solomon, W. and Tjian, R. (1999) Composite co-activator ARC mediates chromatin-directed transcriptional activation. *Nature* **398**: 828–832.

Nagy, L., Kao, H.Y., Chakravarti, D. *et al.* (1997) Nuclear receptor repression mediated by a complex containing SMRT, mSin3A, and histone deacetylase. *Cell* **89**: 373–380.

Nagy, L., Kao, H.Y. and Love, J.D. *et al.* (2000) Mechanism of copressor binding and release from nuclear hormone receptors. *Genes and Development* **13**: 3209–3216.

Nawaz, Z., Tsai, M.J. and O'Malley, B.W. (1995) Specific mutations in the ligand binding domain selectively abolish silencing function of human thyroid hormone receptor beta. *Proc. Natl. Acad. Sci. USA* **92**: 11 691–11 695.

Nayeri, S., Kahlen, J.P. and Carlberg, C. (1996) The high affinity ligand binding conformation of the nuclear 1,25-dihydroxyvitamin D3 receptor is functionally linked to the transactivation domain 2 (AF-2). *Nucleic Acids Res.* **24**: 4513–4518.

Nishigori, H. and Toft, D. (1980) Inhibition of progesterone receptor activation by sodium molydate. *Biochemistry* **19**: 77–83.

Noda, M., Vogel, R.L., Craig, A.M., Prahl, J., DeLuca, A.F. and Denhardt, D.T. (1990) Identification of a DNA sequence responsible for binding of the 1,25-dihydroxyvitamin D3 receptor and 1,25-dihydroxyvitamin D3 enhancement of mouse secreted phosphoprotein 1 (SPP-1 or osteopontin) gene expression. *Proc. Natl. Acad. Sci. USA* **87**: 9995–9999.

Nolte, R.T., Wisely, G.B., Westin, S. *et al.* (1998) Ligand binding and co-activator assembly of the peroxisome proliferator-activated receptor-γ. *Nature* **395**: 137–143.

Noteboom, W.D. and Gorski, J. (1965) Stereospecific binding of estrogens in the rat uterus. *Arch. Biochem. Biophys.* **111**: 559–568.

Nuclear Receptor Nomenclature Committee (1999) A unified nomenclature system for the nuclear receptor superfamily. *Cell* **97**: 161–163.

Oberfield, J.L., Collins, J.L., Holmes, C.P. *et al.* (1999) A peroxisome proliferator-activated receptor γ ligand inhibits adipocyte differentiation. *Proc. Natl. Acad. Sci. USA* **96**: 6102–6106.

Ogawa, H. and Umesono, K. (1998) Intracellular localization and transcriptional activation by the human glucocorticoid receptor-green fluorescent protein (GFP) fusion proteins. *Acta Histochem. Cytochem.* **31**: 303–308.

Ogryzko, V.V., Schiltz, R.L., Russanova, V., Howard, B.H. and Nakatani, Y. (1996) The transcriptional coactivators p300 and CBP are histone acetyltransferases. *Cell* **87**: 953–959.

O'Malley, B.W., Aronow, A., Peacock, A.C. and Dingman, C.W. (1968) Estrogen-dependent increase in transfer RNA during differentiation of the chick oviduct. *Science* **162**: 567–568.

Onate, S.A., Tsai, S.Y., Tsai, M.J. and O'Malley, B.W. (1995) Sequence and characterization of a coactivator for the steroid receptor superfamily. *Science* **270**: 1354–1357.

Onate, S.A., Boonyaratanakornkit, V., Spencer, T.E. *et al.* (1998) The steroid receptor coactivator-1 contains multiple receptor interacting and activation domains that cooperatively enhance the activation function 1 (AF1) and AF2 domains of steroid receptors. *J. Biol. Chem.* **273**: 12 101–12 108.

Ordentlich, P., Downes, M., Xie, W., Genin, A., Spinner, N.B. and Evans, R.M. (1999) Unique forms of human and mouse nuclear receptor corepressor SMRT. *Proc. Natl. Acad. Sci. USA* **99**: 2639–2644.

Orti, E., Bodwell, J.E. and Munck, A. (1992) Phosphorylation of steroid hormone receptors. *Endocr. Rev.* **13**: 105–128.

Ostrowski, M.C., Huang, A.L., Kessel, M., Wolford, R.G. and Hager, G.L. (1984) Modulation of enhancer activity by the hormone responsive regulatory element from mouse mammary tumor virus. *EMBO J.* **3**: 1891–1899.

Ozono, K., Liao, J., Kerner, S.A., Scott, R.A. and Pike, J.W. (1990) The vitamin D-responsive element in the human osteocalcin gene. *J. Biol. Chem.* **265**: 21 881–21 888.

Papamichail, M., Tsokos, G., Tsawdaroglou, N. and Sekeris, C.E. (1980) Immunocytochemical demonstration of glucocorticoid receptors in different cell types and their translocation from the cytoplasm to the cell nucleus in the presence of dexamethasone. *Exp. Cell Res.* **125**: 490–493.

Park, E.J., Schroen, D.J., Yang, M., Li, H., Li, L. and Chen, J.D. (1999) SMRTe, a silencing mediator for retinoid and thyroid hormone receptor-extended isoform that is more related to the nuclear receptor corepressor. *Proc. Natl. Acad. Sci. USA* **96**: 3519–3524.

Parks, D.J., Blanchard, S.G., Bledsoe, R.K. *et al.* (1999) Bile acids: natural ligands for an orphan nuclear receptor. *Science* **284**: 1365–1368.

Parks, W.P., Scolnick, E.M. and Kozikowski, E.H. (1974) Dexamethasone stimulation of murine mammary tumor virus expression: a tissue culture source of virus. *Science* **184**: 158–160.

Patel, P.D., Sherman, T.G., Goldman, D.J. and Watson, S.J. (1989) Molecular cloning of a mineralocorticoid (type I) receptor complementary DNA from rat hippocampus. *Mol. Endocrinol.* **3**: 1877–1885.

Pavletich, N.P. and Pabo, C.O. (1991) Zinc finger–DNA recognition: crystal structure of a Zif268–DNA complex at 2.1 A. *Science* **252**: 809–817.

Pavlik, E.J., Nelson, K., Van Nagell, J.R., Jr. *et al.* (1985) Hydrodynamic characterizations of estrogen receptors complexed with [3H]-4-hydroxytamoxifen: evidence in support of contrasting receptor transitions mediated by different ligands. *Biochemistry* **24**: 8101–8106.

Payvar, F., DeFranco, D., Firestone, G.L. *et al.* (1983) Sequence-specific binding of glucocorticoid receptor to MTV DNA at sites within and upstream of the transcribed region. *Cell* **35**: 381–392.

Perissi, V., Staszewski, L.M., McInerney, E.M. *et al.*. (2000) Molecular determinants of nuclear receptor–corepressor interaction. *Genes and Development* **13**: 3198–3208.

Perlmann, T., Rangarajan, P.N., Umesono, K. and Evans, R.M. (1993) Determinants for selective RAR and TR recognition of direct repeat HREs. *Genes Dev.* **7**: 1411–1422.

Perrot-Applanat, M., Logeat, F., Groyer-Picard, M.T. and Milgrom, E. (1985) Immunocytochemical study of mammalian progesterone receptor using monoclonal antibodies. *Endocrinology* **116**: 1473–1484.

Petkovich, M., Brand, N.J., Krust, A. and Chambon, P. (1987) A human retinoic acid receptor which belongs to the family of nuclear receptors. *Nature* **330**: 444–450.

Petty, K.J., Desvergne, B., Mitsuhaslin, T. and Nikoderm, V.M. (1990) Identification of a thyroid hormone response element in the malic enzyme gene. *J. Biol. Chem.* **265**: 7395–7400.

Pfahl, M. (1982) Specific binding of the glucocorticoid–receptor complex to the mouse mammary tumor proviral promoter region. *Cell* **31**: 475–482.

Picard, D. and Yamamoto, K.R. (1987) Two signals mediate hormone-dependent nuclear localization of the glucocorticoid receptor. *EMBO J.* **6**: 3333–3340.

Picard, D., Salser, S.J. and Yamamoto, K.R. (1988) A movable and regulable inactivation function within the steroid binding domain of the glucocorticoid receptor. *Cell* **54**: 1073–1080.

Picard, D., Khursheed, B., Garabedian, M.J., Fortin, M.G., Lindquist, S. and Yamamoto, K.R. (1990) Reduced levels of hsp90 compromise steroid receptor action *in vivo*. *Nature* **348**: 166–168.

Pike, W.J. and Sleator, N.M. (1985) Hormone-dependent phosphorylation of the 1,25-dihydroxyvitamin D3 receptor is generated through a hormone-dependent phosphorylation. *Biochem. Biophys. Res. Commun.* **131**: 378–385.

Power, R.F., Mani, S.K., Codina, J., Conneely, O.M. and O'Malley, B.W. (1991) Dopaminergic and ligand-independent activation of steroid hormone receptors. *Science* **254**: 1636–1639.

Pratt, W.B., Jolly, D.J., Pratt, D.V. *et al.* (1988) A region in the steroid binding domain determines formation of the non-DNA-binding, 9S glucocorticoid receptor complex. *J. Biol. Chem.* **263**: 267–273.

Predki, P.F., Zamble, D., Sarkar, B. and Giguere, V. (1994) Ordered binding of retinoic acid and retinoid-X receptors to asymmetric response elements involves determinants adjacent to the DNA-binding domain. *Mol. Endocrinol.* **8**: 31–39.

Press, M.F., Xu, S., Wang, J. and Greene, G.L. (1989) Subcellular distribution of estrogen receptor and progesterone receptor with and without specific ligand. *Am. J. Pathol.* **135**: 857–864.

Quack, M., Clarin, A., Binderup, E., Bjorkling, F., Hansen, C.M. and Carberg, C. (1998) Structural variants of the vitamin D analog EB1089 reduce its ligand sensitivity and promoter selectivity. *J. Cell. Biochem.* **71**: 340–350.

Rachez, C., Suldan, Z., Ward, J. *et al.* (1998) A novel protein complex that interacts with the vitamin D3 receptor in a ligand-dependent manner and enhances VDR transactivation in a cell free system. *Genes Dev.* **12**: 1787–1800.

Rachez, C., Lemon, B.D., Suldan, Z. *et al.* (1999) Ligand-dependent transcription activation by nuclear receptors requires the DRIP complex. *Nature* **398**: 824–828.

Rastinejad, F., Perlmann, T., Evans, R.M. and Sigler, P.B. (1995) Structural determinates of nuclear receptor assembly on DNA direct repeats. *Nature* **375**: 203–211.

Reichstein, T. and Shoppee, C.W. (1943) Hormones of the adrenal cortex. *Vitam. Horm.* **1**: 345–413.

Renaud, J.P., Rochel, N., Ruff, M. *et al.* (1995) Crystal structure of the RAR-γ ligand binding domain bound to all-*trans* retinoic acid. *Nature* **378**: 681–689.

Renkawitz, R., Schutz, G., von der Ahe, D. and Beato, M. (1984) Sequences in the promoter region of the chicken lysozyme gene required for steroid regulation and receptor binding. *Cell* **37**: 503–510.

Ringold, G.M., Yamamoto, K.R., Tomkins, G.M., Bishop, J.M. and Varmus, H.E. (1975) Dexamethasone-mediated induction of mouse mammary tumor virus RNA. System for studying glucocorticoid action. *Cell* **6**: 299–305.

Ringold, G.M., Yamamoto, K.R., Bishop, J.M. and Varmus, H.E. (1977) Glucocorticoid-stimulated accumulation of mouse mammary tumor virus RNA: increased rate of synthesis of viral RNA. *Proc. Natl. Acad. Sci. USA* **74**: 2879–2883.

Robins, D.M., Paek, I., Seeburg, P.H. and Axel, R. (1982) Regulated expression of human growth hormone genes in mouse cells. *Cell* **29**: 623–631.

Rochette-Egly, C., Gaub, M., Lutz, Y., Ali, S., Scheuer, I. and Chambon, P. (1992) Retinoic acid receptor-beta: immunodetection and phosphorylation on tyrosine residues. *Mol. Endocrinol.* **6**: 2197–2209.

Rochette-Egly, C., Adams, S., Rossignol, M., Egly, J.M. and Chambon, P. (1997) Stimulation of RARa activation function AF1 through binding to the general transcription factor TFIIH and phosphorylation by CDK. *Cell* **90**: 97–107.

Rodriguez, R., Weigel, N.L., O'Malley, B.W. and Schrader, W.T. (1990) Dimerization of the chicken progesterone receptor *in vitro* can occur in the absence of hormone and DNA. *Mol. Endocrinol.* **4**: 1782–1790.

Rosenfeld, G.C., Comstock, J.P., Means, A.R. and O'Malley, B.W. (1972) Estrogen-induced synthesis of ovalbumin messenger RNA and its translation in a cell free system. *Biochem. Biophys. Res. Commun.* **46**: 1695–1703.

Rusconi, S. and Yamamoto, K.R. (1987) Functional dissection of the hormone and DNA binding activities of the glucocorticoid receptor. *EMBO J.* **6**: 1309–1315.

Rusconi, S., Miesfeld, R., Godowski, P.J., Vaderbilt, J.N., Maler, B.A. and Yamamoto, K.R. (1987) Functional analysis of cloned glucocorticoid receptor sequences. In: W.S. Reznikoff (ed.) *RNA Polmerase and the Regulation of Transcription*, pp. 257–266. Elsevier, New York.

Ruzicka, L. and Wettstein, A. (1935) Uber die kunstliche herstellung des testikelhormons testosterone (androsten-3-on-17-ol). *Helv. Chem. Acta* **18**: 1264.

Saatcioglu, F., Bartunek, P., Deng, T., Zenke, M. and Karin, M. (1993) A conserved C-terminal sequence that is deleted in v-ErbA is essential for the biological activities of c-ErbA (the thyroid hormone receptor). *Mol. Cell. Biol.* **13**: 3675–3685.

Sabbah, M., Redeuilh, G., Secco, C. and Baulieu, E.E. (1987) The binding activity of estrogen receptor to DNA and heat shock protein (Mr 90 000) is dependent on receptor-bound metal. *J. Biol. Chem.* **262**: 8631–8635.

Sabbah, M., Gouilleux, F., Sola, B., Redeuilh, G. and Baulieu, E.E. (1991) Structural differences between the hormone and antihormone estrogen receptor complexes bound to hormone response element. *Proc. Natl. Acad. Sci. USA* **88**: 390–394.

Sadovsky, Y., Webb, P., Lopez, G. *et al.* (1995) Transcriptional activators differ in their responses to overexpression of TATA-box-binding protein. *Mol. Cell. Biol.* **15**: 1554–1563.

Sanchez, E.R., Toft, D.O., Schesinger, M.J. and Pratt, W.B. (1985) Evidence that the 90-kDa phosphoprotein associated with the untransformed L-cell glucocorticoid receptor is a murine heat shock protein. *J. Biol. Chem.* **260**: 12 398–12 401.

Sande, S. and Privalsky, M.L. (1996) Identification of TRAC (T3 receptor-associated cofactors), a family of cofactors that associate with, and modulate the activity of nuclear hormone receptors. *Mol. Endocrinol.* **10**: 813–825.

Sap, J., Munoz, A., Damm, K. *et al.* (1986) The c-erb-A protein is a high-affinity receptor for thyroid hormone. *Nature* **324**: 635–640.

Sap, J., Munoz, A., Schmitt, J., Stunnenberg, H. and Vennstrom, B. (1989) Repression of transcription mediated at a thyroid hormone response element by the v-*erb-A* oncogene product. *Nature* **340**: 242–244.

Sar, M., Lubahn, D.B., French, F.S. and Wilson, E.M. (1990) Immunohistochemical localization of the androgen receptor in rat and human tissues. *Endocrinology* **127**: 3120–3126.

Scheidereit, C. and Beato, M. (1984) Contacts between hormone receptor and DNA double helix within a glucocorticoid regulatory element of mouse mammary tumor virus. *Proc. Natl. Acad. Sci. USA* **81**: 3029–3033.

Scheidereit, C., Geisse, S., Westphal, H.M. and Beato, M. (1983) The glucocorticoid receptor binds to defined nucleotide sequences near the promoter of mouse mammary tumour virus. *Nature* **304**: 749–752.

Schulman, I.G., Chakravarti, D., Juguilon, H., Romo, A. and Evans, R.M. (1995) Interactions between the retinoid X receptor and a conserved region of the TATA-binding protein mediate hormone-dependent transactivation. *Proc. Natl. Acad. Sci. USA* **92**: 8288–8292.

Schulman, I.G., Shao, G. and Heyman, R.A. (1998) Transactivation by retinoid X receptor–peroxisome proliferator-activated receptor γ heterodimers: intermolecular synergy requires only the PPARγ hormone-dependent activation function. *Mol. Cell. Biol.* **18**: 3483–3494.

Schwabe, J.W., Neuhaus D. and Rhodes, D. (1990) Solution structure of the DNA-binding domain of the oestrogen receptor. *Nature* **348**: 458–461.

Schwabe, J.W.R., Chapman, L., Finch, J.T. and Rhodes, D. (1993) The crystal structure of the estrogen receptor DNA-binding domain bound to DNA: how receptors discriminate between their response elements. *Cell* **75**: 567–578.

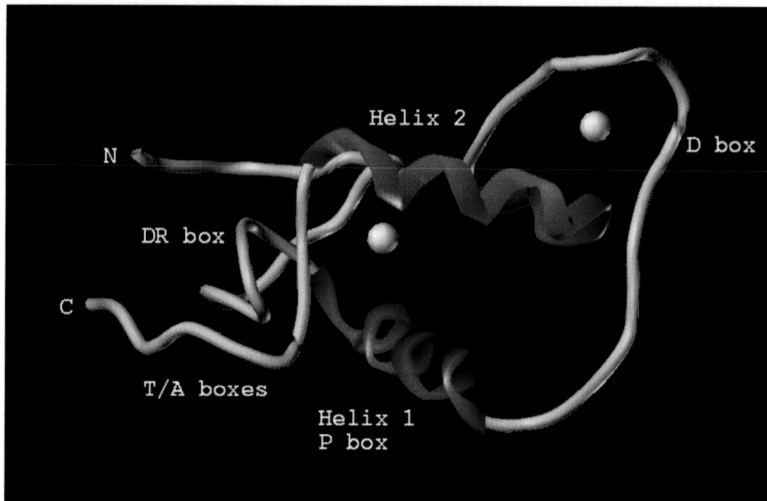

Fig. 1.11 Crystal structure of the DNA-binding domain of the glucocorticoid receptor. The physical structure of DNA-binding domains is very well conserved among the nuclear receptors, and comprises a globular structure organized from two amphipathic α helices arranged perpendicularly to each other. Two zinc ions are coordinated within the structure and are indicated by the gray spheres. Each zinc finger contains a single α helix initiating at the second pair of coordinating cysteines within each respective finger. The relative positions of the DR, D, P and T/A boxes are indicated.

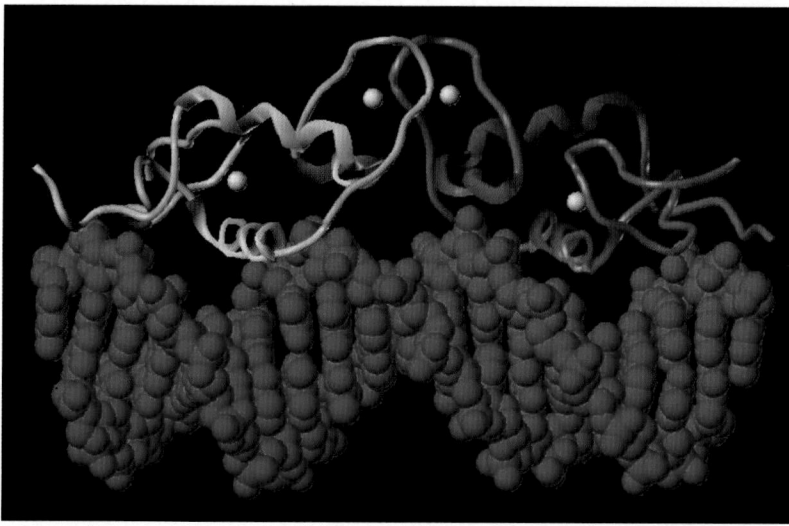

Fig. 1.12 Co-crystal structure of a glucocorticoid receptor homodimer bound to its DNA response element. The D box : D box dimerization interface is clearly visible as well as the direct interaction of the P box and helix 1 with the DNA.

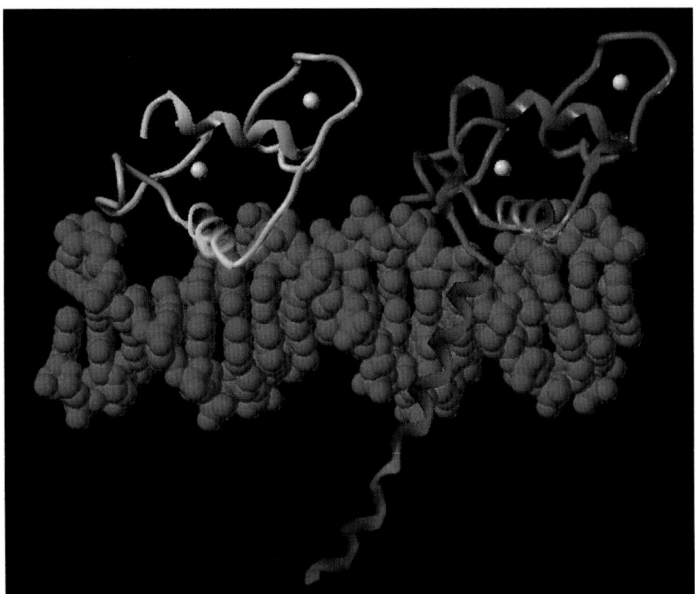

Fig. 1.13 Co-crystal structure of an RXR : TR heterodimer complexed with DNA on a DR4 response element. The retinoid X receptor (RXR) occupies the 5′ half-site while the thyroid hormone receptor (TR) occupies the 3′ site. The head-to-tail dimerization arrangement of the two receptors is facilitated by interactions of the D box of RXR with the DR and T boxes of TR.

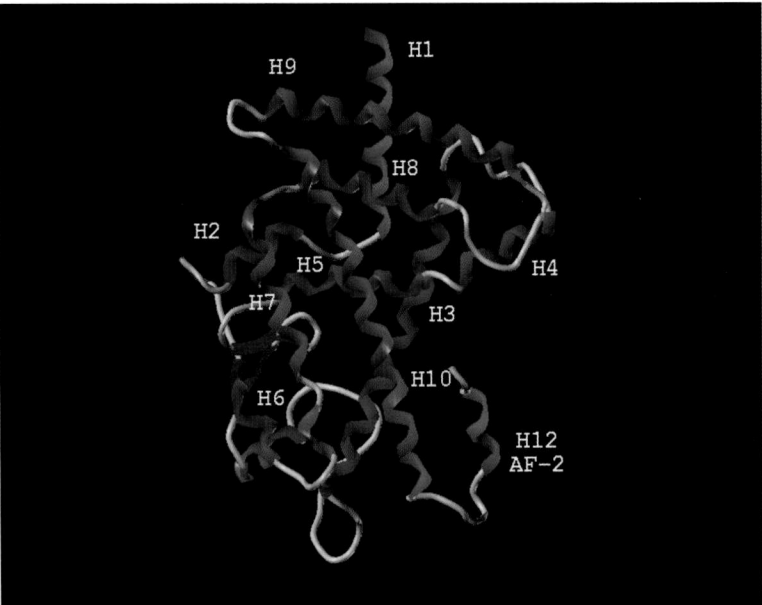

Fig. 1.14 Crystal structure of the ligand-binding domain of peroxisome proliferator activated receptor γ (PPARγ). (a) The structure of the aporeceptor ligand-binding domain is illustrated along with the relative locations of the various helices. Helix 12 (activation function-2 (AF-2) helix; H12) is shown near the bottom right portion of the structure. *Continued.*

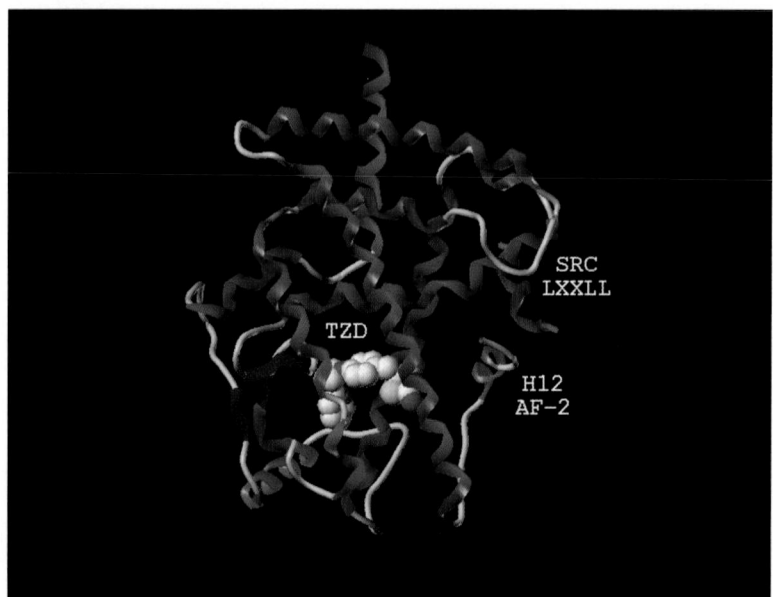

Fig. 1.14 *continued.* (b) The structure of the ligand-bound receptor is illustrated. The thiazolidinedione, rosiglitazone (TZD), occupies the ligand cavity and induces significant conformational changes. Note the repositioning of H12 and the steroid receptor coactivator peptide (SRC LXXLL) which has bound to the surface of the receptor.

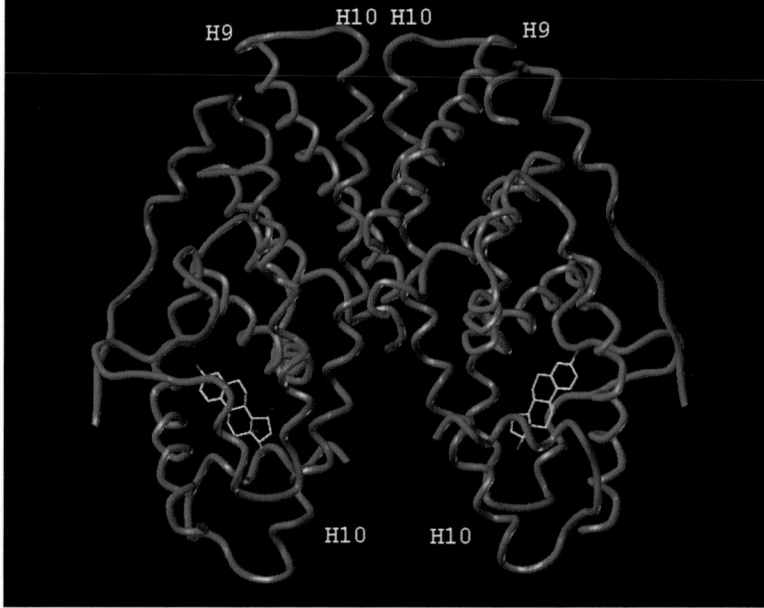

Fig. 1.15 Crystal structure of an estrogen receptor homodimer illustrating the dimerization interfaces. Helix 10 (H10) provides the major dimerization interface; however, helices 9 and 7 as well as the loop between helices 7 and 8 also contribute to the interface.

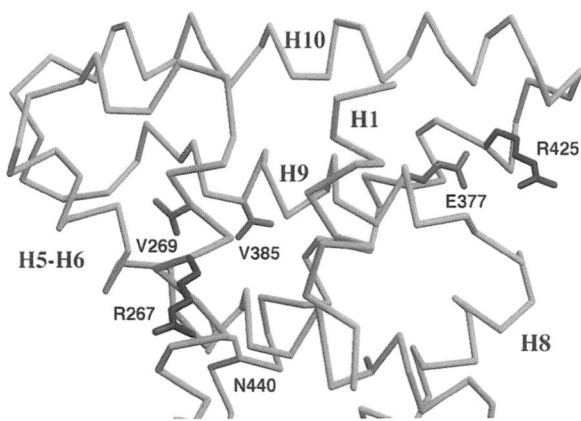

Fig. 5.1 Model of the hydrophobic core of the DAX1 ligand-binding domain in which the seven single amino acid changes map (Zhang *et al.*, 1998). This model was based on sequence homology with rat thyroid hormone receptor α_1 (rTRα_1) and human retinoid X receptor α (hRXRα), for which the structures were known (Bourguet *et al.*, 1995; Wagner *et al.*, 1995).

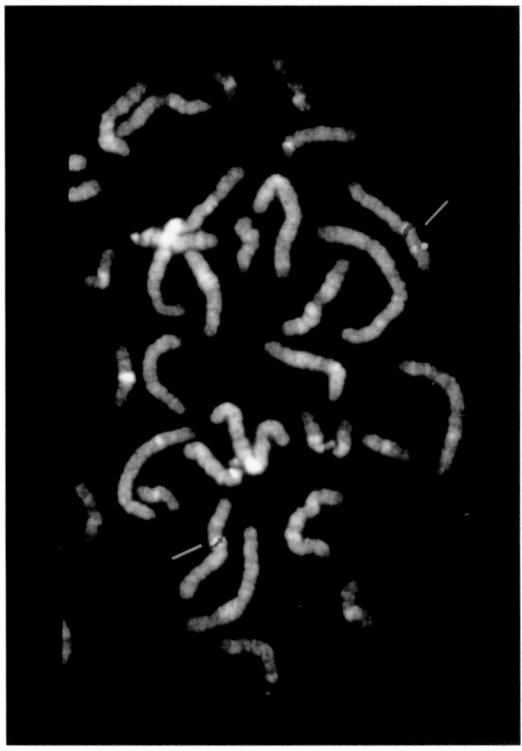

Fig. 5.2 Fluorescence *in situ* hybridization (FISH) analysis demonstrating deletion of the *DAX1* region in one of the X chromosomes from the mother of two boys who were shown to be deleted by polymerase chain reaction. The red signals represent the X-specific centromeric control probe and identify the two X chromosomes. The yellow signal represents cosmid 8E10 containing the *DAX1* gene and is present on only one of the two X chromosomes from this carrier female.

Schwerk, C., Klotzbucher, M., Sachs, M., Ulber, V. and Klein-Hitpass, L. (1995) Identification of a transactivation function in the progesterone receptor that interacts with the TAFII110 subunit of the TFIID complex. *J. Biol. Chem.* **270**: 21 331–21 338.

Scolnick, E.M., Young, H.A. and Parks, W.P. (1976) Biochemical and physiological mechanisms in glucocorticoid hormone induction of mouse mammary tumor virus. *Virology* **69**: 148–156.

Seol, W., Choi, H.S. and Moore, D.D. (1995) Isolation of proteins that interact with the retinoid X receptor: two novel orphan receptors. *Mol. Endocrinol.* **9**: 72–85.

Seol, W., Choi, H.S. and Moore, D.D. (1996) An orphan nuclear hormone receptor that lacks a DNA-binding domain and heterodimerizes with other receptors. *Science* **272**: 1336–1339.

Shao, D. and Lazar, M.A. (1999) Modulating nuclear receptor function: may the phos be with you. *J. Clin. Invest.* **103**: 1617–1618.

Shiau, A.K., Barstad, D., Loria, P.M. *et al.* (1998) The structural basis of estrogen receptor/coactivator recognition and antagonism of this interaction by tamoxifen. *Cell* **95**: 927–937.

Shibata, H., Nawaz, Z., Tsai, S.Y., O'Malley, B.W. and Tsai, M.J. (1997) Gene silencing by chicken ovalbumin upstream promoter transcription factor I (COUP-TFI) is mediated by transcriptional corepressors, nuclear receptor-corepressor (N-CoR) and silencing mediator for retinoic acid receptor and thyroid hormone receptor (SMRT). *Mol. Endocrinol.* **11**: 714–724.

Shyamala, G. and Gorski, J. (1969) Estrogen receptors in the rat uterus. Studies on the interaction of cytosol and nuclear binding sites. *J. Biol. Chem.* **244**: 1097–1103.

Simental, J.A., Sar, M., Lane, M.V., French, F.S. and Wilson, E.M. (1991) Transcriptional activation and nuclear targeting signals of the human androgen receptor. *J. Biol. Chem.* **266**: 510–518.

Simpson, S.A. and Tait, J.F. (1955) Recent progress in methods of isolation, chemistry, and physiology of aldosterone. *Recent Prog. Horm. Res.* **11**: 183–210.

Simpson, S.A., Tait, J.F., Wettstein, A. *et al.* (1953) Isolierung eines neuen kristallisierten hormons aus nebennieren mit besonders hoher wirksamkeit auf den mineralstoffweksel. *Experientia* **9**: 333–335.

Slater, E.P., Rabenau, O., Karin, M., Baxter, J.D. and Beato, M. (1985) Glucocorticoid receptor binding and activation of a heterologous promoter by dexamethasone by the first intron of the human growth hormone gene. *Mol. Cell. Biol.* **5**: 2984–2992.

Slotta, K.H., Rushig, H. and Fels, W. (1934a) Reindarstellung der hormone aus dem corpus luteum (vorlauf mitteil). *Ber. Dtsch. Chem. Ges.* **67**: 1270–1273.

Slotta, K.H., Rushig, H. and Fels, W. (1934b) Reindarstellung der hormone aus dem corpus luteum (II. Mitteil). *Ber. Dtsch. Chem. Ges.* **67**: 1624–1626.

Smith, C.L., Onate, S.A., Tsai, M.J. and O'Malley, B.W. (1996) CREB binding protein acts synergistically with steroid receptor coactivator 1 to enhance steroid receptor dependent transcription. *Proc. Natl. Acad. Sci. USA* **93**: 8884–8888.

Smith, C.L., Nawaz, Z. and O'Malley, B.W. (1997) Coactivator and corepressor regulation of the agonist/antagonist activity of the mixed antiestrogen, 4-hydroxytamoxifen. *Mol. Endocrinol.* **11**: 657–666.

Smith, D.F. and Toft, D.O. (1993) Steroid receptors and their associated proteins. *Mol. Endocrinol.* **7**: 4–11.

Soel, W., Choi, H.S. and Moore, D.D. (1995) Isolation of proteins that interact specifically with the retinoid X receptor: two novel orphan receptors. *Mol. Endocrinol.* **17**: 6131–6138.

Somers, J.P. and DeFranco, D.B. (1992) Effects of okadaic acid a protein phosphatase inhibitor on glucocorticoid receptor-mediated enhancement. *Mol. Endocrinol.* **6**: 26–34.

Sone, T., Kerner, S. and Pike, J.W. (1991) Vitamin D receptor interaction with specific DNA. Association as a 1,25-dihydroxyvitamin D3-modulated heterodimer. *J. Biol. Chem.* **266**: 23 296–23 305.

Spelsberg, T.C., Steggles, A.W. and O'Malley, B.W. (1971) Progesterone-binding components of chick oviduct. III. Chromatin acceptor sites. *J. Biol. Chem.* **246**: 4186–4195.

Spencer, T.E., Jenster, G., Burcin, M.M. *et al.* (1997) Steroid receptor coactivator-1 is a histone acetyltransferase. *Nature* **389**: 194–198.

Sporn, M.B., Dunlop, N.M., Newton, D.L. and Smith, J.M. (1976) Prevention of chemical carcinogenesis by vitamin A and its synthetic analogs (retinoids). *Fed. Proc.* **35**: 1332–1338.

Strahle, U., Klock, G. and Schutz, G. (1987) A DNA sequence of 15 base pairs is sufficient to mediate both glucocorticoid and progesterone induction of gene expression. *Proc. Natl. Acad. Sci. USA* **84**: 7871–7875.

Strahle, U., Boshart, M., Klock, G., Stewart, F. and Schutz, G. (1989) Glucocorticoid- and progesterone-specific effects are determined by differential expression of the respective hormone receptors. *Nature* **339**: 629–632.

Sucov, H.M., Murakami, K.K. and Evans, R.M. (1990) Characterization of an autoregulated response element in the mouse retinoic acid receptor type β gene. *Proc. Natl. Acad. Sci. USA* **87**: 5392–5398.

Suen, C.S., Berrodin, T.S., Mastroeni, R., Cheskis, B.J., Lyttle, C.R. and Frail, D.E. (1998) A transcriptional coactivator, steroid receptor coactivator-3, selectively augments steroid receptor transcriptional activity. *J. Biol. Chem.* **273**: 27 645–27 653.

Takeshita, Z., Cardona, G.R., Koibuchi, N., Suen, C.S. and Chin, W.W. (1997) TRAM-1, a novel 160 kDa thyroid hormone receptor activator molecule exhibits distinct properties from steroid receptor coactivator-1. *J. Biol. Chem.* **272**: 27 629–27 634.

Tanenbaum, D.M., Wang, Y., Williams, S.P. and Sigler, P.B. (1998) Crysallographic comparison of the estrogen and progesterone receptor's ligand binding domains. *Proc. Natl. Acad. Sci. USA* **95**: 5998–6003.

Tasset, D., Tora, L., Fromental, C., Scheer, E. and Chambon, P. (1990) Distinct classes of trancriptional activating domain function by different mechanisms. *Cell* **62**: 1177–1187.

Tate, A.C., Greene, G.L., DeSombre, E.R., Jensen, E.V. and Jordan, V.C. (1984) Differences between estrogen- and antiestrogen-estrogen receptor complexes from human breast tumors identified with an antibody raised against the estrogen receptor. *Cancer Res.* **44**: 1012–1018.

Teng, C.S. and Hamilton, T.H. (1968) The role of chromatin in estrogen action in the uterus, I. The control of template capacity and chemical composition and the binding of ^{3}H-estradiol. *Proc. Natl. Acad. Sci. USA* **60**: 1410–1417.

Tilley, W.D., Marcelli, M., Wilson, J.D. and McPhaul, M.J. (1989) Characterization and expression of cDNA encoding the human androgen receptor. *Proc. Natl. Acad. Sci. USA* **86**: 327–331.

Tini, M., Otulakowski, G., Breitman, M.L., Tsui, L.C. and Giguere, V. (1993) An everted repeat mediates retinoic acid induction of the γf-crystallin gene: evidence of a direct role for retinoids in lens development. *Genes Dev.* **7**: 295–307.

Toft, D. (1972) The interaction of uterine estrogen receptors with DNA. *J. Steroid Biochem.* **3**: 512–522.

Toft, D. and Gorski, J. (1966) A receptor molecule for estrogens: isolation from the rat uterus and preliminary characterization. *Proc. Natl. Acad. Sci. USA* **55**: 1574–1581.

Tora, L., White, J., Brou, C. *et al.* (1989) The human estrogen receptor has two independent nonacidic transcriptional activation functions. *Cell* **59**: 477–487.

Torchia, J., Rose, D.W., Inostroza, J. *et al.* (1997) The transcriptional coactivator p/CIP binds CBP and mediates nuclear receptor function. *Nature* **387**: 677–684.

Trapman, J., Klaassen, P., Kuiper, G.G.J.M. *et al.* (1988) Cloning, structure and expression of a cDNA encoding the human androgen receptor. *Biochem. Biophys. Res. Commun.* **153**: 241–248.

Tsai, M.J. and O'Malley, B.W. (1994) Molecular mechanism of action of steroid/thyroid receptor superfamily members. *Ann. Rev. Biochem.* **63**: 451–486.

Tsai, S.Y., Sagami, I., Wang, L.H., Tsai, M.J. and O'Malley, B.W. (1987) Interactions between a DNA-binding transcription factor (COUP) and a non-DNA binding factor (S300-II). *Cell* **50**: 701–709.

Tsai, S.Y., Carlstedt-Duke, J., Weigel, N.L. *et al.* (1988) Molecular interactions of steroid hormone receptor with its enhancer element: evidence for receptor dimer formation. *Cell* **55**: 361–369.

Tsai, S.Y., Srinivasan, G., Allan, G.F., Thompson, E.B., O'Malley, B.W. and Tsai, M.J. (1990) Recombinant human glucocorticoid receptor induces transcription of hormone response genes *in vitro*. *J. Biol. Chem.* **265**: 17 055–17 061.

Ucker, D.S., Ross, S.R. and Yamamoto, K.R. (1981) Mammory tumor virus DNA contains sequences required for its hormone-regulated transcription. *Cell* **27**: 257–266.

Ueda, H., Sun, G.C., Murata, T. and Hirose, S. (1992) A novel DNA-binding motif abuts the zinc finger domain of insect nuclear hormone receptor FTZ-F1 and mouse embryonal long terminal repeat-binding protein. *Mol. Cell. Biol.* **12**: 5667–5672.

Umesono, K. and Evans, R.M. (1989) Determinants of target gene specificity for steroid/ thyroid hormone receptors. *Cell* **57**: 1139–1146.

Umesono, K., Murakami, K.K., Thompson, C.C. and Evans, R.M. (1991) Direct repeats as selective response elements for the thyroid hormone, retinoic acid, and vitamin D3 receptors. *Cell* **65**: 1255–1266.

Uppenberg, J., Svensson, C., Jaki, M., Bertilsson, G., Jendeberg, L. and Berkenstam, A. (1998) Crystal structure of the ligand binding domain of the human nuclear receptor PPARγ. *J. Biol. Chem.* **47**: 31 108–31 112.

Voegel, J.J., Heine, M.J., Zechel, C., Chambon, P. and Gronemeyer, H. (1996) TIF2, a 160 kDa transcriptional mediator for the ligand-dependent activation function AF-2 of nuclear receptors. *EMBO J.* **15**: 3667–3675.

Voegel, J.J., Heine, M.J.S., Tini, M., Vivat, V., Chambon, P. and Gronemeyer, H. (1998) The coactivator TIF2 contains three nuclear receptor binding motifs and mediates transactivation through CBP binding dependent and independent pathways. *EMBO J.* **17**: 507–519.

von der Ahe, D., Janich, S., Scheidereit, C., Renkawitz, R., Schutz, G. and Beato, M. (1985) Glucocorticoid and progesterone receptors bind to the same sites in two hormonally regulated promoters. *Nature* **313**: 706–709.

Wagner, R.L., Apriletti, J.W., McGrath, M.E., West, B.L., Baxter, J.D. and Fletterick, R.J. (1995) A structural role for hormone in thryoid hormone receptor. *Nature* **378**: 690–697.

Walker, P., Germond, J.E., Brown-Leudi, M., Givel, F. and Wahli, W. (1984) Sequence homologies in the region preceding the transcription initiation site of the liver estrogen-responsive vitellogenin and apo-VLDL II genes. *Nucleic Acids Res.* **12**: 8611–8626.

Walter, P., Green, S., Greene, G. *et al.* (1985) Cloning of the human estrogen receptor cDNA. *Proc. Natl. Acad. Sci. USA* **82**: 7889–7893.

Wang, H., Chen, J., Hollister, K., Sowers, L.C. and Forman, B.M. (1999) Endogenous bile acids are ligands for the nuclear receptor FXR/BAR. *Mol. Cell* **3**: 543–553.

Wang, L.H., Tsai, S.Y., Cook, R.G., Beattie, W.G., Tsai, M.J. and O'Malley, B.W. (1989)

COUP transcription factor is a member of the steroid receptor superfamily. *Nature* **340**: 163–166.

Washburn, T.F., Hocutt, A., Brautigan, D.L. and Korach, K.S. (1991) Uterine estrogen receptor *in vivo*: phosphorylation of nuclear specific forms on serine residues. *Mol. Endocrinol.* **5**: 235–242.

Webster, N.J.G., Green, S., Jin, J.R. and Chambon, P. (1988) The hormone-binding domains of the estrogen receptor and glucocorticoid receptors contain an inducible transactivation function. *Cell* **54**: 199–207.

Weigel, N.L. (1994) Receptor phosphorylation. In: M.J. Tsai and B.W. O'Malley (eds) *Mechanism of Steroid Hormone Regulation of Gene Transcription*, pp. 93–110. R.G. Landes, Austin, Texas.

Weigel, N.L., Beck, C.A., Estes, P.A. *et al.* (1992) Ligands induce conformational changes in the carboxy-terminus of progesterone receptors which are detected by a site-directed antipeptide monoclonal antibody. *Mol. Endocrinol.* **6**: 1585–1597.

Weinberger, C., Hollenberg, S.M., Rosenfeld, M.G. and Evans, R.M. (1985) Domain structure of human glucocorticoid receptor and its relationship to the v-*erb-A* oncogene product. *Nature* **318**: 670–672.

Weinberger, C., Thompson, C.C., Ong, E.S., Lebo, R., Gruol, D.J. and Evans, R.M. (1986) The c-*erb-A* gene encodes a thyroid hormone receptor. *Nature* **324**: 641–646.

Welshons, W.V., Krummel, B.M. and Gorski, J. (1985) Nuclear localization of unoccupied receptors for glucocorticoids, estrogens, and progesterone in GH3 cells. *Endocrinology* **117**: 2140–2147.

Westin, S., Kurokawa, R., Nolte, R.T. *et al.* (1998) Interactions controlling assembly of nuclear-receptor heterodimers and co-activators. *Nature* **395**: 199–202.

Wicks, W.D., Greenman, D.L. and Kenney, F.T. (1965) Stimulation of ribonucleic acid synthesis by steroid hormones. I. Transfer ribonucleic acid. *J. Biol. Chem.* **240**: 4414–4419.

Williams, S.P. and Sigler, P.B. (1998) Atomic structure of progesterone complexed with its receptor. *Nature* **393**: 392–396.

Willy, P.J., Umesono, K., Ong, E.S., Evans, R.M., Heyman, R.A. and Mangelsdorf, D.J. (1995) LXR, a nuclear receptor that defines a distinct retinoid response pathway. *Genes Dev.* **9**: 1033–1045.

Wilson, T.E., Fahrner, T.J., Johnston, M. and Milbrantdt, J. (1991) Identification of the DNA binding site for NGFI-B by genetic selection in yeast. *Science* **252**: 1296–1300.

Wilson, T.E., Paulsen, R.E., Padgett, K.A. and Milbrandt, J. (1992) Participation of non-zinc finger residues in DNA binding by two nuclear orphan receptors. *Science* **256**: 107–110.

Wilson, T.E., Fahrner, T.J. and Milbrandt, J. (1993) The orphan receptors NGFI-B and steroidogenic factor 1 establish monomer binding as a third paradigm of nuclear receptor–DNA interaction. *Mol. Cell. Biol.* **13**: 5794–5804.

Wintersteiner, O. and Allen, W.M. (1934) Crystalline progestin. *J. Biol. Chem.* **107**: 321–336.

Wrange, O. and Gustafsson, J.A. (1978) Separation of the hormone- and DNA-binding sites of the hepatic glucocorticoid receptor by means of proteolysis. *J. Biol. Chem.* **253**: 856–865.

Xu, H.E., Lambert, M.H., Montana, V.G. *et al.* (1999) Molecular recognition of fatty acids by peroxisome proliferator-activated receptors. *Mol. Cell* **3**: 397–403.

Xu, J., Qiu, Y., Demayo, F.J., Tsai. S.Y., Tsai, M.J. and O'Malley, B.W. (1998) Partial hormone resistance in mice with disruption of the steroid receptor coactivator (SRC-1) gene. *Science* **279**: 1922–1925.

Yamamoto, K.R. and Alberts, B. (1974) On the specificity of the binding of the estradiol receptor protein to dexoyribonucleic acid. *J. Biol. Chem.* **249**: 7076–7086.

Yamamoto, K.R. and Alberts, B. (1975) Interaction of estradiol-receptor protein with the genome. Argument for the existence of undetected specific sites. *Cell* **4**: 301–310.

Yamamoto, K.R., Stampfer, M.R. and Tomkins, G.M. (1974) Receptors from glucocorti-coid-sensitive lymphoma cells and two classes of insensitive clones: physical and DNA-binding properties. *Proc. Natl. Acad. Sci. USA* **71**: 3901–3905.

Yamamoto, K.R., Payvar, F., Firestone, G.L. *et al.* (1983) Biological activity of cloned mammary tumor virus DNA fragments that bind purified glucocorticoid receptor protein *in vitro. Cold Spring Harbor Symp. Quant. Biol.* **47**: 977–984.

Yao, T.P., Segraves, W.A., Oro, A.E., McKeown, M. and Evans, R.M. (1992) *Drosophila* ultraspiracle modulates ecdysone receptor function via heterodimerization. *Cell* **71**: 63–72.

Yao, T.P., Ku, G., Zhou, N., Scully, R. and Livingston, D.M. (1996) The nuclear hormone receptor coactivator SRC-1 is a specific target of p300. *Proc. Natl. Acad. Sci. USA* **93**: 10 626–10 631.

Yen, P.M., Darling, D.S. and Chin, W.W. (1991) Basal and thyroid hormone receptor auxiliary protein-enhanced binding of thyroid hormone receptor isoforms to native thyroid hormone response elements. *Endocrinology* **129**: 3331–3336.

Ylikomi, T., Bocquel, M.T., Berry, M., Gronemeyer, H. and Chambon, P. (1992) Coopera-tion of proto-signals for nuclear accumulation of estrogen and progesterone receptors. *EMBO J.* **11**: 3681–3694.

Young, H.A., Shih, T.Y., Scolnick, E.M. and Parks, W.P. (1977) Steroid induction of mouse mammary tumor virus: effect upon synthesis and degradation of viral RNA. *J. Virol.* **21**: 139–146.

Yu, K., Bayona, W., Kallen, C.B. *et al.* (1995) Differential activation of peroxisome proliferator-activated receptors by eicosanoids. *J. Biol. Chem.* **270**: 23 975–23 983.

Yuan, C.X., Ito, M., Fondell, J.D., Fu, Z.Y. and Roeder, R.G. (1998) The TRAP 220 component of a thyroid hormone receptor-associated protein (TRAP) coactivator com-plex interacts directly with nuclear receptors in a ligand-dependent fashion. *Proc. Natl. Acad. Sci. USA* **95**: 7939–7944.

Zamir, I., Harding, H.P., Atkins, G.B. *et al.* (1996) A nuclear hormone receptor corepressor mediates transcriptional silencing by receptors with distinct repression domains. *Mol. Cell. Biol.* **16**: 5458–5465.

Zeng, Z., Allan, G.F., Thaller, C. *et al.* (1994) Detection of potential ligands for nuclear receptors in cellular extracts. *Endocrinology* **135**: 248–252.

Zhang, J. and Fondell, J.D. (1999) Identification of mouse TRAP110: a transcriptional coregulatory factor for thyroid hormone and vitamin D receptors. *Mol. Endocrinol.* **13**: 1130–1140.

Zhang, X., Kun, H., Birgit, T., Paul, B.V., Graupner, G. and Pfahl, M. (1992) Retinoid X receptor is an auxiliary protein for thyroid hormone and retinoic acid receptor. *Nature* **355**: 441–446.

Zhao, Q., Khorasanizadeh, S., Miyoshi, Y., Lazar, M.A. and Rastinejad, F. (1998) Structural elements of an orphan nuclear receptor–DNA complex. *Mol. Cell* **1**: 849–861.

Zhou, G., Cummings, R., Li, Y. *et al.* (1998) Nuclear receptors have distinct affinities for coactivators: characterization by fluorescence resonance energy transfer. *Mol. Endocrinol.* **12**: 1594–1604.

Chapter 2

Thyroid Hormone Receptors

Uwe Dressel and Aria Baniahmad

INTRODUCTION AND HISTORICAL PERSPECTIVE

The thyroid gland produces two thyroid hormones: thyroxine (T_4) and L-3,5,3′-triiodothyronine (T_3). T_4 is converted in cells into the biologically active T_3. Thyroid hormone regulates a large number of cellular functions including growth, metabolic rate and myocardial contractility. Furthermore, thyroid hormone is critical in the development and maturation of the central nervous system. The correlation of loss of thyroid gland function and cretinism in humans and rats has been known for decades (for review see Lebel and Dussault, 1994; Bernal and Nunez, 1995; Oppenheimer and Schwartz, 1997). The first receptor gene for thyroid hormone, the α receptor (TRα, c-erbAα, NR1A1; Nuclear Receptor Nomenclature Committee, 1999) was identified in 1986 (Sap *et al.*, 1986; Weinberger *et al.*, 1986). Subsequently, a more tissue-specific TR variant (TRβ, c-erbβ, NR1A2) was isolated (Hodin *et al.*, 1989), which is encoded by a different gene. Later, splice variants (Chassande *et al.*, 1997) of both receptor genes were identified (Fig. 2.1).

Interestingly, both the α and β variants of TR genes are involved in disease. The oncogene v-*erbA* represents a mutated form of TRα (Fig. 2.2). The oncoprotein is derived from the avian erythroleukemia virus and causes erythroleukemia by arresting erythroid differentiation and sarcomas (reviewed in Rascle *et al.*, 1997).

In humans, the syndrome of resistance to thyroid hormone (RTH) is caused by mutations of the TRβ gene (Usala *et al.*, 1991a; Refetoff, 1982, 1994a,b; McDermott and Ridgway, 1993; Kopp *et al.*, 1996; Chatterjee, 1997). The disease was first defined in 1967 (Refetoff *et al.*, 1967) and the close association of human TRβ receptor mutants was originally described in 1988 (Usala *et al.*, 1988). The RTH syndrome is, in the majority of cases, an autosomal dominant inherited disorder in which body tissues are resistant to the effects of thyroid hormones.

Recently, mouse model systems for RTH syndrome have been developed successfully (Hayashi *et al.*, 1996; Wong *et al.*, 1997; Abel *et al.*, 1999; Gloss *et al.*, 1999), and can be used to analyze the molecular basis for the syndrome and to search for target genes that may cause the disease when dysregulated.

This chapter gives a short overview of mechanisms of transcriptional properties of

Nuclear Receptors and Genetic Disease
ISBN 0-12-146160-2

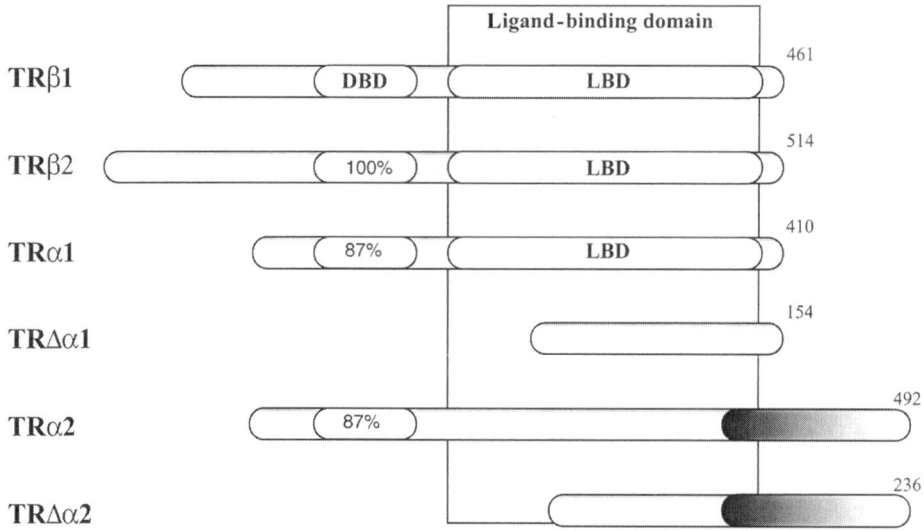

Fig. 2.1   Thyroid hormone receptor (TR) isoforms. Schematic representation of the known TR isoforms. Two genes (c-*erbA* α and β) encode the α and β form of TR. Alternative splice variants and promoter usage generate the multiplicity of different TR isoforms. Amongst the nuclear receptors the highly conserved DNA-binding domain (DBD) separates the variable N-terminus from the conserved C-terminus, which contains the ligand-binding domain (LBD). Note that TR$\Delta\alpha$1 as well as TRα2 and TR$\Delta\alpha$2 are unable to bind ligand. The N-termini of TRβ1, TRβ2 and TRα are completely divergent.

TR and provides an extensive list of known functional properties of mutant receptors involved in the RTH syndrome.

PHYSIOLOGY

The synthesis of thyroid hormones (T_3 and T_4) in the thyroid gland is induced by thyroid-simulating hormone (thyrotropin; TSH) produced in the pituitary, which in turn is activated by the hypothalamic thyrotropin-releasing hormone (TRH). In a negative feed-back mechanism, T_3 downregulates TSH secretion by means of reduced TSH synthesis mediated by repression of the genes encoding for both TSH α and β subunits. Examples of genes that are positively controlled by T_3 are malic enzyme, myosin heavy chain, Na–Ca adenosine triphosphatase (ATPase) in myocardium, and myelin basic protein in brain. Interestingly, both positive and negative regulation of genes by T_3 is mediated by the receptors for thyroid hormone. T_3 deficiency in the brain leads to irreversible abnormalities, such as deficiency in myelination, impairment of proliferation and migration of nerve cells, retardation in synapse formation and alterations in neurotransmitter levels, such as norephedrine and dopamine (Lebel and Dussault, 1994).

In the heart, thyroid hormones increase myocardial inotropy and heart rate, and dilate peripheral arteries to increase cardiac output (Aronow, 1995; Broderick and Wechsler, 1997; Gomberg-Maitland and Frishman, 1998). Thus, hypothyroidism is associated with diminished cardiac output. Hyperthyroidism, on the other hand, is associated with

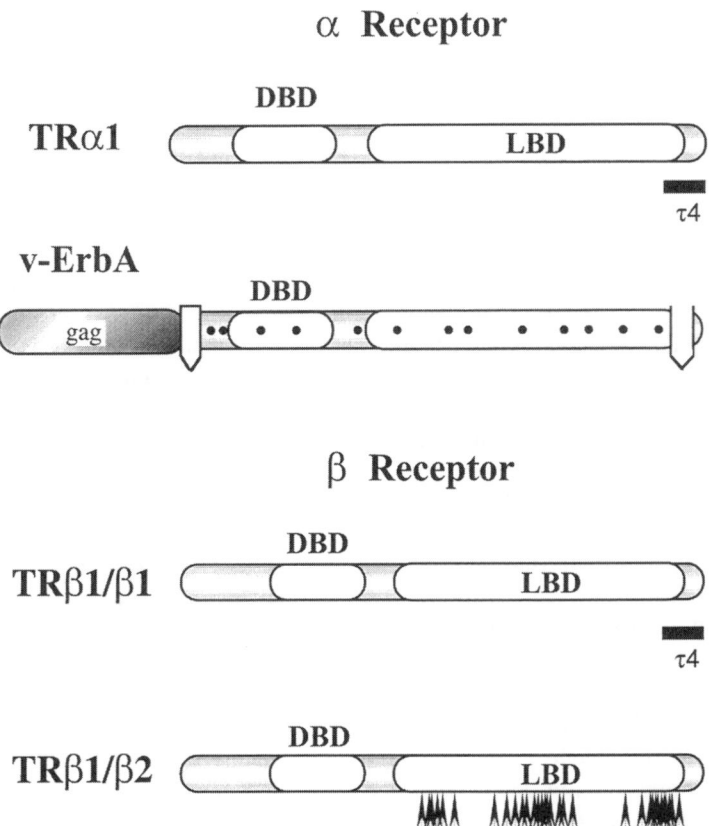

Fig. 2.2 Naturally occurring mutant forms of thyroid hormone receptors (TRs). The oncogene v-*erbA* represents a mutated form of the α receptor. v-*erbA* differs by two deletions, a few amino acid exchanges (filled dots) as well as the gag fusion from the α receptor. For the β receptor a large battery of different, mostly point, mutations has been described, derived from patients with resistance to thyroid hormone (RTH) syndrome. Schematically, the arrows indicate the clustering of observed mutations in the receptor ligand-binding domain (LBD). The detailed location of each mutation is shown in Fig. 2.5. The small conserved activation domain τ4, essential for corepressor release and coactivator recruitment, is lost in the oncogene product v-ErbA and in some β-receptor mutants. DBD, DNA-binding domain.

hyperactivity and cardiovascular symptoms which include palpitations, angina pectoris, orthopnea or paroxysomal nocturnal dyspnea (Aronow, 1995). Future experiments will shed some light on to which effects are mediated by control of gene expression or by fast nongenomic effects. So far, there is no evidence for a TR other than the known ones (Gothe *et al.*, 1999), which implies that hormone effects are mediated by the α and β TRs and their isoforms.

Mice model systems devoid of either α or β TR will elucidate the detailed role of each receptor in development and homeostasis (Forrest *et al.*, 1996c; Fraichard *et al.*, 1997; Johansson *et al.*, 1998; Wikstrom *et al.*, 1998). Recently, mice have been generated that are devoid of known TRs. They exhibit a very hyperactive pituitary–thyroid axis, poor female fertility and retarded growth with associated bone maturation (Gothe *et al.*, 1999).

STRUCTURE AND MECHANISM OF ACTION OF THYROID HORMONE RECEPTORS

The TRs belong to the superfamily of nuclear receptors with a characteristic tripartite structure (Mangelsdorf and Evans, 1995; Mangelsdorf *et al.*, 1995). The DNA-binding domain consisting of a zinc finger motif separates the amino- (N) terminus and the carboxy- (C) terminus. This DNA-binding domain enables DNA recognition as homodimer or as a heterodimer with the receptor for 9-*cis* retinoic acid (retinoid X receptor; RXR). On most DNA-binding sites, TR homodimerizes in the absence of ligand, and in the presence of ligand heterodimerization with RXR is preferred. In contrast to the receptors for steroids, TRs have a short N-terminus. The C-terminus harbors multiple functions such as hormone binding, dimerization, nuclear localization, gene silencing and hormone-dependent transactivation (Fig. 2.3). A very small stretch of amino acids localized at the extreme C-terminus of TR is essential for hormone-dependent transactivation (Barettino *et al.*, 1994; Baniahmad *et al.*, 1995). This sequence, called activation function-2-activation domain (AF-2-AD), τ4 or τc, represents helix 12 of the TR ligand-binding domain (LBD) crystal structure (Wagner *et al.*, 1995) and is necessary for coactivator binding. In addition, this sequence is essential for corepressor release (Baniahmad *et al.*, 1995). Thus, mutations within this region, observed for the oncogene v-*erbA* and some TRβ receptors from patients with RTH, result in receptors lacking corepressor release and coactivator binding. Most of the functions of TR can be transferred to other proteins, and thus the receptor parts consist of functional domains (Baniahmad *et al.*, 1992a, 1994, 1997; Tenbaum and Baniahmad, 1997).

Gene activation is mediated by a complex of the receptor and so-called coactivators,

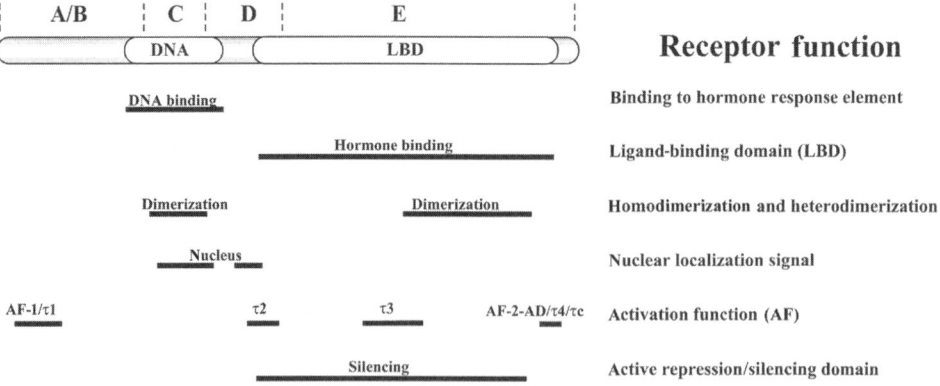

Fig. 2.3 Functional domains of nuclear receptors. Based on sequence homologies among nuclear receptors the receptor parts have been divided into regions A–F. The thyroid hormone receptor does not have an F region. Functional properties and their localization are shown by filled bars. Most of these functions can be transferred to other proteins and thus represent functional domains. AD, activation domain.

whereas gene silencing is mediated by a receptor–corepressor complex (Fig. 2.3). Although several coactivator classes are known that bind in a hormone-dependent manner to nuclear receptors, only two corepressor classes, the SMRT nuclear receptor corepressor (NCoR) (silencing mediator for retinoic acid and thyroid hormone receptors/ nuclear receptor corepressor) class (Chen and Evans, 1995; Hörlein *et al.*, 1995) and a novel corepressor called Alien (Dressel *et al.*, 1999), have been characterized, which bind to the receptor in a hormone-sensitive manner.

It is thought that one mechanism by which coactivators enhance receptor-mediated gene transcription is through the recruitment of histone acetylase function to the receptor, which leads to an opening of the chromatin. *In vitro* gene activation in the absence of chromatin is shown to be mediated by the TR (thyroid hormone receptor) associated protein (AP)–Vitamin D3 receptor interacting protein (DRIP) protein complex (Ito *et al.*, 1999; Rachez *et al.*, 1999). On the other hand, gene silencing is mediated by binding of corepressors and recruitment of histone deacetylase function, which is thought to transform the chromatin into a repressed state. *In vitro* gene silencing mediated by TR in the absence of chromatin was shown to occur in the presence of purified basal transcription factors and may involve the inhibition of the preinitiation complex formation through interacting with transcription factor (TF) IIB and TATA-binding protein (TBP) (Baniahmad *et al.*, 1993; Fondell *et al.*, 1996). Thus, the role of hormone is to shift the receptor from a silencer into an activator, which leads to dissociation of corepressors and binding of coactivators (Fig. 2.4).

There are other mechanisms by which TRs regulate gene transcription. Gene activation mediated by associated protein-1 (AP-1), a heterodimer of the proto-

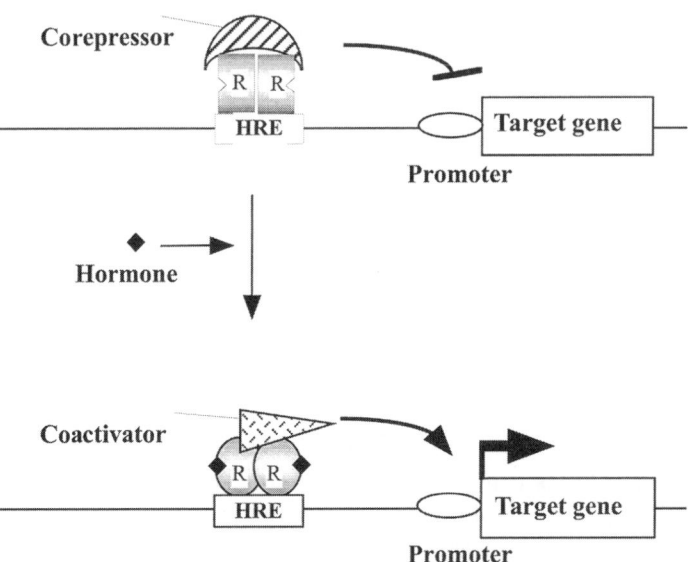

Fig. 2.4 Mechanism of hormone-induced changes of the wild-type thyroid hormone receptor (TR). In the absence of hormone, TR is associated with corepressors and this protein complex silences gene expression. Ligand binding leads to a conformational change in the receptor, dissociation of corepressors and recruitment of coactivators, which are responsible for subsequent target gene activation. R, receptor; HRE, hormone response element.

oncogenes Fos and Jun, can be inhibited by liganded TR as well as by other nuclear receptors (Kamei *et al.*, 1996). There may be multiple mechanisms including squelching of the co-integrator CREB (cAMP response element binding factor) binding protein P300 and/or inhibition of Jun N-terminal kinase (JNK) activity (Caelles *et al.*, 1997). This crosstalk represents a pathway by which nuclear receptors control important pharmacologic actions including immunosuppressive, antiinflammatory and antineoplastic pathways.

Also, some TR isoforms regulate overall TR action. The splice variant TRα2, for example, inhibits the transactivation of wild-type TR (Katz and Lazar, 1993). Similarly, the recently identified shorter forms of TRα (TRαΔ1 and TRαΔ2; Fig. 2.1) are inhibitors of TR transactivation (Chassande *et al.*, 1997). Thus, there is the interesting feature that the gene locus of TR also encodes the receptor's own inhibitors.

The inhibition of myosin heavy chain gene, however, is mediated by a different mechanism involving so-called negative thyroid hormone response elements (TREs). Thereby, unliganded TR activates the gene, while hormone-bound TR is able to repress the activated status (Glass *et al.*, 1989). The detailed mechanism is still under investigation and is poorly understood.

Various TRβ mutant receptors from patients with RTH have been analyzed for their role in transcriptional regulation. In general, the mutant receptors have reduced affinity or lost completely the ability to bind to T_3. Consequently, all the hormone-dependent effects mediated by TR are reduced or completely absent. Thereby, TRβ mutant receptors also show an abberrant effect on the AP-1 inhibition. Mutant receptors with significant impairment of hormone binding are unable to repress AP-1-activated gene expression (Ways *et al.*, 1993).

Similarly, the silencing function of TR mutants is not relieved by the actual hormone concentrations in a cell (Baniahmad *et al.*, 1992b). This is in accordance with reduced dissociation of the corepressors (Yoh *et al.*, 1997; Clifton-Bligh *et al.*, 1998; Nagaya *et al.*, 1998; Safer *et al.*, 1998) and reduced or lacking recruitment of coactivators (Fig. 2.5). Thus, genes that should be activated by hormone remain repressed. The repressive status mediated by TRβ mutants is even more enforced by their dominant negative activity. Thereby wild-type receptor activity is also inhibited by mutant receptors. However, not only the remaining wild-type TRβ and TRα are inhibited by the mutant receptors but also the closely related retinoic acid receptors (Baniahmad *et al.*, 1992b). The underlying mechanism presumably includes formation of inactive heterodimers (Collingwood *et al.*, 1994; Yen and Chin, 1994).

GENETICS

The syndrome of RTH is caused mostly by point mutations that affect the coding sequences of the LBD of TRβ (Fig. 5) on chromosome 3. Thus, both TRβ1 and TRβ2 are affected by the mutations. Since it is autosomally inherited, both sexes are affected. A list of known receptor mutants with the respective amino acid change is shown in Fig. 2.5. Mutations causing RTH are localized in exons 7 to 10 of human TRβ (hTRβ). Thereby, three clusters of mutations can be observed (Fig. 2.5). These mutations lead to a reduced hormone-binding affinity. With one exception, the disease is inherited in a dominant manner. Only one case with recessive inheritance has been reported (Refetoff *et al.*, 1967). This patient, also called the Refetoff patient (from kindred G), was homozygous for a major deletion of both alleles in the TRβ gene which resulted in the

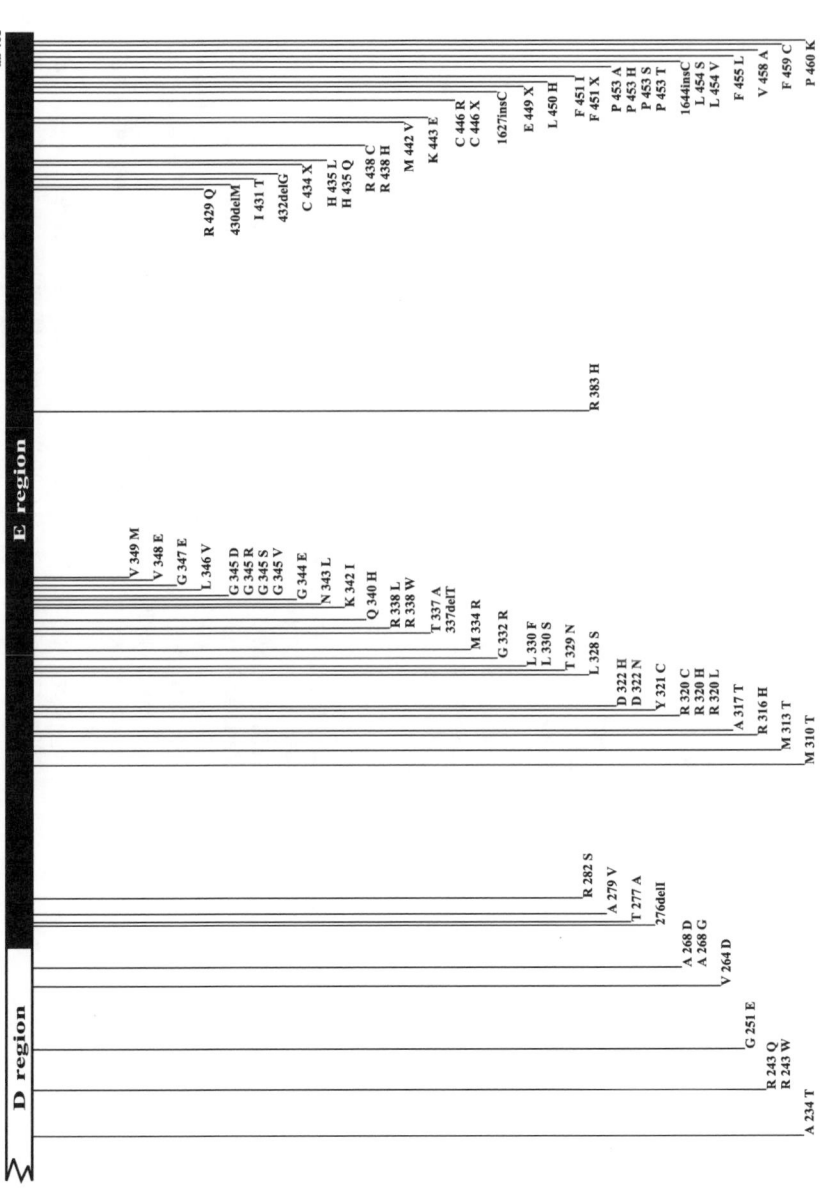

Fig. 2.5 Overview of the localization of each mutant within the C terminus of the β receptor. The localization of known mutations of thyroid hormone receptor β identified in patients with resistance to thyroid hormone is shown. Changes in the amino acid sequence are indicated. Mutations found, so far, are localized only in the C terminus (regions D and E) of the β receptor. Nomenclature is according to Beck-Peccoz et al. (1994).

Table 2.1 Thyroid hormone receptor β mutations

Mutation	Codon change	References
Exons 4–10Δ (family III)		Takeda et al. (1991, 1992a)
Exon 7		
A234T (A229T, kindred Td)	GCC → ACC	Behr and Loos (1992), Mixson et al. (1992), Collingwood et al. (1998), Safer et al. (1998)
R243Q	CGG → CAG	Onigata et al. (1995), Yagi et al. (1997), Collingwood et al. (1998), Safer et al. (1998)
R243W	CGG → TGG	Pohlenz et al. (1996), Yagi et al. (1997), Collingwood et al. (1998), Safer et al. (1998)
Exon 8		
G251E	GGA → GAA	Macchia et al. (1998)
V264D (kindred GP)	GTT → GAT	Adams et al. (1994), Collingwood et al. (1994)
A268D	GCC → GAC	Collingwood et al. (1998)
A268G		Jezequel et al. (1996)
Δ276I	ΔATC	Collingwood et al. (1998)
T277A	ACA → GCA	Collingwood et al. (1998)
A279V	GCA → GTA	Collingwood et al. (1998)
R282S	AGA → AGT	Collingwood et al. (1998)
Exon 9		
M310T	ATG → ACG	Takeda et al. (1992b)
M313T	ATG → ACG	Refetoff et al. (1996), di Fulvio et al. (1997)
R316H (R311H, kindreds GH, deG)	CGC → CAC	Geffner et al. (1993), Ways et al. (1993), Adams et al. (1994), Collingwood et al. (1994), Weiss et al. (1994a), Kitajima et al. (1995), Takeda et al. (1997)
A317T (A312T, kindreds ED, CM, PC)	GCT → ACT	Parilla et al. (1991), Mixson et al. (1992), Takeda et al. (1992b), Weiss et al. (1993), Adams et al. (1994), Cheng et al. (1994), Brucker-Davis et al. (1995), Pohlenz et al. (1995), Seto and Weintraub (1996), Sunthornthepvarakul et al. (1997)
R320C (R315C, kindred WR)	CGC → TGC	Burman et al. (1992), Mixson et al. (1992), Weiss et al. (1993, 1997), Pohlenz and Knobl (1996)
R320H (R315H, kindreds CL, PM)	CGC → CAC	Cugini et al. (1992), Mixson et al. (1992), Ways et al. (1993), Weiss et al. (1993), Adams et al. (1994), Collingwood et al. (1994), Liu et al. (1998)

R320L (kindreds SC, GM)	CGC → CTC	Adams *et al.* (1994), Collingwood *et al.* (1994), Erichsen *et al.* (1998)
Y321C (kindred ST)	TAT → TGT	Adams *et al.* (1994), Collingwood *et al.* (1994)
D322H (D317H, kindred IR)	GAC → CAC	Mixson *et al.* (1992), Cheng *et al.* (1994), Brucker-Davis *et al.* (1995)
D322N (TRβ-CN)	GAC → AAC	Behr and Loos (1996)
L328S	TTA → TCA	Brucker-Davis *et al.* (1995)
T329N	ACC → AAC	Sarkissian *et al.* (1999)
L330F		Usala *et al.* (1992)
L330S		Pohlenz *et al.* (1997)
G332R (G327R, kindreds BB, FW)	GGG → AGG	Parilla *et al.* (1991), Mixson *et al.* (1992), Takeda *et al.* (1992b), Adams *et al.* (1994)
M334R	ATG → AGG	Collingwood *et al.* (1994)
T337A	ACA → GCA	Asteria *et al.* (1999)
Δ337T (332del, kindred S)	ΔCAC	Ono *et al.* (1991), Usala *et al.* (1991b), Baniahmad *et al.* (1992b), Takeda *et al.* (1992a), Ways *et al.* (1993), Abel *et al.* (1999)
R338L (kindred NM)	CGG → CTG	Adams *et al.* (1994), Safer *et al.* (1997), Menzaghi *et al.* (1999)
R338W (kindreds KT, LF, JM, RM, LM)	CGG → TGG	Mixson *et al.* (1992, 1993), Sasaki *et al.* (1993), Weiss *et al.* (1993), Adams *et al.* (1994), Cheng *et al.* (1994), Collingwood *et al.* (1994), Brucker-Davis *et al.* (1995), Sasaki *et al.* (1995), Ando *et al.* (1996), Seto and Weintraub (1996), Safer *et al.* (1997)
Q340H (Q335H)	CAG → CAC	Usala *et al.* (1991c), Mixson *et al.* (1992), Brucker-Davis *et al.* (1995)
K342I	AAA → ATA	Brucker-Davis *et al.* (1995), Seto and Weintraub (1996)
N343L		Brucker-Davis *et al.* (1995)
G344E (kindred BK)	GGG → GAG	Adams *et al.* (1994), Collingwood *et al.* (1994)
G345D	GGT → GAT	Takeda *et al.* (1992b)
G345R (G340R, hTRβ-Mf)	GGT → CGT	Sakurai *et al.* (1989), Chatterjee *et al.* (1991), Takeda *et al.* (1991, 1997), Adams *et al.* (1992), Mixson *et al.* (1992), Takeda *et al.* (1992b), Nagaya and Jameson (1993), Kitajima *et al.* (1995), Ando *et al.* (1996), Yagi *et al.* (1997), Hayashi *et al.* (1998), Liu *et al.* (1998)
G345S (G340S)	GGT → AGT	Adams *et al.* (1992), Collingwood *et al.* (1994)
G345V (G340V)	GGT → GTT	Parilla *et al.* (1991), Mixson *et al.* (1992), Brucker-Davis *et al.* (1995)
L346V	CTT → GTT	Macchia *et al.* (1997)
G347E (G342E)	GGG → GAG	Parilla *et al.* (1991), Mixson *et al.* (1992), Cheng *et al.* (1994)
V348E (kindred JA)	GTG → GAG	Seto and Weintraub (1996)
V349M (kindred SS)	CTG → ATG	Adams *et al.* (1994), Safer *et al.* (1997)
R383H	CGC → CAC	Clifton-Bligh *et al.* (1998)

continued overleaf

Table 2.1 *Continued*

Mutation	Codon change	References
Exon 10		
R429Q (kindreds LL, MS, MA)	CGG → CAG	Adams et al. (1994), Flynn et al. (1994), Hayashi et al. (1995), Safer et al. (1997), Clifton-Bligh et al. (1998)
Δ430M	ΔATG	Collingwood et al. (1994)
I431T (kindred LO)	ATA → ACA	Adams et al. (1994), Safer et al. (1997)
Δ432G	ΔGGA	Collingwood et al. (1994)
C434X	TGC → TGA	Behr et al. (1997)
H435L	CAT → CTT	Tsukaguchi et al. (1995), Nomura et al. (1996)
H435Q	CAT → CAA	Tsukaguchi et al. (1995), Nomura et al. (1996)
R438C (kindred CMa)	CGC → TGC	Adams et al. (1994), Collingwood et al. (1994)
R438H (R433H; kindreds BW, CS, GS, JH)	CGC → CAC	Boothroyd et al. (1991), Mixson et al. (1992), Takeda et al. (1992b), Sakurai et al. (1993), Weiss et al. (1993), Adams et al. (1994), Collingwood et al. (1994), Gurnell et al. (1998)
M442V (M437V, kindred OK)	AGT → GTG	Parrilla et al. (1991), Mixson et al. (1992), Cheng et al. (1994), Brucker-Davis et al. (1995)
K443E (K438E)	AAG → GAG	Sasaki et al. (1992, 1995), Ando et al. (1996)
C446R	TGC → TGA	Weiss et al. (1994b)
C446X (Δ446–461)	TGC → TGA	Groenhout and Dorin (1994), Miyoshi et al. (1998)
1627insC (frameshift 443 to >458, kindred PV)	C insertion	Parrilla et al. (1991), Mixson et al. (1992), Brucker-Davis et al. (1995)
E449X	GAA → TAA	Taniyama et al. (1996), Miyoshi et al. (1998)
L450H (L445H)	CTC → CAC	Mixson et al. (1992), Cheng et al. (1994), Brucker-Davis et al. (1995)
F451I	TTC → ATC	Tsukaguchi et al. (1995)
F451X	TTC → TAA	Miyoshi et al. (1998), Nishiyama et al. (1998)
P453A (kindred TB)	CCT → GCT	Adams et al. (1994), Collingwood et al. (1994)
P453H (P448H, hTRβ-Mh, kindred A)	CCT → CAT	Usala et al. (1990a,b), Takeda et al. (1991), Baniahmad et al. (1992b), Mixson et al. (1992), Takeda et al. (1992b), Nagaya and Jameson (1993), Cheng et al. (1994), Collingwood et al. (1994), Brucker-Davis et al. (1995), Zavacki et al. (1996), Safer et al. (1997)
P453S	CTT → TCT	Takeda et al. (1992b), Adams et al. (1994), Collingwood et al. (1994), Refetoff et al. (1994), Ozata et al. (1995)

Mutation	Codon change	References
P453T (P448T, kindred QW; kindreds SH, PA, MC)	CCT → ACT	Parrilla et al. (1991), Mixson et al. (1992), Shuto et al. (1992a,b), Adams et al. (1994), Collingwood et al. (1994), Brucker-Davis et al. (1995), Radetti et al. (1997)
1644insC (frsh452, L454F + frameshift, kindred S, 1638i)	C insertion	Takeda et al. (1992b), Adams et al. (1994), Collingwood et al. (1994)
L454S (kindred MW)	TTG → TCG	Tagami et al. (1998)
L454V	TTG → GTG	Collingwood et al. (1997)
F455L (F450L)	TTC → TTA	Hiramatsu et al. (1994)
V458A	GTG → GCG	Weiss et al. (1996)
F459C (F459C)	TTC → TGC	Mixson et al. (1992), Cheng et al. (1994), Brucker-Davis et al. (1995)
P460K (kindred MP)	GAG → AAG	Adams et al. (1994)

complete loss of the functional β receptor (deletion of exons 4 to 10). The patient had only mild symptoms, such as short stature. It is thought that the α receptor is, at least in part, able to overcome the TRβ deficiency. In this particular patient there is no dominant negative-acting receptor which can inhibit wild-type receptors. This is presumably the reason for the recessive inheritance in this particular case.

Interestingly, the Bercu patient, another homozygous patient, had, on both alleles, a TRβ deletion of only one amino acid, called TRΔ337 or kindred S receptor, which results in complete loss of hormone binding and strong dominant negative activity. The patient had profound symptoms, such as severe delay in bone maturation and intellectual impairment (Refetoff, 1994b; Refetoff *et al.*, 1967; Usala, 1991).

Mutations causing resistance to thyroid hormone

Below, known receptor mutants from patients with RTH and their effects on the molecular function of TR are described. A comprehensive overview of the mutant receptors is also provided, citing both the nucleotides and amino acids as well as the citations for each receptor mutant (Table 2.1), according to the nomenclature (Beck-Peccoz *et al.*, 1994). As hTRβ was later shown to have 461 amino acids rather than 456 (Sakurai *et al.*, 1990), it may be confusing to compare the original citations concerning the localization of TRβ mutations. For example, A234T was originally termed A229T (Behr and Loos, 1992). Termination signals of translation, caused by a mutation, are abbreviated with 'X'. A summary of the localization of each mutant within the C-terminus of the receptor is shown in Fig. 2.5.

Mutations within exon 7

A234T

The first report of a mutation within the hinge region of TRβ describes a G to A transition in codon 234 in one TRβ allele of three affected members of one family, designated family Td (Behr and Loos, 1992). The *in vitro* expressed mutant TRβ was shown to bind with high affinity to various TREs. However, the affinity of the mutant receptor for T_3 is reduced threefold, indicating that the hinge domain of TRβ is important for full ligand-binding activity. On a molecular level, DNA binding, hetero-dimerization and corepressor recruitment were preserved in several mutations, including A234T and others, extending from codon 234 to 282 (Collingwood *et al.*, 1998). However, Safer *et al.* (1998) reported that hinge mutants with near-normal T_3 binding showed an impaired dissociation of the corepressor NCoR and recruitment of the steroid receptor coactivator (SRC-1).

R243Q/W

Onigata *et al.* (1995) described a Japanese family with RTH that had a G to A transition in codon 243 in one allele. This resulted in substitution of the normal arginine with glutamine within the hinge region. Pohlenz *et al.* (1996) reported a C to T transition in the same codon in a family with generalized resistance to thyroid hormone (GRTH), resulting in a substitution of the arginine with tryptophan. Both R243 mutants exhibited normal silencing activity in the absence of T_3 but fail to transactivate in the presence of T_3 (Yagi *et al.* 1997). However, the affinity for T_3 is not significantly reduced *in vitro*. Similar to A234T, DNA binding, heterodimerization and corepressor recruitment are preserved, with near-normal T_3 binding, while an impaired dissociation of the corepressor NCoR and recruitment of the coactivator SRC-1 was observed (Collingwood et al., 1998; Safer et al., 1998).

Mutations within exon 8

G251E

Macchia et al. (1998) identified a heterozygous G to A transition at nucleotide 1037 in codon 251, resulting in a glycine to glutamic acid substitution in a patient affected by RTH. This mutation probably represents a de novo mutation since both parents of the patient were unaffected and did not have this genomic mutation. The mutant protein exhibited a markedly decreased affinity for T_3, which presumably leads to impaired corepressor dissociation.

V264D

This mutation, resulting from a T to A transition in codon 264, was found in a kindred (GP) with GRTH (Adams et al., 1994; Collingwood et al. 1994). The mutant receptor acts as a dominant negative in transient transfection assays and exhibits almost no affinity for T_3. However, the dominant negative action is slightly decreased at high T_3 concentrations.

A268G

Jezequel et al. (1996) reported this mutation in a large family with RTH. However, the molecular mechanisms by which this mutant receptor causes RTH is unknown.

A268D

A transition from C to A, found in two patients with RTH, leads to a mutated protein with almost no affinity for T_3 (Collingwood et al., 1998). Accordingly, A268D retains dominant negative activity even at high T_3 levels. Also, interaction with the corepressor SMRT is retained even in the presence of T_3. These data clearly indicate that the unconserved hinge region of TR is important for hormone binding. DNA binding and heterodimerization with RXR are preserved.

Δ276I

An inframe deletion of one entire codon, found in a 7-year-old girl with RTH, results in a mutant receptor with almost no affinity for T_3 and which retains dominant negative activity even at high T_3 levels. Also, interaction with the corepressor SMRT is retained in the presence of T_3, indicating again that the unconserved hinge region of TR is important for hormone binding. DNA binding, silencing in the absence of T_3, and heterodimerization with RXR are preserved.

T277A

A substitution from G to A was found in a 31-year-old woman with RTH (Collingwood et al., 1998). The mutant receptor showed just a threefold reduced affinity for T_3 in vitro but interestingly retained dominant negative activity even at high T_3 levels. The authors showed that this effect might be due to an impaired recruitment of the coactivator SRC-1. This further correlates with the observation that interaction with the corepressor SMRT is retained in the presence of T_3, while DNA binding, silencing function in the absence of T_3 and heterodimerization with RXR are preserved.

A279V

A transition from C to T, found in two patients with RTH, leads to a mutant receptor with almost no affinity for T_3 (Collingwood *et al.*, 1998). Also interaction with the corepressor SMRT and dominant negative activity are retained even at high T_3 levels, while DNA binding and heterodimerization with RXR are preserved. This indicates a strong relationship between lack of corepressor dissociation and manifestion of the RTH syndrome.

R282S

A transition from A to T, found in five patients with RTH, leads to a mutant receptor with only a threefold reduced affinity for T_3 (Collingwood *et al.*, 1998). Accordingly, the dominant negative activity is abolished at high T_3 levels. Interestingly, however, interaction with the corepressor SMRT is retained even in the presence of T_3, while DNA binding and heterodimerization with RXR are preserved.

Mutations within exon 9

M310T

Takeda *et al.* (1992b) reported this mutation in a family with GRTH. However, the molecular mechanisms by which this mutant receptor causes RTH are not known.

M313T

Refetoff *et al.* (1996) reported a heterozygous T to C transition in two unrelated families with RTH. Whereas *attention deficit hyperactivity disorder* (ADHD) was associated with the mutant receptor in seven of the nine affected individuals, these symptoms were also present in two family members without RTH. Therefore, the ADHD syndrome is not necessarily associated with RTH, and is also seen in another mutation (R316H; Weiss *et al.*, 1994a). However, the molecular mechanisms by which this mutant receptor causes RTH are not known.

R316H

Geffner *et al.* (1993) reported a G to A transition in one allele of a female patient (kindred GH) with a severe pituitary form of RTH (PRTH). Interestingly, this mutation was also found in her unaffected father and half-sister. The mutant receptor revealed significantly defective T_3-binding activity. Both alleles, the mutant and the wild-type, showed no differential expression in leukocytes and fibroblasts, neither in the patient nor in her unaffected father. Also the mutant receptor exhibited no dominant negative activity in transfection assays. Thus, the authors suggested that the R316H mutation may contribute to PRTH in patient GH, but cannot be the sole cause of the disease.

In contrast, Collingwood *et al.* (1994) and Kitajima *et al.* (1995) reported that R316H does act as a dominant negative in transient transfection assays, but only at physiologic concentrations of T_3. At saturating concentrations of T_3, the dominant negative function is abolished. Both groups also found that, in contrast to the wild-type receptor, the R316H mutation leads to loss of homodimerization on selected TREs, but heterodimerization with RXR is preserved in all cases. The selective loss of homodimerization is independent of the orientation of the half-sites within the TRE. The latter group also showed that introduction of the R316H mutation into a dominant negative mutation (G345R) causes a reduction of the dominant negative activity to the level of

R316H alone. Therefore, they suggested that mutations that alter homodimerization have a reduced dominant negative activity for some target genes. However, it still remains unclear why some members of kindred GH, harboring this mutation, show no RTH phenotype (Geffner *et al.*, 1993).

Another family with RTH, designated F120 (Weiss *et al.*, 1994a), that harbors this mutation was analyzed for the association of ADHD with the RTH syndrome. The authors found that their data do not support a genetic linkage between ADHD and RTH, which is also seen for another mutation (M313T), but suggested that RTH is associated with lower IQ scores and so may confer a higher likelihood of exhibiting ADHD symptoms.

A317T

Parrilla *et al.* (1991) described this mutation in one allele of a patient (kindred ED). Both parents had two wild-type alleles, suggesting a spontaneous mutation. Weiss *et al.* (1993) reported this mutation in two other unrelated families, designated F52 and F100. Another family harboring this mutation was identified through a child with goiter and raised thyroid hormone levels (Pohlenz *et al.*, 1995). The father and a brother were later identified as affected. The phenotype of the affected family members differed from that previously described (Parrilla *et al.*, 1991), concerning 'significant articulation problems'. The first report of a TRβ mutation causing RTH in an individual of Thai origin showed that this patient also harbored this mutation (Sunthornthepvarakul *et al.*, 1997).

R320C

This mutation, due to a C to T transition, was first reported in one allele of a kindred (WR) with GRTH (Burman *et al.*, 1992). The mutant showed a 50% reduced T_3 binding affinity. Weiss *et al.* (1993) reported this mutation in another unrelated family, designated F88 (Pohlenz and Knobl, 1996). In another patient harboring this mutation, an 8-year-old boy with PRTH, treatment with D-thryroxin (DT_4) diminished the symptoms of thyrotoxicosis (Burman *et al.*, 1992). However in another case with the same mutation the affected family members reacted differently on treatment with D-T_4 (Weiss *et al.*, 1997).

R320H

Cugini *et al.* (1992) reported a kindred (CL) with a G to A transition in codon 320 in a patient with mild GRTH. The affinity of the mutant receptor for T_3 was only slightly reduced. Liu *et al.* (1998) showed that the slightly reduced affinity to T_3 correlates with a reduced ability of the mutant receptor to release the corepressor NCoR and to recruit the coactivator SRC-1.

R320L

Adams *et al.* (1994) reported a G to T transition in two kindreds (SC and GM). Interestingly, SC was affected by GRTH, while GM showed more effect on the pituitary. The mutant receptor exhibited a 10-fold decrease in T_3 binding. Interestingly, they also detected two other mutations, R429Q and R338W, where there was no correlation between genotype and phenotype. R320L shows a reduced loss of homodimerization in the presence of T_3 but no effect on heterodimerization with RXR (Collingwood *et al.*, 1994). In transient transfection assays the mutant receptor acts as a dominant negative. High concentrations of T_3 abrogate this function.

Y321C

This mutation, resulting from an A to G transition in a kindred (ST) with PRTH, shows almost no T_3 binding and acts as a dominant negative in transient transfection assays (Adams *et al.*, 1994; Collingwood *et al.*, 1994). Both homodimerization and hetero-dimerization with RXR are not affected significantly. Presumably, the lack of T_3 binding abrogates corepressor dissociation and/or coactivator binding, which leads to dominant negative activity.

D322H

This mutation, due to a G to C transition in codon 322 in a kindred termed IR, leads to a mutant receptor with a threefold reduced affinity for T_3 (Mixson *et al.*, 1992). Presumably the relatively slight reduction of T_3 binding abrogates corepressor dissociation and/or coactivator binding at physiologic T_3 concentrations, which then leads to dominant negative activity. This effect might be overcome by using high T_3 concentrations, as shown for several mutant receptors with just slightly reduced affinity for T_3.

D322N

Behr and Loos (1996) reported this mutation in a patient, termed CN, and his daughter. The mutant receptor revealed strongly decreased T_3-binding activity. At low T_3 levels, the mutant shows only limited transactivation function, whereas full activity is regained at a high T_3 concentration. At low T_3 levels, the mutant exerts a dominant negative effect on wild-type TRβ, whereas this effect is diminished in the presence of high T_3 concentrations. The mutant, however, cannot be activated through RXR in the presence of T_3, whereas addition of 9-*cis*-retinoic acid results in transactivation through RXR independently of the presence of the mutant receptor. This indicates that the time-dependent, variable THR phenotype of patient CN might be influenced by the differential expression of RXRs and the T_3 and 9-*cis*-retinoic acid hormonal status.

L328S

This mutation, due to a T to C transition at codon 328, was found in a kindred termed VO (Brucker-Davis *et al.*, 1995). However, the molecular mechanisms by which this mutant receptor causes RTH is not known.

T329N

Sarkissian *et al.* (1999) reported a C to A transition at nucleotide 1271 in a 30-year-old woman who was investigated after recurrent spontaneous abortions and was found to have RTH. D-T_4 treatment led to the delivery of a healthy boy and an RTH-affected girl. The mutant receptor reveals a markedly reduced affinity to T_3, functions in a dominant negative manner, and is transcriptional inactive. Formation of mutant homodimers and heterodimers with RXR is not affected, but the ability of T_3 to interrupt mutant wild-type homodimerization is markedly reduced.

L330F

Usala *et al.* (1992) reported this mutation in two unrelated patients with selective PRTH. However, the molecular mechanisms by which this mutant receptor causes RTH are not yet known.

L330S

At the same codon, where a mutation to phenylalanine was reported previously (Usala *et al.*, 1992), Pohlenz *et al.* (1997) identified a T to C transition in an 11-year-old boy who presented with symptoms suggestive of both hyperthyroidism and hypothyroidism.

G332R

This mutation was identified in two unrelated families (Parrilla *et al.*, 1991; Takeda *et al.*, 1992b). However, the molecular mechanisms by which this mutant receptor causes RTH are not known.

M334R

This mutation was identified in a patient with PRTH (Collingwood *et al.*, 1994). The mutant receptor exhibits a 20-fold reduced affinity for T_3 and acts as a dominant negative in transient transfection assays even at high doses of T_3. Homodimerization and heterodimerization with RXR depend on the type of TRE, but the presence of T_3 does not lead to a decrease in homodimerization.

T337A

Asteria *et al.* (1999) reported this mutation in a 29-year-old woman with PRTH, successfully treated with TRIAC (triiodothyroacetic acid) until pregnancy. Therapy was discontinued to prevent harm to the fetus. Prenatal diagnostic testing revealed that the fetus also harbored the mutant allele. TRIAC treatment was restarted at low doses. At week 29 mild fetal growth retardation and goiter were observed. The TRIAC dose was then increased, and after 5 weeks the goiter was reduced. The baby was delivered by cesarean section and showed the biochemical features of RTH. This is the first report of successful treatment during pregnancy. However, the molecular mechanisms by which this mutant receptor causes RTH are not known.

T337del

The kindred S receptor has a CAC deletion in codon 337 which results in loss of threonine at this position (Usala *et al.*, 1991b). Heterozygotes of kindred S displayed raised levels of free T_4 and inappropriately normal TSH levels. However, patient S1, homozygous for the S receptor, had markedly increased TSH and free T_4 levels, and displayed profound abnormalities in brain development and linear growth. The mutant receptor does not bind T_3, but binds to TREs with the same affinity as wild-type hTRβ. Ono *et al.* (1991) described the clinical features of the homozygous patient and reported that the phenotype was markedly different from recently reported patients with no TRβ, suggesting that TRβ mutants can inhibit at least some thyroid hormone action mediated by TRα. The kindred S receptor acts as a constitutive silencer in transient transfection assays (Baniahmad *et al.*, 1992b), suggesting that this feature enforces the dominant negative activity of the receptor. Abel *et al.* (1999) developed a transgenic mouse model with pituitary-specific expression of T337del and reported findings indicating that pituitary expression of this mutant TR impairs both T_3-mediated suppression and T_3-independent activation of TSH production *in vivo*. Also, alteration of the cardiac phenotype has been observed (Gloss *et al.*, 1999).

R338L

Adams *et al.* (1994) reported this G to T transition in codon 338 in a kindred (NM) with PRTH. R338L shows a fourfold decrease in T_3 binding. Presumably the reduction of T_3

affinity abrogates corepressor dissociation and/or coactivator binding at physiologic T_3 concentrations, which then leads to dominant negative activity. This effect might be overcome by using high T_3 concentrations, as shown for several mutant receptors with just slightly reduced affinity for T_3.

R338W

A transition from C to T at codon 338 was first found in kindred KT (Mixson *et al.*, 1992). The mutant receptor exhibits a significant reduced affinity for T_3. This receptor was also found in families termed F29 and F106 (Weiss *et al.*, 1993), as well as LF (Mixson *et al.*, 1993). Whereas LF suffered from selective PRTH, kindred KT was reported to have GRTH, while Sasaki *et al.* (1993) reported another family harboring this mutation where affected members suffered from PRTH. Adams *et al.* (1994) reported that this mutation was associated with selective PRTH in four of five kindreds, but found a further patient who had GRTH. They also reported another two mutations (R429Q and R338W), where there was no stringent correlation between genotype and phenotype. The receptor exhibits relatively mild transcriptional activation, homodimerization and dominant negative activity. High doses of T_3 abrogate the dominant negative activity, which correlates with a still detectable T_3 binding capacity (Collingwood *et al.*, 1994; Sasaki *et al.*, 1995).

Accordingly, Ando *et al.* (1996) showed that introducing this mutation into severe dominant negative mutant receptors (K443E and F451X) significantly reduces their dominant negative activity.

Q340H

The kindred D receptor, a G to C transition, was described in a family with GRTH (Usala *et al.*, 1991c). However, the molecular mechanisms by which this mutant receptor causes RTH is not yet understood.

K342I

Brucker-Davis *et al.* (1995) reported this mutation in a kindred termed YH. However, the molecular mechanisms by which this mutant causes RTH are not known.

N343L

Brucker-Davis *et al.* (1995) reported this mutation in a kindred termed ZP. However, the molecular mechanism by which this mutant causes RTH is not yet known.

G344E

Adams *et al.* (1994) and Collingwood *et al.* (1994) reported this mutation, resulting from a G to A transition, in a kindred (BK) with GRTH. The mutant receptor shows no detectable T_3 binding, correlating with the observation that dominant negative function in transient transfection assays is retained even at extremely high T_3 concentrations, and that homodimerization is not disrupted by the addition of ligand.

G345D

Takeda *et al.* (1992b) reported this mutation in one family with RTH. However, the molecular mechanism by which this mutant receptor causes RTH is not known.

G345R/S

These two mutations in the same codon exhibit more or less the same biochemical features. A transition from G to C in one of two alleles in affected members of a family (Mf) with inherited GRTH, results in the replacement of glycine by arginine (Sakurai *et al.*, 1989). This was the first mutation in the TRβ gene found to cause RTH. The *in vitro* translated mutant receptor fails to bind T_3. The same mutation was found in other, unrelated families with GRTH (Takeda *et al.*, 1991, 1992b). The mutant functions neither as a transactivator on a positive TRE nor as a repressor on a negative TRE in the presence of T_3 (Chatterjee *et al.*, 1991). Furthermore, the mutant receptor exhibits dominant negative activity on the wild-type receptor. Consistent with this, the DNA-binding activity of the mutant is not affected. Another mutation leads to an exchange of the same glycine with serine, and exhibits the same properties (G345S; Adams *et al.*, 1992). Nagaya and Jameson (1993) reported that G345R, as well as P453H, leads to the formation of homodimers as well as heterodimers with RXR. Introduction of an artificial amino acid substitution (L428R) impairs heterodimerization (but not homodimerization) and also results in loss of heterodimerization of the double mutants.

In transient transfection assays, these double mutants lose their dominant activity on the wild-type receptor. This supports the idea that a possible mechanism of dominant negative activity is the formation of functionally inactive heterodimers on DNA which inhibits the access by wild-type receptors. Also, introduction of a naturally occurring RTH mutation, R316H, known to impair heterodimerization, into G345R, reduces the dominant negative activity to the level of R316H alone (Kitajima *et al.*, 1995). Furthermore, introduction of R338W, another mutant affecting heterodimerization, but not G345R, into dominant negative mutants results in a reduction of dominant negative activity.

Liu *et al.* (1998) reported that G345R binds the corepressor NCoR in both the presence and the absence of T_3 but does not form a complex with the coactivator SRC-1 when T_3 is added. Additionally, when N-CoR and SRC-1 are incubated together with the receptor at various T_3 concentrations, only TR–NCoR or TR–SRC-1 complexes, but no ternary complexes, are observed. This suggests that NCoR release is necessary before SRC-1 binds to TR.

These data provide a new insight into the molecular mechanisms of dominant negative activity in RTH, suggesting that the inability of mutant TRs to dissociate corepressors and to interact with coactivators is a determinant of dominant negative activity.

G345V

Parrilla *et al.* (1991) reported a G to T transition in codon 345 resulting in a replacement of the original glycine by valine. However, the molecular mechanisms by which this mutant receptor causes RTH are not known.

L346V

This mutant receptor shows a marked reduction in affinity for T_3, impaired ligand-dependent transactivation and potent dominant negative activity. This functional impairment cannot be alleviated, even at high concentrations of T_3, suggesting that the mutation might interfere with the intrinsic ligand-dependent transactivation function (AF-2) (Macchia *et al.*, 1997). The receptor mutant was found in a woman and her son. The presence of the mutation in the son, who died from complications associated with congenital heart disease, raises the possibility that RTH might have contributed to the pathogenesis or severity of the disease.

G347E

Parrilla *et al.* (1991) reported a T to A transition in codon 345 resulting in replacement of the original glycine by glutamic acid. This receptor exhibits reduced affinity for T_3 and thyroid hormone analogs (Cheng *et al.*, 1994).

V348E

Seto and Weintraub (1996) reported a T to A transition in a kindred termed JA. However, the molecular mechanism by which this mutant receptor causes RTH is not known.

V349M

A transition from C to A was found in a kindred (SS) with PRTH (Adams *et al.*, 1994). T_3 binding was reduced fourfold. This and other mutant receptors, derived from PRTH cases, function mainly as dominant negatives for the TRβ2 isoform, but not (or less) for the TRβ1 isoform. RTH caused by these mutations is therefore limited to tissues that express the TRβ2 isoform, most notably the anterior pituitary and hypothalamus (Safer *et al.*, 1997).

R383H

Clifton-Bligh *et al.* (1998) reported this RTH mutation caused by a G to A transition in codon 383. No other mutations are described for this region (Fig. 2.5). Interestingly, R383H has impaired negative trancriptional regulation, but transactivation of positively regulated genes is not affected. The unliganded activation of negative TREs is not reversed by T_3 compared with the wild-type receptor. The R383H mutation impairs ligand-induced corepressor release but does not affect coactivator recruitment. The homologous amino acid of R383 in TRα (R329) interacts with R375 of TRα (homologous to R429 in TRβ). The R429Q mutant of TRβ is also associated with RTH and has been shown to exhibit functional impairment of negative transcriptional regulation (Flynn *et al.*, 1994; Hayashi *et al.*, 1995). This suggests that the physical interaction of R383 and R429 is critical for the release of corepressors.

Mutations within exon 10

R429Q

Adams *et al.* (1994) reported the same G to A transition in two kindreds (MS and MA). Interestingly MS suffered from PRTH, while MA had GRTH. The mutant receptor exhibits a fourfold decrease in T_3 binding. Two other mutations (R320L and R338W) are described, where there is no correlation between genotype and phenotype. This example shows that the distinction between PRTH and GRTH is, in terms of a mechanical view, misleading and presumably not dependent on the genetic background.

M430del

This mutation, which results from a frame deletion of codon 430, was identified in a patient with PRTH. The mutant receptor shows no detectable T_3 binding, which correlates with the observation that dominant negative activity is retained even at high T_3 concentrations (Collingwood *et al.*, 1994). M430del forms homodimers and heterodimers with RXR, but the addition of T_3 does not lead to a decrease of homodimers.

I431T

This mutation, resulting from a T to C transition, was identified in a kindred, termed LO, with PRTH. The mutant receptor shows extremely reduced T_3 binding, which correlates with the observation that dominant negative function in transient transfection assays is slightly retained at high T_3 concentration. Homodimerization is slightly increased on several TREs, but addition of T_3 does not lead to decreased homodimerization (Adams et al., 1994; Collingwood et al., 1994). This and other mutant receptors derived from patients with PRTH function mainly as dominant negative for the TRβ2 isoform, but less for the TRβ1 isoform. RTH caused by these mutations is therefore limited to tissues that express the TRβ2 isoform, most notably the anterior pituitary and hypothalamus (Safer et al., 1997).

G432del

Collingwood et al. (1994) reported this mutation in a patient with PRTH. The mutant receptor shows no detectable T_3 binding, which correlates with the observation that dominant negative activity in transient transfection assays is retained even at high concentrations of T_3. G432del also forms homodimers and heterodimers with RXR, but the addition of T_3 does not lead to a decrease of homodimerization.

H435L/Q

Tsukaguchi et al. (1995) reported these two point mutations (A to T and T to A transition, respectively) in two unrelated Japanese families. These mutations, which lie at the C-terminus of the dimerization domain near the ninth heptad, are associated with different subtypes of RTH: the patient harboring H435L suffered from GRTH, while the H435Q mutant was reported in a patient with PRTH. Interestingly, a crystallography study suggests that H435 is critical for direct contact with T_3 (Wagner et al., 1995). The analysis of H435L, H435Q and two artificial mutants (H435R and H435E) revealed that these point mutations show no T_3 binding, are inactive in T_3-dependent transactivation and exhibit dominant negative activity (Nomura et al., 1996). Heterodimerization with RXR is not affected by mutations of H435, but the ability for homodimer formation is decreased when histidine is exchanged for a neutral (L) or acidic (E) amino acid, suggesting that H435 is essential for homodimer formation. However, homodimerization of H435Q is less affected than that of H435L, providing a link between different biochemical features and different phenotypes.

R438C

A transition from C to T was identified in a kindred (CMa) with GRTH. T_3 binding of the mutant receptor was slightly reduced (threefold), which correlates with the observation that dominant negative activity in transient transfection assays can be abolished at higher T_3 concentrations. Heterodimerization with RXR and homodimerization are not affected (Adams et al., 1994; Collingwood et al., 1994).

R438H

Boothroyd et al. (1991) first reported a G to A transition at codon 438. The same mutation has since been found in several other, unrelated families (Takeda et al., 1992a; Sakurai et al., 1993; Weiss et al., 1993; Adams et al., 1994). The mutant receptor shows a reduced, but still detectable, affinity for T_3, and remains able to bind DNA (Sakurai et al., 1993; Adams et al., 1994). Accordingly, the dominant negative function in transient transfection assays is abolished at high T_3 levels (Collingwood et al., 1994).

M442V

This mutation is a result of an A to T transition and was identified in a kindred, designated OK, with GRTH (Parrilla *et al.*, 1991). The mutant receptor exhibits a sixfold reduction of T_3-binding activity.

K443E

A transition from A to G in codon 443 was identified in a kindred with GRTH (Sasaki *et al.*, 1992). The mutant receptor has reduced T_3 binding affinity and acts in a dominant negative manner in transient transfection assays (Sasaki *et al.*, 1995).

C446R

This mutant receptor, identified from a patient with RTH, shows no T_3 binding, fails to exhibit transactivation, and shows dominant negative activity in transient transfection assays (Weiss *et al.*, 1994b). Presumably this reduction in T_3 binding abrogates corepressor dissociation and/or coactivator binding at physiologic T_3 concentrations, which then leads to dominant negative activity. This effect might be overcome by using high T_3 concentrations, as shown for several mutant receptors with slightly reduced affinity for T_3.

L450H

Mixson *et al.* (1992) reported a T to A transition in a kindred (TP) with GRTH. The mutant receptor shows a threefold reduction in T_3 binding. Presumably this reduction in T_3 binding abrogates corepressor dissociation and/or coactivator binding at physiologic T_3 concentrations, which then leads to dominant negative activity. This effect might be overcome by using high T_3 concentrations, as shown for several mutant receptors with slightly reduced affinity for T_3.

F451I

Tsukaguchi *et al.* (1995) reported a T to A transition in codon 451 in a patient with RTH. The parents did not harbor this mutation, indicating that this is a *de novo* mutation. However, the molecular mechanisms by which this mutant receptor causes RTH are unknown.

P453A

A transition from C to G at codon 453 was identified in a kindred (TB) with PRTH (Adams *et al.*, 1994; Collingwood *et al.*, 1994). The mutant receptor exhibits a fivefold reduction in T_3 binding and acts as a dominant negative at low concentrations of T_3. Extremely high concentrations of T_3 abrogate this inhibition. Interestingly, another mutation affecting the same codon (P453H; see below), from patients with GRTH, involves loss of dominant negative activity at a tenfold higher T_3 concentration, providing a link between different biochemical features and different phenotypes. Features of homodimerization and heterodimerization with RXR are comparable to the wild-type receptor.

P453H

The kindred A receptor was found in patients with GRTH and derives from a C to A transition at codon 453 (Usala *et al.*, 1990a,b). Takeda *et al.* (1991) found this mutation in another, unrelated, family. The mutant receptor exhibits a tenfold decrease in T_3 binding affinity (Chatterjee *et al.*, 1991) and acts as a dominant negative in transient transfection assays (Collingwood *et al.*, 1994). Under saturating T_3 concentrations, this dominant negative activity is partially relieved. P453H represses transcription in the absence of T_3, similar to the wild-type receptor (Baniahmad *et al.*, 1992b).

P453S

This point mutation, due to a T to C transition in codon 453, was first reported by Takeda *et al.* (1992b) and has since been found in another family (kindred MC; Adams *et al.*, 1992). The mutant receptor shows almost no T_3 binding affinity and acts as a dominant negative in transient transfections, even at high T_3 levels (Collingwood *et al.*, 1994). Heterodimerization with RXR and homodimerization are preserved, but formation of homodimers was less decreased, compared with the wild-type.

P453T

A substitution of C with an A in codon 453 caused this mutation in a kindred (QW; Parrilla *et al.*, 1991). The mutant receptor showed only a relatively mild decrease in binding affinity to T_3 (twofold to fivefold). Transactivation in transient transfection assays occurs only at high levels of T_3 (Shuto *et al.*, 1992a,b). Furthermore, P453T acts as a dominant negative receptor when overexpressed, but a high molar excess of the mutant is necessary. Collingwood *et al.* (1994) showed that the dominant negative function varies depending on the type of TRE, and that increasing T_3 concentrations decrease the ability to inhibit transactivation of wild-type receptor. Heterodimerization with RXR is not affected, while homodimerization was less decreased by addition of T_3 than the wild-type receptor.

L454S

A replacement of T with C at codon 454, within the $\tau 4$/AF2-AD activation domain, was found in a patient with RTH (MW) by Tagami *et al.* (1998). The parents did not have the mutation, indicating that this is a *de novo* mutation. The mutant receptor binds T_3 but with a fivefold decreased affinity. The ability to activate positive TREs and to repress negative TREs is markedly impaired. As anticipated from the location of the amino acid substitution, only a weak interaction with coactivators (glucocorticoid receptor interacting protein 1 (GRIP-1) and SRC-1), even at high T_3 levels, is observed. In the absence of T_3, L454S interacts much more strongly with the corepressor NCoR than does wild-type receptor, and the T_3-dependent release of NCoR is markedly impaired. In comparison, the NCoR interaction and T_3-dependent dissociation of an adjacent artificial AF-2 domain mutant (E457A) is normal, suggesting that L454 may be involved directly, or indirectly, in the release of corepressors as well as in the recruitment of coactivators.

L454V

A transition from T to G at codon 454 leads to this mutation in the kindred JA (Collingwood *et al.*, 1997). The mutant receptor reveals unusual biochemical properties in that it binds T_3 comparably to the wild-type receptor and forms homodimers and heterodimers comparable to the wild-type. However, the ability to transactivate is markedly impaired. The authors found that this is probably due to a reduced interaction with coactivators (SRC-1 and receptor interacting protein (RIP) 140). This hydrophobic leucine residue lies within an amphipathic α helix at the C-terminus of TRβ, and the position of the homologous residue in the crystal structure of TRα (Wagner *et al.*, 1995) indicates that the side-chain is solvent exposed and might be responsible for interaction with other cofactors.

F455L

This mutation was found in one allele of two GRTH-affected members of a Japanese family (Hiramatsu *et al.*, 1994). The *in vitro* translation product of this mutant has a significantly reduced T_3 binding affinity.

V458A

This mutation, involving a T to C transition at codon 458, was reported in a patient with RTH whose daughter was also affected (Weiss *et al.*, 1996). However, the molecular mechanism by which this mutation causes RTH is unknown.

F459C

A replacement of a T by a G at codon 459 in a kindred (RL) leads to this mutation, first reported by Mixson *et al.* (1992). The mutant receptor shows a threefold decreased affinity for T_3 and several analogs (Cheng *et al.*, 1994). Presumably, the relatively slight reduction of T_3 binding abrogates corepressor dissociation and/or coactivator binding at physiologic T_3 concentrations, which then leads to dominant negative activity. This effect might be overcome by using high T_3 concentrations, as shown for several mutant receptors with slightly reduced affinity for T_3.

P460K

A transition from G to A at codon 460 was observed in two members of a kindred termed MP (Adams *et al.*, 1994). The affinity of the mutant receptor for T_3 is decreased fourfold. Presumably, the relatively slight reduction of T_3 binding abrogates corepressor dissociation and/or coactivator binding at physiologic T_3 concentrations, which then leads to dominant negative activity. This effect might be overcome by using high T_3 concentrations, as shown for several mutant receptors with slightly reduced affinity for T_3.

τ4/AF-2 activation domain truncation mutants (C434X, C446X, E449X and F451X)

Four truncation mutants of TRβ, associated with RTH, have been described so far. These mutations result from base substitutions which lead to premature translation termination signals (stop codons) before or within the τ4/AF-2 activation domain at the extreme C-terminus of the receptor. The TRβ mutant C434X (TRβ-del434–461), resulting from a C to A transition at codon 434 exhibits no T_3 binding (Behr *et al.*, 1997). The truncated receptor shows strong dominant negative activity. The other truncated receptor, C446X (TRβ-del446–461), resulting from a C to A transition at codon 446, was found in one allele of a patient with GRTH (Groenhout and Dorin, 1994). Both parents are homozygous for the wild-type receptor, suggesting a *de novo* mutation in the patient. Taniyama *et al.* (1996) first reported E449X (TRβ-del449–461) in a patient with GRTH with mild resistance for thyroid hormone. F451X, resulting from two consecutive base substitutions, T to A and C to T at nucleotides 1637 and 1638 respectively, was found in a patient with GRTH (Miyoshi *et al.*, 1995). This truncated TR shows very low T_3 binding activity, and transactivating activity, and a very strong dominant negative effect, even at high T_3 concentrations. Comparing the latter three mutants, F451X and C446X were found in patients with severe intellectual impairment, whereas the patient with E449X showed no remarkable clinical symptoms, except for goiter (Miyoshi *et al.*, 1998).

Transient expression studies revealed that all three mutants have negligible T$_3$ binding and transactivation function. Interestingly, the dominant negative activity and silencing activity are significantly stronger for F451X than for E449X or C446X. Electrophoretic-mobility shift assay (EMSA) experiments revealed no apparent differences in homo-dimer formation between wild-type or mutant TRβ proteins and in heterodimer formation with RXR. These observations may indicate that τ4/AF-2 activation domain affects diverse TR functions. Thus, the position between the last 13th and 11th C-terminal amino acids strongly influences ligand-independent TR functions, such as dominant negative and silencing activities.

Frameshift mutants

Three different kindreds harboring frameshift mutations have been reported: kindred PV (Parrilla *et al.*, 1991) (1627insC) with a cytosine insertion at nucleotide 1627 (codon 448 with a stop codon at 463); family XI (Takeda *et al.* 1992b) and kindred SN (Adams *et al.*, 1994; Collingwood *et al.*, 1994). The latter two harbor a C insertion next to or within a cluster of six cytosine residues. The mutation described by Takeda *et al.* (1992b) was designated (1644insC; codon 454 with a stop codon at 463) by Beck-Peccoz *et al.* (1994), while Adams *et al.* (1994) reported their mutation to have a cytosine insertion at nucleotide 1638 and therefore could be named (1638insC; codon 452 with a stop codon at 463). However, no matter where the cytosine is inserted, it leads to the same mutant receptor (Fig. 2.6). Therefore, we refer here to the earlier report by Takeda *et al.* (1992b):

Wild-type

```
T    E    L    F    P    P    L    F    L    E    V    F    E    D    *
ACA  GAA  CTC  TTC  CCC  CCT  TTG  TTC  CTG  GAA  GTG  TTC  GAG  GAT  TAG  ACT  GAC
```

1627insC

```
T    R    T    L    P    P    F    V    P    G    S    V    R    G    L    D    *
CAC  AGA  ACT  CTT  CCC  CCC  TTT  GTT  CCT  GGA  AGT  GTT  CGA  GGA  TTA  GAC  TGA
```

1638insC

```
T    E    L    F    P    P    F    V    P    G    S    V    R    G    L    D    *
ACA  GAA  CTC  TTC  CCC  CCC  TTT  GTT  CCT  GGA  AGT  GTT  CGA  GGA  TTA  GAC  TGA
```

1644insC

```
T    E    L    F    P    P    F    V    P    G    S    V    R    G    L    D    *
ACA  GAA  CTC  TTC  CCC  CCC  TTT  GTT  CCT  GGA  AGT  GTT  CGA  GGA  TTA  GAC  TGA
```

Fig. 2.6 Frameshift mutations in the β-receptor C terminus. Naturally inserted nucleotides (bold and underlined) result in frameshift mutants (bold amino acids) that have lost the τ4 activation domain at the extreme C terminus. Stars represent stop codons.

1644insC, as suggested by Beck-Peccoz et al. (1994). Both mutations destroy the extreme C-terminus of TR, containing the $\tau 4$/AF-2 activation domain, and therefore it is not surprising that these receptors show similar biochemical features to the $\tau 4$/AF-2 truncation mutants (C434X, C446X, E449X and F451X). Both receptors show negligible T_3 binding and transcriptional activities. They also have, as far as has been tested, strongly dominant negative activity, even at high T_3 concentrations. 1644insC forms homo and heterodimers with RXR, but addition of T_3 does not lead to decreased homodimerization.

Deletion of exons 4–10

This deletion of seven of eight coding exons (exons 4–10) of the TRβ gene was found in a consanguineous family (designated family III) with GRTH inherited recessively (Takeda et al., 1991, 1992a). Heterozygous members of the family were clinically normal, indicating that in particular this mutant does not act as a dominant negative and that the presence of a single wild-type allele is sufficient for the normal phenotype. The affected homozygous members of that family, one of whom is known as the Refetoff patient (Refetoff et al., 1967), had severe hyposensitivity to thyroid hormone and exhibited congenital deafness, epiphyseal dysgenesis and other minor somatic abnormalities.

PATHOPHYSIOLOGY

The syndrome of RTH is characterized by raised serum levels of free thyroid hormones and inappropriate levels of TSH. In patients, higher levels of thyroid hormones are unable to downregulate the secretion of pituitary TSH and to produce signs of hyperthyroidism (Refetoff et al., 1967; Refetoff, 1994a,b). Rather, symptoms correspond to peripheral hypothyroidism. In most cases the thyroid gland is enlarged. A high prevalence of hyperactivity, hearing defects, learning difficulties and language abnormalities, as well as intellectual impairment of varying degree is observed (Refetoff, 1982; Mixson et al., 1992; Hauser et al., 1993; McDermott and Ridgway, 1993; Forrest et al., 1996a,b). Also, short stature and an associated delay in bone maturation, as well as tachycardia, have been reported (Usala, 1991; Refetoff, 1994b; Brucker-Davis et al., 1995). However, the overall clinical manifestations of the symptoms of patients with RTH syndrome show a high degree of variation. Presumably in cases of partial resistance some patients can nearly fully compensate for some symptoms. In general, the response of target tissue to supplied thyroid hormone, which normally would be excessive, is reduced significantly in patients with RTH.

According to the early findings and description of clinical manifestations, RTH is distinguished into a pituitary (PRTH) and a generalized (GRTH) form. In contrast to a general tissue resistance (GRTH), PRTH is a pituitary selective resistance to thyroid hormone without exhibiting resistance of peripheral tissue. However, on one hand there is in general a great variation of the symptoms, and on the other, since the identical mutations exhibit different phenotypes (PRTH or GRTH) in different families, it is difficult or even impossible to state that a specific mutation is the cause of either of these two forms of RTH.

Through analysis of a battery of different TRβ mutants from patients with RTH, Mixson et al. (1992) reported that patients bearing mutations in exon 9 had severe difficulties in reading, reversing words and numbers, whereas patients with mutations in exon 10 usually had no language problems. As more than 400 individuals with RTH are now known (Refetoff, 1994a,b), it is important to repeat these studies with more patients in order to generalize this mechanistically interesting correlation.

Several functional defects of mutant receptors have been identified at the molecular level. They exhibit decreased or absent homone-binding affinity. Thus, the hormone-induced effects are reduced or completely abolished, including the T_3-induced dissociation of homodimers and, thus, the receptors remain as transcriptional silencers (Baniahmad *et al.*, 1992b; Collingwood *et al.*, 1994). Furthermore, the mutant receptors show a dominant negative effect on wild-type TR. Thereby, the hormone-activated functions of wild-type receptors are impaired. Analysis of mutant receptors with virtually no hormone-binding capacity and surprisingly weak dominant negative effects showed that homodimerization is impaired, but not RXR heterodimerization (Collingwood *et al.*, 1994, 1998; Yen and Chin, 1994; Nishiyama *et al.*, 1998). Although the role of other potential dimerization partners of TR, such as retinoic acid receptor and presumably some orphan receptors, has to be elucidated, the dominant negativity of mutant receptors associates with homodimerization. This implies that mutant receptors act in a dominant negative fashion through homodimerization with wild-type receptor, forming heterodimers that are unable to activate genes. Target genes that should be activated are even actively repressed, since reduced or complete lack of hormone binding results in a reduction or lack of both corepressor dissociation and coactivator recruitment. Therefore, the mutant receptors act as transcriptional silencers of target genes in the presence of ligand. Thus, the dominant negative action is a combination of both silencing function, even in the presence of hormone, and homodimerization with wild-type α and β receptor (Fig. 2.7).

The dominant negative effect of mutant receptors leads to the failure of negative feedback in the pituitary–thyroid axis. This results in non suppressed TSH secretion and stimulated thyroid hormone production, which is one of the key features of the disorder. However, the dysregulated genes involved in the associated RTH symptoms are still largely unknown.

DIAGNOSIS

Diagnostic procedures must be initiated only after persistence of raised serum thyroid hormone concentrations in association with nonsuppressed TSH levels has been confirmed on samples obtained several weeks apart (Refetoff, 1994a). However, it is hard for a clinician to distinguish between the RTH syndrome and TSH-induced hyperthyroidism caused by TSH-producing adenomas. Testing of peripheral tissues, such as mononuclear cells or fibroblasts, from patients with RTH for reduced hormone-binding capacity was unsuccessful (Refetoff, 1994a). There are other altered hormonal responses in peripheral tissues such as ferritin, cholesterol, triglycerides and creatinine phosphokinase, urine hydroxyproline and carnitine, basal metabolic rate, cardiac contractility by echography, and deep tendon relaxation time. However, these have not yet developed to diagnostic value because they are not specific and lack the sensitivity necessary to discriminate patients with RTH from subjects with mild thyrotoxicosis, or are altered by factors unrelated to thyroid gland function. Thus, for diagnostic purposes the search for mutations by polymerase chain reaction sequencing of the TRβ gene is important.

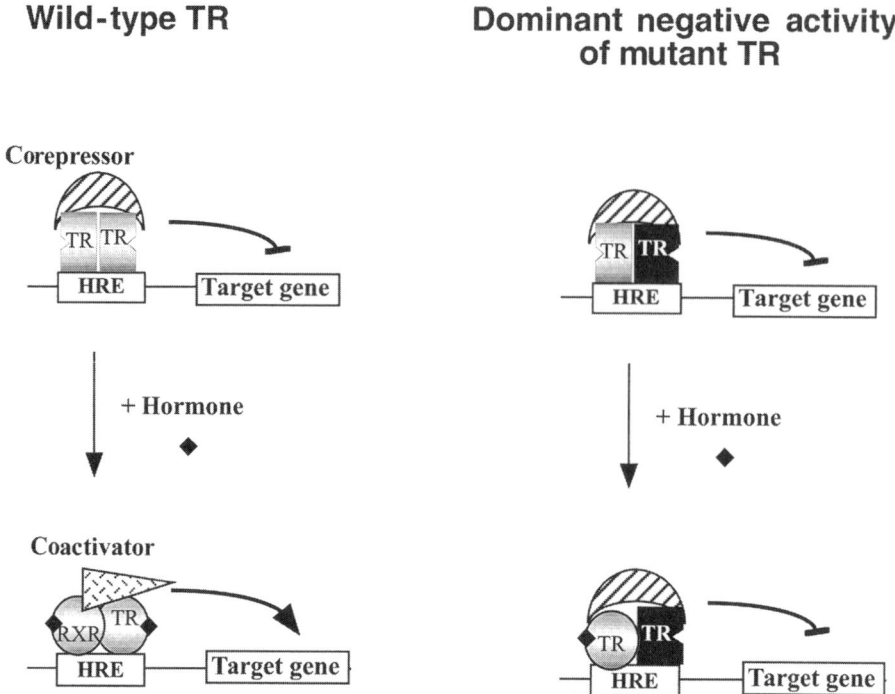

Fig. 2.7   Comparison of the hormonal effects of wild-type and mutant thyroid hormone receptors (TRs). Dominant negative-acting mutant receptors (black) have both reduced or completely absent hormone-binding affinity and coactivator recruitment. Hormone induces heterodimerization with the retinoid X receptor (RXR), while a large number of the mutant receptors prefer homodimerization in the presence of hormone. This leads to the formation of homodimers that repress genes even in the presence of ligand. HRE, hormone response element.

PHARMACOLOGY

No specific treatment is available to correct the defect of RTH fully and specifically (Refetoff, 1994a). Owing to tissue resistance, most patients are compensated by means of endogenously increased levels of thyroid hormone. There have been reports of the successful treatment of individuals with a T_3 analog, TRIAC (Darendeliler and Bas, 1997; Radetti *et al.*, 1997; Asteria *et al.*, 1999). One 2-year study with one patient bearing the TRβ mutation P453T showed that TRIAC reduced TSH and thyroid hormone levels, but maintained normal heart rate after discontinuation of atenolol (Radetti *et al.*, 1997).

However, therapy of patients with RTH by using thyroid hormone and analogs has several limitations. One limitation is the receptor mutant itself. Thyroid hormone treatment should be considered only when the receptor has remaining hormone-binding capability. Another limitation involves the occurrence of tachycardia as a result of higher doses of thyroid hormone.

In general, a major goal would be specifically to inactivate the dominant negative-acting TRβ mutants. This can be pursued via the search for drugs that are able to bind to the receptor and either reactivate the mutant receptor or at least inactivate its dominant negativity. Since dominant negativity is thought to be mediated by lack of corepressor dissociation, tests could be developed to find substances that release corepressors. Another goal would be to activate the repressed target genes of mutant TRβ. As some response elements can also be bound and regulated by the retinoic acid receptor, it would be a possibility to test retinoids or rexinoids for their ability to reverse at least some of the symptoms of resistance. Transgenic mouse model systems for RTH (Wong *et al.*, 1997; Abel *et al.*, 1999; Gloss *et al.*, 1999) would therefore be the first choice for testing candidate substances and a very suitable model for future treatment of patients with RTH.

Acknowledgments

Owing to space limitations, the authors apologize to all authors in this area for not citing their contributions. They are grateful to Dr Les Burke for critically reading this chapter. This work was supported by grants from the Sonderforschungsbereich SFB 397 of the Deutsche Forschungsgemeinschaft.

REFERENCES

Abel, E.D., Kaulbach, H.C., Campos-Barros, A. *et al.* (1999) Novel insight from transgenic mice into thyroid hormone resistance and the regulation of thyrotropin. *J. Clin. Invest.* **103**: 271–279.

Adams, M., Nagaya, T., Tone, Y., Jameson, J.L. and Chatterjee, V.K. (1992) Functional properties of a novel mutant thyroid hormone receptor in a family with generalized thyroid hormone resistance syndrome. *Clin. Endocrinol. (Oxf.)* **36**: 281–289.

Adams, M., Matthews, C., Collingwood, T.N., Tone, Y., Beck-Peccoz, P. and Chatterjee, K.K. (1994) Genetic analysis of 29 kindreds with generalized and pituitary resistance to thyroid hormone. Identification of thirteen novel mutations in the thyroid hormone receptor beta gene. *J. Clin. Invest.* **94**: 506–515.

Ando, S., Nakamura, H., Sasaki, S. *et al.* (1996) Introducing a point mutation identified in a patient with pituitary resistance to thyroid hormone (Arg338 to Trp) into other mutant thyroid hormone receptors weakens their dominant negative activities. *J. Endocrinol.* **151**: 293–300.

Aronow, W.S. (1995) The heart and thyroid disease. *Clin. Geriatr. Med.* **11**: 219–229.

Asteria, C., Rajanayagam, O., Collingwood, T.N. *et al.* (1999) Prenatal diagnosis of thyroid hormone resistance. *J. Clin. Endocrinol. Metab.* **84**: 405–410.

Baniahmad, A., Kohne, A.C. and Renkawitz, R. (1992a) A transferable silencing domain is present in the thyroid hormone receptor, in the v-*erbA* oncogene product and in the retinoic acid receptor. *EMBO J.* **11**: 1015–1023.

Baniahmad, A., Tsai, S.Y., O'Malley, B.W. and Tsai, M.J. (1992b) Kindred S thyroid hormone receptor is an active and constitutive silencer and a repressor for thyroid hormone and retinoic acid responses. *Proc. Natl. Acad. Sci. USA* **89**: 10633–10637.

Baniahmad, A., Ha, I., Reinberg, D., Tsai, S., Tsai, M.J. and O'Malley, B.W. (1993) Interaction of human thyroid hormone receptor β with transcription factor TFIIB may mediate target gene derepression and activation by thyroid hormone. *Proc. Natl. Acad. Sci. USA* **90**: 8832–8836.

Baniahmad, A., Burris, T.P. and Tsai, M.-J. (1994) The nuclear hormone receptor superfamily. In: M.-J. Tsai and B.W. O'Malley (eds) *Mechanism of Steroid Hormone Regulation of Gene Transcription*, pp. 1–24. CRC Press, Austin, Texas.

Baniahmad, A., Leng, X., Burris, T.P., Tsai, S.Y., Tsai, M.J. and O'Malley, B.W. (1995) The tau 4 activation domain of the thyroid hormone receptor is required for release of a putative corepressor(s) necessary for transcriptional silencing. *Mol. Cell. Biol.* **15**: 76–86.

Baniahmad, A., Eggert, M. and Renkawitz, R. (1997) Steroid, thyroid and retinoid receptors as transcription factors. In: A.G. Papavassiliou (ed.) *Transcription Factors in Eukaryotes*, pp. 95–123. CRC Press, Austin, Texas.

Barettino, D., Vivanco Ruiz, M.M. and Stunnenberg, H.G. (1994) Characterization of the ligand-dependent transactivation domain of thyroid hormone receptor. *EMBO J.* **13**: 3039–3049.

Beck-Peccoz, P., Chatterjee, V.K., Chin, W.W. *et al.* (1994) Nomenclature of thyroid hormone receptor β-gene mutations in resistance to thyroid hormone: consensus statement from the first workshop on thyroid hormone resistance, 10–11 July 1993, Cambridge, UK. *J. Clin. Endocrinol. Metab.* **78**: 990–993.

Behr, M. and Loos, U. (1992) A point mutation (Ala229 to Thr) in the hinge domain of the c-*erbA* β thyroid hormone receptor gene in a family with generalized thyroid hormone resistance. *Mol. Endocrinol.* **6**: 1119–1126.

Behr, M. and Loos, U. (1996) Periodically hyperthyroid phenotype in thyroid hormone resistance is associated with mutation D322N in the thyroid hormone receptor β gene: transcriptional properties of the mutant and the role of retinoid X receptor. *Exp. Clin. Endocrinol. Diabetes* **104**: 111–116.

Behr, M., Ramsden, D.B. and Loos, U. (1997) Deoxyribonucleic acid binding and transcriptional silencing by a truncated c-*erbA* β1 thyroid hormone receptor identified in a severely retarded patient with resistance to thyroid hormone. *J. Clin. Endocrinol. Metab.* **82**: 1081–1087.

Bernal, J. and Nunez, J. (1995) Thyroid hormones and brain development. *Eur. J. Endocrinol.* **133**: 390–398.

Boothroyd, C.V., Teh, B.T., Hayward, N.K., Hickman, P.E., Ward, G.J. and Cameron, D.P. (1991) Single base mutation in the hormone binding domain of the thyroid hormone receptor β gene in generalised thyroid hormone resistance demonstrated by single stranded conformation polymorphism analysis. *Biochem. Biophys. Res. Commun.* **178**: 606–612.

Broderick, T.J. and Wechsler, A.S. (1997) Triiodothyronine in cardiac surgery. *Thyroid* **7**: 133–137.

Brucker-Davis, F., Skarulis, M.C., Grace, M.B. *et al.* (1995) Genetic and clinical features of 42 kindreds with resistance to thyroid hormone. The National Institutes of Health Prospective Study. *Ann. Intern. Med.* **123**: 572–583.

Burman, K.D., Djuh, Y.Y., Nicholson, D. *et al.* (1992) Generalized thyroid hormone resistance: identification of an arginine to cystine mutation in codon 315 of the c-*erbA* β thyroid hormone receptor. *J. Endocrinol. Invest.* **15**: 573–579.

Caelles, C., Gonzalez-Sancho, J.M. and Munoz, A. (1997) Nuclear hormone receptor antagonism with AP-1 by inhibition of the JNK pathway. *Genes Dev.* **11**: 3351–3364.

Chassande, O., Fraichard, A., Gauthier, K. *et al.* (1997) Identification of transcripts initiated from an internal promoter in the c-*erbA* α locus that encode inhibitors of retinoic acid receptor-α and triiodothyronine receptor activities. *Mol. Endocrinol.* **11**: 1278–1290.

Chatterjee, V.K. (1997) Resistance to thyroid hormone. *Horm. Res.* **484**: 43–46.

Chatterjee, V.K., Nagaya, T., Madison, L.D., Datta, S., Rentoumis, A. and Jameson, J.L. (1991) Thyroid hormone resistance syndrome. Inhibition of normal receptor function by mutant thyroid hormone receptors. *J. Clin. Invest.* **87**: 1977–1984.

Chen, J.D. and Evans, R.M. (1995) A transcriptional co-repressor that interacts with nuclear hormone receptors. *Nature* **377**: 454–457.

Cheng, S.Y., Ransom, S.C., McPhie, P., Bhat, M.K., Mixson, A.J. and Weintraub, B.D. (1994) Analysis of the binding of 3,3′,5-triiodo-L-thyronine and its analogues to mutant human β1 thyroid hormone receptors: a model of the hormone binding site. *Biochemistry* **33**: 4319–4326.

Clifton-Bligh, R.J., de Zegher, F., Wagner, R.L. *et al.* (1998) A novel TRβ mutation (R383H) in resistance to thyroid hormone syndrome predominantly impairs corepressor release and negative transcriptional regulation. *Mol. Endocrinol.* **12**: 609–621.

Collingwood T.N., Adams M., Tone Y. and Chatterjee V.K. (1994) Spectrum of transcriptional, dimerization, and dominant negative properties of twenty different mutant thyroid hormone β-receptors in thyroid hormone resistance syndrome. *Mol. Endocrinol.* **8**: 1262–1277.

Collingwood, T.N., Rajanayagam, O., Adams, M. *et al.* (1997) A natural transactivation mutation in the thyroid hormone β receptor: impaired interaction with putative transcriptional mediators. *Proc. Natl. Acad. Sci. USA* **94**: 248–253.

Collingwood, T.N., Wagner, R., Matthews, C.H. *et al.* (1998) A role for helix 3 of the TRβ ligand-binding domain in coactivator recruitment identified by characterization of a third cluster of mutations in resistance to thyroid hormone. *EMBO J.* **17**: 4760–4770.

Cugini, C.D., Jr., Leidy, J.W., Jr., Chertow, B.S. *et al.* (1992) An arginine to histidine mutation in codon 315 of the c-erbA β thyroid hormone receptor in a kindred with generalized resistance to thyroid hormones results in a receptor with significant 3,5,3′-triiodothyronine binding activity. *J. Clin. Endocrinol. Metab.* **74**: 1164–1170.

Darendeliler, F. and Bas, F. (1997) Successful therapy with 3,5,3′-triiodothyroacetic acid (TRIAC) in pituitary resistance to thyroid hormone. *J. Pediatr. Endocrinol. Metab.* **10**: 535–538.

di Fulvio, M., Chiesa, A.E., Baranzini, S.E., Gruniero-Papendieck, L., Masini-Repiso, A.M. and Targovnik, H.M. (1997) A new point mutation (M313T) in the thyroid hormone receptor β gene in a patient with resistance to thyroid hormone. *Thyroid* **7**: 43–44.

Dressel, U., Thormeyer, D., Altincicek, B. *et al.* (1999) Alien, a highly conserved protein with characteristics of a corepressor for members of the nuclear hormone receptor superfamily. *Mol. Cell. Biol.* **19**: 3383–3394.

Erichsen, K.E., Berg, J.P., Torjesen, P.A., Haug, E. and Johannesen, O. (1998) Thyroid hormone resistance. Clinical, biochemical and genetic study of a family (in Norwegian). *Tidsskr. Nor. Laegeforen.* **118**: 525–529.

Flynn, T.R., Hollenberg, A.N., Cohen, O. *et al.* (1994) A novel C-terminal domain in the thyroid hormone receptor selectively mediates thyroid hormone inhibition. *J. Biol. Chem.* **269**: 32713–32716.

Fondell, J.D., Brunel, F., Hisatake, K. and Roeder R.G. (1996) Unliganded thyroid hormone receptor α can target TATA-binding protein for transcriptional repression. *Mol. Cell. Biol.* **16**: 281–287.

Forrest, D., Erway, L.C., Ng, L., Altschuler, R. and Curran, T. (1996a) Thyroid hormone receptor β is essential for development of auditory function. *Nat. Genet.* **13**: 354–357.

Forrest, D., Golarai, G., Connor, J. and Curran, T. (1996b) Genetic analysis of thyroid hormone receptors in development and disease. *Recent Prog. Horm. Res.* **51**: 1–22.

Forrest, D., Hanebuth, E., Smeyne, R.J. *et al.* (1996c) Recessive resistance to thyroid hormone in mice lacking thyroid hormone receptor β: evidence for tissue-specific modulation of receptor function. *EMBO J.* **15**: 3006–3015.

Fraichard, A., Chassande, O., Plateroti, M. *et al.* (1997) The T3R α gene encoding a thyroid hormone receptor is essential for post-natal development and thyroid hormone production. *EMBO J.* **16**: 4412–4420.

Geffner, M.E., Su, F., Ross, N.S. *et al.* (1993) An arginine to histidine mutation in codon 311 of the c-*erbA* β gene results in a mutant thyroid hormone receptor that does not mediate a dominant negative phenotype. *J. Clin. Invest.* **91**: 538–546.

Glass, C.K., Lipkin, S.M., Devary, O.V. and Rosenfeld, M.G. (1989) Positive and negative regulation of gene transcription by a retinoic acid–thyroid hormone receptor heterodimer. *Cell* **59**: 697–708.

Gloss, B., Sayen, M.R., Trost, S.U. *et al.* (1999) Altered cardiac phenotype in transgenic mice carrying the Δ337 threonine thyroid hormone receptor β mutant derived from the S family. *Endocrinology* **140**: 897–902.

Gomberg-Maitland, M. and Frishman, W.H. (1998) Thyroid hormone and cardiovascular disease. *Am. Heart. J.* **135**: 187–196.

Gothe, S., Wang, Z., Ng, L. *et al.* (1999) Mice devoid of all known thyroid hormone receptors are viable but exhibit disorders of the pituitary–thyroid axis, growth, and bone maturation. *Genes Dev.* **13**: 1329–1341.

Groenhout, E.G. and Dorin, R.I. (1994) Generalized thyroid hormone resistance due to a deletion of the carboxy terminus of the c-*erbA* β receptor. *Mol. Cell. Endocrinol.* **99**: 81–88.

Gurnell, M., Rajanayagam, O., Barbar, I., Jones, M.K. and Chatterjee, V.K. (1998) Reversible pituitary enlargement in the syndrome of resistance to thyroid hormone. *Thyroid* **8**: 679–682.

Hauser, P., Zametkin, A.J., Martinez, P. *et al.* (1993) Attention deficit-hyperactivity disorder in people with generalized resistance to thyroid hormone. *N. Engl. J. Med.* **328**: 997–1001.

Hayashi, Y., Weiss, R.E., Sarne, D.H. *et al.* (1995) Do clinical manifestations of resistance to thyroid hormone correlate with the functional alteration of the corresponding mutant thyroid hormone-β receptors? *J. Clin. Endocrinol. Metab.* **80**: 3246–3256.

Hayashi, Y., Mangoura, D. and Refetoff, S. (1996) A mouse model of resistance to thyroid hormone produced by somatic gene transfer of a mutant thyroid hormone receptor. *Mol. Endocrinol.* **10**: 100–106.

Hayashi, Y., Xie, J., Weiss, R.E., Pohlenz, J. and Refetoff, S. (1998) Selective pituitary resistance to thyroid hormone produced by expression of a mutant thyroid hormone receptor β gene in the pituitary gland of transgenic mice. *Biochem. Biophys. Res. Commun.* **245**: 204–210.

Hiramatsu, R., Abe, M., Morita, M., Noguchi, S. and Suzuki, T. (1994) Generalized resistance to thyroid hormone: identification of a novel c-*erbA* β thyroid hormone receptor variant (Leu450) in a Japanese family and analysis of its secondary structure by the Chou and Fasman method. *Jpn. J. Hum. Genet.* **39**: 365–377.

Hodin, R.A., Lazar, M.A., Wintman, B.I. *et al.* (1989) Identification of a thyroid hormone receptor that is pituitary-specific. *Science* **244**: 76–79.

Hörlein, A.J., Naar, A.M., Heinzel, T. *et al.* (1995) Ligand-independent repression by the thyroid hormone receptor mediated by a nuclear receptor co-repressor. *Nature* **377**: 397–404.

Ito, M., Yuan, C.X., Malik, S. *et al.* (1999) Identity between TRAP and SMCC complexes indicates novel pathways for the function of nuclear receptors and diverse mammalian activators. *Mol. Cell* **3**: 361–370.

Jezequel, P., Guilhem, I., Hespel, J.P. *et al.* (1996) Identification of a novel mutation (A268G) in exon 8 of the HTR β gene in a large family with thyroid hormone resistance. *Hum. Mutat.* **8**: 396.

Johansson, C., Vennstrom, B. and Thoren, P. (1998) Evidence that decreased heart rate in thyroid hormone receptor-α1-deficient mice is an intrinsic defect. *Am. J. Physiol.* **275**: 640–646.

Kamei, Y., Xu, L., Heinzel, T. *et al.* (1996) A CBP integrator complex mediates transcriptional activation and AP-1 inhibition by nuclear receptors. *Cell* **85**: 403–414.

Katz, D. and Lazar, M.A. (1993) Dominant negative activity of an endogenous thyroid hormone receptor variant (α2) is due to competition for binding sites on target genes. *J. Biol. Chem.* **268**: 20904–20910.

Kitajima, K., Nagaya, T. and Jameson, J.L. (1995) Dominant negative and DNA-binding properties of mutant thyroid hormone receptors that are defective in homodimerization but not heterodimerization. *Thyroid* **5**: 343–353.

Koenig, R.J. (1998) Thyroid hormone receptor coactivators and corepressors. *Thyroid* **8**: 703–713.

Kopp, P., Kitajima, K. and Jameson, J.L. (1996) Syndrome of resistance to thyroid hormone: insights into thyroid hormone action. *Proc. Soc. Exp. Biol. Med.* **211**: 49–61.

Lebel, J.-M. and Dussault, J.H. (1994) The thyroid and brain development. *Curr. Opin. Endocrinol. Diabetes* **1994**: 167–174.

Liu, Y., Takeshita, A., Misiti, S., Chin, W.W. and Yen, P.M. (1998) Lack of coactivator interaction can be a mechanism for dominant negative activity by mutant thyroid hormone receptors. *Endocrinology* **139**: 4197–4204.

Macchia, E., Gurnell, M., Agostini, M. *et al.* (1997) Identification and characterization of a novel *de novo* mutation (L346V) in the thyroid hormone receptor β gene in a family with generalized thyroid hormone resistance. *Eur. J. Endocrinol.* **137**: 370–376.

Macchia, E., Agostini, M., Sarkissian, G. *et al.* (1998) Detection of a new *de novo* mutation at codon 251 of exon 8 of thyroid hormone receptor β gene in an Italian kindred with resistance to thyroid hormone. *J. Endocrinol. Invest.* **21**: 226–233.

Mangelsdorf, D.J. and Evans, R.M. (1995) The RXR heterodimers and orphan receptors. *Cell* **83**: 841–850.

Mangelsdorf, D.J., Thummel, C., Beato, M. *et al.* (1995) The nuclear receptor superfamily: the second decade. *Cell* **83**: 835–839.

McDermott, M.T and Ridgway, E.C. (1993) Thyroid hormone resistance syndromes. *Am. J. Med.* **94**: 424–432.

Menzaghi, C., Balsamo, A., Di Paola, R. *et al.* (1999) Association between an R338L mutation in the thyroid hormone receptor-β gene and thyrotoxic features in two unrelated kindreds with resistance to thyroid hormone. *Thyroid* **9**: 1–6.

Mixson, A.J., Parrilla, R., Ransom, S.C. *et al.* (1992) Correlations of language abnormalities with localization of mutations in the β-thyroid hormone receptor in 13 kindreds with generalized resistance to thyroid hormone: identification of four new mutations. *J. Clin. Endocrinol. Metab.* **75**: 1039–1045.

Mixson, A.J., Renault, J.C., Ransom, S., Bodenner, D.L. and Weintraub, B.D. (1993) Identification of a novel mutation in the gene encoding the β-triiodothyronine receptor in a patient with apparent selective pituitary resistance to thyroid hormone. *Clin. Endocrinol. (Oxf.)* **38**: 227–234.

Miyoshi, Y., Nakamura, H., Sasaki, S. *et al.* (1995) Two consecutive nucleotide substitutions resulting in the T$_3$ receptor β gene resulting in an 11-amino acid truncation in a patient with generalized resistance to thyroid hormone. *Mol. Cell. Endocrinol.* **114**: 9–17.

Miyoshi, Y., Nakamura, H., Tagami, T. *et al.* (1998) Comparison of the functional properties of three different truncated thyroid hormone receptors identified in subjects with resistance to thyroid hormone. *Mol. Cell. Endocrinol.* **137**: 169–176.

Nagaya, T. and Jameson, J.L. (1993) Thyroid hormone receptor dimerization is required for dominant negative inhibition by mutations that cause thyroid hormone resistance. *J. Biol. Chem.* **268**: 15 766–15 771.

Nagaya, T., Fujieda, M. and Seo, H. (1998) Requirement of corepressor binding of thyroid hormone receptor mutants for dominant negative inhibition. *Biochem. Biophys. Res. Commun.* **247**: 620–623.

Nishiyama, K., Andoh, S., Kitahara, A. *et al.* (1998) Difference in dominant negative activities between mutant thyroid hormone receptors $\alpha 1$ and $\beta 1$ with an identical truncation in the extreme carboxyl-terminal τ4 domain. *Mol. Cell. Endocrinol.* **138**: 95–104.

Nomura, Y., Nagaya, T., Tsukaguchi, H., Takamatsu, J. and Seo, H. (1996) Amino acid substitutions of thyroid hormone receptor-β at codon 435 with resistance to thyroid hormone selectively alter homodimer formation. *Endocrinology* **137**: 4082–4086.

Nuclear Receptor Nomenclature Committee (1999) A unified nomenclature system for the nuclear receptor superfamily. *Cell* **97**: 161–163.

Onigata, K., Yagi, H., Sakurai, A. *et al.* (1995) A novel point mutation (R243Q) in exon 7 of the c-*erbA* β thyroid hormone receptor gene in a family with resistance to thyroid hormone. *Thyroid* **5**: 355–358.

Ono, S., Schwartz, I.D., Mueller, O.T., Root, A.W., Usala, S.J. and Bercu, B.B. (1991) Homozygosity for a dominant negative thyroid hormone receptor gene responsible for generalized resistance to thyroid hormone. *J. Clin. Endocrinol. Metab.* **73**: 990–994.

Oppenheimer, J.H. and Schwartz, H.L. (1997) Molecular basis of thyroid hormone-dependent brain development. *Endocr. Rev.* **8**: 462–475.

Ozata, M., Suzuki, S., Takeda, T. *et al.* (1995) Functional analysis of a proline to serine mutation in codon 453 of the thyroid hormone receptor $\beta 1$ gene. *J. Clin. Endocrinol. Metab.* **80**: 3239–3245.

Parrilla, R., Mixson, A.J., McPherson, J.A., McClaskey, J.H. and Weintraub, B.D. (1991) Characterization of seven novel mutations of the c-*erbA* β gene in unrelated kindreds with generalized thyroid hormone resistance. Evidence for two 'hot spot' regions of the ligand binding domain. *J. Clin. Invest.* **88**: 2123–2130.

Pohlenz, J. and Knobl, D. (1996) Treatment of pituitary resistance to thyroid hormone (PRTH) in an 8-year-old boy. *Acta Paediatr.* **85**: 387–890.

Pohlenz, J., Wirth, S., Winterpacht, A., Wemme, H., Zabel, B. and Schonberger, W. (1995) Phenotypic variability in patients with generalised resistance to thyroid hormone. *J. Med. Genet.* **32**: 393–395.

Pohlenz, J., Schonberger, W., Wemme, H., Winterpacht, A., Wirth, S. and Zabel, B. (1996) New point mutation (R243W) in the hormone binding domain of the c-*erbA* $\beta 1$ gene in a family with generalized resistance to thyroid hormone. *Hum. Mutat.* **7**: 79–81.

Pohlenz, J., Wildhardt, G., Zabel, B. and Willgerodt, H. (1997) Resistance to thyroid hormone in a family caused by a new point mutation L330S in the thyroid receptor (TR) β gene. *Thyroid* **7**: 39–41.

Rachez, C., Lemon, B.D., Suldan, Z. *et al.* (1999) Ligand-dependent transcription activation by nuclear receptors requires the DRIP complex. *Nature* **398**: 824–828.

Radetti, G., Persani, L., Molinaro, G. *et al.* (1997) Clinical and hormonal outcome after two years of triiodothyroacetic acid treatment in a child with thyroid hormone resistance. *Thyroid* **7**: 775–778.

Rascle, H., Gandrillon, O., Cabello, G. and Samarut, J. (1997) The V-*erbA* oncogene. Oncogenes as transcriptional regulators 1. M. Taniv and J. Ghysael, *Retroviral oncogenes*, 117–163.

Refetoff, S. (1982) Syndromes of thyroid hormone resistance. *Am. J. Physiol.* **243**: 88–98.

Refetoff, S. (1994a) Resistance to thyroid hormone and its molecular basis. *Acta Paediatr. Jpn.* **36**: 1–15.

Refetoff, S. (1994b) Resistance to thyroid hormone: an historical overview. *Thyroid* **4**: 345–349.

Refetoff, S. (1997) Resistance to thyroid hormone. *Curr. Ther. Endocrinol. Metab.* **6**: 132–134.

Refetoff, S., DeWind, L.T. and DeGroot, L.J. (1967) Familial syndrome combining deaf-mutism, stuppled epiphyses, goiter and abnormally high PBI: possible target organ refractoriness to thyroid hormone. *J. Clin. Endocrinol. Metab.* **27**: 279–294.

Refetoff, S., Weiss, R.E., Wing, J.R., Sarne, D., Chyna, B. and Hayashi, Y. (1994) Resistance to thyroid hormone in subjects from two unrelated families is associated with a point mutation in the thyroid hormone receptor β gene resulting in the replacement of the normal proline 453 with serine. *Thyroid* **1**: 249–254.

Refetoff, S., Tunca, H., Wilansky, D.L., Mussey, V.C. and Weiss, R.E. (1996) Mutation in the thyroid hormone receptor (TR) β gene (M313T) not previously reported in two unrelated families with resistance to thyroid hormone (RTH). *Thyroid* **6**: 571–573.

Safer, J.D., Langlois, M.F., Cohen, R. *et al.* (1997) Isoform variable action among thyroid hormone receptor mutants provides insight into pituitary resistance to thyroid hormone. *Mol. Endocrinol.* **11**: 16–26.

Safer, J.D., Cohen, R.N., Hollenberg, A.N. and Wondisford, F.E. (1998) Defective release of corepressor by hinge mutants of the thyroid hormone receptor found in patients with resistance to thyroid hormone. *J. Biol. Chem.* **273**: 30 175–30 182.

Sakurai, A., Takeda, K., Ain, K. *et al.* (1989) Generalized resistance to thyroid hormone associated with a mutation in the ligand-binding domain of the human thyroid hormone receptor β. *Proc. Natl. Acad. Sci. USA* **86**: 8977–8981.

Sakurai, A., Nakai, A. and DeGroot, L.J. (1990) Structural analysis of human thyroid hormone receptor β gene. *Mol. Cell. Endocrinol.* **71**: 83–91.

Sakurai, A., Miyamoto, T., Hughes, I.A. and DeGroot, L.J. (1993) Characterization of a novel mutant human thyroid hormone receptor β in a family with hereditary thyroid hormone resistance. *Clin. Endocrinol. (Oxf.)* **38**: 29–38.

Sap, J., Munoz, A., Damm, K. *et al.* (1986) The c-*erb*-*A* protein is a high-affinity receptor for thyroid hormone. *Nature* **324**: 635–640.

Sarkissian, G., Dace, A., Mesmacque, A. *et al.* (1999) A novel resistance to thyroid hormone associated with a new mutation (T329N) in the thyroid hormone receptor β gene. *Thyroid* **9**: 165–171.

Sasaki, S., Nakamura, H., Tagami, T., Miyoshi, Y., Tanaka, K. and Imura, H. (1992) A point mutation of the T_3 receptor β1 gene in a kindred of generalized resistance to thyroid hormone. *Mol. Cell. Endocrinol.* **84**: 159–166.

Sasaki, S., Nakamura, H., Tagami, T. *et al.* (1993) Pituitary resistance to thyroid hormone associated with a base mutation in the hormone-binding domain of the human 3,5,3'-triiodothyronine receptor-β. *J. Clin. Endocrinol. Metab.* **76**: 1254–1258.

Sasaki, S., Nakamura, H., Tagami, T., Miyoshi, Y. and Nakao, K. (1995) Functional properties of a mutant T_3 receptor β (R338W) identified in a subject with pituitary resistance to thyroid hormone. *Mol. Cell. Endocrinol.* **113**: 109–117.

Seto, D. and Weintraub, B.D. (1996) Rapid molecular diagnosis of mutations associated with generalized thyroid hormone resistance by PCR-coupled automated direct sequencing of genomic DNA: detection of two novel mutations. *Hum. Mutat.* **8**: 247–257.

Shuto, Y., Wakabayashi, I., Amuro, N., Minami, S. and Okazaki, T. (1992a) A point mutation in the 3,5,3′-triiodothyronine-binding domain of thyroid hormone receptor-β associated with a family with generalized resistance to thyroid hormone. *J. Clin. Endocrinol. Metab.* **75**: 213–217.

Shuto, Y., Okazaki, T. and Wakabayashi, I. (1992b) Transcriptional activity of a mutant thyroid hormone receptor β in a family with generalized resistance to thyroid hormone. *Mol. Cell. Endocrinol.* **90**: 111–115.

Sunthornthepvarakul, T., Angsusingha, K., Likitmaskul, S. Ngowngarmratana, S. and Refetoff, S. (1997) Mutation in the thyroid hormone receptor β gene (A317T) in a Thai subject with resistance to thyroid hormone. *Thyroid* **7**: 905–907.

Tagami, T., Gu, W.X., Peairs, P.T., West, B.L. and Jameson, J.L. (1998) A novel natural mutation in the thyroid hormone receptor defines a dual functional domain that exchanges nuclear receptor corepressors and coactivators. *Mol. Endocrinol.* **12**: 1888–1902.

Takeda, K., Balzano, S., Sakurai, A., DeGroot, L.J. and Refetoff, S. (1991) Screening of nineteen unrelated families with generalized resistance to thyroid hormone for known point mutations in the thyroid hormone receptor β gene and the detection of a new mutation. *J. Clin. Invest.* **87**: 496–502.

Takeda, K., Weiss, R.E. and Refetoff, S. (1992a) Rapid localization of mutations in the thyroid hormone receptor-β gene by denaturing gradient gel electrophoresis in 18 families with thyroid hormone resistance. *J. Clin. Endocrinol. Metab.* **74**: 712–719.

Takeda, K., Sakurai, A., DeGroot, L.J. and Refetoff, S. (1992b) Recessive inheritance of thyroid hormone resistance caused by complete deletion of the protein-coding region of the thyroid hormone receptor-β gene. *J. Clin. Endocrinol. Metab.* **74**: 49–55.

Takeda, T., Nagasawa, T., Miyamoto, T., Hashizume, K. and DeGroot, L.J. (1997) The function of retinoid X receptors on negative thyroid hormone response elements. *Mol. Cell. Endocrinol.* **128**: 85–96.

Taniyama, M., Kusano, S., Miyoshi, Y. *et al.* (1996) Mild resistance to thyroid hormone with a truncated thyroid hormone receptor β. *Exp. Clin. Endocrinol. Diabetes* **104**: 339–343.

Tenbaum, S. and Baniahmad, A. (1997) Nuclear receptors: structure, function and involvement in disease. *Int. J. Biochem. Cell. Biol.* **29**: 1325–1341.

Tsukaguchi, H., Yoshimasa, Y., Fujimoto, K. *et al.* (1995) Three novel mutations of thyroid hormone receptor β gene in unrelated patients with resistance to thyroid hormone: two mutations of the same codon (H435L and H435Q) produce separate subtypes of resistance. *J. Clin. Endocrinol. Metab.* **80**: 3613–3616.

Usala, S.J. (1991) Molecular diagnosis and characterization of thyroid hormone resistance syndromes. *Thyroid* **1**: 361–367.

Usala, S.J., Bale, A.E., Gesundheit, N. *et al.* (1988) Tight linkage between the syndrome of generalized thyroid hormone resistance and the human c-*erbA* β gene. *Mol. Endocrinol.* **2**: 1217–1220.

Usala, S.J., Wondisford, F.E., Watson, T.L., Menke, J.B. and Weintraub, B.D. (1990a) Thyroid hormone and DNA binding properties of a mutant c-*erbA* β receptor associated with generalized thyroid hormone resistance. *Biochem. Biophys. Res. Commun.* **171**: 575–580.

Usala, S.J., Tennyson, G.E., Bale, A.E. *et al.* (1990b) A base mutation of the c-*erbA* β thyroid hormone receptor in a kindred with generalized thyroid hormone resistance. Molecular heterogeneity in two other kindreds. *J. Clin. Invest.* **85**: 93–100.

Usala, S.J., Bercu, B.B. and Refetoff, S. (1991a) Diverse abnormalities of the c-*erbA* β thyroid hormone receptor gene in generalized thyroid hormone resistance. *Adv. Exp. Med. Biol.* **299**: 251–258.

Usala, S.J., Menke, J.B., Watson, T.L. *et al.* (1991b) A homozygous deletion in the c-*erbA β* thyroid hormone receptor gene in a patient with generalized thyroid hormone resistance: isolation and characterization of the mutant receptor. *Mol. Endocrinol.* **5**: 327–335.

Usala, S.J., Menke, J.B., Watson, T.L. *et al.* (1991c) A new point mutation in the 3,5,3′-triiodothyronine-binding domain of the c-*erbA β* thyroid hormone receptor is tightly linked to generalized thyroid hormone resistance. *J. Clin. Endocrinol. Metab.* **72**: 32–38.

Usala, S.J. Menke, J.B., Hao, E.H. *et al.* (1992) Mutations in the c-*erbA β* gene in two different patients with selective pituitary resistance to thyroid hormones. 74th Annual Meeting of the Endocrine Society, San Antonio, Texas 135. Abstract 335.

Wagner, R.L., Apriletti, J.W., McGrath, M.E., West, B.L., Baxter, J.D. and Fletterick, R.J. (1995) A structural role for hormone in the thyroid hormone receptor. *Nature* **378**: 690–697.

Ways, D.K., Qin, W., Cook, P. *et al.* (1993) Dominant and nondominant negative c-*erbA β*1 receptors associated with thyroid hormone resistance syndromes augment 12-*O*-tetradecanoyl-phorbol-13-acetate induction of the collagenase promoter and exhibit defective 3,5,3′-triiodothyronine-mediated repression. *Mol. Endocrinol.* **7**: 1112–1120.

Weinberger, C., Thompson, C.C., Ong, E.S., Lebo, R., Gruol, D.J. and Evans, R.M. (1986) The c-*erb-A* gene encodes a thyroid hormone receptor. *Nature* **324**: 641–646.

Weiss, R.E., Weinberg, M. and Refetoff, S. (1993) Identical mutations in unrelated families with generalized resistance to thyroid hormone occur in cytosine-guanine-rich areas of the thyroid hormone receptor *β* gene. Analysis of 15 families. *J. Clin. Invest.* **91**: 2408–2415.

Weiss, R.E., Stein, M.A., Duck, S.C. *et al.* (1994a) Low intelligence but not attention deficit hyperactivity disorder is associated with resistance to thyroid hormone caused by mutation R316H in the thyroid hormone receptor *β* gene. *J. Clin. Endocrinol. Metab.* **78**: 1525–1528.

Weiss, R.E., Chyna, B., Duell, P.B., Hayashi, Y., Sunthornthepvarakul, T. and Refetoff, S. (1994b) A new point mutation (C446R) in the thyroid hormone receptor-*β* gene of a family with resistance to thyroid hormone. *J. Clin. Endocrinol. Metab.* **78**: 1253–1256.

Weiss, R.E., Tunca, H., Gerstein, H.C. and Refetoff, S. (1996) A new mutation in the thyroid hormone receptor (TR) *β* gene (V458A) in a family with resistance to thyroid hormone (RTH). *Thyroid* **6**: 311–312.

Weiss, R.E., Tunca, H., Knapple, W.L., Faas, F.H. and Refetoff, S. (1997) Phenotype differences of resistance to thyroid hormone in two unrelated families with an identical mutation in the thyroid hormone receptor *β* gene (R320C). *Thyroid* **7**: 35–38.

Wikstrom, L., Johansson, C., Salto, C. *et al.* (1998) Abnormal heart rate and body temperature in mice lacking thyroid hormone receptor α1. *EMBO J.* **17**: 455–461.

Wong, R., Vasilyev, V.V., Ting, Y.T. *et al.* (1997) Transgenic mice bearing a human mutant thyroid hormone *β*1 receptor manifest thyroid function anomalies, weight reduction, and hyperactivity. *Mol. Med.* **3**: 303–314.

Yagi, H., Pohlenz, J., Hayashi, Y., Sakurai, A. and Refetoff, S. (1997) Resistance to thyroid hormone caused by two mutant thyroid hormone receptors *β*, R243Q and R243W, with marked impairment of function that cannot be explained by altered *in vitro* 3,5,3′-triiodothyroinine binding affinity. *J. Clin. Endocrinol. Metab.* **82**: 1608–1614.

Yen, P.M. and Chin, W.W. (1994) Molecular mechanisms of dominant negative activity by nuclear hormone receptors. *Mol. Endocrinol.* **8**: 1450–1454.

Yoh, S.M., Chatterjee, V.K. and Privalsky, M.L. (1997) Thyroid hormone resistance syndrome manifests as an aberrant interaction between mutant T$_3$ receptors and transcriptional corepressors. *Mol. Endocrinol.* **11**: 470–480.

Zavacki, A.M., Harney, J.W., Brent, G.A. and Larsen, P.R. (1996) Structural features of thyroid hormone response elements that increase susceptibility to inhibition by an RTH mutant thyroid hormone receptor. *Endocrinology* **137**: 2833–2841.

Chapter 3

Estrogen and Progesterone Receptors

George F. Allan

INTRODUCTION AND HISTORICAL PERSPECTIVE

This chapter provides an overview of the biology, pharmacology and genetics of the estrogen and progesterone receptors. The physiologic regulation of these two steroid receptors overlaps to a degree that justifies their inclusion together in any discussion. At the same time, as will be shown, their target tissues and temporal modes of action are distinct enough to allow pharmaceutical intervention for different purposes. A section of the chapter will describe the genetic mutations and variants that have been defined for both of the receptors. The majority of these have been described for the α form of the estrogen receptor; indeed, it is fair to say that, with regard to the occurrence of mutations in nuclear receptor genes, the estrogen receptor has been by far the most intensively investigated.

The estrogen and progesterone receptors were two of the earliest nuclear receptors to be biochemically isolated. Together, they play central roles in regulating the reproductive systems of female mammals, as well as the menstrual cycles of higher primates.

The field of endocrinology can be said to have begun at the very beginning of the twentieth century with the demonstration by Knauer (1900) that grafting ovarian tissue on to ovariectomized guinea-pigs prevented the symptoms of castration. This was followed by a similar study in dogs, where estrus in an ovariectomized animal was induced using ovarian extracts from another animal (Marshall and Jolly, 1905). Together, these two studies introduced the concept of a diffusible substance required for normal maintenance of reproductive tissues. Moreover, the work of the latter investigators indicated that the ovary was producing two hormones, one producing estrus and the other originating from the corpora lutea. Subsequently, the two substances were isolated as the steroids estrone (Allen and Doisy, 1923, 1924; Butenandt, 1929; Doisy *et al.*, 1929) and progesterone (Corner and Allen, 1929; Allen and Wintersteiner, 1934; Butenandt and Dannenbaum, 1934; Hartmann and Wettstein, 1934; Slotta *et al.*, 1934). Progesterone was shown to cause endometrial proliferation in the rabbit uterus (Corner and Allen, 1929).

It would be another three and a half decades before these substances would be established as receptor ligands. In the meantime, the commonly held view was that,

Nuclear Receptors and Genetic Disease
ISBN 0-12-146160-2

because estrogens had stimulatory effects on metabolic activity, they must act as substrates for enzymes to produce biosynthetic precursors (Talalay, 1962; Villee, 1962). This viewpoint began to change with the demonstration that labeled estrogen accumulated in and could be recovered unaltered from stimulated uterus (Glasscock and Hoekstra, 1959; Jensen and Jacobsen, 1962). Taken together with work demonstrating that estrogens and progesterone enhanced protein and RNA synthesis in target tissues (Mueller, 1960; Hamilton, 1962; Noteboom and Gorski, 1963; Wilson, 1963; Hamilton, 1968; O'Malley *et al.*, 1969), the concept began to evolve that these steroids were ligands for a cellular receptor protein that interacted with DNA in the nucleus (King *et al.*, 1965). Jensen (1965) showed that an estrogen antagonist, nafoxidine, inhibited the uterotropic activity of estradiol in parallel with uterine uptake of the steroid, providing indirect evidence for a binding moiety in this tissue.

Finally, the estrogen receptor was defined biochemically in the Gorski and Davidson laboratories (Talwar *et al.*, 1964; Toft and Gorski, 1966). Subsequent work showed that another receptor for progesterone was present in rat, rabbit, chicken and human tissues (Laumas and Farooq, 1966; Sherman *et al.*, 1970; Wiest and Rao, 1971; Schrader and O'Malley, 1972). Both of the receptors were molecularly cloned in the mid 1980s (Walter *et al.*, 1985; Conneely *et al.*, 1986). Since then, the receptors have been found in all vertebrates, including mammals, reptiles, amphibians, birds and fish.

Recently, a second estrogen receptor (estrogen receptor β) was cloned (Kuiper *et al.*, 1996; Mosselman *et al.*, 1996). We now know that the progesterone and estrogen receptors are members of the nuclear receptor superfamily comprising 50 or more mammalian members. As two of the five known steroid receptors, they are minor members of this family in abundance, although not in importance. Today, the original 'two-step' model of steroid action (Gorski *et al.*, 1968; Jensen *et al.*, 1968), invoking a cytoplasmic binding event followed by a nuclear translocation event, forms the basis of a detailed, though by no means complete, appreciation of the complexity of hormone action. Some of this complexity is addressed in the next two sections.

PHYSIOLOGY

The effects of estrogen and progesterone on female physiology are intimately related. Both are required for proper development and functioning of the reproductive system and of secondary sexual characteristics, and both are required to prepare for and maintain pregnancy. In addition, estrogen plays a crucial role in maintaining the health of nonreproductive tissues including bone and those of the cardiovascular system.

Ovulation and pregnancy

In higher primates, the development and release of an egg from the ovary is exquisitely tied to the levels of key hormones in the blood. In humans, the rise and fall of these hormones follows a typical 28-day cycle known as the menstrual cycle (Figs 3.1 and 3.2) (Mishell *et al.*, 1971). This cycle can be divided into two phases. The first, estrogen-dominated or follicular phase begins, by convention, on the first day of menstrual bleeding. During these 14 or so days, the ovarian follicle destined to release the mature egg develops and its granulocytes produce estradiol, the secreted levels of the hormone gradually rising towards the latter part of the phase. Progesterone, at this point, is essentially undetectable in the circulation. Between days 10 and 13 there is a rapid rise in

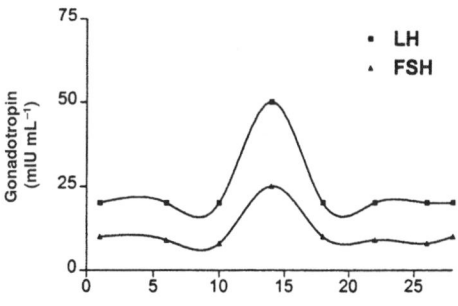

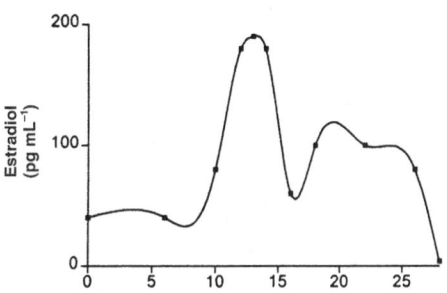

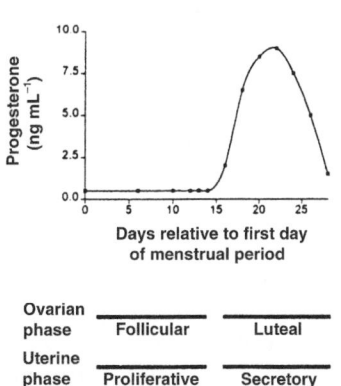

Fig. 3.1 Hormone fluctuations during the menstrual cycle. The phases of the cycle at the ovary and uterus are shown at the bottom. Redrawn from Thorneycroft *et al.* (1971), with permission.

estrogen levels, followed by a similarly rapid decrease over the succeeding 2 days. This peak in estradiol production is thought to stimulate a sudden surge in gonadotropin release from the pituitary gonadotrope cells – the so-called luteinizing hormone (LH) surge – 1 day later, on or about day 14. The LH surge marks the transition from the follicular phase to the luteal phase, and is followed by ovulation some 20–44 hours later (Kirton *et al.*, 1970; Monroe *et al.*, 1970; Pauerstein *et al.*, 1978).

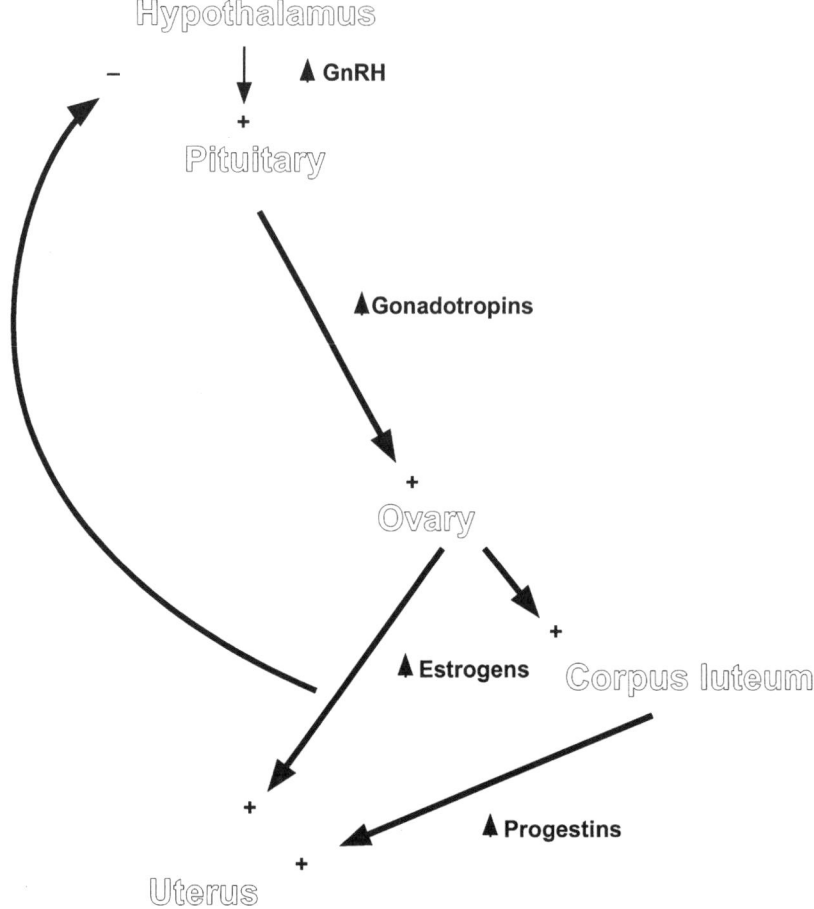

Fig. 3.2 Simplified illustration of the positive and negative regulatory pathways involving estrogen and progesterone in the female primate reproductive system. GnRH, gonadotropin-releasing hormone.

Meanwhile, under the influence of rising estrogen levels, the uterine lining begins to recover from the previous menstrual period during the second week of the phase. The lining of the uterus, known as the endometrium, consists chiefly of relatively undifferentiated stromal and epithelial cells at this point. Estrogen, a mitogen, stimulates these cells to divide and proliferate. It also stimulates the cellular synthesis of progesterone receptors via its own receptor by binding to and activating transcription of the progesterone receptor gene.

Following rupture and release of the egg, the follicle develops into a corpus luteum, whose chief function appears to be to produce large amounts of progesterone and estradiol, as well as peptide hormones such as inhibin. Progesterone stimulates differentiation of the estrogen-primed uterine endometrium into a complex glandular structure capable of accepting a fertilized egg. In this context, progesterone acts as an estrogen antagonist. Towards the second half of a nonpregnant cycle, progesterone and estrogen levels begin to wane as the luteum deteriorates. Lacking the continued support provided by progesterone, the endometrium begins to break down and menstrual flow begins. In the rhesus monkey this is thought to occur about 3 days after progesterone levels fall below 1 ng mL^{-1} (Goodman and Hodgen, 1983).

In the event of pregnancy, another hormone, chorionic gonadotropin, which is produced by the implanting blastocyst, acts to maintain the integrity of the corpus luteum. Thus, progesterone levels are maintained and the normal cycle is suppressed. Progesterone, the 'pregnancy hormone', remains at high levels throughout gestation, although in humans its main site of synthesis is the placenta after the second month of pregnancy. It appears that, at least in animals such as the rabbit, one of the main functions of the hormone at this time is to maintain the quiescence of the uterus (Csapo, 1973). Towards term, estrogen levels begin to rise, coinciding with and probably promoting enhanced uterine contractility by activating genes for contraction-associated proteins such as oxytocin (Riemer et al., 1986).

Humans and monkeys are unique in their capacity periodically to shed their endometria. Although lower mammals do not menstruate, the dominance of estrogen and progesterone in their reproductive systems is no less obvious. For example, in the 4-day cycle of the laboratory rat, circulating estradiol levels rise during metestrus and diestrus to peak during proestrus, a subsequent fall-off in levels coinciding with the preovulatory LH surge. In these animals, progesterone is produced by both the corpus luteum (with peak production during the night between metestrus and diestrus) and the preovulatory follicle (peaking at proestrus). In rabbits, mature females are in constant estrus, with low levels of circulating estrogen and progesterone. Immediately after coitus the ovary begins to secrete both hormones in sufficient quantity to support follicular maturation and differentiation of the endometrium. Ovulation follows 9–12 hours later. Likewise, in seasonal breeders such as the sheep and dog, follicular and uterine development and pregnancy maintenance require one or both of the hormones.

Mice lacking a functional estrogen receptor α gene are sterile due to defective ovarian folliculogenesis (Lubahn et al., 1993). It appears that folliculogenesis is impaired because of disruption of the hypothalamic–pituitary–gonadal regulatory axis, suggesting a role for the α receptor in the function of the gonadotrope. A similar disruption of the estrogen receptor β gene has been generated (Krege et al., 1998). Although capable of producing litters, female estrogen receptor β mutants are subfertile owing to defective ovarian function and reduced oocyte production. Finally, and somewhat surprisingly, disruption of the mouse progesterone receptor gene produced female animals that were infertile due to a failure of preovulatory follicles to rupture (Lydon et al., 1995). The uteri of the progesterone receptor mutants were grossly and histologically abnormal following treatment (but not before treatment) with exogenous estradiol and progesterone. It thus seems likely that there is interplay between the hormones at the level of the ovary as well as at the level of the uterus.

Mammary gland development and lactogenesis

In the human female, estrogen is required for the development of one of the major secondary sexual characteristics associated with puberty, breast development. As the ovary matures and approaches menarche, serum estrogen levels increase and stimulate lobuloalveolar development and fat deposition. The mammary glands of estrogen receptor α 'knockout' animals fail to mature (Lubahn et al., 1993). Prolonged exposure to the proliferative effect of estrogen on breast tissue is associated with an increased risk of breast cancer as a woman ages, especially if she experienced early menarche or late menopause, or had no children.

It has long been known that progesterone acts with estrogen to bring about proliferation of the acini of the mammary gland during pregnancy and to a lesser extent during the luteal phase of the cycle. Thus, progesterone may not be an estrogen 'antagonist' in breast, as it is in uterus. Recent gene disruption data obtained with progesterone receptor mutant mice have revealed a hitherto unappreciated role for progesterone in mammary gland development (Lydon *et al.*, 1995). It appears that progesterone acts as a mitogen in this tissue, as ᵗthe mutant animals exhibited rudimentary ductal structures with the complete absence of lobuloalveoles.

Lactogenesis, the cell developmental process by which the mammary gland is prepared for postnatal lactation, requires a complex interplay of hormones during pregnancy. Estrogen is one of those players, possibly acting to increase prolactin and glucocorticoid secretion, both of which subsequently directly stimulate milk production (Tucker, 1974). Progesterone, on the other hand, inhibits lactogenesis (Liu and Davis, 1967). Decreasing levels of the hormone in late pregnancy, combined with stimulatory levels of estrogen, prolactin and glucocorticoids, act to facilitate the preparation for milk production.

Postmenopausal estrogen deficiency

At the menopause, the human ovary ceases ovulation and serum estrogen levels drop precipitously. Over the ensuing 2–10 years, estradiol levels continue to decline, dropping, on average, to a final concentration of $15\,\mathrm{pg\,mL}^{-1}$, compared with a premenopausal early follicular phase high of $50\,\mathrm{pg\,mL}^{-1}$ (Yen, 1977). Women who have had their ovaries surgically removed have estradiol levels of $10\,\mathrm{pg\,mL}^{-1}$ or less. Estrogen does not disappear completely, possibly because of peripheral aromatization of androgens.

However, the residual hormone is not sufficient to prevent the appearance of some of the distressing symptoms of the postmenopausal state in many women (Table 3.1). The symptoms, more than likely in all cases, are a direct consequence of the loss of estrogen, as they can be relieved successfully by the administration of exogenous estrogens as part of a hormone replacement therapy regimen. Studies of the postmenopausal state have established that estrogen plays a role in the homeostasis of every important organ in the body. This is perhaps not surprising, given the widespread distribution of estrogen receptors α and β throughout most mammalian tissues (Kuiper *et al.*, 1997).

Osteoporosis

Bone mineral density decreases as a function of estrogen loss. Indications that this was the case came from a large number of studies showing that bone loss accelerated following the menopause or ovariectomy (Cosman *et al.*, 1997). More detailed

Table 3.1 Established and likely consequences of estrogen loss after the menopause

Osteoporosis
Hot flushes
Vaginal dryness
Skin dryness
Increased susceptibility to cardiovascular disease
Urinary incontinence
Cognitive deficiencies

longitudinal studies confirmed these preliminary findings (Richelson *et al.*, 1984; Nilas and Christiansen, 1988). In a more recent study examining midshaft radius bone mineral density over a year or more, both before and after the menopause, bone loss showed a strong correlation with estradiol levels (Rannevik *et al.*, 1995). Women with the lowest postmenopausal estrogen levels ($15 \, pg \, mL^{-1}$) lost between 10 and 40% of bone density in this region.

Supplementation with estrogen has been shown clearly to reduce bone loss and the risk of osteoporosis. This is true whether it is administered orally in the form of conjugated hormones prepared from the urine of pregnant mares (Lindsay *et al.*, 1984), as micronized estradiol (Christiansen and Lindsay, 1991), as estrone (Harris *et al.*, 1991), as a synthetic derivative such as mestranol (Lindsay *et al.*, 1976), or whether it is administered transdermally (Lufkin *et al.*, 1992) or percutaneously (Riis *et al.*, 1987).

The postmenopausal decrease in bone density appears to be due to an imbalance in the activities of bone-forming osteoblast and bone-resorbing osteoclast cells. However, the way in which estrogen loss leads to this situation is by no means clear (Turner *et al.*, 1994). In the animal, estrogen is a potent inhibitor of the differentiation of both types of cell, and may also inhibit their activities. *In vitro*, the published effects of estrogens on proliferation and gene transcription in osteoblast-like cells and osteoclasts have been contradictory, probably influenced by the cell type used, a lack of complex cell-to-cell interactions in clonal cell systems, and low to undetectable levels of estrogen receptor in these cells. There is some consensus in the field that the effects of estrogen are indirect, possibly involving regulation of paracrine and/or autocrine interactions between cells in the bone marrow. For example, estrogen inhibits interleukin 6 (IL-6) release from osteoblast-like cells (Girasole *et al.*, 1992), and, in turn, this cytokine has been reported to stimulate osteoclast activity *in vitro* (Ishimi *et al.*, 1990). Moreover, administration of neutralizing antibodies against IL-6 to mice prevented the proliferation of osteoclasts associated with ovariectomy (Jilka *et al.*, 1992), and disruption of the IL-6 gene *in vivo* prevented estrogen-deficient bone loss (Poli *et al.*, 1994). However, given the complexity of estrogen's effects at different skeletal sites (Turner *et al.*, 1994) and the number of other signaling molecules that have been implicated in estrogen deficiency-dependent bone loss, including transforming growth factor β and insulin-like growth factor 1 (Ernst *et al.*, 1989; Westerlind *et al.*, 1994), it seems likely that IL-6 is but one player in a network of regulatory signaling molecules.

Cardiovascular disease and insulin resistance

It is widely accepted that estrogen replacement after the menopause dramatically reduces the incidence of coronary heart disease (Stampfer *et al.*, 1991). Although the beneficial effects of estrogen on cardiovascular diseases in general are unproven, there is a wealth of observational data supporting such a contention (Stampfer and Colditz, 1991). Moreover, estrogen is known to have a number of physiologic effects on the cardiovascular system, although the extent and nature of the effect (that is, whether it is positive or negative) depends on the type of estrogen used for replacement (Cooper and Stevenson, 1997). Beneficial effects include increased plasma levels of high-density lipoprotein, decreased levels of low-density lipoprotein, and decreased levels of coagulation factors (fibrinogen, plasminogen activator inhibitor 1). Estrogen receptors are present in the liver (where they might influence the activity of enzymes regulating lipoproteins) and in the endothelial cells lining the blood vessel wall. The latter cells regulate vascular tone by releasing vasoactive and vasoconstricting factors, including acetylcholine. The vaso-dilator response of the latter in women has been shown to be improved by estrogen administration (Gilligan *et al.*, 1994). Raised activity of serum angiotensin-converting enzyme has been associated with an increased risk of coronary artery disease, and hormone replacement therapy has been shown to reduce the activity of this enzyme by 20% (Proudler *et al.*, 1995).

Insulin resistance and diabetes often lead to heart disease, probably by a combination of factors including changes in lipid profile, direct damage to the arterial wall and effects on coagulation and fibrinolytic factors. Postmenopausal women show a redistribution of body fat towards the male pattern, and exhibit changes in carbohydrate metabolism and insulin resistance that may increase their risk of cardiovascular disease (Proudler *et al.*, 1992; Walton *et al.*, 1992). Hormone replacement therapy, possibly with a dependence on the regimen, has been reported to improve insulin sensitivity (Brussaard *et al.*, 1997; Crook *et al.*, 1997). However, more work is needed to prove a link between estrogen and insulin resistance, and to define a mechanism by which estrogen influences insulin secretion.

Skin and urogenital atrophy

Estrogen receptors are present in skin fibroblasts. The collagen content of skin decreases after the menopause, and estrogen has been reported to prevent this decrease (Brincat *et al.*, 1987; Maheux *et al.*, 1994).

The vaginal epithelium is extremely sensitive to estrogen. Indeed, the two most common bioassays for estrogenicity examine the effects of test compounds on uterine weight and on vaginal cornification in rats. Estrogen receptors are present in the vagina, urethra and bladder (Iosif *et al.*, 1981). Menopausal women suffer from an increased incidence of urogenital disorders that include urinary incontinence, urinary tract infections, vaginal dryness and dyspareunia. Hormone replacement relieves some of these symptoms (Semmens and Wagner, 1982; Fantl *et al.*, 1994). As in skin, collagen is lost from urogenital tissues in the absence of estrogen. In addition, estrogen appears to stimulate α-adrenergic-responsive urethral smooth muscle by increasing the number of α_2-adrenergic receptors (Larsson *et al.*, 1984).

Central nervous system effects

Sex steroids and their receptors are present in the brain, and both estrogen and progesterone are active in this tissue. This section summarizes some of the more important actions, particularly as they relate to postmenopausal women. For a more detailed review of the area, see the review by Smith (1994).

The loss of estrogen at the menopause is associated with a number of deleterious symptoms that are related to the central nervous system. Most immediate is the hot flush, a physiologic response to declining hormone whereby an increase in peripheral body temperature and a decrease in core body temperature dissipate heat. Episodes can last 20 minutes or more, and can result in sleep-disturbing night sweats. Because hot flushes often appear during perimenopause, they are the leading reason for women to visit a physician for the relief of menopausal symptoms. The mechanism of the flush is unknown, but it appears to center on the hypothalamus.

There is tantalizing evidence linking a reduced incidence of dementia with estrogen use in older women (Paganini-Hill and Henderson, 1996). Women who had been on estrogen were 10–50% less likely to have had Alzheimer's disease or dementia, with the likelihood increasing the longer the hormone replacement regimen had been used. Estrogen may also have beneficial effects on cognition, an umbrella term that includes attention, concentration, language, memory and abstract reasoning (Sherwin, 1997).

Finally, estrogen has positive effects on mood and well-being (Sherwin and Gelfand, 1985; Ditkoff *et al.*, 1991). The mood disorders that sometimes occur postpartum and after the menopause are probably linked to hormonal fluctuations. Although unlikely to be effective in clinical depression, estrogen treatment at the doses used during hormone replacement therapy enhances mood in nondepressed women.

Progesterone and the premenstrual syndrome

Plasma progesterone levels plummet even more dramatically than those of estrogen at menstruation and after parturition. As with estrogen, the sudden loss of progesterone may underlie the mood changes associated with the premenstrual and postnatal periods. 5α-reduced analogs of progesterone (e.g. allopregnanolone) are neuroactive and have a sedative action that mimics that of the benzodiazepines (Bitran *et al.*, 1991; Brot *et al.*, 1997). They appear to sedate by enhancing the function of γ-aminobutyric acid (GABA), the brain's major inhibitory neurotransmitter. It is likely that allopregnanolone binds directly to the $GABA_A$ receptor, rather than to its own receptor, to affect this (Puia *et al.*, 1990). In an animal model of progesterone withdrawal that mimicked the premenstrual state, chronically treated rats were more prone to seizures and became insensitive to benzodiazepines following removal of the steroid (Smith *et al.*, 1998). These data suggest that severe premenstrual syndrome may be due to the disruption of anxiolytic circuits in the brain as plasma progesterone levels drop.

Estrogen action in the male

Estrogen's status as the 'female hormone' has recently begun to change with the recognition that the hormone also has important roles in the male (likewise, androgens can no longer be regarded solely as 'male hormones'). This has come about through a combination of scientific advances from a number of directions. The realization of estrogen's activity in nonreproductive tissues such as bone and in the cardiovascular system proved that it was not merely a 'sex' steroid. And the development of male estrogen receptor knockout mice with reproductive abnormalities proved that estrogen action was important for male fertility (Lubahn *et al.*, 1993). Finally, the discovery of an adult human male with a mutation of the estrogen receptor gene that disrupted bone development clearly showed that our view of its action had been restricted (Smith *et al.*, 1994).

Before these recent findings, it had been known that estrogens derived from aromatization in the testis could feedback inhibit follicle-stimulating hormone secretion from the pituitary (Finkelstein *et al.*, 1991; Bagatell *et al.*, 1994). Estrogen was also thought to inhibit the differentiation of Leydig cells before puberty and the production of testosterone in adult Leydig cells (Ge *et al.*, 1996).

Male mice in which the estrogen receptor α gene has been disrupted are infertile, due, apparently, to a defect in sperm production (Lubahn *et al.*, 1993; Eddy *et al.*, 1996; Hess *et al.*, 1997). Sperm counts decline in these animals as they age, and the efferent ducts fail to reabsorb seminiferous tubule fluid properly as the sperm develop. Thus far, male mice lacking a functional estrogen receptor β gene are fertile (Krege *et al.*, 1998), suggesting that the α isoform has a more prominent role in regulating male fertility.

The finding of an essentially normal adult man with a mutated estrogen receptor α gene proved that the receptor is not required for male development (Smith *et al.*, 1994). Reproductive organs were apparently unaffected. The man had raised plasma levels of gonadotropins, impaired glucose tolerance, hyperinsulinemia, incomplete closing of the bone epiphyses, and abnormally low bone density. With only one subject, it is impossible to know the significance of these defects. Nevertheless, the bone defects are probably real, because other male subjects have been found in whom the aromatase gene has been mutated, preventing the conversion of androgens to estrogens (Morishima *et al.*, 1995; Carani *et al.*, 1997). Like the estrogen receptor-mutated subject, these men exhibit a juvenile bone age, open epiphyses and tall stature. Some authors (e.g. Sharpe, 1998) now suggest that, in men, estrogen has a stimulatory role in inducing bone growth at puberty and an inhibitory role in the adult.

STRUCTURE AND MECHANISM OF ACTION OF ESTROGEN AND PROGESTERONE RECEPTORS

Structure

Like all members of the nuclear receptor superfamily, the estrogen and progesterone receptors consist of discrete protein domains (see Chapter 1 and Fig. 3.3); that is, an N-terminal A/B domain containing a transactivation function (activation function-1, AF-1), a central C domain containing two 'Cysteine2, Cysteine2'-type zinc fingers that has DNA-binding activity, a short hinge (D) domain, and a C-terminal E domain containing ligand-binding, dimerization, nuclear translocation and transactivation functions (AF-2). Of these domains, the DNA-binding domain is the most conserved across the receptor superfamily, the N-terminal domain is the least conserved, and the ligand-binding domain is moderately conserved. The estrogen receptors (α and β) contain an additional F domain (Fig. 3.3) C-terminal of the ligand-binding domain that is present in only one other nuclear receptor, the retinoic acid receptor. Its function is unknown.

Ten years after the initial cloning of the estrogen receptor (Walter *et al.*, 1985), a second receptor was discovered in rat prostate (Kuiper *et al.*, 1996) and human leukocyte complementary DNA (cDNA) libraries (Mosselman *et al.*, 1996). The receptor has a similar domain structure to the first receptor (now called estrogen receptor α), including an F domain, a highly conserved DNA-binding domain and a homologous ligand-

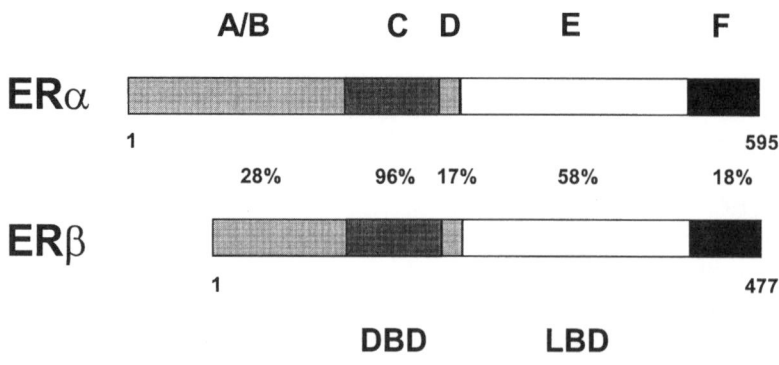

Fig. 3.3 Domain homology comparison of the human estrogen receptors α and β. ER, estrogen receptor; DBD, DNA-binding domain; LBD, ligand-binding domain.

binding domain (58% conserved at the amino acid level) (Fig. 3.3). Estrogen receptor β is the product of a separate gene from α (Enmark *et al.*, 1997). It is functionally similar to α, binding to an estrogen response element (see below) and recognizing all the same ligands (Kuiper *et al.*, 1997). Its tissue distribution, as determined by its RNA level, overlaps substantially with that of the α receptor, being present in most major mammalian tissues. However, there are differences; for example, the β subtype is more highly expressed in certain regions of the brain (Osterlund *et al.*, 1998), while the α receptor is the only subtype found in the liver (Kuiper *et al.*, 1997, 1998). In tissues where both are expressed it is possible that heterodimers form between the two subtypes (Pettersson *et al.*, 1997). However, the different phenotypes of the α and β receptor knockout mice (Lubahn *et al.*, 1993; Krege *et al.*, 1998) suggest that heterodimers are not absolutely required *in vivo*.

Two forms of the progesterone receptor also exist, although in this case both arise from the same gene (Conneely *et al.*, 1987). The A form is identical to the B form except that it lacks 165 amino acids at the N-terminus (see Fig. 3.6), due either to initiation of translation at an alternate internal methionine (Conneely *et al.*, 1987) or to differential splicing of the messenger RNA (mRNA) (Kastner *et al.*, 1990). Functionally, the two forms are identical in their DNA and ligand-binding characteristics. Heterodimers and homodimers of each form probably exist *in vivo*. The A form is substantially less active in transient cotransfection assays, and may even act as a repressor of B form activity in the appropriate promoter and cell environment (Vegeto *et al.*, 1993).

Nuclear receptor domains are modular and can be switched between different nuclear receptors and on to heterologous DNA-binding domains with retention of activity (Green and Chambon, 1987; Oro *et al.*, 1988). Thus, crystallization of purified nuclear receptor domains, rather than of the entire protein, has been a reasonable, and indeed necessary, approach to determining the structures of these receptors. Thus far, the structures of the DNA-binding and ligand-binding domains of human estrogen receptor α (Schwabe *et al.*, 1993; Brzozowski *et al.*, 1997; Shiau *et al.*, 1998), and the ligand-binding domain of the human progesterone receptor have been determined (Williams and Sigler, 1998). The DNA-binding domain of the estrogen receptor, and of other analyzed nuclear receptors, consists of two helices, one lying perpendicular to the other. At the base of each helix, a zinc ion is chelated by four conserved cysteine residues. When bound to DNA, two domains dimerize, with one helix from each monomer lying in the major groove of the hexameric estrogen response element half-site.

The ligand-binding domains of the estrogen and progesterone receptors consist of three-layered α-helical 'sandwiches', the middle helices lying perpendicular to those of the two outer layers. The ligand-binding pocket lies at one end of the structure. Ligand binding is stabilized by a mixture of electrostatic interactions with the A ring of the steroid (plus the D ring in the case of estradiol) and hydrophobic interactions with the entire molecule. Dimerization is accomplished differently by the two receptors (Tanenbaum *et al.*, 1998). The estrogen receptor has a larger dimerization interface, dominated by a hydrophobic region in helix 10 of each monomer. The dimerization interface of the progesterone receptor is one-half the size of the estrogen receptor's, and is probably less stable; it requires mutual interactions between helix 11 of one monomer and helix 12 of the other (Williams and Sigler, 1998). Finally, at least for the estrogen receptor (Shiau *et al.*, 1998), transactivation is probably accomplished by the formation of a coactivator interaction surface involving helices 3, 4, 5 and 12.

Mechanism of action

Many of the details of the molecular mechanism of action of estrogen and progesterone were elucidated during the 1970s and 1980s. This apparently central mechanism applies to all nuclear receptors with few modifications, and can be called the 'classic' mode of action. However, in the 1990s we began to understand how other regulatory signaling pathways could influence steroid action.

Classic mechanism

A schematic view of our current understanding of estrogen and progestin action in the cell is shown in Fig. 3.4. This mechanism can be applied to all type 1 members of the nuclear receptor superfamily with little modification (Chapter 1). The classic pathway of steroid action consists of the following steps:

1. Passive diffusion of the hydrophobic ligand across the plasma cell membrane.
2. Binding of ligand to the inactive receptor.
3. Ligand-induced activation via a conformational change.
4. Dissociation of associated proteins, chiefly heat shock proteins.
5. Dimerization.
6. Tight association of the activated receptor to nuclear DNA at DNA-binding sites adjacent to hormone-responsive genes.
7. Binding of coactivators.
8. Acetylation of histones and consequent loosening of the adjacent chromatin.
9. Binding of other gene-specific transcription factors and/or members of the basal transcription regulatory complex (including transcription activation factors and TATA-binding protein).
10. Gene activation.

Of all these steps, step 1 is perhaps the only one for which there is no good biochemical evidence. It is conceivable, although unlikely, that steroids are actively transported into the cell. Also, many coactivators, including SRC-1 (steroid receptor coactivator 1), ACTR (activator), CBP (cAMP response element binding [CREB] protein) and P/CAF (p300/CBP activation factor), possess histone acetyltransferase activity (Chen et al., 1997; Spencer et al., 1997; Blanco et al., 1998; Martinez-Balbas et al., 1998), but evidence that all possess it is lacking. It is possible that some coactivators help to recruit histone acetyltransferases to the transcription complex, much as nuclear receptor corepressors such as SMRT (silencing mediator of retinoid and thyroid receptors) and NCoR (nuclear receptor corepressor) recruit histone deacetylases (Heinzel et al., 1997; Nagy et al., 1997).

One of the differences between the mechanisms of action of steroid receptors and most other members of the nuclear receptor superfamily is in their association with heat shock proteins before ligand binding. Sedimentation analysis of purified progesterone receptors had revealed an 'untransformed' 8–10S form and, after heat, salt or ligand treatment, a 'transformed' 4–6S form (Baulieu et al., 1971; Pratt, 1987; Denis et al., 1988). Subsequent purification of the components of the complexes showed that one of them was heat shock protein 90 (Sanchez et al., 1985; Schuh et al., 1985). Today, it is clear that the 8S and 4S complexes contain a variety of proteins besides the progesterone receptor, including heat shock proteins and immunophilins such as FKPB56 (Pratt and Toft, 1997). Although we have known of their presence for 15 years, there is currently no accepted role defined for these associations. In vitro transcription assays with heat shock protein-free progesterone receptor indicated that they play no role in gene activation

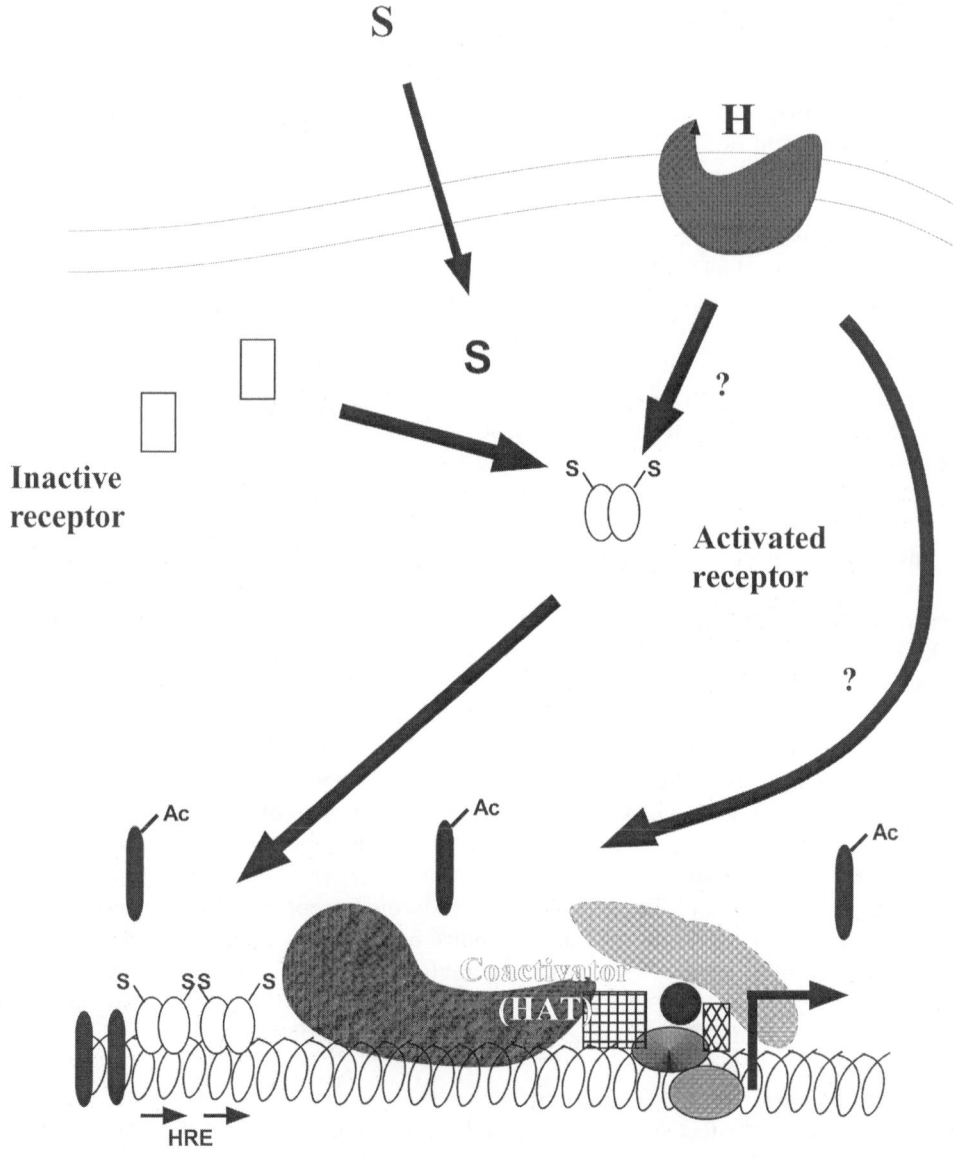

Fig. 3.4 Schematic illustration of the mechanism of action of the estrogen and progesterone receptors. S, steroid; H, peptide hormone or neurotransmitter; Ac, acetyl group; HAT, histone acetyltransferase; HRE, hormone response element.

(Bagchi *et al.*, 1990); in contrast, a study in yeast led to the conclusion that activation is compromised (Picard *et al.*, 1990). Although it still cannot be ruled out that protein association is an artefact of biochemical isolation of the steroid receptor, it seems more likely that it acts in some way to help stabilize the ligand-free receptor against degradation.

Steroid receptors are also unique in the nature of the DNA response elements that they recognize. Unlike most nuclear receptor DNA-binding sites, the estrogen and progesterone receptors bind to inverted repeats of a six-base-pair half-site separated by three base pairs. The consensus half-site for the progesterone receptor is AGAACA (Scheidereit et al., 1983; Scheidereit and Beato, 1984), and the consensus half-site for the estrogen receptor is AGGTCA (Klein-Hitpass et al., 1986, 1988). However, the consensus is rarely followed absolutely; most binding sites have one half-site with close similarity to it and one half-site with less similarity. Examples of progesterone-binding sites are in the chicken lysozyme gene promoter and the mouse mammary tumor virus long terminal repeat (von der Ahe et al., 1985). Estrogen-binding sites have been described upstream of the frog vitellogenin and human pS2 genes (Klein-Hitpass et al., 1986; Berry et al., 1989). Surprisingly, the progesterone response element is also bound by the glucocorticoid, androgen and mineralocorticoid receptors (von der Ahe et al., 1985; Darbre et al., 1986; Strahle et al., 1987). As a consequence, how specificity in transcriptional regulation is achieved in the cell is unclear. Tissue-specific expression of receptors probably contributes (Strahle et al., 1989). In addition, there may be subtle differences in the natural hormone response elements that favor binding of one receptor over another, or the milieu of transcription factors bound to the promoter of a target gene may stabilize binding of one receptor and/or inhibit binding of others.

Functional dissection of the glucocorticoid and estrogen receptor DNA-binding domains has revealed that a short sequence of amino acids overlapping the most C-terminal cysteine residue of the first zinc finger (the 'P box') specifies discrimination between the estrogen and the glucocorticoid/progesterone response elements (Green and Chambon, 1987; Danielsen et al., 1989; Mader et al., 1989; Umesono and Evans, 1989). Similar analyses have shown that the sequence between the first pair of cysteines in the second zinc finger (the 'D box') recognizes the length of the spacer between the response element's half-sites (Umesono and Evans, 1989). The orientation of the D boxes in each monomer of a receptor dimer appears to control whether the receptor will bind to an inverted repeat of two half-sites (such as in an estrogen or progesterone response element), a direct repeat, or an everted repeat, all of which are variously bound by thyroid hormone and retinoic acid receptor homodimers and heterodimers with retinoid X receptor (Cooney and Tsai, 1994). In the steroid receptor subfamily, interactions between adjacent D boxes in the dimer stabilize binding to the inverted repeat (Luisi et al., 1991; Schwabe et al., 1993), while interactions with and between another DNA-binding domain motif, the 'DR box', determine binding of other superfamily members to direct and everted DNA-binding sites (Perlmann et al., 1993).

Interactions with other signaling pathways

Treatment of cotransfected mammalian cells with inducers of phosphorylation or the neurotransmitter dopamine induces transcriptional activation by progesterone, estrogen and other nuclear receptors (Denner et al., 1990; Power et al., 1991). This was the first serious challenge to the dogma that nuclear receptors were transcription factors with a sole dependence on ligand binding for transcriptional activation. Subsequent work showed that insulin-like growth factor 1, epidermal growth factor and transforming growth factor α act in synergy with progesterone or estradiol to induce reporter genes via their respective receptors (Aronica and Katzenellenbogen, 1991, 1993; Ignar-Trowbridge et al., 1993). Animal studies strengthened the link between peptide and non-peptide hormonal signaling pathways. Korach, MacLachlan and coworkers showed that the estrogen-like stimulatory effects of epidermal growth factor on rat uterine growth were inhibited by an estrogen antagonist (Ignar-Trowbridge et al., 1992). And O'Malley and coworkers provided evidence that progesterone-dependent sexual behavior in rats could be mimicked by dopamine agonists. In turn, dopamine activation was inhibited by a progesterone antagonist or by antisense oligonucleotides against the progesterone receptor (Mani *et al.*, 1994).

Together, the *in vitro* and *in vivo* data make a compelling argument for the existence of 'crosstalk', as the phenomenon has been called. Its mechanism is less well understood, and its relative physiologic importance has not been established. Because nuclear receptors are phosphoproteins (Weigel, 1994), and because intracellular signaling pathways often consist of protein kinase and phosphatase cascades, it is thought that phosphorylation (or dephosphorylation) of a common factor is responsible for synergistic and/or ligand-independent activation of nuclear receptors. That factor could be the receptor itself; however, mutagenesis of known phosphorylation sites in the estrogen and progesterone receptors has only modest effects on transactivation (Ali *et al.*, 1993; Bai *et al.*, 1994; Bai and Weigel, 1996). It seems more likely that a coactivator, corepressor or other protein with which the receptor interacts is the target of an activated kinase or phosphatase. One candidate is the cyclic adenosine monophosphate (cAMP) response element-binding protein (CBP/p300), a target of adenyl cyclase-stimulated second messenger pathways that has been shown to be a receptor coactivator (Chakravarti *et al.*, 1996). Clearly, there is still much to learn about this interesting aspect of steroid hormone action.

GENETICS

As mentioned above, estrogen plays an important role in the development of mammary gland and endometrial tissues. It is also involved in the growth and development of uterine and breast cancers. Tamoxifen, an estrogen antagonist that inhibits the proliferation of breast cancer cells, is a commonly prescribed breast cancer therapeutic agent. Likewise, progesterone antagonists inhibit breast tumor growth. Based on the work of McGuire and coworkers (Knight *et al.*, 1977; McGuire *et al.*, 1979), tumors are often classified based on the presence of estrogen and progesterone receptors that are capable of binding hormone, with those that possess binding activity for both receptors having the best prognostic forecast. Breast tumors often become resistant to drug treatment over time, although they may retain receptor protein. Thus, possession of a receptor does not by itself ensure responsiveness of a cancer cell to tamoxifen therapy. How resistance occurs is unclear, but one possibility is that cancer cells in which the receptor(s) is mutated gain predominance in the tumor in the presence of the drug. Over the past decade, dozens of naturally occurring estrogen receptor α RNA variants have been described. Estrogen receptor β and progesterone receptor variants have also been described more recently. Some of these variants are probably true mutants, although it is not clear that any of them emerged in response to drug therapy; many others are likely to be naturally occurring variants with a yet-to-be established physiologic role.

Estrogen receptor α

In 1989, Murphy and Dotzlaw noted the presence of multiple truncated RNA species in Northern blots from breast tumor biopsy RNA samples. The first estrogen receptor variants were described in the laboratories of Horwitz and McGuire in RNA from the established breast cancer cell lines T47D and MCF7 (Graham *et al.*, 1990; McGuire *et al.*, 1991). Subsequently, new variants began to appear with increasing frequency, mainly in tissue samples rather than cell lines; this was facilitated by the establishment of the polymerase chain reaction as a standard laboratory technique, which allowed the analysis of RNA from limiting tumor samples. Most variants have been obtained from RNA derived from breast tumor samples, although uterus, testis, ovary, cervix, pituitary, hypothalamus, heart, liver and bone have also been sources.

Variant sequence deviations affect all domains of the estrogen receptor, with those that coincide with the DNA- and ligand-binding domains having the most dramatic effects on receptor function. Point mutations, small deletions, multiple exon deletions, exon duplications, *trans*-splice variants and insertions have all been described (Table 3.2). Only those variants that alter amino acid sequence are discussed here.

Some of the more dramatic variants include the exon-skipping and exon duplication variants. Most are presumably the result of alternate splicing, and are found in both normal and neoplastic tissues. Thus, they probably do not have anything to do with drug resistance in breast cancer. Instead, they may act to regulate the response of the wild-type receptor to normal physiologic stimuli. Indeed, at least one variant, truncated estrogen receptor product 1 (TERP-1), appears to be tissue specific, present only in and regulating estrogenic responses of pituitary cells from the rat (Friend *et al.*, 1995; Schreihofer *et al.*, 1999).

The human estrogen receptor α gene has eight protein-coding exons, all of which (with the exception of exon 8) have been deleted in one variant or another, either alone or in combination with other exons (Fig. 3.5, Table 3.2). In most cases these variants are incapable of binding DNA and/or ligand, and are functionally inert. Some of the variant receptors (deletions of exon 3, exon 4, exon 7, or exons 4 and 5; and duplication of exons

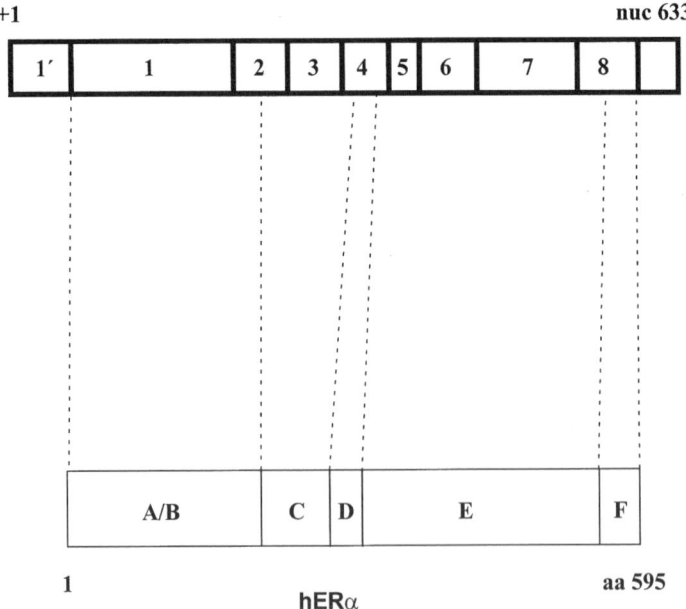

Fig. 3.5 Schematic diagram of the human estrogen receptor α (hERα) messenger RNA and its encoded protein. Numbering is according to Green *et al.* (1986). The gene structure of human estrogen receptor β is virtually identical (Enmark *et al.*, 1997). aa, Amino acid; nuc, nucleotide.

Table 3.2 Variations in the estrogen receptor α gene. Nucleotides are numbered relative to the initiator methionine codon (+1 ATG+3)

RNA	Domain(s) or residue affected	Functional effect	Tissue	References
Exon 2 o-f del	All	No activity	Uterus, PBMC, pituitary, breast	Wang and Miksicek, 1991; Madsen et al., 1995; Pfeffer et al., 1995; Gotteland et al., 1995; Leygue et al., 1996a; Chaidarun et al., 1997; Moutsatsou et al., 1998
Exon 3 i-f del	DBD	Dominant negative	Uterus, pituitary	Wang and Miksicek, 1991; Miksicek et al., 1993; Chaidarun et al., 1997; Rey et al., 1998
Exon 4 i-f del	DBD, hinge, LBD	Dominant negative	Pituitary, uterus, heart, hypothalamus, liver, ovary, testes, VSMC, bone	Pfeffer et al., 1993, 1995; Skipper et al., 1993; Koehorst et al., 1994; Gotteland et al., 1995; Madsen et al., 1995; Hoshino et al., 1995; Hu et al., 1996; Inoue et al., 1996; Park et al., 1996; Pfeffer, 1996; Friend et al., 1997
Exon 5 o-f del	LBD	Dominant positive	Uterus, ovary, cervix, pituitary	Fuqua et al., 1991, 1993; Castles et al., 1993; Zhang et al., 1993; Madsen et al., 1995; Villa et al., 1995, 1996; Hu et al., 1996; Wilson et al., 1996; Chaidarun and Alexander, 1998; Mito et al., 1998
Exon 6 o-f del	LBD	Unknown	Breast	Miksicek et al., 1993
Exon 7 o-f del	LBD	Dominant negative	Uterus, pituitary	Wang and Miksicek, 1991; Fuqua et al., 1992; Madsen et al., 1995; Hu et al., 1996; Leygue et al., 1996b; Chaidarun et al., 1997; Rey et al., 1998
Exon 2 + 3 del	All	Unknown	Breast	Leygue et al., 1996a,b
Exon 3 + 4 del	DBD, hinge	No activity	Pituitary, uterus, heart, hypothalamus, liver, testes, VSMC, bone	Gotteland et al., 1995; Hoshino et al., 1995; Hu et al., 1996; Inoue et al., 1996; Leygue et al., 1996a,b; Friend et al., 1997
Exon 4 + 5 del	DBD, hinge, LBD	Dominant negative	Uterus, VSMC	Hu et al., 1996; Inoue et al., 1996
Exon 4 + 7 del	DBD, hinge, LBD	Unknown	Breast	Madsen et al., 1995; Pfeffer et al., 1995; Leygue et al., 1996b

continued overleaf

Table 3.2 *Continued*

RNA	Domain(s) or residue affected	Functional effect	Tissue	References
Exon 5 + 6 del	LBD	Unknown	Pituitary, uterus, heart, hypothalamus, liver, testes	Friend et al., 1997
Exon 2 + 3 + 4 del	All	Unknown	Breast	Leygue et al., 1996b
Exon 4 + 5 + 6 del	DBD, hinge, LBD	No E binding; no dimerization, no transaction	Breast	Chan and Dowsett, 1997
Exon 5 + 6 + 7 del	LBD	Unknown	Breast	Pfeffer et al., 1995
Exon 3 + 7 partial del	DBD, hinge, LBD	Unknown	Breast	Leygue et al., 1996b
Exon 4 + 5 partial del	DBD, hinge, LBD	No E binding; no nuclear localization	Breast cell line	Graham et al., 1990
Exon 4 + 7 partial del	DBD, hinge, LBD	Unknown	Uterus, breast	Daffada and Dowsett, 1995
Exon 2/intron *trans*-splice	DBD, hinge, LBD	Unknown	Breast	Dotzlaw et al., 1992
Exon 3/intron *trans*-splice	DBD, hinge, LBD	Unknown	Breast	Dotzlaw et al., 1992
Exon 1 + 2 + 3 + 4 del (TERP-2)	All	Biphasic dominant negative, dominant positive	Pituitary, uterus, heart, hypothalamus, liver, testes	Friend et al., 1995, 1997
Exon 1 + 2 + 3 + 4 del (TERP-1)	All	Biphasic dominant negative, dominant positive	Pituitary	Friend et al., 1995
Duplication of exon 6	LBD	Unknown	Breast	Murphy et al., 1996
Duplication of exons 3 + 4	DBD, hinge	Unknown	Breast	Murphy et al., 1996
Duplication of exons 6 + 7	LBD	Dominant negative	Breast cell line	Pink et al., 1995, 1996
69 bp insert between exons 5 and 6	LBD	Unknown	Breast	Murphy et al., 1996; Wang et al., 1997
480GC > TG	160G > C	Unknown	Breast	Andersen et al., 1997; Zelada-Hedman et al., 1997
207C > G	69N > K	Unknown	Breast	Roodi et al., 1995
471C > T	R157X	Decreased E binding	Lymphocytes	Smith et al., 1994

750–751insTT	250M > I,X	No E binding; no nuclear localization	Breast cell line	Graham et al., 1990; Leslie et al., 1992
888T > C	296L > P	Unknown	Breast	McGuire et al., 1991
1056G > T	352D > Y	Activated by tamoxifen	Breast cell line	Wolf and Jordan, 1994; Catherino et al., 1995
1059A > T	353E > V	Unknown	Breast	Karnik et al., 1994
1188A > G	396M > V	Unknown	Breast	Roodi et al., 1995
1233delG	D411T, + 6novel	No E binding	Breast cell line	Graham et al., 1990
1296delT	S432H, + 4novel	Tamoxifen resistance	Breast	Karnik et al., 1994
1272–1319 replaced with 1105–1147	I424R, + 28novel	No E binding	Breast	Karnik et al., 1994

Silent mutations are not listed. o-f, out-of-frame; i-f, in-frame; del, deletion; ins, insert; DBD, DNA binding domain; LBD, ligand-binding domain; E, estrogen; PBMC, peripheral blood mononuclear cells; VSMC, vascular smooth muscle cells; TERP, truncated estrogen receptor product; bp, base pairs.

6 and 7) are still capable of binding DNA or of dimerizing with wild-type receptors, although they have no activity by themselves (Table 3.2). Such variants have dominant negative effects and would be likely to interfere with the action of normal receptor present in the same cell. Another variant, lacking exon 5 and therefore most ligand-binding domain functions, has a dominant positive effect (Chaidarun and Alexander, 1998). However, it must be borne in mind that it has yet to be shown that exon-skipping and duplication variants actually exist as protein *in vivo*; for example, it is possible that the translation products are unstable. Even if they are not, there is as yet no convincing evidence that they play a functional role outside transfected cells.

The same caveats apply to point mutants and small deletions of estrogen receptor α (Table 3.2). However, in their cases it seems more likely that they are true somatic mutants rather than products of inefficient processing. Evidence for altered responses to tamoxifen have been provided for two such variants (Karnik *et al.*, 1994; Catherino *et al.*, 1995). In a rare comprehensive analysis of large numbers of patients and controls for estrogen receptor α gene mutations, a germline mutation converting a glycine in the DNA-binding domain into a cysteine was described in three women with a family history of breast cancer (Andersen *et al.*, 1997; Zelada-Hedman *et al.*, 1997). Although not conclusive, the possibility exists that the mutation might increase the risk of breast cancer. No somatic mutations of the receptor gene were detected in breast tumors.

Currently, the only estrogen receptor α mutation that seems certain to have had a physiologic consequence is the arginine conversion to a stop codon in a man with unfused epiphyses described in the discussion of estrogen action in males (Smith *et al.*, 1994).

Estrogen receptor β

Although it was cloned relatively recently, some splicing variants that are similar to the α versions have already been described for estrogen receptor β (Table 3.3). Two non-exon-skipping mutations that alter the receptor protein have also been discovered (Rosenkranz *et al.*, 1998). Since the original receptor cDNA was cloned, two to four longer versions with an extra coding sequence at the 3′ end have been described (Moore *et al.*, 1998; Leygue *et al.*, 1999). Which of these represents the 'true' receptor in the cell, or whether they are all viable differentially regulated variants, remains to be determined.

Table 3.3 Variations in the estrogen receptor β gene

RNA	Domain or residue affected	Functional effect	Tissue	References
Exon 5 i-f del	LBD	Unknown	Breast, ovary, uterus	Lu *et al.*, 1998; Vladusic *et al.*, 1998
Exon 6 i-f del	LBD	Unknown	Breast, ovary, uterus	Lu *et al.*, 1998
Exons 5 + 6 i-f del	LBD	Unknown	Breast, ovary, uterus	Lu *et al.*, 1998
809–830del	Q238-K244del	Unknown		Rosenkranz *et al.*, 1998
846G > A	250G > S	Unknown		Rosenkranz *et al.*, 1998

Silent mutations are not listed. See Table 3.2 for abbreviations.

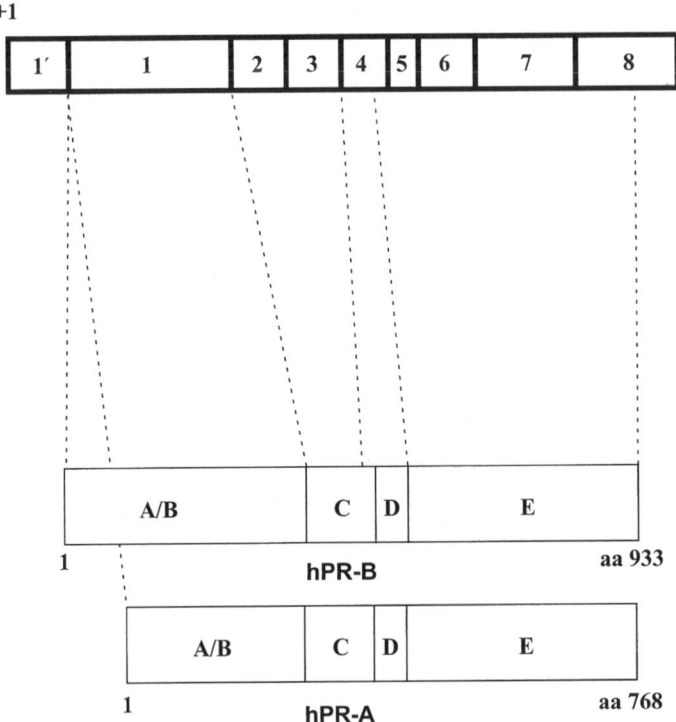

Fig. 3.6 Schematic diagram of the human progesterone receptor (hPR) messenger RNA and its encoded proteins (A and B forms). aa, Amino acid.

Progesterone receptor

Progesterone antagonists have potential for the treatment of breast cancer, but they have yet to be widely used clinically. Thus, their susceptibility to resistance is unknown. Several variants of the progesterone receptor have been described (Fig. 3.6, Table 3.4); they are all exon-skipping splice variants. As with the similar estrogen receptor splice variants, their physiologic significance is unproved. Assuming that they are functional *in vivo*, it is likely that they would act in some way to regulate the progestational response in the cell.

PHARMACOLOGY

Because of their central roles in ovulation and pregnancy maintenance, the estrogen and progesterone receptors have been the predominant targets of hormonal contraceptives and abortifacients. In addition, recognition of the primacy of estrogen loss in complications of the postmenopausal state has led to the widespread use of supplemental estrogens for hormone replacement therapy. Estrogen antagonists are used for preventing and treating breast cancer. With our growing appreciation of the pleiotropic physiologic actions of estrogen, other uses can be envisaged for estrogen and its pharmaceutical analogs. Likewise, the importance of progesterone in uterine physiology and breast development will allow the development of new therapies for disorders in those tissues.

Table 3.4 Variations in the progesterone receptor gene

RNA	Domain(s) affected	Functional effect	Tissue	References
Exon 2 i-f del	DBD	Unknown	Breast	Richer *et al.*, 1998
Exon 4 i-f del	DBD, hinge	Unknown	Uterus, breast	Leygue *et al.*, 1996c; Misao *et al.*, 1998; Richer *et al.*, 1998
Exon 6 i-f del	LBD	Dominant negative	Uterus, breast	Misao *et al.*, 1998; Richer *et al.*, 1998
Exon 6 del	LBD	Unknown	Breast	Leygue *et al.*, 1996c
Exon 3, 6 del	DBD, LBD	Unknown	Breast	Leygue *et al.*, 1996c
Exon 5, 6 i-f del	LBD	Dominant negative	Uterus, breast	Leygue *et al.*, 1996c; Misao *et al.*, 1998; Richer *et al.*, 1998
Exon 4, 5, 6 i-f del	DBD, LBD, hinge	Unknown	Uterus	Misao *et al.*, 1998
Exon 4, 6 i-f del	DBD, LBD, hinge	Unknown	Uterus	Misao *et al.*, 1998

See Table 3.2 for abbreviations.

Contraception

Oral contraceptives are the most effective type of contraceptive available, short of surgical sterilization. The chance of pregnancy drops to 1–2% per cycle when used correctly. Over 60 million women are currently taking oral contraceptives worldwide.

Most oral contraceptives contain both an estrogen and a progestin together. Two common analogs of each, ethinylestradiol and norgestimate respectively, are shown in Figs 3.7 and 3.8. Either component alone is an effective contraceptive, but each helps to balance out some of the adverse effects of the other. For example, estrogen by itself (so-called 'unopposed estrogen') would hyperstimulate the endometrium and increase the risk of uterine tumor formation; introduction of a progestin in low amounts, or at the correct point in the monthly cycle, allows appropriate endometrial turnover. A progestin by itself is slightly less effective than an estrogen, and tends to produce irregular cycles, reducing its popularity.

Oral contraceptives appear to work by inhibiting the secretion of gonadotropins from pituitary gonadotropes (Briggs, 1976). As a consequence, ovulation is inhibited. There may be additional effects on the endometrium, which is unlikely to be capable of accepting a blastocyst under the altered hormonal conditions engendered by contraceptive use.

Progesterone antagonists (RU486, also known as mifepristone) (Fig. 3.8) have been shown to be effective as postcoital contraceptives (Glasier *et al.*, 1992; Webb *et al.*, 1992). They act by inhibiting ovulation in the proliferative phase and by inducing menses in the secretory phase of the menstrual cycle (van Uem *et al.*, 1989; Kekkonen *et al.*, 1993).

Hormone replacement

As discussed in the Physiology section, the deleterious effects of ovarian shutdown that appear during and after the menopause can be alleviated by estrogen replacement. Perimenopausal complaints such as hot flushes and vaginitis disappear (Ernster *et al.*, 1988). In the long term, the increased risk of bone fractures due to osteoporosis is

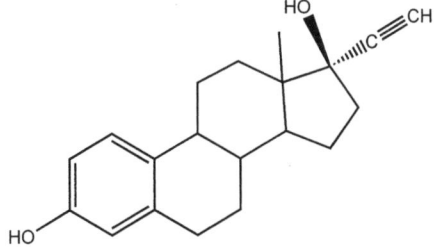

Estradiol

Ethinylestradiol

ICI 182 780

Fig. 3.7 Steroidal estrogen agonists and antagonists.

reduced (Kiel *et al.*, 1987). Estrogens may also reduce the risk of coronary heart disease (Stampfer and Colditz, 1991; Stampfer *et al.*, 1991).

Estrogen is administered in a variety of forms as part of a hormone replacement regimen. Naturally occurring sulfated estrogen mixtures, estrone sulfates and pure estradiol can be taken orally, and estradiol can be applied in a patch. Ethinylestradiol, a synthetic estrogen that is a common component of oral contraceptives (see above), can also be used for hormone replacement therapy.

Although highly effective, long-term treatment with unopposed estrogen increases the risk of developing endometrial carcinoma (Shapiro *et al.*, 1985). Although the epidemiologic data are less clear, there is also concern among practitioners and patients that estrogen use increases the risk of breast cancer in postmenopausal women (Bergkvist and Persson, 1996). Certainly, estrogens have a proliferative effect on breast cells *in vitro* and

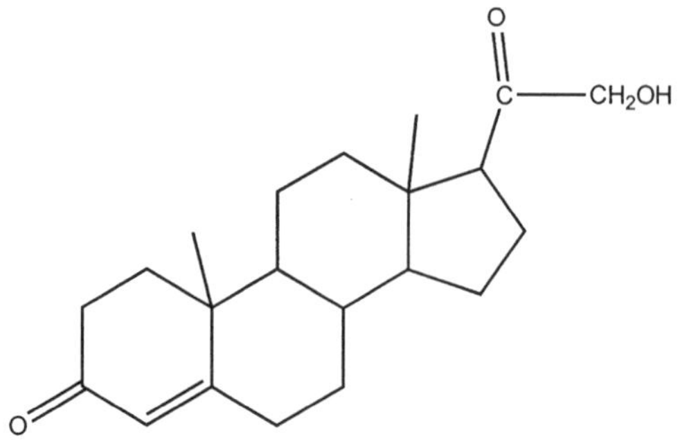

Progesterone

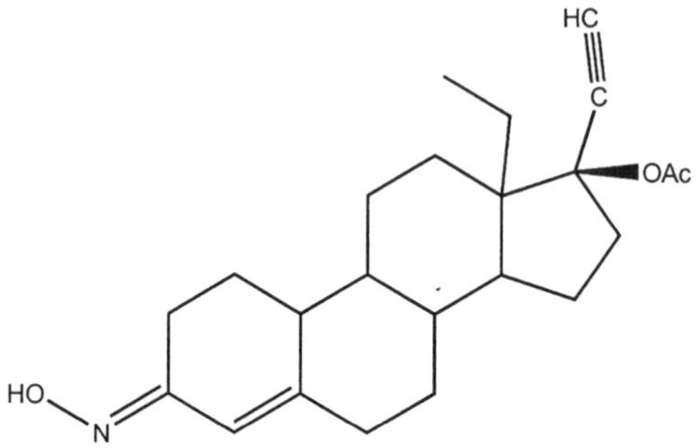

Norgestimate

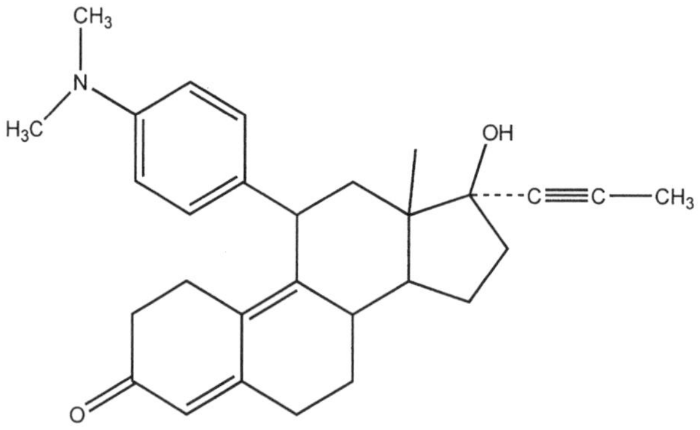

RU 38486

Fig. 3.8 Steroidal progestin agonists and antagonists.

on breast tumors in animals, and an estrogen antagonist is an effective treatment for breast cancer (Sledge and McGuire, 1983; Pritchard, 1997). Thus, alternatives to estrogen alone for hormone replacement have been sought. Early on, following the rationale that led to the development of combination oral contraceptives, a progestin was occasionally included with the estrogen. However, doing so led to the resumption of menstrual bleeding, an unfavorable situation for most postmenopausal women. In addition, the progestin might counteract the beneficial effects of the estrogen on cardiovascular health.

These considerations led to the ongoing search for synthetic tissue-specific estrogens that have the bone- and cardiovascular system-preserving properties of the natural hormone without its stimulatory effects on breast and uterus. The first of these selective estrogen receptor modulators, raloxifene, is now in use (Fig. 3.9). Others are currently in clinical trials. Raloxifene is effective in preventing bone loss in women and does not stimulate the breast or uterus in animal models. Its effects on the cardiovascular system in women have yet to be determined. Its drawbacks are reduced potency relative to estrogen and a failure to prevent hot flushes (Khovidhunkit and Shoback, 1999). With our new appreciation of the benefits of estrogen use to ageing women, for example in slowing the development of cognitive defects and perhaps in preventing Alzheimer's

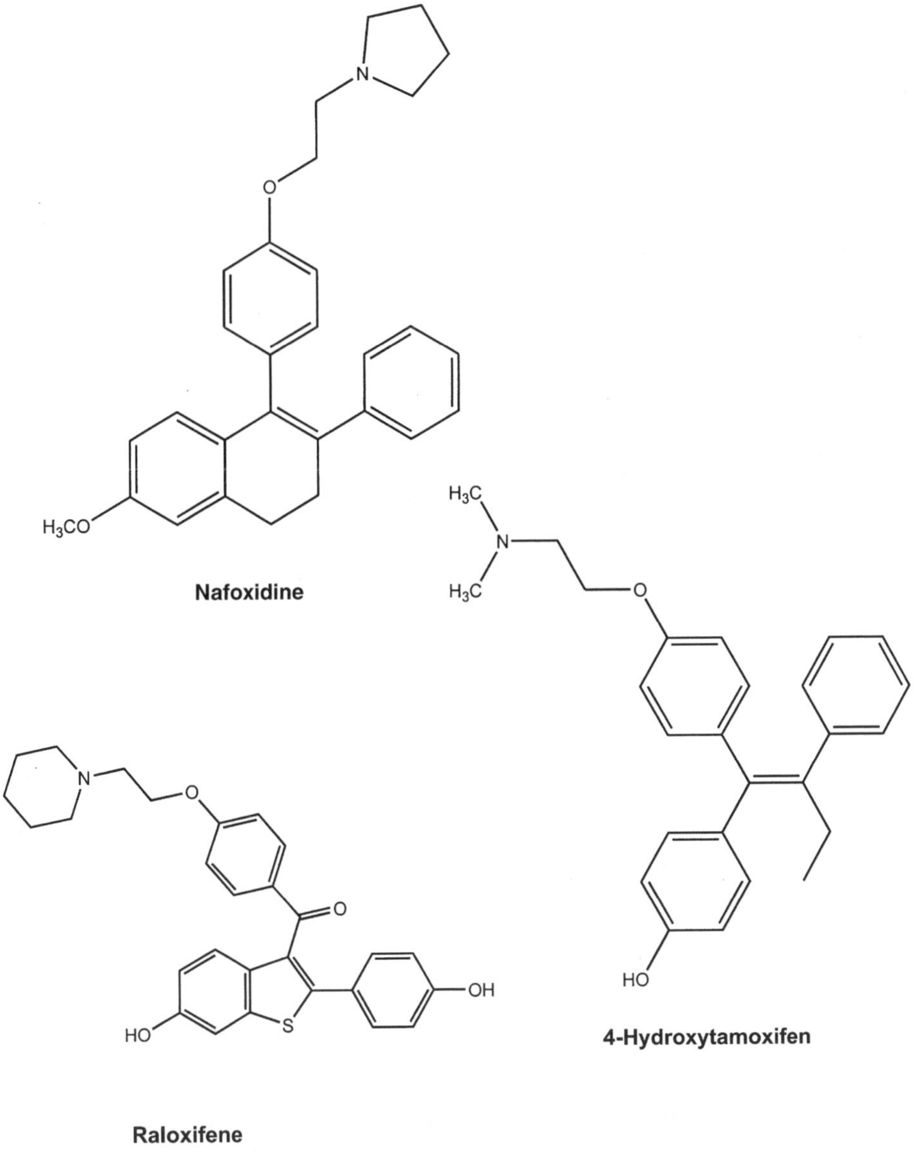

Nafoxidine

4-Hydroxytamoxifen

Raloxifene

Fig. 3.9 Some currently marketed nonsteroidal estrogen antagonists.

disease, there is a great deal of excitement about the potential of tissue-specific estrogens for treating many of the negative effects of ageing.

Breast cancer

Today, the estrogen antagonist tamoxifen (Fig. 3.9) is one of the major therapeutic weapons used against breast cancer. Initially developed in the early 1960s for emergency contraception, it was later shown to inhibit metastatic breast tumor growth in postmenopausal women (Cole *et al.*, 1971; Ward, 1978). Tamoxifen is now used in postmenopausal and premenopausal women, either as an adjuvant to surgery, or alone. Many other triphenylethylenes, such as nafoxidine (Fig. 3.9), have been synthesized, but none has gained the clinical success of tamoxifen.

The use of tamoxifen as a cancer preventive has been more controversial. One study of 13 000 subjects showed a dramatic (49%) decrease in the incidence of contralateral breast cancer in women aged over 60 years and in younger high-risk women who took tamoxifen for 5 years (Fisher *et al.*, 1998). In contrast, a second study of 2500 high-risk women who received the drug for 6 years showed no decrease in disease incidence (Powles *et al.*, 1998). Tamoxifen appears to increase the risk of endometrial cancer in women on extended therapy (Assikis and Jordan, 1995) and has been shown to induce liver tumors in animals (Williams *et al.*, 1993).

As mentioned in the Genetics section above, one of the major problems with tamoxifen use is the appearance of tumor resistance over time. One response to this problem has been the development of so-called 'pure' estrogen antagonists, compounds that, unlike tamoxifen and raloxifene, possess no estrogenic activity. ICI 182 780, a steroidal compound (Fig. 3.7), is currently under clinical investigation for the treatment of advanced breast cancer that has not responded to tamoxifen (Howell *et al.*, 1996).

More recently, the possibility that progesterone antagonists could be used to treat breast cancer has been explored. RU486 inhibits the proliferation of breast cancer cell lines *in vitro* (Bardon *et al.*, 1985; Horwitz, 1985) and of hormone-dependent mammary tumors *in vivo* (Bakker *et al.*, 1987). However, an initial clinical trial has not shown a significant benefit of RU486 therapy (Perrault *et al.*, 1996). Other antiprogestins are currently under investigation.

Dysfunctional uterine bleeding

This condition occurs at menarche or menopause, and is a result of the continuous secretion of estrogen and insufficient secretion of progesterone, with resultant incomplete sloughing of a hyperplastic endometrium at menstruation. Excessive bleeding results. High, regulated doses of a progestin such as norethindrone will both stop the bleeding and reinstate the normal cycle.

Pregnancy termination

The progesterone antagonist RU486 is used in some European countries for abortion, and is nearing approval in the United States. The compound works by interfering with the suppressive effect of progesterone on uterine contractility and by disrupting the stable environment provided to the fetus by the endometrial layer (deciduum) of the uterus. When RU486 is taken with a prostaglandin, which enhances smooth muscle contractions, surgical abortion can be avoided in over 90% of cases (Baulieu, 1989).

Other therapeutic uses of progesterone antagonists

The social and political controversy surrounding abortion and RU486 is unfortunate because of the clear potential of progesterone antagonists to satisfy some major unmet medical needs. RU486 has been shown to relieve the pain associated with endometriosis, and at high doses can reduce the size of endometriotic lesions (Kettel *et al.*, 1998). Endometriosis affects 5–15% of all women and is currently best treated nonsurgically with a gonadotropin-releasing hormone agonist, a regimen that can be followed for only 6 months due to the bone demineralization caused by extended exposure to these agents. Likewise, uterine fibroids (leiomyomata) are common (present in 30–40% of all women), but are amenable to surgery only when clinical symptoms appear. Fibroid volumes were reduced following treatment of women with RU486 (Murphy *et al.*, 1993).

Progesterone antagonists may also have potential for cervical ripening (Stiemer and Elger, 1990) and for treating brain meningiomas, which contain progesterone receptors (Grunberg *et al.*, 1991). Clearly, the potential uses of progesterone antagonists have only begun to be tapped.

REFERENCES

Ali, S., Metzger, D., Bornert, J.M. and Chambon, P. (1993) Modulation of transcriptional activation by ligand-dependent phosphorylation of the human estrogen receptor A/B region. *EMBP J.* **12**: 1153–1160.

Allen, E. and Doisy, E.A. (1923) An ovarian hormone. *JAMA* **81**: 819–821.

Allen, E. and Doisy, E.A. (1924) The induction of a sexually mature condition in immature females by injection of the ovarian follicular hormone. *Am. J. Physiol.* **69**: 577–588.

Allen, W.M. and Wintersteiner, O.S. (1934) Crystalline progestin. *Science* **80**: 190–191.

Andersen, T.I., Wooster, R., Laake, K. *et al.* (1997) Screening for ESR mutations in breast and ovarian cancer patients. *Hum. Mutat.* **9**: 531–536.

Aronica, S.M. and Katzenellenbogen, B.S. (1991) Progesterone receptor regulation in uterine cells: stimulation by estrogen, cyclic adenosine 3′,5-monophosphate, and insulin-like growth factor 1 and suppression by antiestrogens and protein kinase inhibitors. *Endocrinology* **128**: 2045–2052.

Aronica, S.M. and Katzenellenbogen, B.S. (1993) Stimulation of estrogen receptor-mediated transcription and alteration in the phosphorylation state of the rat uterine estrogen receptor by estrogen, cyclic adenosine monophosphate, and insulin-like growth factor-1. *Mol. Endocrinol.* **7**: 743–752.

Assikis, V.J. and Jordan, V.C. (1995) Gynecologic effect of tamoxifen and the association with endometrial carcinoma. *Int. J. Gynecol. Obstet.* **49**: 241–257.

Bagatell, C.J., Dahl, K.D. and Bremner, W.J. (1994) The direct pituitary effect of testosterone to inhibit gonadotropin secretion in men is partially mediated by aromatization to estradiol. *J. Androl.* **15**: 15–21.

Bagchi, M.K., Tsai, S.Y., Tsai, M.-J. and O'Malley, B.W. (1990) Identification of a functional intermediate in receptor activation in progesterone-dependent cell-free transcription. *Nature* **345**: 547–550.

Bai, W. and Weigel, N.L. (1996) Phosphorylation of Ser211 in the chicken progesterone receptor modulates its transcriptional activity. *J. Biol. Chem.* **271**: 12 801–12 806.

Bai, W., Tullos, S. and Weigel, N.L. (1994) Phosphorylation of Ser530 facilitates hormone-dependent transcriptional activation of the chicken progesterone receptor. *Mol. Endocrinol.* **8**: 1465–1473.

Bakker, G.H., Setyono-Han, B., Henkelman, M.S. *et al.* (1987) Comparison of the actions of the antiprogestin mifepristone (RU486), the progestin megestrol acetate, the LH–RH-analog buserelin, and ovariectomy in treatment of rat mammary tumors. *Cancer Treat. Rep.* **71**: 1021–1027.

Bardon, S., Vignon, F., Chalbos, D. and Rochefort, H. (1985) RU486, a progestin and glucocorticoid antagonist, inhibits the growth of breast cancer cells via the progesterone receptor. *J. Clin. Endocrinol. Metab.* **60**: 692–697.

Baulieu, E.-E. (1989) Contragestion and other clinical applications of RU 486, an antiprogesterone at the receptor. *Science* **245**: 1351–1357.

Baulieu, E.-E., Alberga, A., Jung, I. *et al.* (1971) Metabolism and protein binding of sex steroids in target organs. Approach to the mechanism of hormone action. *Recent Prog. Horm. Res.* **27**: 351–419.

Bergkvist, L. and Persson, I. (1996) Hormone replacement therapy and breast cancer: a review of current knowledge. *Drug Saf.* **15**: 360–370.

Berry, M., Nunez, A.M. and Chambon, P. (1989) Estrogen-responsive element of the human pS2 gene is an imperfectly palindromic sequence. *Proc. Natl. Acad. Sci. USA.* **86**: 1218–1222.

Bitran, D., Hilvers, R.J. and Kellogg, C.K. (1991) Anxiolytic effects of 3α-hydroxy-5α[β]-pregnan-20-one: endogenous metabolites of progesterone that are active at the $GABA_A$ receptor. *Brain Res.* **561**: 157–161.

Blanco, J.C.G., Minucci, S., Lu, J. *et al.* (1998) The histone acetylase PCAF is a nuclear receptor coactivator. *Genes Dev.* **12**: 1638–1651.

Briggs, M. (1976) Biochemical effects of oral contraceptives. In: M.H. Briggs and G.A. Christie (eds) *Advances in Steroid Biochemistry and Pharmacology*, Vol. 5, pp. 66–160. Academic Press, London.

Brincat, M., Versi, E., O'Dowd, T. *et al.* (1987) Skin collagen changes in postmenopausal women receiving estradiol gel. *Maturitas* **9**: 1–5.

Brot, M.D., Akwa, Y., Purdy, R.H., Koob, G.F. and Britton, K.T. (1997) The anxiolytic-like effects of the neurosteroid allopregnanolone: interactions with $GABA_A$ receptors. *Eur. J. Pharmacol.* **325**: 1–7.

Brussaard, H.E., Leuven, G.J.A., Frolich, M., Kluft, C. and Krans, H.M.J. (1997) Short-term estrogen replacement therapy improves insulin resistance, lipids and fibrinolysis in postmenopausal women with NIDDM. *Diabetologia* **40**: 843–849.

Brzozowski, A.M., Pike, A.C.W. *et al.* (1997) Molecular basis of agonism and antagonism in the estrogen receptor. *Nature* **389**: 753–758.

Butenandt, A. (1929) Untersuchungen uber das weibliche sexual hormon. *Dtsch. Med. Wochenschr.* **55**: 2171–2173.

Butenandt, A. and Dannenbaum, H. (1934) Uber androsteron: isolierung eines neuen, physiologisch unwirksamen steroiderivatives ans mannernarn, seine verknutsung mit dehydroandrosterone und androsterone bin beitrag vurk constitution des androsterone. *Z. Physiol. Chem.* **229**, 192–208.

Carani, C., Qin, K., Simoni, M. *et al.* (1997) Effect of testosterone and estradiol in a man with aromatase deficiency. *N. Engl. J. Med.* **337**: 91–95.

Castles, C.G., Fuqua, S.A.W., Klotz, D.M. and Hill, S.M. (1993) Expression of a constitutively active estrogen receptor variant in the estrogen receptor-negative BT-20 human breast cancer cell line. *Cancer Res.* **53**: 5934–5939.

Catherino, W.H., Wolf, D.M. and Jordan, V.C. (1995) A naturally occurring estrogen receptor mutation results in increased estrogenicity of a tamoxifen analog. *Mol. Endocrinol.* **9**: 1053–1063.

Chaidarun, S.S. and Alexander, J.M. (1998) A tumor-specific truncated estrogen receptor splice variant enhances estrogen-stimulated gene expression. *Mol. Endocrinol.* **12**: 1355–1366.

Chaidarun, S.S., Klibanski, A. and Alexander, J.M. (1997) Tumor-specific expression of alternatively spliced estrogen receptor messenger ribonucleic acid variants in human pituitary adenomas. *J. Clin. Endocrinol. Metab.* **82**: 1058–1065.

Chakravarti, D., LaMorte, V.J., Nelson, M.C. *et al.* (1996) Role of CBP/P300 in nuclear receptor signalling. *Nature* **383**: 99–103.

Chan, C.M.W. and Dowsett, M. (1997) A novel estrogen receptor variant mRNA lacking exons 4 to 6 in breast carcinoma. *J. Steroid Biochem. Mol. Biol.* **62**: 419–430.

Chen, H., Lin, R.J., Schiltz, R.L. *et al.* (1997) Nuclear receptor coactivator ACTR is a novel histone acetyltransferase and forms a multimeric activation complex with P/CAF and CBP/p300. *Cell* **90**: 569–580.

Christiansen, C. and Lindsay, R. (1991) Estrogens, bone loss and preservation. *Osteoporosis Int.* **1**: 7–13.

Cole, M.P., Jones, C.T.A. and Todd, I.D.H. (1971) A new antioestrogenic agent in late breast cancer: an early clinical appraisal of ICI 46 474. *Br. J. Cancer* **25**: 270–275.

Conneely, O.M., Sullivan, W.P., Toft, D.O. *et al.* (1986) Molecular cloning of the chicken progesterone receptor. *Science* **233**: 767–770.

Conneely, O.M., Maxwell, B.L., Toft, D.O., Schrader, W.T. and O'Malley, B.W. (1987) The A and B forms of the chicken progesterone receptor arise by alternate initiation of translation of a unique mRNA. *Biochem. Biophys. Res. Commun.* **149**: 493–501.

Cooney, A.J. and Tsai, S.Y. (1994) Nuclear receptor–DNA interactions. In: M.-J. Tsai and B.W. O'Malley (eds) *Mechanism of Steroid Hormone Regulation of Gene Transcription*, pp. 25–59. R.G. Landes, Austin, Texas.

Cooper, A.J. and Stevenson, J.C. (1997) Effects on the cardiovascular system: clinical aspects. In: R. Lindsay, D.W. Dempster and V.C. Jordan (eds) *Estrogens and Antiestrogens. Basic and Clinical Aspects*, pp. 119–128. Lippincott-Raven, Philadelphia.

Corner, G.W. and Allen, W.M. (1929) Physiology of the corpus luteum. *Am. J. Physiol.* **88**: 326–339.

Cosman, F., Dempster, D. and Lindsay, R. (1997) Clinical effects of estrogens and antiestrogens on the skeleton and skeletal metabolism. In: R. Lindsay, D.W. Dempster and V.C. Jordan (eds) *Estrogens and Antiestrogens. Basic and Clinical Aspects*, pp. 151–164. Lippincott-Raven, Philadelphia.

Crook, D., Godsland, I.F., Hull, J. and Stevenson, J.C. (1997) Hormone replacement therapy with dydrogesterone and 17β-estradiol: effects on serum lipoproteins and glucose tolerance during 24 month follow up. *Br. J. Obstet. Gynaecol.* **104**: 298–304.

Csapo, A.I. (1973) Regulatory interplay of progesterone and $PGF_{2\alpha}$ in the control of the pregnant uterus. In: J.B. Josimovich (ed.) *Uterine Contractions: Side Effects of Steroid Contraceptives*, pp. 223–232. Wiley, London.

Daffada, A.I. and Dowsett, M. (1995) Tissue-dependent expression of a novel splice variant of the human estrogen receptor. *J. Steroid Biochem. Mol. Biol.* **55**: 413–421.

Danielsen, M., Hinck, L. and Ringold, G.M. (1989) Two amino acids within the knuckle of the first zinc finger specify DNA response element activation by the glucocorticoid receptor. *Cell* **57**: 1131–1138.

Darbre, P., Page, M. and King, R.J.B. (1986) Androgen regulation by the long terminal repeat of mouse mammary tumor virus. *Mol. Cell. Biol.* **6**: 2847–2854.

Denis, M., Poellinger, L., Wikstrom, A.C. and Gustafsson, J.-A. (1988) Requirement of hormone for thermal conversion of the glucocorticoid receptor to a DNA-binding state. *Nature* **333**: 686–688.

Denner, L.A., Weigel, N.L., Maxwell, B.L., Schrader, W.T. and O'Malley, B.W. (1990) Regulation of progesterone receptor-mediated transcription by phosphorylation. *Science* **250**: 1740–1743.

Ditkoff, E.C., Crary, W.G., Crisito, M. and Lobo, R.A. (1991) Estrogen improves psychological function in asymptomatic post-menopausal women. *Obstet. Gynecol.* **78**: 991–995.

Doisy, E.A, Veler C.D. and Thayer, S. (1929) Folliculin from urine of pregnant women. *Am. J. Physiol.* **90**: 329–330.

Dotzlaw, H., Alkhalaf, M. and Murphy, L.C. (1992) Characterization of estrogen receptor variant mRNAs from human breast cancers. *Mol. Endocrinol.* **6**: 773–785.

Eddy, E.M., Washburn, T.F., Bunch, D.O. *et al.* (1996) Targeted disruption of the estrogen receptor gene in male mice causes alteration of spermatogenesis and infertility. *Endocrinology* **137**: 4796–4805.

Enmark, E., Pelto-Huikko, M., Grandien, K. *et al.* (1997) Human estrogen receptor β-gene structure, chromosomal localization, and expression pattern. *J. Clin. Endocrinol. Metab.* **82**: 4258–4265.

Ernst, M., Heath, J.K. and Rodan, G.A. (1989) Estradiol effects on proliferation, messenger ribonucleic acid for collagen and insulin-like growth factor-1, and parathyroid hormone-stimulated adenylate cyclase activity in osteoblastic cells from calvariae and long bones. *Endocrinology* **125**: 825–833.

Ernster, V.L., Huggins, G.R., Hulka, B.S., Kelsey, J.L. and Schottenfeld, F. (1988) Benefits and risks of menopausal estrogen and/or progestin hormone use. *Prev. Med.* **17**: 201–233.

Fantl, J.A., Cardozo, L.D., Ekberg, J., McClish, D.K. and Heimer, G. (1994) Estrogen therapy in the management of urinary incontinence in postmenopausal women. A meta-analysis. *Obstet. Gynecol.* **83**: 12–18.

Finkelstein, J.S., O'Dea, L.S.L., Whitcomb, R.W. and Crowley, W.F. (1991) Sex steroid control of gonadotropin secretion in the human male. II. Effects of estradiol administration in normal and gonadotropin-releasing hormone-deficient men. *J. Clin. Endocrinol. Metab.* **70**: 621–628.

Fisher, B., Costantino, J.P., Wickerham, D.L. *et al.* (1998) Tamoxifen for prevention of breast cancer: report of the National Surgical Adjuvant Breast and Bowel Project P-1 study. *J. Natl. Cancer Inst.* **90**: 1371–1388.

Friend, K.E., Ang, L.W. and Shupnik, M.A. (1995) Estrogen regulates the expression of several different estrogen receptor mRNA isoforms in rat pituitary. *Proc. Natl. Acad. Sci. USA* **92**: 4367–4371.

Friend, K.E., Resnick, E.M., Ang, L.W. and Shupnik, M.A. (1997) Specific modulation of estrogen receptor mRNA isoforms in rat pituitary throughout the estrous cycle and in response to steroid hormones. *Mol. Cell. Endocrinol.* **131**: 147–155.

Fuqua, S.A.W., Fitzgerald, S.D., Chamness, G.C. *et al.* (1991) Variant human breast tumor estrogen receptor with constitutive transcriptional activity. *Cancer Res.* **51**: 105–109.

Fuqua, S.A.W., Fitzgerald, S.D., Allred, D.C. *et al.* (1992) Inhibition of estrogen receptor action by a naturally occurring variant in human breast tumors. *Cancer Res.* **52**: 482–486.

Fuqua, S.A.W., Allred, D.C. and Auchus, R.J. (1993) Expression of estrogen receptor variants. *J. Cell. Biochem.* **51**: 135–139.

Ge, R.-S., Shan, L.-X. and Hardy, M.P. (1996) Pubertal development of Leydig cells. In:

A.H. Payne, M.P. Hardy and L.D. Russell (eds) *The Leydig Cell*, pp. 159–172. Cache River Press, Clearwater, Florida.

Gilligan, D.M., Badar, D.M., Panza, J.A., Quyyumi, A.A. and Cannon, R.O. (1994) Acute vascular effects of estrogen in postmenopausal women. *Circulation* **90**: 786–791.

Girasole, G., Jilka, R.L., Passeri, G. *et al.* (1992) 17β-Estradiol inhibits interleukin-6 production by bone marrow-derived stromal cells and osteoblasts *in vitro*: a potential mechanism for the antiosteoporotic effect of estrogens. *J. Clin. Invest.* **89**: 883–891.

Glasier, A., Thong, K.J., Dewar, M., Mackie, M. and Baird, D.T. (1992) Mifepristone (RU 486) compared with high-dose estrogen and progestogen for emergency postcoital contraception. *N. Engl. J. Med.* **327**: 1041–1044.

Glasscock, R.F. and Hoekstra, W.G. (1959) Selective accumulation of tritium-labeled hexosterol by the reproductive organs of immature female goats and sheep. *Biochem. J.* **72**: 673–682.

Goodman, A.L. and Hodgen, G.D. (1983) The ovarian triad of the primate menstrual cycle. *Recent Prog. Horm. Res.* **39**: 1–73.

Gorski, J., Toft, D., Shyamala, G., Smith, D. and Notides, A. (1968) Hormone receptors: studies on the interaction of estrogen with the uterus. *Recent Prog. Horm. Res.* **24**: 45–80.

Gotteland, M., Desauty, G., Delarue, J.C., Liu, L. and May, E. (1995) Human estrogen receptor messenger RNA variants in both normal and tumor breast tissues. *Mol. Cell. Endocrinol.* **112**: 1–13.

Graham, M.L., II, Krett, N.L., Miller, L.A. *et al.* (1990) T47Dco cells, genetically unstable and containing estrogen receptor mutations, are a model for the progression of breast cancers to hormone resistance. *Cancer Res.* **50**: 6208–6217.

Green, S. and Chambon, P. (1987) Estradiol induction of a glucocorticoid-responsive gene by a chimeric receptor. *Nature* **325**: 75–78.

Green, S., Walter, P., Kumar, V. *et al.* (1986) Human estrogen receptor cDNA: sequence, expression and homology to v-*erb-A*. *Nature* **320**: 134–139.

Grunberg, S.M., Weiss, M.H., Spitz, I.M. *et al.* (1991) Treatment of unresectable meningioma with the anti-progestational agent mifepristone. *J. Neurosurg.* **74**: 861–866.

Hamilton, H. (1962) Inhibition of protein synthesis and some quantifications of early estrogen action and response. *Science* **138**: 989.

Hamilton, T.H. (1968) Control by estrogen of genetic transcription and translation. Binding to chromatin and stimulation of nucleolar RNA synthesis are primary events in the early estrogen action. *Science* **161**: 649–661.

Harris, S.T., Genant, H.K., Baylink, D.J. *et al.* (1991) The effects of estrone (Ogen) on spinal bone density of postmenopausal women. *Arch. Intern. Med.* **151**: 1980–1984.

Hartmann, M. and Wettstein, A. (1934) Zur Kenntnis der corpus-luteum hormone. *Helv. Chim. Acta* **17**: 1365–1372.

Heinzel, T., Lavinsky, R.M., Mullen, T.-M. *et al.* (1997) A complex containing N-CoR, mSin3 and histone deacetylase mediates transcriptional repression. *Nature* **387**: 43–48.

Hess, R.A., Bunick, D., Lee, K.-H. *et al.* (1997) A role of estrogens in the male reproductive system. *Nature* **390**: 509–512.

Horwitz, K.B. (1985) The antiprogestin RU 38 486: receptor-mediated progestin versus antiprogestin actions screened in estrogen-insensitive T47Dco human breast cancer cells. *Endocrinology* **116**: 2236–2245.

Hoshino, S., Inoue, S., Hosoi, T. *et al.* (1995) Demonstration of isoforms of the estrogen receptor in the bone tissues and in osteoblastic cells. *Calcif. Tissue Int.* **57**: 466–468.

Howell, A., DeFriend, D., Robertson, J. *et al.* (1996) Pharmacokinetics, pharmacological and anti-tumor effects of the specific anti-estrogen ICI 182 780 in women with advanced breast cancer. *Br. J. Cancer* **74**: 300–308.

Hu, C., Hyder, S.M., Needleman, D.S. and Baker, V.V. (1996) Expression of estrogen receptor variants in normal and neoplastic human uterus. *Mol. Cell. Endocrinol.* **118**: 173–179.

Ignar-Trowbridge, D.M., Nelson, K.G., Bidwell, M.C. *et al.* (1992) Coupling of dual signaling pathways: epidermal growth factor action involves the estrogen receptor. *Proc. Natl. Acad. Sci. USA* **89**: 4658–4662.

Ignar-Trowbridge, D.M., Teng, C.T., Ross, K.A., Parker, M.G., Korach, K.S. and McLachlan, J.A. (1993) Peptide growth factors elicit estrogen receptor-dependent transcriptional activation of an estrogen-responsive element. *Mol. Endocrinol.* **7**: 992–998.

Inoue, S., Hoshino, S., Miyoshi, H. *et al.* (1996) Identification of a novel isoform of estrogen receptor, a potential inhibitor of estrogen action, in vascular smooth muscle cells. *Biochem. Biophys. Res. Commun.* **219**: 766–772.

Iosif, S., Batra, S., Ek, A. and Astedt, B. (1981) Oestrogen receptors in the human female lower urinary tract. *Am. J. Obstet. Gynecol.* **141**: 817–820.

Ishimi, Y., Miyaura, C., Jin, C.H. *et al.* (1990) IL-6 is produced by osteoblasts and induces bone resorption. *J. Immunol.* **145**: 3297–3303.

Jensen, E.V. (1965) Mechanism of estrogen action in relation to carcinogenesis. *Canadian Cancer Conf.* **6**: 143–165.

Jensen, E.V. and Jacobsen, H.I. (1962) Basic guides to the mechanism of estrogen action. *Recent Prog. Horm. Res.* **18**: 387–414.

Jensen, E.V., Suzuki, T., Kawashima, T., Stumpf, W.E., Jungblut, P.W. and De Sombre, E. R. (1968) Two-step mechanism for the interaction of estradiol with rat uterus. *Proc. Natl. Acad. Sci. USA* **59**: 632–638.

Jilka, R.L., Hangoc, G., Girasole, G. *et al.* (1992) Increased osteoclast development after estrogen loss: mediation by interleukin-6. *Science* **257**: 88–91.

Karnik, P.S., Kulkarni, S., Liu, X.-P., Budd, G.T. and Bukowski, R.M. (1994) Estrogen receptor mutations in tamoxifen-resistant breast cancer. *Cancer Res.* **54**: 349–353.

Kastner, P., Bocquel, M.T., Turcotte, B. *et al.* (1990) Transient expression of human and chicken progesterone receptors does not support alternative translational initiation from a single mRNA as the mechanism generating two receptor isoforms. *J. Biol. Chem.* **265**: 12 163–12 167.

Kekkonen, R., Lahteenmaki, P., Luukkainen, T. and Tuominen, J. (1993) Sequential regimen of the antiprogesterone RU486 and synthetic progestin for contraception. *Fertil. Steril.* **60**: 610–615.

Kettel, L.M., Murphy, A.A., Morales, A.J. and Yen, S.S.C. (1998) Preliminary report on the treatment of endometriosis with low-dose mifepristone (RU 486). *Am. J. Obstet. Gynecol.* **178**: 1151–1156.

Khovidhunkit, W. and Shoback, D.M. (1999) Clinical effects of raloxifene hydrochloride in women. *Ann. Intern. Med.* **130**: 431–439.

Kiel, D.P., Felson, D.T., Anderson, J.J., Wilson, P.W.F. and Moskowitz, M.A. (1987) Hip fracture and the use of estrogen in postmenopausal women: the Framingham Study. *N. Engl. J. Med.* **317**: 1169–1174.

King, R.J.B., Gordon, J. and Inman D.R. (1965) The intracellular localization of oestrogen in rat tissues. *J. Endocrinol.* **32**: 9–15.

Kirton, K.T., Niswender, G.G., Midgley, A.R., Jr., Jaffe, R.B. and Forbes, A.D. (1970) Serum luteinizing hormone and progesterone concentration during the menstrual cycle of the rhesus monkey. *J. Clin. Endocrinol. Metab.* **30**: 105–110.

Klein-Hitpass, L., Schorpp, M., Wagner, U. and Ryffel, G.U. (1986) An estrogen-responsive element derived from the 5' flanking region of the *Xenopus* vitellogenin A2 gene functions in transfected human cells. *Cell* **46**: 1053–1061.

Klein-Hitpass, L., Ryffel, G.U., Heitlinger, E. and Cato, A.C.B. (1988) A 13 bp palindrome is a functional estrogen responsive element and interacts specifically with estrogen receptor. *Nucleic Acids Res.* **16**: 647–663.

Knauer, E. (1900) Die ovarientransplantation. *Arch. Gynakol.* **60**: 322–376.

Knight, W.A., III, Livingston, R.B., Gregory, E.J. and McGuire, W.L. (1977) Estrogen receptor as an independent prognostic factor for early recurrence in breast cancer. *Cancer Res.* **37**: 4669–4671.

Koehorst, S.G.A., Cox, J.J., Donker, G.H. *et al.* (1994) Functional analysis of an alternatively spliced estrogen receptor lacking exon 4 isolated from MCF-7 breast cancer cells and meningioma tissue. *Mol. Cell. Endocrinol.* **101**: 237–245.

Krege, J.H., Hodgin, J.B., Couse, J.F. *et al.* (1998) Generation and reproductive phenotypes of mice lacking estrogen receptor β. *Proc. Natl. Acad. Sci. USA* **95**: 15 677–15 682.

Kuiper, G.G.J.M., Enmark, E., Pelto-Huikko, M., Nilsson, S. and Gustafsson, J.-A. (1996) Cloning of a novel estrogen receptor expressed in rat prostate and ovary. *Proc. Natl. Acad. Sci. USA* **93**: 5925–5930.

Kuiper, G.G.J.M., Carlsson, B., Grandien, K. *et al.* (1997) Comparison of the ligand binding specificity and transcript tissue distribution of estrogen receptors α and β. *Endocrinology* **138**: 863–870.

Kuiper, G.G.J.M., Carlquist, M. and Gustafsson, J.-A. (1998) Estrogen is a male and female hormone. *Sci. Med.* **5**: 36–45.

Larsson, B. Andersson, K.E., Batra, S., Mattiasson, A. and Sjoegren, C. (1984) Effects of estradiol on norepinephrine-induced contraction, alpha-adrenoceptor number and nor-epinephrine content in the female rabbit urethra. *J. Pharmacol. Exp. Ther.* **229**: 557–563.

Laumas, K.R. and Farooq, A. (1966) The uptake *in vivo* of [1,2-^{3}H]-progesterone by the brain and genital tract of the rat. *J. Endocrinol.* **36**: 95–96.

Leslie, K.K., Tasset, D.M. and Horwitz, K.B. (1992) Functional analysis of a mutant estrogen receptor isolated from T47Dco breast cancer cells. *Am. J. Obstet. Gynecol.* **166**: 1053–1061.

Leygue, E.R., Watson, P.H. and Murphy, L.C. (1996a) Estrogen receptor variants in normal human mammary tissue. *J. Natl. Cancer Inst.* **88**: 284–290.

Leygue, E., Huang, A., Murphy, L.C. and Watson, P.H. (1996b) Prevalence of estrogen receptor variant messenger RNAs in human breast cancer. *Cancer Res.* **56**: 4324–4327.

Leygue, E., Dotzlaw, H., Watson, P.H. and Murphy, L.C. (1996c) Identification of novel exon-deleted progesterone receptor variant mRNAs in human breast tissue. *Breast Cancer Res. Treat.* **228**: 63–68.

Leygue, E., Dotzlaw, H., Watson, P.H. and Murphy, L.C. (1999) Expression of estrogen receptor β1, β2, and β5 messenger RNAs in human breast tissue. *Cancer Res.* **59**: 1175–1179.

Lindsay, R., Hart, D.M., Aitken, J.M., MacDonald, E.B., Anderson, J.B. and Clarke, A.C. (1976) Long-term prevention of postmenopausal osteoporosis by oestrogen. *Lancet* **307**: 1038–1041.

Lindsay, R., Hart, D.M. and Clark, D.M. (1984) The minimum effective dose of estrogen for prevention of postmenopausal bone loss. *Obstet. Gynecol.* **63**: 759–763.

Liu, T.M.Y. and Davis, J.W. (1967) Induction of lactation by ovariectomy of pregnant rats. *Endocrinology* **80**: 1043–1050.

Lu, B., Leygue, E., Dotzlaw, H., Murphy, L.J., Murphy, L.C. and Watson, P.H. (1998)

Estrogen receptor-β mRNA variants in human and murine tissues. *Mol. Cell. Endocrinol.* **138**: 199–203.

Lubahn, D.B., Moyer, J.S., Golding, T.S., Couse, J.F., Korach, K.S. and Smithies, O. (1993) Alteration of reproductive function but not prenatal sexual development after insertional disruption of the mouse estrogen receptor gene. *Proc. Natl. Acad. Sci. USA* **90**: 11 162–11 166.

Lufkin, E.G., Wahner, H.W., O'Fallon, W.M. *et al.* (1992) Treatment of postmenopausal osteoporosis with transdermal estrogen. *Ann. Intern. Med.* **117**: 1–9.

Luisi, B.F., Xu, W.X., Otwinowski, Z., Freedman, L.P., Yamamoto, K.R. and Sigler, P.B. (1991) Crystallographic analysis of the interaction of the glucocorticoid receptor with DNA. *Nature* **352**: 497–505.

Lydon, J.P., DeMayo, F.J., Funk, C.R. *et al.* (1995) Mice lacking progesterone receptor exhibit pleiotropic reproductive abnormalities. *Genes Dev.* **9**: 2266–2278.

Mader, S., Kumar, V., de Verneuil, H. and Chambon, P. (1989) Three amino acids of the estrogen receptor are essential to its ability to distinguish an estrogen from a glucocorticoid-responsive element. *Nature* **338**: 271–274.

Madsen, M.W., Reiter, B.E. and Lykkesfeldt, A.E. (1995) Differential expression of estrogen receptor mRNA splice variants in the tamoxifen resistant human breast cancer cell line, MCF-7/TAMR-1 compared to the parental MCF-7 cell line. *Mol. Cell. Endocrinol.* **109**: 197–207.

Maheux, R., Naud, F., Rioux, M. *et al.* (1994) A randomized, double-blind, placebo-controlled study on the effect of conjugated estrogens on skin thickness. *Am. J. Obstet. Gynecol.* **170**: 642–649.

Mani, S.K., Allen, J.M.C., Clark, J.H., Blaustein, J.D. and O'Malley, B.W. (1994) Convergent pathways for steroid hormone- and neurotransmitter-induced rat sexual behavior. *Science* **265**: 1246–1249.

Marshall, F.H.A. and Jolly, W.A. (1905) Contributions to the physiology of mammalian reproduction. *Philos. Trans. R. Soc. Lond. [Biol.]* **198**: 99–142.

Martinez-Balbas, M.A., Bannister, A.J., Martin, K., Haus-Seuffert, P., Meisterernst, M. and Kouzarides, T. (1998) The acetyltransferase activity of CBP stimulates transcription. *EMBO J.* **17**: 2886–2893.

McGuire, W.L., Horwitz, K.B., Zava, D.T. and Garola, R.E. (1979) Estrogen and progesterone receptors in human breast cancer. *Steroid Recept. Manag. Cancer* **1**: 31–40.

McGuire, W.L., Chamness, G.C. and Fuqua, S.A.W. (1991) Estrogen receptor variants in clinical breast cancer. *Mol. Endocrinol.* **5**: 1571–1577.

Miksicek, R.J., Lei, Y. and Wang, Y. (1993) Exon skipping gives rise to alternatively spliced forms of the estrogen receptor in breast tumor cells. *Breast Cancer Res. Treat.* **26**: 163–174.

Misao, R., Sun, W.-S., Iwagaki, S., Fujimoto, J. and Tamaya, T. (1998) Identification of various exon-deleted progesterone receptor mRNAs in human endometrium and ovarian endometriosis. *Biochem. Biophys. Res. Commun.* **252**: 302–306.

Mishell, D.R., Jr., Nakamura, R.M., Crosignani, P.G. *et al.* (1971) Serum gonadotropin and steroid patterns during the normal menstrual cycle. *Amer. J. Obstet. Gynecol.* **111**: 60–65.

Mito, K., Tamura, T., Hosokawa, K., Kondo, T., Yamamoto, T. and Honjo, H. (1998) Expression of exon 5 deleted estrogen receptor variant messenger RNA in human uterine myometrium and leiomyoma. *J. Steroid Biochem. Mol. Biol.* **67**: 9–15.

Monroe, S.E., Atkinson, L.E. and Knobil, E. (1970) Patterns of circulating luteinizing hormone and their relation to plasma progesterone levels during the menstrual cycle of the rhesus monkey. *Endocrinology* **87**: 453–455.

Moore, J.T., McKee, D.D., Slentz-Kesler, K. *et al.* (1998) Cloning and characterization of human estrogen receptor β isoforms. *Biochem. Biophys. Res. Commun.* **247**: 75–78.

Morishima, A., Grumbach, M.M., Simpson, E.R., Fisher, C. and Qin, K. (1995) Aromatase deficiency in male and female siblings caused by a novel mutation and the physiological role of estrogens. *J. Clin. Endocrinol. Metab.* **80**: 3689–3698.

Mosselman, S., Polman, J. and Dijkema, R. (1996) ERβ: identification and characterization of a novel human estrogen receptor. *FEBS Lett.* **392**: 49–53.

Moutsatsou, P., Kassi, E., Creatsas, G., Coulocheri, S., Scheller, K. and Sekeris, C. E. (1998) Detection of estrogen receptor variants in endometrium, myometrium, leiomyoma, and peripheral blood mononuclear cells. Comparison to variants present in breast cancer. *J. Cancer Res. Clin. Oncol.* **124**: 478–484.

Mueller, G.C. (1960) Biochemical parameters of estrogen action. In: G. Pincus and E.P. Vollmer (eds) *Biological Activities of Steroids in Relation to Cancer*, pp. 129–145. Academic Press, New York.

Murphy, A.A., Kettel, L.M., Morales, A.J., Roberts, V.J. and Yen, S.C. (1993) Regression of uterine leiomyomata in response to the antiprogesterone RU-486. *J. Clin. Endocrinol. Metab.* **76**: 513–517.

Murphy, L.C. and Dotzlaw, H. (1989) Variant estrogen receptor mRNA species detected in human breast cancer biopsy samples. *Mol. Endocrinol.* **3**: 687–693.

Murphy, L.C., Wang, M., Coutts, A. and Dotzlaw, H. (1996) Novel mutations in the estrogen receptor messenger RNA in human breast cancers. *J. Clin. Endocrinol. Metab.* **81**: 1420–1427.

Murphy, L.C., Dotzlaw, H., Leygue, E., Douglas, D., Coutts, A. and Watson, P.H. (1997) Estrogen receptor variants and mutations. *J. Steroid Biochem. Mol. Biol.* **62**: 363–372.

Nagy, L., Kao, H.-Y., Chakravarti, D. *et al.* (1997) Nuclear receptor repression mediated by a complex containing SMRT, mSin3A, and histone deacetylase. *Cell* **89**: 373–380.

Nilas, L. and Christiansen, C. (1988) Rates of bone loss in normal women: evidence of accelerated trabecular bone loss after the menopause. *Eur. J. Clin. Invest.* **18**: 529–534.

Noteboom, W.D. and Gorski, J. (1963) An early effect of estrogen on protein synthesis. *Proc. Natl. Acad. Sci. USA* **50**: 250–255.

O'Malley, B.W., McGuire, W.L., Kohler, P.O. and Korenman, S.G. (1969) Studies on the mechanism of steroid hormone regulation of synthesis of specific proteins. *Recent Prog. Horm. Res.* **25**: 105–160.

Oro, A.E., Hollenberg, S.M. and Evans, R.M. (1988) Transcriptional inhibition by a glucocorticoid receptor-β-galactosidase fusion protein. *Cell* **55**: 1109–1114.

Osterlund, M., Kuiper, G.G.J.M., Gustafsson, J.-A. and Hurd, Y.L. (1998) Differential distribution and regulation of estrogen receptor-α and -β mRNA within the female rat brain. *Mol. Brain Res.* **54**: 175–180.

Paganini-Hill, A. and Henderson, V.W. (1996) Estrogen replacement therapy and risk of Alzheimer disease. *Arch. Intern. Med.* **156**: 2213–2217.

Park, W., Choi, J.-J., Hwang, E.-S. and Lee, J.-H. (1996) Identification of a variant estrogen receptor lacking exon 4 and its coexpression with wild-type estrogen receptor in ovarian carcinomas. *Clin. Cancer Res.* **2**: 2029–2035.

Pauerstein, C.J., Eddy, C.A., Croxatto, H.D., Hess, R., Siler-Khodr, T.M. and Croxatto, H.B. (1978) Temporal relationships of estrogen, progesterone, and luteinizing hormone levels to ovulation in women and infrahuman primates. *Am. J. Obstet. Gynecol.* **130**: 876–886.

Perlmann, T., Rangarajan, P.N., Umesono, K. and Evans, R.M. (1993) Determinants for selective RAR and TR recognition of direct repeat HREs. *Genes Dev.* **7**: 1411–1422.

Perrault, D., Eisenhauer, E.A., Pritchard, K.I. *et al.* (1996) Phase II study of the progesterone antagonist mifepristone in patients with untreated metastatic breast carcinoma: a National Cancer Institute of Canada Clinical Trials Group study. *J. Clin. Oncol.* **14**: 2709–2712.

Pettersson, K., Grandien, K., Kuiper, G.G.J.M. and Gustafsson, J.-A. (1997) Mouse estrogen receptor β forms estrogen response element-binding heterodimers with estrogen receptor α. *Mol. Endocrinol.* **11**: 1486–1496.

Pfeffer, U. (1996) Estrogen receptor mRNA variants; do they have a physiological role? *Ann. N. Y. Acad. Sci.* **784**: 304–313.

Pfeffer, U., Fecarotta, E., Castagnetta, L. and Vidali, G. (1993) Estrogen receptor variant messenger RNA lacking exon 4 in estrogen-responsive human breast cancer cell lines. *Cancer Res.* **53**: 741–743.

Pfeffer, U., Fecarotta, E. and Vidali, G. (1995) Coexpression of multiple estrogen receptor variant messenger RNAs in normal and neoplastic breast tissues and in MCF-7 cells. *Cancer Res.* **55**: 2158–2165.

Picard, D., Khursheed, B., Garabedian, M.J., Fortin, M.G., Lindquist, S. and Yamamoto, K.R. (1990) Reduced levels of hsp90 compromise steroid receptor action *in vivo*. *Nature* **348**: 166–168.

Pink, J.J., Jiang, S.-Y., Fritsch, M. and Jordan, V.C. (1995) An estrogen-independent MCF-7 breast cancer cell line which contains a novel 80-kilodalton estrogen receptor-related protein. *Cancer Res.* **55**: 2583–2590.

Pink, J.J., Wu, S.-Q., Wolf, D.M., Bilimoria, M.M. and Jordan, V.C. (1996) A novel 80 kDa human estrogen receptor containing a duplication of exons 6 and 7. *Nucleic Acids Res.* **24**: 962–969.

Poli, V., Balena, R., Fattori, E. *et al.* (1994) Interleukin-6 deficient mice are protected from bone loss caused by estrogen depletion. *EMBO J.* **13**: 1189–1196.

Power, R.F., Mani, S.K., Codina, J., Conneely, O.M. and O'Malley, B.W. (1991) Dopaminergic and ligand-independent activation of steroid hormone receptors. *Science* **254**: 1636–1639.

Powles, T., Eeles, R., Ashley, S. *et al.* (1998) Interim analysis of the incidence of breast cancer in the Royal Marsden Hospital tamoxifen randomized chemoprevention trial. *Lancet* **352**: 98–101.

Pratt, W.B. (1987) Transformation of glucocorticoid and progesterone receptors to the DNA-binding state. *J. Cell. Biochem.* **35**: 51–68.

Pratt, W.B. and Toft, D.O. (1997) Steroid receptor interactions with heat shock protein and immunophilin chaperones. *Endocr. Rev.* **18**: 306–360.

Pritchard, K. (1997) Effects on breast cancer: clinical aspects. In: R. Lindsay, D.W. Dempster and V.C. Jordan (eds) *Estrogens and Antiestrogens. Basic and Clinical Aspects*, pp. 175–210. Lippincott-Raven, Philadelphia.

Proudler, A.J., Felton, C.V. and Stevenson, J.C. (1992) Aging and the response of plasma insulin, glucose and C-peptide concentrations to intravenous glucose in postmenopausal women. *Clin. Sci.* **83**: 489–494.

Proudler, A.J., Hasib Ahmed, A.I., Crook, D., Fogelman, I., Rymer, J.M. and Stevenson, J.C. (1995) Hormone replacement therapy and serum angiotensin-converting-enzyme activity in postmenopausal women. *Lancet* **346**: 89–90.

Puia, G., Santi, M., Vicini, S., Pritchett, D.B., Purdy, R.H. and Costa, E. (1990) Neurosteroids act on recombinant human GABAA receptors. *Neuron* **4**: 759–765.

Rannevik, G., Jeppsson, S., Johnell, O., Bjerre, B., Laurell-Borulf, Y. and Svanberg, L.

(1995) A longitudinal study of the perimenopausal transition: altered profiles of steroid and pituitary hormones, SHBG and bone mineral density. *Maturitas* **21**: 103–113.

Rey, J.M., Pujol, P., Dechaud, H., Edouard, E., Hedon, B. and Maudelonde, T. (1998) Expression of estrogen receptor-α splicing variants and estrogen receptor-β in endometrium of infertile patients. *Mol. Hum. Reprod.* **4**: 641–647.

Richelson, L.S., Wahner, H.W., Melton, J.L. and Riggs, B.L. (1984) Relative contributions of aging and estrogen deficiency to post-menopausal bone loss. *N. Engl. J. Med.* **311**: 1273–1275.

Richer, J.K., Lange, C.A., Wierman, A.M. *et al.* (1998) Progesterone receptor variants found in breast cells repress transcription by wild-type receptors. *Breast Cancer Res. Treat.* **48**: 231–241.

Riemer, R.K., Goldfien, A.C., Goldfien, A. and Roberts, J.M. (1986) Rabbit uterine oxytocin receptors and *in vitro* contractile response: abrupt changes at term and the role of eicosanoids. *Endocrinology* **119**: 699–709.

Riis, B., Thomsen, K., Strom, V. and Christiansen, C. (1987) The effect of percutaneous estradiol and natural progesterone on postmenopausal bone loss. *Am. J. Obstet. Gynecol.* **156**: 61–65.

Roodi, N., Bailey, L.R., Kao, W.-Y. *et al.* (1995) Estrogen receptor gene analysis in estrogen receptor-positive and receptor-negative primary breast cancer. *J. Natl. Cancer Inst.* **87**: 446–451.

Rosenkranz, K., Hinney, A., Ziegler, A. *et al.* (1998) Systematic mutation screening of the estrogen receptor beta gene in probands of different weight extremes: identification of several genetic variants. *J. Clin. Endocrinol. Metab.* **83**: 4524–4527.

Sanchez, E.R., Toft, D.O., Schlesinger, M.J. and Pratt, W.B. (1985) Evidence that the 90-kDa phosphoprotein associated with the untransformed L-cell glucocorticoid receptor is a murine heat shock protein. *J. Biol. Chem.* **260**: 12398–12401.

Scheidereit, C. and Beato, M. (1984) Contacts between hormone receptor and DNA double helix within a glucocorticoid regulatory element of mouse mammary tumor virus. *Proc. Natl. Acad. Sci. USA* **81**: 3029–3033.

Scheidereit, C., Geisse, S., Westphal, H.M. and Beato, M. (1983) The glucocorticoid receptor binds to defined nucleotide sequences near the promoter of mouse mammary tumor virus. *Nature* **304**: 749–752.

Schrader, W.T. and O'Malley, B.W. (1972) Progesterone-binding components of chick oviduct. IV. Characterization of purified subunits. *J. Biol. Chem.* **247**: 51–59.

Schreihofer, D.A., Resnick, E.M., Soh, A.Y. and Shupnik, M.A. (1999) Transcriptional regulation by a naturally occurring truncated rat estrogen receptor (ER), truncated ER product-1 (TERP-1). *Mol. Endocrinol.* **13**: 320–329.

Schuh, S., Yonemoto, W., Brugge, J. *et al.* (1985) A 90,000-dalton binding protein common to both steroid receptors and the Rous sarcoma virus transforming protein, pp60v-src. *J. Biol. Chem.* **260**: 14292–14296.

Schwabe, J.W.R., Chapman, L., Finch, J.T. and Rhodes, D. (1993) The crystal structure of the estrogen receptor DNA-binding domain bound to DNA: how receptors discriminate between their response elements. *Cell* **75**: 567–578.

Semmens, J.P. and Wagner, G. (1982) Estrogen deprivation and vaginal function in postmenopausal women. *JAMA* **248**: 445–448.

Shapiro, S., Kelly, J.P., Rosenberg, L. *et al.* (1985) Risk of localized and widespread endometrial cancer in relation to recent and discontinued use of conjugated estrogens. *N. Engl. J. Med.* **313**: 969–972.

Sharpe, R.M. (1998) The roles of estrogen in the male. *Trends Endocrinol. Metab.* **9**: 371–377.

Sherman, M.R., Corvol, P.L. and O'Malley, B.W. (1970) Progesterone-binding components of chick oviduct. I. Preliminary characterization of cytoplasmic components. *J. Biol. Chem.* **245**: 6085–6096.

Sherwin, B.B. (1997) Estrogenic effects on the central nervous system: clinical aspects. In: R. Lindsay, D.W. Dempster and V.C. Jordan (eds) *Estrogens and Antiestrogens: Basic and Clinical Aspects*, pp. 75–87. Lippincott-Raven, Philadelphia.

Sherwin, B.B. and Gelfand, M.M. (1985) Sex steroids and affect in the surgical menopause: a double-blind cross-over study. *Psychoneuroendocrinology* **10**: 325–335.

Shiau, A.K., Barstad, D., Loria, P.M. *et al.* (1998) The structural basis of estrogen receptor/coactivator recognition and the antagonism of this interaction by tamoxifen. *Cell* **95**: 927–937.

Skipper, J.K., Young, L.J., Bergeron, J.M., Tetzlaff, M.T., Osborn, C.T. and Crews, D. (1993) Identification of an isoform of the estrogen receptor messenger RNA lacking exon four and present in the brain. *Proc. Natl. Acad. Sci. USA* **90**: 7172–7175.

Sledge, G.W., Jr. and McGuire, W.L. (1983) Steroid hormone receptors in human breast cancer. *Adv. Cancer Res.* **38**: 61–75.

Slotta, K.H., Ruschig, H. and Fels, E. (1934) Die reindarstellung der hormone aus dem corpus luteum. *Bev. Dtsch. Chem. Ges.* **67**: 1207–1208.

Smith, E.P., Boyd, J., Frank, G.R. *et al.* (1994) Estrogen resistance caused by a mutation in the estrogen-receptor gene in a man. *N. Engl. J. Med.* **331**: 1056–1061.

Smith, S.S. (1994) Female sex steroid hormones: from receptors to networks to performance-actions on the sensorimotor system. *Prog. Neurobiol.* **44**: 55–86.

Smith, S.S., Gong, Q.H., Hsu, F.-C., Markowitz, R.S., French-Mullen, J.M.H. and Li, X. (1998) $GABA_A$ receptor $\alpha 4$ subunit suppression prevents withdrawal properties of an endogenous steroid. *Nature* **392**: 926–930.

Spencer, T.E., Jenster, G., Burcin, M.M. *et al.* (1997) Steroid receptor coactivator-1 is a histone acetyltransferase. *Nature* **389**: 194–198.

Stampfer, M.J. and Colditz, G. (1991) Estrogen replacement therapy and coronary heart disease: a quantitative assessment of the epidemiologic evidence. *Prev. Med.* **20**: 47–63.

Stampfer, M.J., Colditz, G.A., Willett, W.C. *et al.* (1991) Postmenopausal estrogen therapy and cardiovascular disease. Ten-year follow-up from the Nurses' Health Study. *N. Engl. J. Med.* **325**: 756–762.

Stiemer, B. and Elger, W. (1990) Cervical ripening of the rat in dependence on endocrine milieu; effects of antigestagens. *J. Perinat. Med.* **18**: 419–429.

Strahle, U., Klock, G. and Schutz, G. (1987) A DNA sequence of 15 base pairs is sufficient to mediate both glucocorticoid and progesterone induction of gene expression. *Proc. Natl. Acad. Sci. USA* **84**: 7871–7875.

Strahle, U., Boshart, M., Klock, G., Stewart, F. and Schutz, G. (1989) Glucocorticoid- and progesterone-specific effects are determined by differential expression of the respective hormone receptors. *Nature* **339**: 629–632.

Talalay, P. (1962) Studies on the placental 17β-hydroxy steroid dehydrogenase. In: *On Cancer and Hormones: Essays in Experimental Biology*, pp. 271–289. University of Chicago Press, Chicago.

Talwar, G.P., Segal, S.J., Evans, A. and Davidson, O.W. (1964) The binding of estradiol in the uterus: a mechanism for derepression of RNA synthesis. *Proc. Natl. Acad. Sci. USA* **52**: 1059–1066.

Tanenbaum, D.M., Wang, Y., Williams, S.P. and Sigler, P.B. (1998) Crystallographic comparison of the estrogen and progesterone receptor's ligand binding domains. *Proc. Natl. Acad. Sci. USA* **95**: 5998–6003.

Thorneycroft, I.H., Mishell, D.R., Jr., Stone, S.C., Kharma, K.M. and Nakamura, R.M. (1971) Relation of serum 17-hydroxyprogesterone and 17β-estradiol levels during the human menstrual cycle. *Am. J. Obstet. Gynecol.* **111**: 947–951.

Toft, D. and Gorski, J. (1966) A receptor molecule for estrogens: isolation from the rat uterus and preliminary characterization. *Proc. Natl. Acad. Sci. USA* **55**: 1574–1581.

Tucker, H.A. (1974) General endocrinological control of lactation. In: B.L. Larson and V.R. Smith (eds) *Lactation: A Comprehensive Treatise*, pp. 277–326. Academic Press, New York.

Turner, R.T., Riggs, B.L. and Spelsberg, T.C. (1994) Skeletal effects of estrogen. *Endocr. Rev.* **15**: 275–300.

Umesono, K. and Evans, R.M. (1989) Determinants of target gene specificity for steroid/thyroid hormone receptors. *Cell* **57**: 1139–1146.

van Uem, J.F.H.M., Hsiu, J.G., Chillik, C.F. *et al.* (1989) Contraceptive potential of RU 486 by ovulation inhibition: I. Pituitary versus ovarian action with blockade of estrogen-induced endometrial proliferation. *Contraception* **40**: 171–184.

Vegeto, E., Shahbaz, M.M., Wen, D.X., Goldman, M.E., O'Malley, B.W. and McDonnell, D.P. (1993) Human progesterone receptor A form is a cell- and promoter-specific repressor of human progesterone receptor B function. *Mol. Endocrinol.* **7**: 1244–1255.

Villa, E., Camellini, L., Dugani, A. *et al.* (1995) Variant estrogen receptor messenger RNA species detected in human primary hepatocellular carcinoma. *Cancer Res.* **55**: 498.

Villa, E., Dugani, A., Fantoni, E. *et al.* (1996) Type of estrogen receptor determines response to antiestrogen therapy. *Cancer Res.* **56**: 3883–3885.

Villee, C.A. (1962) The role of steroid hormones in the control of metabolic activity. In: J.M. Allen (ed.) *The Molecular Control of Cellular Activity*, pp. 297–313. McGraw-Hill, New York.

Vladusic, E.A., Hornby, A.E., Guerra-Vladusic, F.K. and Lupu, R. (1998) Expression of estrogen receptor β messenger RNA variant in breast cancer. *Cancer Res.* **58**: 210–214.

von der Ahe, D., Janich, S., Scheidereit, C., Renkawitz, R., Schuetz, G. and Beato, M. (1985) Glucocorticoid and progesterone receptors bind to the same sites in two hormonally regulated promoters. *Nature* **313**: 706–709.

Walter, P., Green, S., Greene, G. *et al.* (1985) Cloning of the human estrogen receptor cDNA. *Proc. Natl. Acad. Sci. USA* **82**: 7889–7893.

Walton, C., Godsland, I.F., Proudler, A.J., Felton, C.V. and Wynn, V. (1992) Effect of body mass index and fat distribution on insulin sensitivity, secretion, and clearance in nonobese healthy men. *J. Clin. Endocrinol. Metab.* **75**: 170–175.

Wang, M., Dotzlaw, H., Fuqua, S.A.W. and Murphy, L.C. (1997) A point mutation in the human estrogen receptor gene is associated with the expression of an abnormal estrogen receptor mRNA containing a 69 novel nucleotide insertion. *Breast Cancer Res. Treat.* **44**: 145–151.

Wang, Y. and Miksicek, R.J. (1991) Identification of a dominant negative form of the human estrogen receptor. *Mol. Endocrinol.* **5**: 1707–1715.

Ward, H.W.C. (1978) Anti-estrogen therapy for breast cancer: a report on 300 patients treated with tamoxifen. *Clin. Oncol.* **4**: 11–17.

Webb, A.M.C., Russell, J. and Elstein, M. (1992) Comparison of Yuzpe regimen, danazol, and mifepristone (RU486) in oral postcoital contraception. *BMJ* **305**: 927–931.

Weigel, N.L. (1994) Receptor phosphorylation. In: M.-J. Tsai and B.W. O'Malley (eds) *Mechanism of Steroid Hormone Regulation of Gene Transcription*, pp. 93–110. R.G. Landes, Austin, Texas.

Westerlind, K.C., Wronski, T.J., Evans, G.L. and Turner, R.T. (1994) The effect of long-term ovarian hormone deficiency on transforming growth factor-β and bone matrix protein mRNA expression in rat femora. *Biochem. Biophys. Res. Commun.* **200**: 283–289.

Wiest, W.G. and Rao, B.R. (1971) Progesterone binding proteins in rabbit uterus and human endometrium. In: G. Raspe (ed.) *Advances in the Biosciences*, No. 7, pp. 251–266. Pergamon Press, New York.

Williams, G.M., Iatropoulos, M.J., Djordjevic, M.V. and Kaltenberg, O.P. (1993) The triphenylethylene drug tamoxifen is a strong liver carcinogen in the rat. *Carcinogenesis* **14**: 315–317.

Williams, S.P. and Sigler, P.B. (1998) Atomic structure of progesterone complexed with its receptor. *Nature* **393**: 392–396.

Wilson, J.D. (1963) The nature of the RNA response to estradiol administration by the uterus of the rat. *Proc. Natl. Acad. Sci. USA* **50**: 93–100.

Wilson, K.B., Evans, M. and Abdou, N.I. (1996) Presence of a variant form of the estrogen receptor in peripheral blood mononuclear cells from normal individuals and lupus patients. *J. Reprod. Immunol.* **31**: 199–208.

Wolf, D.M. and Jordan, V.C. (1994) The estrogen receptor from a tamoxifen stimulated MCF-7 tumor variant contains a point mutation in the ligand binding domain. *Breast Cancer Res. Treat.* **31**: 129–138.

Yen, S.S.C. (1977) The biology of menopause. *J. Reprod. Med.* **18**: 287–296.

Zelada-Hedman, M., Borresen-Dale, A.-L. and Lindblom, A. (1997) Screening of 229 family cancer patients for a germline estrogen receptor gene (ESR) base mutation. *Hum. Mutat.* **9**: 289.

Zhang, Q.X., Borg, A. and Fuqua, S.A.W. (1993) An exon 5 deletion variant of the estrogen receptor frequently coexpressed with wild-type estrogen receptor in human breast cancer. *Cancer Res.* **53**: 5882–5884.

Chapter 4

The Androgen Receptor

Guido Jenster, Jan Trapman and Albert O. Brinkmann

INTRODUCTION

The androgens testosterone and dihydrotestosterone (DHT) are crucial for normal male sexual development and are essential for male reproductive function. Organs such as the testis and prostate are dependent on androgens for development to a functional gland and for maintenance of their function, while many other cells and tissues, including muscle, bone, facial hair follicles, vocal cords and many more, are responsive to testosterone and DHT exposure. Androgen action at the target cell level is mediated through the intracellular androgen receptor (AR; NR3C4), which is a member of the superfamily of nuclear receptors. This family includes receptors for steroids (androgen, estrogen, progesterone, glucocorticoid, mineralocorticoid), retinoids, thyroid hormone, vitamin D, fatty acids and other small hydrophobic molecules (Laudet *et al.*, 1992; Mangelsdorf *et al.*, 1995). In addition, there is a large group of orphan receptors for which the existence or identity of a ligand has not yet been determined (Enmark and Gustafsson, 1996). Upon ligand binding, the AR is able to recognize specific DNA sequences, so-called androgen response elements (AREs). These AREs are typically located in the promoter or enhancer regions of target genes. Binding to an ARE triggers the upregulation or downregulation of transcription of these genes.

The complementary DNA (cDNA) encoding the human AR was cloned in 1988 by different groups (Chang *et al.*, 1988; Lubahn *et al.*, 1988; Trapman *et al.*, 1988; Tilley *et al.*, 1989). Due to a polymorphic variation in the length of a CAG repeat encoding glutamine residues and a GGN repeat encoding a stretch of glycines in the N-terminal domain of the AR, the different groups have reported different AR sizes varying from 910 to 919 amino acids. In this review, the 919-amino-acid AR will be the standard (Fig. 4.1). The molecular mass of the human AR is approximately 110 kDa. Two major messenger RNA (mRNA) transcripts of 11 and 8 kilobases (kb) and one minor 4.7-kb mRNA are transcribed from the AR gene.

Based on the inheritance pattern of diseases associated with AR dysfunction, the AR gene was predicted to be located on the X chromosome. With the cloning of the AR cDNA and subsequent chromosomal localization to Xq11-12, this was shown to be

Nuclear Receptors and Genetic Disease
ISBN 0-12-146160-2

```
  1   MEVQLGLGRVYPRPPSKTYRGAFQNLFQSVREVIQNPGPRHPEAASAAPP
 51   GASLLLLQQQQQQQQQQQQQQQQQQQQQQQQETSPRQQQQQQGEDGSPQAHRR
101   GPTGYLVLDEEQQPSQPQSALECHPERGCVPEPGAAVAASKGLPQQLPAP
151   PDEDDSAAPSTLSLLGPTFPGLSSCSADLKDILSEASTMQLLQQQQQEAV
201   SEGSSSGRAREASGAPTSSKDNYLGGTSTISDNAKELCKAVSVSMGLGVE
251   ALEHLSPGEQLRGDCMYAPLLGVPPAVRPTPCAPLAECKGSLLDDSAGKS
301   TEDTAEYSPFKGGYTKGLEGESLGCSGSAAAGSSGTLELPSTLSLYKSGA
351   LDEAAAYQSRDYYNFPLALAGPPPPPPPPHPHARIKLENPLDYGSAWAAA
401   AAQCRYGDLASLHGAGAAGPGSGSPSAAASSSWHTLFTAEEGQLYGPCGG
451   GGGGGGGGGGGGGGGGGGGGGGGGEAGAVAPYGYTRPPQGLAGQESDFTAPD
501   VWYPGGMVSRVPYPSPTCVKSEMGPWMDSYSGPYGDMR LETARDHVLPI
550   DYYFPPQKTCLICGDEASGCHYGALTCGSCKVFFKRAAEGKQKYLCASRN
600   DCTIDKFRRKNCPSCRLRKCYE AGMTLGARKLKKLGNLKLQEEGEASST
649   TSPTEETTQKLTVSHIEGYECQ PIFLNVLEAIEPGVVCAGHDNNQPDSF
698   AALLSSLNELGERQLVHVVKWAKALPGFRNLHVDDQMAVIQYSWMGLMVF
748   AMGWRSFTNVNSRMLYFAPDLVFNEYRMHKSRMYSQCVRMRHLSQEFGWL
798   QITPQEFLCMKALLLFSIIPVDGLKNQKFFDELRMNYIKELDRIIACKRK
848   NPTSCSRRFYQLTKLLDSVQPIARELHQFTFDLLIKSHMVSVDFPEMMAE
898   IISVQVPKILSGKVKPIYFHTQ    919
```

Fig. 4.1 Amino acid sequence of the human androgen receptor. The residues encoding the DNA-binding domain are depicted by the white letters in the black box. The ligand-binding domain is shaded gray. The homopolymeric amino acid stretches in the N-terminal domain (residues 1–537) are shown in bold type. In addition, the variable glutamine (CAG) and glycine (GGN) repeats are underlined.

correct (Lubahn *et al.*, 1988; Brown *et al.*, 1989; Mahtani *et al.*, 1991). It is important to realize that 46XY individuals have only one allele of the AR gene.

The AR is unique owing to its direct involvement in three unrelated diseases. AR gene aberrations can cause androgen insensitivity syndrome (AIS) and Kennedy disease, which is also referred to as spinal and bulbar muscular atrophy (SBMA). In addition, AR gene aberrations can contribute to the progression of prostate cancer.

PHYSIOLOGY

Androgen action

Testosterone is produced and secreted mainly by the Leydig cells in the testis and converted into DHT by the two 5α-reductase enzymes (Russell and Wilson, 1994).

5α-Reductase type I is expressed largely in sebaceous glands and liver, while the type II enzyme is expressed in the male urogenital tract. DHT is the more potent androgen with a higher binding affinity to the AR (Wilbert et al., 1983). The importance of DHT is revealed in patients with 5α-reductase type II deficiency who have ambiguous external genitalia and a highly underdeveloped and unpalpable prostate (Thigpen et al., 1992; Wilson et al., 1993). Both the testis and the adrenals secrete large amounts of inactive precursor steroids dehydroepiandrosterone (DHEA), its sulfate form (DHEAS) and androstenedione. DHEA and androstenedione can be converted into testosterone and eventually to DHT in most peripheral tissues.

The role of androgens and the AR in males is very obvious, and knockout of the AR in males results in AIS (Quigley et al., 1995). Although females express the AR protein and produce small amounts of androgens in their adrenals and ovaries, the role of this pathway in their development and reproductive function is less well defined. Female mice with both AR alleles mutated were found to be fertile but proved to have impaired reproductive performance and premature cessation of reproduction (Lyon and Glenister, 1980).

The AR is expressed in many different tissues. Using techniques such as [^{3}H]-androgen radiolabeling, AR antibody staining and RNA expression analysis, AR protein and AR mRNA expression have been detected in the male and female reproductive systems as well as kidney, liver, adrenal cortex, pituitary gland, muscle, central nervous system, skin, bone and other tissues (reviewed in Chang et al., 1995).

ANDROGEN RECEPTOR STRUCTURE AND FUNCTION

Androgen receptor gene and protein structure

The AR gene has a length of more than 90 kb and is organized into eight exons and seven introns (Kuiper et al., 1989; Lubahn et al., 1989). The genomic structure is characteristic for the nuclear receptor gene family with different exons contributing to independent functional receptor domains: (1) an N-terminal domain, also referred to as the A/B domain harboring transcription activation functions; (2) a centrally located DNA-binding domain (DBD), also referred to as the C domain; (3) a hinge or D region; and (4) a C-terminal E domain, which is the ligand-binding domain (LBD) (Fig. 4.2). Both the DBD and the LBD share homology among the different family members. The N-terminal domain and the hinge region both vary in size and primary sequence, and can be considered receptor specific. The sequence encoding the AR N-terminal domain is present in the large first exon (Faber et al., 1989). The DBD, which consists of two zinc cluster structures, is encoded by exons 2 and 3. The first zinc cluster is responsible for direct DNA binding and harbors a so-called P box (residues 577-GSCKV-781) for the specific recognition of the ARE (Danielsen et al., 1989; Mader et al., 1989; Umesono and Evans, 1989; Zilliacus et al., 1991). The second zinc cluster is involved in protein–protein interaction of the AR homodimer via the so-called D box (residues 596-ASRND-600) (Hard et al., 1990; Schwabe et al., 1990; Luisi et al., 1991). The first half of exon 4 encodes the hinge region, while the second half and exons 5–8 encode the LBD (Fig. 4.2).

Androgen receptor function

In its inactive state, the unliganded AR is associated with heat shock proteins and cannot perform its transactivating function (reviewed in Smith and Toft, 1993). This large 8–10S

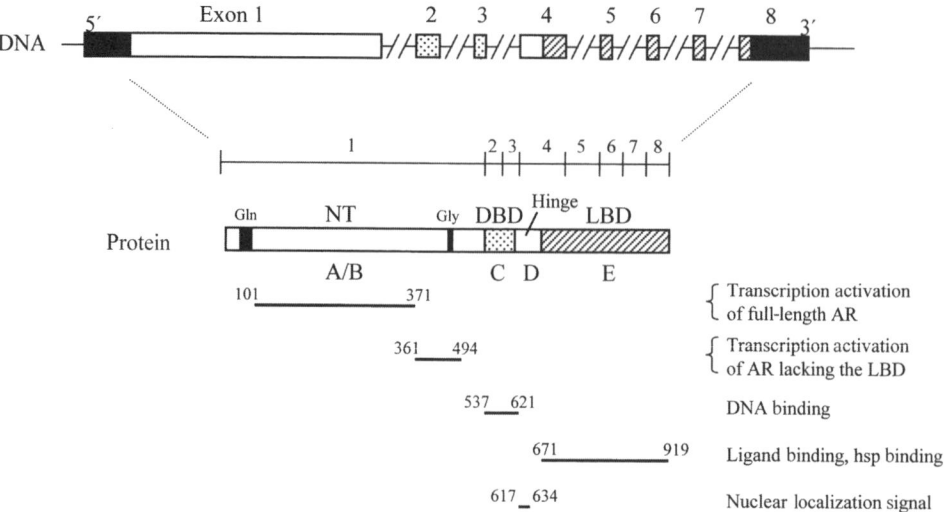

Fig. 4.2 Gene organization and functional domain structure of the human androgen receptor (AR). The AR gene contains eight exons encoding the different protein domains. The black boxes in the DNA diagram represent 5' and 3' untranslated regions. The AR N-terminal domain (NT), also referred to as the A/B domain, harbors the receptor's main transcription activation functions. The DNA-binding domain (DBD), or C domain, is located in the middle of the molecule and N-terminal to the hinge region (D domain) and ligand-binding domain (LBD), or E domain. The signal responsible for nuclear import is located at the junction of the DBD and hinge region.

complex may be present in the cytoplasm or nucleus. The different heat shock proteins interact with the receptor's LBD. Upon testosterone or DHT binding, the AR dissociates from the heat shock proteins and, as a homodimer, recognizes the AREs. These elements are typically located in the promoter or enhancer regions of AR target genes. The consensus DNA-binding site for the AR is comprised of two imperfect palindromic 6-base-pair (bp) elements separated by a 3-bp spacer: GG(A/T)ACAnnnTGTTCT (Roche *et al.*, 1992). Interestingly, this type of binding site can also be recognized by the glucocorticoid, progesterone and mineralocorticoid receptors. Besides a consensus ARE, more AR-specific response elements have been identified in typical androgen-regulated genes (Adler *et al.*, 1993; Rundlett and Miesfeld, 1995; Claessens *et al.*, 1996). Binding to the DNA element and recruitment of cofactor proteins trigger the upregulation or downregulation of transcription of these genes (Tsai and O'Malley, 1994; Quigley *et al.*, 1995; Brinkmann *et al.*, 1996).

The role of coactivators and corepressors in androgen receptor function

The cloning and characterization of cofactor proteins, called coactivators and corepressors, have greatly enhanced our understanding of how steroid receptors activate or inhibit transcription of their target genes. Multiple coactivators have been identified that fit the definition of a protein that connects or bridges the DNA-bound receptor to proteins in the preinitiation complex, and thereby enhance transcription (Horwitz *et al.*,

1996; Glass *et al.*, 1997; Shibata *et al.*, 1997; Chen and Li, 1998; Jenster, 1998; Kingston, 1999; McKenna *et al.*, 1999). Besides this bridging function, some coactivators can modify chromatin by histone acetylation and make promoters more accessible for the binding of other transcription factors. Both chromatin remodeling and recruitment of basal transcription factors are essential processes for full target gene expression (Archer *et al.*, 1992; Jenster *et al.*, 1997).

The steroid receptor domains involved in coactivator binding include both the N-terminal domain and the LBD (Aarnisalo *et al.*, 1998; Ding *et al.*, 1998; Onate *et al.*, 1998). The AR N-terminus harbors the main activation functions, while the LBD by itself is not transcriptionally active (Jenster *et al.*, 1995). This suggests that the LBD and N-terminus work in concert to recruit coactivators into the DNA-bound AR dimer. This hypothesis is supported by multiple observations: (1) the AR N-terminal domain interacts with its own LBD, potentially creating a novel protein–protein interacting interface (Langley *et al.*, 1995; Doesburg *et al.*, 1997; Ikonen *et al.*, 1997; Berrevoets *et al.*, 1998; Sui *et al.*, 1999), and (2) a different AR N-terminal activation domain is responsible for transactivation when the LBD has been deleted, compared with the full-length AR (Jenster *et al.*, 1995; Ikonen *et al.*, 1997).

The role of corepressors in AR function is poorly defined. Some nuclear receptors (including retinoid receptors, thyroid hormone receptors, vitamin D receptor and certain orphan receptors) are not associated with heat shock proteins in their unliganded state and subsequently bind DNA and repress transcription. These receptors recruit corepressors that indirectly bind histone deacetylases, which will 'lock' the nucleosomes in a closed conformation excluding transcription factors from binding to the promoter template (Alland *et al.*, 1997; Heinzel *et al.*, 1997; Nagy *et al.*, 1997). Although the AR and other steroid receptors will not be able to bind DNA in their unliganded state, specific antagonists have been shown to induce corepressor binding to the progesterone receptor and estrogen receptor, creating the same scenario. Both tamoxifen and RU486 (mifepristone) are able to induce active repression by estrogen receptor α and progesterone receptor respectively (Jackson *et al.*, 1997; Smith *et al.*, 1997; Lavinsky *et al.*, 1998; Wagner *et al.*, 1998; Zhang *et al.*, 1998). Although there are no reports yet, it is possible that binding of certain antiandrogens to the AR will also result in corepressor recruitment and active target gene repression.

Androgen receptor phosphorylation

Immediately after translation, the AR becomes phosphorylated resulting in the appearance of two isoforms separable by sodium dodecyl sulfate (SDS)–polyacrylamide gel electrophoresis (Kuiper *et al.*, 1991; Blok *et al.*, 1996). The nonphosphorylated, faster migrating, 110-kDa isoform is converted into a 112-kDa phosphoisoform. Mutational analysis of serine 81 or 94 in the AR N-terminal domain abolishes this upshift, indicating that phosphorylation of these serine residues probably contributes to the phosphorylation of the 112-kDa AR isoform (Jenster *et al.*, 1994; Zhou *et al.*, 1995). Three other AR phosphorylation sites have been identified by means of mutational analysis and trypsin digestion of ^{32}P-labeled AR followed by high-performance liquid chromatography (HPLC) analysis and Edman degradation (Zhou *et al.*, 1995; Blok *et al.*, 1998). These include the serine residues at position 515, 650 and 662. The importance of the potential phosphorylation of serines 81, 94 and 650 in AR function has been studied using AR mutants in which these serine residues were replaced by an alanine or glycine. Only substitution of serine 650 reduced AR activity up to 30%. Mutation of serines 81 and 94 had little or no effect on AR function (Jenster *et al.*, 1994; Zhou *et al.*, 1995). Interestingly, the protein kinase A stimulator forskolin rapidly induces the dephosphorylation of serines 650 and 662 in the AR hinge region (Blok *et al.*, 1998). This results in the impairment of AR target gene transcription, probably as a result of decreased ligand binding. These data in combination with the mutational analysis strongly suggest an important role for phosphorylation in optimal AR functioning.

Besides the basal phosphorylation resulting in the 110–112-kDa doublet, addition of androgen induces another shift and the 110–112–114-kDa AR triplet (Jenster *et al.*, 1994). This triplet is the result of both an addition and a redistribution of phosphorylated sites; however, it is not known which exact residues are involved (Kuiper *et al.*, 1993; Zhou *et al.*, 1995). Interestingly, mutations that inactivate AR function, such as mutations resulting in loss of DNA binding or transactivation, inhibit the formation of the 114-kDa isoform. This suggests that part of the androgen-induced phosphorylation occurs during or after AR transcription regulation (Jenster *et al.*, 1994; Brüggenwirth *et al.*, 1997).

Androgen receptor N-terminal domain: transcription activation and interaction with the ligand-binding domain

The AR N-terminal domain harbors the major transcription activation functions. Within its 537 amino acids, two independent activation domains have been identified: (1) activation function 1 (located between residues 102 and 371), which is essential for transactivity of the full-length AR; and (2) activation function 5 (located between residues 361 and 494), which is required for transactivity of a constitutively active AR, which lacks its LBD (Jenster *et al.*, 1995; Ikonen *et al.*, 1997). Activation function 1 has been further delineated into two distinct units in the rat AR (Chamberlain *et al.*, 1996). One important message from these studies is that only large deletions and/or multiple amino acid substitutions within the AR N-terminal domain will affect transcription activity. Single amino acid substitutions within the AR N-terminus have not been observed to impair AR function significantly. This could explain why there have been only a few reports of these types of mutation in the AR of individuals with the AIS (discussed below).

The AR N-terminal domain is characterized by the presence of several homopolymeric amino acid stretches (Fig. 4.1). Two of these repeats, the long glutamine repeat and the glycine repeat, vary in length among the normal human population (La Spada *et al.*, 1991; Edwards *et al.*, 1992; Sleddens *et al.*, 1992, 1993; Irvine *et al.*, 1995; Choong and Wilson, 1998; Kantoff *et al.*, 1998). The exact function of these repeats is unknown although variation in the length of the glutamine stretch has been linked to AR activity and stability: the shorter the repeat the higher the AR transcription activity (Mhatre *et al.*, 1993; Chamberlain *et al.*, 1994; Jenster *et al.*, 1994; Kazemi-Esfarjani *et al.*, 1995; Choong *et al.*, 1996a). This observation is important as it may explain the potential correlation of the length of the CAG repeat with changes in AR function in both individuals with Kennedy disease and patients with prostate cancer (discussed below).

Another function of the AR N-terminal domain is its binding to the C-terminal AR LBD (Langley *et al.*, 1995; Doesburg *et al.*, 1997; Ikonen *et al.*, 1997; Berrevoets *et al.*, 1998; Sui *et al.*, 1999). The N-terminal regions required for the binding of the LBD have been mapped to two essential units: the first 36 amino acids and residues 371–503 (Berrevoets *et al.*, 1998). This same type of C–N interaction has also been observed for the estrogen receptor α and the progesterone receptor, but the exact function and importance is unclear (Kraus *et al.*, 1995; Tetel *et al.*, 1997, 1999). Moreover, it has not yet been established whether the C–N interaction occurs intramolecularly or intermolecularly when the AR occupies its ARE as a homodimer.

Nuclear import and subcellular distribution

The AR can both be cytoplasmic or nuclear in the absence of ligand. In the presence of androgens, however, it is exclusively nuclear. The signal responsible for nuclear import has been mapped to the last part of the second zinc clusters and the first part of the hinge region (Jenster *et al.*, 1993; Zhou *et al.*, 1994) (Fig. 4.2). This type of nuclear localization signal (NLS) is characterized by two clusters of basic residues spaced by 10 amino acids, and was first identified in the nucleoplasmin protein (Robbins *et al.*, 1991). This bipartite NLS is highly conserved among the members of the steroid receptor family (Jenster *et al.*, 1993). The sequence of the AR NLS is: 616-L**RK**CYEAGMTLGA**RK**L**KK**L-634. In underlined bold are the basic amino acid residues that play an intricate part in the NLS function.

Androgen receptor ligand independent activation

In addition to androgens, specific growth factors and cytokines have also been shown to induce AR activation in *in vitro* cell line systems (Zhu and Liu, 1997; Weigel and Zhang, 1998). These include insulin-like growth factor 1 (IGF-1), keratinocyte growth factor (KGF), epidermal growth factor (EGF), luteinizing hormone-releasing hormone (LHRH) and interleukin 6 (IL-6) (Culig *et al.*, 1994, 1997; Reinikainen *et al.*, 1996; Hobisch *et al.*, 1998). The growth factor/cytokine reporter gene activation is mediated by the AR because antiandrogens block growth factor-induced transcription. All the growth factors and cytokines transduce signaling through protein kinase pathways, and it is not surprising that manipulation of these different pathways can mimic or block the ligand-independent AR transactivity (Ikonen *et al.*, 1994; de Ruiter *et al.*, 1995; Nazareth and Weigel, 1996; Reinikainen *et al.*, 1996; Culig *et al.*, 1997; Blok *et al.*, 1998; Hobisch *et al.*, 1998; Sadar 1999).

It is still unclear how manipulation of kinases results in the activation of the AR. Although it is well established that the AR is a phosphoprotein, it remains to be proven that induction of protein kinase pathways directly affects the phosphorylation status, resulting in receptor activation. Besides the receptor itself, phosphorylation of other proteins (including coactivators and basal transcription factors) might contribute to the AR ligand-independent transactivity. Importantly, ligand-independent AR activation is synergistic with minimal levels of androgens, thus lowering the androgen concentration required for full AR function. This observation is particularly relevant for androgen ablation treatment of patients with prostate cancer. The minimal amount of androgens (mainly produced by the adrenals) present in these patients might still be utilized by the cancer cells to synergize with an activated growth factor or cytokine pathway, resulting in the survival and growth of the prostatic tumors. However, all of the studies revealing ligand-independent mechanisms to activate the AR have been performed *in vitro*. To date, there is no direct *in vivo* evidence for AR target gene activation in a ligand-independent fashion.

ANDROGEN RECEPTOR AND DISEASE

Androgen insensitivity syndrome

It has been known for quite some time that defects in male sexual differentiation in 46XY individuals has an X-linked pattern of inheritance. This was reported in 1947, in families with severe hypospadias, infertility and gynecomastia (Reifenstein, 1947). The end-organ resistance to androgens has been designated as AIS, and is distinct from other forms of male pseudohermaphroditism such as 17β-hydroxysteroid dehydrogenase type 3 deficiency or 5α-reductase type 2 deficiency (Wilson *et al.*, 1974, 1993; Geissler *et al.*, 1994). It is generally accepted that defects in the AR gene can prevent the normal development of both internal and external male structures in 46XY individuals, and information on the molecular structure of the human AR gene has facilitated the study of molecular defects associated with androgen insensitivity. Owing to the X-linked character of the syndrome, only 46XY individuals are affected; in female carriers only sporadic reports are available on delayed menarche (Sai *et al.*, 1990). Naturally occurring mutations in the AR gene are an interesting source for the investigation of receptor structure–function relationships. In addition, the variation in clinical pheno-types provides the opportunity to correlate a mutation in the AR structure with the impairment of a specific physiologic function.

The main phenotypic characteristics of individuals with the complete androgen insensitivity syndrome (CAIS) are: female external genitalia; a short, blind-ending vagina; the absence of wolffian duct-derived structures such as epididymides, vas deferens and seminal vesicles; the absence of a prostate; development of gynecomastia; and the absence of pubic and axillary hair (Quigley *et al.*, 1995). Müllerian duct-derived structures are usually absent because anti-müllerian hormone is synthesized in the testes, which are in the abdomen or inguinal canals. Usually, testosterone levels are raised at puberty, and increased luteinizing hormone (LH) levels are also found, indicating androgen resistance at the hypothalamic–pituitary level. The high testosterone levels are also the substrate for aromatase, resulting in substantial amounts of estrogens, which are responsible for further feminization in individuals with CAIS.

In the partial androgen insensitivity syndrome (PAIS), several phenotypes are found ranging from individuals with predominantly a female appearance (e.g. external female genitalia and pubic hair at puberty, or with mild cliteromegaly, and some fusion of the labia) to persons with ambiguous genitalia or individuals with a predominantly male phenotype (also called Reifenstein syndrome). Patients from this latter group can present with a micropenis, perineal hypospadias and cryptorchidism. In the group of individuals with PAIS, wolffian-duct derived structures may be partially to fully developed, depending on the biochemical phenotype of the AR. At puberty, raised LH, testosterone and estradiol levels are observed, but in general the degree of feminization is less than among those with CAIS. Individuals with mild symptoms of undervirilization (mild androgen insensitivity syndrome, MAIS) and infertility have been described as well.

Phenotypic variation between individuals in different families has been described for several mutations (Batch *et al.*, 1993; Imasaki *et al.*, 1994; Evans *et al.*, 1997) (Table 4.1). However, in cases of CAIS no phenotypic variation has been described within one single family, in contrast to families with PAIS.

Since the cloning of the AR cDNA in 1988 by four independent research groups and the subsequent elucidation of the genomic organization of the AR gene, molecular biology tools have been available for the molecular analysis of the AR gene of individuals with AIS. In addition to endocrinologic data such as levels of testosterone, LH, androstenedione and 5α-DHT, which may vary substantially among individuals with AIS, the most reliable approach is the sequencing of each individual AR exon. In general, AIS can be analyzed routinely and differential diagnosis is possible with syndromes presenting with similar phenotypes but with a completely different molecular cause (e.g. testicular enzyme deficiencies, 5α-reductase deficiency, Leydig cell hypoplasia due to inactivating LH receptor mutations). Furthermore, in pedigree analysis intragenic polymorphisms such as the highly polymorphic CAG repeat encoding a glutamine stretch, the polymorphic GGN repeat encoding a glycine stretch, and the *Hin*dIII polymorphism (Brown *et al.*, 1989) or *Stu*I polymorphism (Lu and Danielsen, 1996) can be used as X-chromosomal markers (Ris-Stalpers *et al.*, 1994a; Davies *et al.*, 1995).

In the AR gene, four different types of mutation have been detected in DNA from individuals with AIS: (1) single point mutations resulting in amino acid substitutions or premature stop codons, (2) nucleotide insertions or deletions most often leading to a frameshift and premature termination, (3) complete or partial gene deletions (more than 10 nucleotides), and (4) intronic mutations in either splice donor or acceptor sites, which affect the splicing of AR RNA.

Single point mutations, insertions and deletions resulting in amino acid substitutions or premature stop codons

Mutations in the N-terminal (A/B) domain   Mutations in the N-terminal domain (exon 1 of the gene) do not occur frequently and the vast majority of the mutations result in a stop codon or in premature termination due to frameshifts caused by nucleotide insertions or deletions. An interesting mutation has been described by Choong *et al.* (1996b) (Table 4.1) in the fourth nucleotide, which results in a decreased translational efficiency of the AR mRNA in an individual with PAIS. Three other missense mutations were reported in combination with mosaicism or with a mutation in another region of the gene.

In a family with PAIS associated with severe hypospadias, the length of the AR N-terminal polyglutamine repeat has been reported to be shortened to only 12 glutamine residues (McPhaul *et al.*, 1991a). The shortened glutamine stretch, as such, is not the cause for the androgen resistance but seems to increase the thermolability of the AR in combination with a point mutation in exon 5 (Y763C) (see Table 4.1) in the LBD. This point mutation causes rapid dissociation but no thermolability. These data support a functional interaction of the two separated regions in the AR and indicate further that the defect becomes critical in only part of the androgen target tissues because of the partial character of the androgen resistance found in this family (see Table 4.1) (McPhaul *et al.*, 1991a).

Mutations in the DNA-binding (C) domain   The DBD of nuclear receptors is the most conserved part of the receptor molecule and consists of two zinc clusters. The first zinc cluster is involved in direct DNA binding and contains the so-called P box for the specific recognition of the ARE. The second zinc cluster is involved in protein–protein interactions and serves as stabilization unit for the dimerization of two receptor molecules. In general, mutations (e.g. single nucleotide substitutions) result in a normal hormone-binding protein, which is defective in DNA binding/dimerization and consequently in transcription activation. In total, 32 different mutations have been published and are illustrated in Table 4.1. Fifteen mutations were observed in the first zinc cluster

Table 4.1 Androgen receptor mutations in individuals with complete (CAIS), partial (PAIS) or mild (MAIS) androgen insensitivity syndrome

Type of mutation					
Nucleotide	Amino acid	Domain	Biochemical phenotype	Clinical phenotype	Reference	Comment
4G > A	E2K	A/B	Thermolabile	PAIS	Choong et al. (1996b)	Stop in codon 172
125–126insC	Frameshift	A/B	Zero binding	CAIS	Brüggenwirth et al. (1996)	
178C > T	Q60X	A/B	Low binding	CAIS	Zoppi et al. (1993)	Stop in codon 172
381delA	Frameshift	A/B	Zero binding	CAIS	Batch et al. (1992)	
418–444del		A/B	–	CAIS	Hiort et al. (1996)	
515T > G	L172X	A/B	Zero/low binding	PAIS	Holterhus et al. (1997)	Mosaicism
581A > G	Q194R	A/B	–	CAIS	Komori et al. (1997)	Also 1790delC
603–604insATCC	Frameshift	A/B	Zero binding	CAIS	Batch et al. (1992)	Stop in codon 232
644–645insC	Frameshift	A/B	–	CAIS	Hiort et al. (1994b)	Stop in codon 232
764T > C	L255P	A/B	Thermolabile	CAIS	Tanaka et al. (1998)	Also G820A
816delA	Frameshift	A/B	Zero binding	CAIS	Brüggenwirth et al. (1996)	Stop in codon 301
1111C > T	G371X	A/B	–	CAIS	Davies et al. (1995)	
1505G > A	W502X	A/B	–	CAIS	Brüggenwirth et al. (1996)	
1602C > G	Y534X	A/B	Zero binding	CAIS	McPhaul et al. (1991a)	
1642C > T	P548S	C	–	MAIS	Sutherland et al. (1996)	
1676G > A	C559Y	C	Normal binding	CAIS	Zoppi et al. (1992)	
1702G > T	G568W	C	Normal binding	PAIS	Lobaccaro et al. (1994)	
1703G > T	G568V	C	Normal binding	PAIS	Allera et al. (1995)	
1712A > G	Y571C	C	–	CAIS	Komori et al. (1998)	
1718C > A	A573D	C	Normal binding	CAIS	Brüggenwirth et al. (1996)	Defective DNA binding
1726T > C	C576R	C	Normal binding	CAIS	Zoppi et al. (1992)	
1736G > A	C579Y	C	–	CAIS	Sultan et al. (1993)	
1736G > T	C579F	C	Normal binding	CAIS	Imasaki et al. (1996)	Defective DNA binding
1737delC	Frameshift	C	–	CAIS	Imai et al. (1995)	Stop in codon 619
1741G > T	V581F	C	Normal binding	CAIS	Lumbroso et al. (1993)	
1742–1744delTCT		C	Low binding	CAIS	Beitel et al. (1994a)	Deletion of F582

Nucleotide change		Amino acid change	Binding	Phenotype	Reference	Comment
1745T > C	C	F582S	Zero binding	PAIS	Hiort et al. (1994b)	
1745T > A	C	F582Y	Normal binding	PAIS	Imasaki et al. (1996)	Reduced DNA binding
1754G > A	C	R585K	–	CAIS	Sultan et al. (1993)	
1768A > T	C	K590X	Zero binding	CAIS	Marcelli et al. (1990b)	
1786G > C	C	A596T	Normal binding	CAIS	Gast et al. (1995)	Abolishes dimerization
1789A > G	C	S597G	Normal binding	PAIS	Zoppi et al. (1992)	Also R617P mutation
1802G > T	C	C601F	–	CAIS	Baldazzi et al. (1994)	
1810G > T	C	D604Y	–	PAIS	Hiort et al. (1994b)	
1819C > T	C	R607X	Zero binding	CAIS	Brown et al. (1993)	
1820G > A	C	R607Q	–	PAIS	Wooster et al. (1992)	Also breast cancer
1823G > A	C	R608K	Normal binding	PAIS	Lobaccaro et al. (1993c)	Also breast cancer
1829A > C	C	N610T	Normal binding	PAIS	Weidemann et al. (1996)	
1842–1844delTCG	C		Normal binding	CAIS	Beitel et al. (1994a)	Deletion of R615
1844G > A	C	R615H	Low binding	CAIS	Ris-Stalpers et al. (1994a)	
1844G > A	C	R615H	–	PAIS	Hiort et al. (1996)	
1844G > C	C	R615P	–	PAIS	Hiort et al. (1996)	
1847T > G	C	L616R	Normal binding	PAIS	De Bellis et al. (1994)	
1847T > C	C	L616P	Normal binding	CAIS	Lobaccaro et al. (1996)	
1850G > C	C	R617P	Normal binding	CAIS	Marcelli et al. (1991)	
1850G > C	C	R617P	Normal binding	PAIS	Zoppi et al. (1992)	Also S597G mutation
1934C > A	D	A645D	–	PAIS	Hiort et al. (1996)	
1991T > A	D	I664N	Low binding	PAIS	Pinsky et al. (1992)	
2012C > A	E	P671H	–	PAIS	Hiort et al. (1996)	
2030T > C	E	L677P	Zero binding	CAIS	Belsham et al. (1995)	
2041G > A	E	E681K	–	CAIS	Hiort et al. (1993)	
2056T > C	E	C686R	–	PAIS	Hiort et al. (1996)	
2060C > T	E	A687V	–	PAIS	Hiort et al. (1996)	
2063G > A	E	G688E	–	CAIS	Hiort et al. (1998)	
2068–2070delGAC	E		–	PAIS	Schwartz et al. (1994)	D690 deleted
2074–2076delAAC	E		Normal binding	CAIS	Batch et al. (1992)	N692 deleted
2083G > C	E	D695H	Low binding	CAIS	Ris-Stalpers et al. (1991)	
2083G > A	E	D695N	Normal binding	CAIS	Ris-Stalpers et al. (1991)	

continued overleaf

Table 4.1 *continued*

Type of mutation

Nucleotide	Amino acid	Domain	Biochemical phenotype	Clinical phenotype	Reference	Comment
2083G > A	D695N	E	–	PAIS	Hiort et al. (1998)	
2084A > T	D695V	E	–	CAIS	Dork et al. (1998)	
2104T > G	S702A	E	Zero binding	CAIS	Pinsky et al. (1992)	
2107A > G	S703G	E	Low binding	PAIS	Radmayr et al. (1997)	
2114A > G	N705S	E	Zero binding	CAIS	Pinsky et al. (1992)	
2120T > G	L707R	E	–	CAIS	Lumbroso et al. (1996)	
2123G > C	G708A	E	–	PAIS	Hiort et al. (1994b)	
2134C > G	L712F	E	–	PAIS	Hiort et al. (1996)	
2154G > A	W718X	E	Zero binding	CAIS	Sai et al. (1990)	
2166G > C	L722F	E	–	CAIS	Hiort et al. (1996)	
2173T > C	F725L	E	Normal binding	PAIS	Quigley et al. (1995)	Oligospermia
2181C > G	N727K	E	–	MAIS	Yong et al. (1994b)	
2183T > C	L728S	E	Low binding	PAIS	McPhaul et al. (1992)	
2194G > A	D732N	E	High K_d	CAIS	Ko et al. (1997)	
2194G > T	D732Y	E	Zero binding	CAIS	Pinsky et al. (1992)	
2199G > T	Q733H	E	–	PAIS	Hiort et al. (1998)	Mosaicism
2210T > C	I737T	E	Low binding	PAIS	Quigley et al. (1995)	
2221T > C	W741R	E	Low binding	CAIS	Marcelli et al. (1994)	
2224A > G	M742V	E	High dissociation	PAIS	Ris-Stalpers et al. (1994a)	
2226G > A	M742I	E	High K_d	PAIS	Bevan et al. (1996)	
2228G > T	G743V	E	Zero binding	CAIS	Lobaccaro et al. (1993b)	
2228G > T	G743V	E	Normal binding	PAIS	Nakao et al. (1993)	
2230C > T	L744F	E	–	PAIS	Brinkmann et al. (1995)	
2234T > C	M745T	E	Zero binding	PAIS	Ris-Stalpers et al. (1994a)	
2236G > A	V746M	E	–	PAIS	Hiort et al. (1996)	
2243C > A	A748D	E	Low binding	PAIS	Marcelli et al. (1994)	
2245A > G	M749V	E	–	CAIS	De Bellis et al. (1992)	Abnormal dissociation

Nucleotide	Amino acid	Exon	Binding	Phenotype	Reference	Comment
2249G > A	G750D	E	Low binding	CAIS	Bevan et al. (1997)	
2251T > A	W751R	E	–	CAIS	Brinkmann et al. (1995)	
2254C > T	R752X	E	Zero binding	CAIS	Brinkmann et al. (1995)	
2255G > A	R752Q	E	Zero binding	CAIS	Evans (1992)	
2260T > G	F754V	E	–	CAIS	Lobaccaro et al. (1993d)	
2262C > A	F754L	E	High K_d	PAIS	Weidemann et al. (1996)	
2264A > G	N756S	E	–	PAIS	Hiort et al. (1996)	
2273A > C	N758T	E	Normal binding	PAIS	Yong et al. (1998)	High dissociation
2276C > T	S759F	E	Zero binding	CAIS	De Bellis et al. (1992)	
2284C > T	L762F	E	Zero binding	CAIS	Bevan et al. (1997)	
2287T > C	Y763H	E	–	CAIS	Quigley et al. (1995)	
2288A > G	Y763C	E	High K_d	PAIS	McPhaul et al. (1991b)	
2290T > C	F764L	E	Zero binding	CAIS	Ris-Stalpers et al. (1994a)	
2290T > C	F764L	E	Low binding	CAIS	Pinsky et al. (1992)	
2290T > C	F764L	E	High dissociation	CAIS	Marcelli et al. (1994)	
2293G > A	A765T	E	Zero binding	CAIS	Bevan et al. (1997)	
2294C > T	A765V	E	Zero binding	CAIS	Pinsky et al. (1992)	
2296C > T	P766S	E	High K_d	CAIS	Marcelli et al. (1994)	
2298delT	Frameshift	E	–	CAIS	Baldazzi et al. (1994)	Stop in codon 807
2301T > G	D767E	E	Low binding	CAIS	Lobaccaro et al. (1993a)	
2311A > C	N771H	E	–	PAIS	Hiort et al. (1994b)	
2314G > T	E772X	E	Zero binding	CAIS	Imasaki et al. (1995)	
2315A > C	E772A	E	High dissociation	PAIS	Pinsky et al. (1992)	
2315A > G	E772G	E	High K_d	PAIS	Tincello et al. (1997)	
2320C > T	R774C	E	Zero binding	CAIS	Prior et al. (1992)	
2321G > A	R774H	E	High K_d	CAIS	Prior et al. (1992)	
2321G > A	R774H	E	–	PAIS	Quigley et al. (1995)	
2335C > T	R779W	E	–	CAIS	Hiort et al. (1994b)	
2340G > A	M780I	E	–	PAIS	Brinkmann et al. (1995)	
2340G > A	M780I	E	–	CAIS	Rodien et al. (1996)	
2356C > T	R786X	E	Zero binding	CAIS	Pinsky et al. (1992)	
2359A > G	M787V	E	Zero binding	CAIS	Nakao et al. (1992)	

continued overleaf

Table 4.1 *continued*

Type of mutation

Nucleotide	Amino acid	Domain	Biochemical phenotype	Clinical phenotype	Reference	Comment
2368C > T	L790F	E	Low K_d	MAIS	Tsukada et al. (1994)	
2379G > C	E793D	E	Normal binding	MAIS	Pinsky et al. (1992)	
2381T > C	F794S	E	–	CAIS	Hiort et al. (1996)	
2388G > A	W796X	E	Low binding	CAIS	Marcelli et al. (1990a)	
2392C > A	Q798E	E	Normal binding	PAIS	Bevan et al. (1996)	
2417G > A	C806Y	E	–	PAIS	Brown et al. (1993)	
2419A > G	M807V	E		CAIS	Murono et al. (1995)	
2420T > G	M807R	E	Zero binding	CAIS	Adeyemo et al. (1993)	
2441G > A	S814N	E	Normal binding	PAIS	Pinsky et al. (1992)	Altered binding specificity
2459G > C	G820A	E	High K_d	CAIS	Tanaka et al. (1998)	Also L257P
2461C > G	L821V	E	Normal binding	PAIS	Pinsky et al. (1992)	
2491C > T	R831X	E	Zero binding	CAIS	De Bellis et al. (1992)	
2492G > A	R831Q	E	Low binding	CAIS	Brown et al. (1990)	
2492G > T	R831L	E	Zero binding	CAIS	Pinsky et al. (1992)	
2501A > G	Y834C	E	Zero binding	CAIS	Wilson et al. (1992)	
2518C > T	R840C	E	High K_d	PAIS	Beitel et al. (1994b)	
2519G > A	R840H	E	High K_d	PAIS	Marcelli et al. (1994)	
2522T > G	I841S	E	–	PAIS	Hiort et al. (1996)	
2525T > C	I842T	E	High K_d	PAIS	Weidemann et al. (1996)	
2543–2544insA	Frameshift	E	Zero binding	CAIS	Brinkmann et al. (1995)	Stop in codon 879
2558C > G	S853X	E	Zero binding	CAIS	Wilson et al. (1992)	
2561G > A	R854K	E	Low binding	PAIS	McPhaul et al. (1992)	
2563C > T	R855C	E	Zero binding	CAIS	Brinkmann et al. (1995)	
2564G > A	R855H	E	High K_d	PAIS	Boehmer et al. (1997)	
2564G > A	R855H	E	Zero binding	CAIS	Weidemann et al. (1996)	
2571C > G	Y857X	E	–	CAIS	Hiort et al. (1998)	
2588T > G	L863R	E	–	CAIS	Brown et al. (1993)	

Nucleotide	Amino acid	Domain	Binding	Phenotype	Reference
2590G > A	D864N	E	Low binding	CAIS	Bevan et al. (1996)
2591A > G	D864G	E	Zero binding	CAIS	De Bellis et al. (1992)
2596G > T	V866L	E	High K_d	PAIS	Saunders et al. (1992)
2596G > A	V866M	E	High K_d	CAIS	Lubahn et al. (1989)
2596G > A	V866M	E	—	PAIS	McPhaul et al. (1992)
2597T > A	V866E	E	—	CAIS	McPhaul et al. (1992)
2607T > G	I869M	E	High K_d	PAIS	Bevan et al. (1996)
2609C > A	A870E	E	—	CAIS	Knoke et al. (1997)
2609C > G	A870G	E	—	CAIS	Hiort et al. (1998)
2609C > T	A870V	E	—	PAIS	Hiort et al. (1994a)
2611A > G	R871G	E	Normal binding	MAIS	Pinsky et al. (1992)
2641C > G	L881V	E	—	CAIS	Davies et al. (1995)
2647A > T	K883X	E	Zero binding	CAIS	Trifiro et al. (1991)
2665G > A	V889M	E	Zero binding	CAIS	Pinsky et al. (1992)
2665G > A	V889M	E	Low binding	PAIS	De Bellis et al. (1994)
2693T > C	L898T	E	—	CAIS	Hiort et al. (1998)
2707G > A	V903M	E	Low binding	PAIS	McPhaul et al. (1992)
2710C > T	P904S	E	High K_d	CAIS	Pinsky et al. (1992)
2711C > A	P904H	E	Zero binding	CAIS	McPhaul et al. (1992)
2719C > T	L907F	E	Low binding	CAIS	Bevan et al. (1997)
2725G > A	G909L	E	Low binding	PAIS	Choong et al. (1996c)
2737C > T	P913S	E	—	PAIS	Ghirri and Brown (1993)
2748C > G	F916L	E	High K_d	CAIS	Radmayr et al. (1997)

Nucleotide substitutions and small nucleotide insertions or deletions are shown. Nucleotide numbering is according to the recommendations by Antonarakis (1998) (the A of the initiator Met codon is denoted nucleotide 1) and the amino acid residue numbering is according to Lubahn et al. (1989). The amino acid nomenclature is the one-letter code and the wild-type amino acid is given before and the mutant amino acid after the codon number. Stop codons are designated by X. The domains are designated as follows: exon 1: A/B = N-terminal domain; exons 2 and 3: C = DNA-binding domain; first part of exon 4: D = hinge region; rest of exon 4 and exons 5–8: E = ligand-binding domain. The biochemical phenotype reflects either information on the number of hormone binding sites or on the thermolability of the complex or on the K_d. Only one reference is given per mutation. For a complete overview of all reported mutations in the androgen receptor gene, see also the AR database at http://www.mcgill.ca/androgendb/ or Gottlieb et al. (1998).

and 17 in the second zinc cluster. As the three-dimensional structures of the DBD of several nuclear receptors have been published, the consequence of mutations in the AR DBD can be predicted on basis of the structure of the glucocorticoid receptor DBD (Luisi *et al.*, 1991). This is illustrated in two studies in which three-dimensional modeling of the mutated DBD of the AR predicts the functional activity of mutant receptors (Lobaccaro *et al.*, 1996; Brüggenwirth *et al.*, 1998). An interesting observation was made with respect to the second zinc cluster in which either one of two adjacent arginine residues (Arg607 and Arg608) was found to be mutated in individuals with PAIS who developed breast cancer (Wooster *et al.*, 1992; Lobaccaro *et al.*, 1993a). It is speculated that a decrease in androgen action within the breast cells could account for the development of male breast cancer by the loss of a protective effect of androgens. Unfortunately, information is lacking about whether activation of estrogen-regulated genes by the change of DNA-binding characteristics of the mutant AR had occurred.

The mutation Ala596Thr in the second zinc cluster in the so-called D box abolished dimerization in a patient with PAIS (Gast *et al.*, 1995). A similar mutation at an identical position in the second zinc cluster of the glucocorticoid receptor DBD has been created to discriminate between dimerization/DNA binding of the glucocorticoid receptor and protein–protein interactions with other transcription factors such as the activation protein-1 (AP-1) transcription complex (Reichardt *et al.*, 1998). It appeared that the dimerization mutant did not affect the crosstalk with other transcription factors. In this way, a tissue-specific response can be influenced by a single amino acid change and, if this is also true for the mutant AR, then the partial phenotype can be explained.

Mutations in the hinge (D) region   In the so-called hinge region, located between amino acid residues 622 and 670, no mutations have been reported yet that result in CAIS, which indicates that this region might be very flexible and that some variation in composition and length of this region is not detrimental for AR function. Two amino acid substitutions within the hinge region have been described that resulted in PAIS (Table 4.1). The I664N substitution on the border of the hinge region and the LBD resulted in decreased hormone binding (Pinsky *et al.*, 1992).

Mutations in the ligand-binding (E) domain   Elucidation of the three-dimensional crystallographic structure of the LBD of several nuclear receptors has established that a variable number (10–12, depending on the type of receptor) of α helices and an antiparallel β sheet arranged in a helical sandwich are involved in formation of the ligand-binding pocket and the formation of an interaction surface for binding proteins such as coactivators and corepressors (Bourguet *et al.*, 1995; Renaud *et al.*, 1995; Wagner *et al.*, 1995; Brzozowski *et al.*, 1997; Williams and Sigler, 1998).

For the AR LBD no information is available with respect to three-dimensional structure, but based on their high homology it can be predicted that it resembles to a large extent that of the progesterone receptor. Information from the three-dimensional structure of the LBD in the progesterone receptor can be applied for the AR. In the progesterone receptor, helices 3, 5, 7, 11 and 12 and the β turn are predominantly involved in the formation of the hydrophobic binding pocket, and amino acid residues in these helices are in close contact with the ligand (Williams and Sigler, 1998). Upon ligand binding, an interaction surface is formed, which allows interactions with other proteins. Similar considerations can be held for the AR LBD. It is therefore not surprising that helix 5 and the β turn in particular have a relatively high number of reported mutations which all affect ligand binding in a variable way (either low binding and normal dissociation constant (K_d), normal binding and raised K_d, or no binding) (Table 4.1).

It is remarkable that in the activation function (AF) 2 core region (893-EMMAEIIS-900) of the AR LBD a relatively low number of mutations has been reported. Only at position Ile898 has a mutation been described in an individual with the complete syndrome (Hiort *et al.*, 1998). It can be speculated that, in this part of helix 12, mutations in the AR LBD are less deleterious for AR function than those in helix 5 and in the β turn, where almost every amino acid residue has been found to be mutated in individuals with AIS.

Deletions of the androgen receptor gene
Only a few cases have been reported with partial or complete AR gene deletions, indicating the relatively low frequency of this type of AR defect (Brown *et al.*, 1988; Akin *et al.*, 1991; Trifiro *et al.*, 1991; Quigley *et al.*, 1992; MacLean *et al.*, 1993; Ris-Stalpers *et al.*, 1994b; Jakubiczka *et al.*, 1997). All cases reported are found in individuals with CAIS, with the exception of two cases, one in which an exon 4 deletion was found in a person with MAIS and azoospermia (Akin *et al.*, 1991) and another one in which a large intron 2 deletion of at least 6 kb was reported involving a branch point site, which resulted in a partial exon 3 skipping during the splicing process (Ris-Stalpers *et al.*, 1994b).

Deletion of either exon 3 or exon 4 occurs both in-frame and results in a nonfunctional protein lacking either the second zinc cluster or the hinge region and the N-terminal part of the LBD. In case of an exon 3 deletion, an intact and functional LBD is present. So far, functional significant mutations in the AR promoter region or in the 5' and 3' untranslated regions of the gene have not been reported.

Splice site mutations affecting androgen receptor RNA splicing
A special group of interesting but rare mutations are the splice donor and splice acceptor site mutations in the AR gene in individuals with AIS. For all splice donor sites in the AR gene, the consensus splice donor site sequence GTAAG/A is present. The five reported mutations in donor splice sites are all substitutions, either at position $+1$ (G $\rightarrow$ A or G $\rightarrow$ T) or at position $+3$ (A $\rightarrow$ T), and result in defective splicing with the consequence of one or more exons lost or the use of a cryptic splice donor site within the preceding exon (Ris-Stalpers *et al.*, 1990; Evans *et al.*, 1991; Pinsky *et al.*, 1992; Yong *et al.*, 1994a; Hiort *et al.*, 1998). In all of the five reported cases, the phenotype is complete androgen insensitivity. In one case, an insertion of one nucleotide (T) at position $+3$ in the splice donor site of intron 1 has been reported, resulting in a complete androgen-insensitive phenotype (Trifiro *et al.*, 1997). Finally, a substitution at position -11 (T $\rightarrow$ G) has been found in the pyrimidine-rich region of the splice acceptor site of intron 2, resulting in the activation of a cryptic splice acceptor site at position $-70/-69$ and consequently in the insertion of 69 nucleotides (corresponding to 23 additional amino acid residues) in the mRNA between exons 2 and 3 (Brüggenwirth *et al.*, 1997). The corresponding protein is defective in DNA binding because the insertion has occurred between the first and second zinc cluster.

Spinal and bulbar muscular atrophy

SBMA or Kennedy disease is an X-linked recessive disease characterized by the selective loss of anterior horn neuronal cells in the spinal cord, depletion of sensory neurons in the dorsal root ganglia, and selective degeneration of motor neurons in the brainstem (motor neurons in the lower cranial nerves) (Harding *et al.*, 1982; Arbizu *et al.*, 1983;

Warner *et al.*, 1992). Motor neuron loss in the spinal cord in the bulbar region results in muscle weakness and atrophy. These symptoms are often preceded by cramps. The clinical symptoms usually manifest in the third to fifth decade (Nance, 1997). In addition, patients with SBMA frequently exhibit endocrinologic abnormalities including testicular atrophy, reduced or absent fertility, gynecomastia, and raised levels of LH, follicle-stimulating hormone (FSH) and estradiol. These symptoms are also observed in MAIS. Sex differentiation proceeds normally and characteristics of the mild androgen insensitivity appear later in life.

The disease is caused by an abnormal increase in the length of a polymorphic polyglutamine stretch, encoded by a CAG repeat. This repeat is located in exon 1, which encodes the N-terminal part of the AR (Faber *et al.*, 1989; Sleddens *et al.*, 1992) (Fig. 4.1). In normal individuals, the CAG repeat contains 9–33 CAGs, whereas 38–75 CAGs are associated with SBMA (La Spada *et al.*, 1991; Nance, 1997). The severity of the symptoms is inversely correlated with length of the CAG repeat (Igarashi *et al.*, 1992; La Spada *et al.*, 1992; Doyu *et al.*, 1993). Similar polyglutamine stretch elongations have been found in a variety of proteins (huntingtin, atrophin, ataxins) which are involved in the pathogenesis of other neurodegenerative diseases (Huntington disease and several types of spinocerebellar ataxia) (Wilmot and Warren, 1998). This indicates the existence of a common neurotoxic mechanism of the expanded CAG repeat.

From the clinical signs, it may be concluded that SBMA is likely a result from a combination of a gain-of-function mechanism in motor neurons and a loss-of-function mechanism, causing partial loss of receptor function in androgen target tissues. The fact that neurologic symptoms of SBMA are not observed in patients with AIS, which is caused by inactivating mutations of the AR gene, points to a gain-of-function mechanism in SBMA.

The intragenic expanded CAG repeat could be pathogenic at either the DNA, RNA or protein level. Increased binding of RNA-binding proteins to RNAs containing expanded (CAG) repeats has been observed, suggesting that RNAs might disrupt normal transport in the cell, or squelch cytoplasmic proteins that bind to CAGs (McLaughlin *et al.*, 1996). It has also been suggested that proteins with an expanded polyglutamine stretch function as better substrates for transglutaminase, an ubiquitously expressed enzyme that catalyzes the coupling of glutamine to lysine residues (Green, 1993). Furthermore, it was shown that peptides containing expanded glutamine stretches are indeed better *in vitro* substrates (Kahlem *et al.*, 1996). The subsequent accumulation of isopeptides and protein aggregates might be toxic for the cells. Although information on a possible mechanism of the AR in neuronal cell death is sparse, recent evidence points in the direction of involvement of the AR with the expanded repeat in aggregate formation and aberrant processing. Evidence has been presented that aggregate formation and proteolytic processing of the AR protein can occur in a polyglutamine repeat length-dependent manner, and aberrant metabolism of the expanded repeat AR is coupled to cellular toxicity (Butler *et al.*, 1998; Merry *et al.*, 1998; Stenoien *et al.*, 1999). Furthermore, in neuronal tissues from patients with SBMA ubiquitinated nuclear inclusions containing AR protein fragments have been detected (Li *et al.*, 1998). Finally, recent evidence has been provided that caspase-3 cleavage of an AR displaying an expanded polyglutamine tract may play a role in the induction of neural cell death (Kobayashi *et al.*, 1998; Ellerby *et al.*, 1999).

Endocrine abnormalities observed in patients with SBMA are indicative of reduced receptor functioning. Cotransfection studies demonstrated that the polymorphic polyglutamine stretch influences the transcription activation capacity, although promoter-dependent differences have been observed (Mhatre *et al.*, 1993; Chamberlain *et al.*, 1994; Jenster *et al.*, 1994; Kazemi-Esfarjani *et al.*, 1995). The length of the CAG repeat is inversely related to transcription activity. Ligand binding (K_d) is not affected (Mhatre *et al.*, 1993; Chamberlain *et al.*, 1994; Kazemi-Esfarjani *et al.*, 1995). From these studies, it was concluded that the length of the CAG repeat might influence intrinsic properties of the receptor. In contrast to these *in vitro* findings, androgen-binding abnormalities were detected in cells derived from patients with SBMA (Warner *et al.*, 1992; Danek *et al.*, 1994; MacLean *et al.*, 1995; Lumbroso *et al.*, 1997). In contrast to spinal cord of controls, AR mRNA expression in the spinal cord of a patient with SBMA was decreased as well (Nakamura *et al.*, 1994). Therefore, it was proposed that the CAG repeat might interfere with transcription of the AR gene, indicating that the expanded repeat also plays a role at the DNA level.

Prostate cancer

The prostate develops late during embryogenesis from the urogenital sinus. Ample evidence exists to show that, at early steps of prostate development, epithelial cell differentiation is directed by androgen-regulated processes in the mesenchymal cells (Cunha *et al.*, 1987). Later steps in prostate development might depend on the AR in the epithelial cells, eventually leading to the production of prostate-specific proteins (Donjacour and Cunha, 1993).

According to the current hypotheses, a prostate tumor cell develops from an aberrantly growing progenitor cell of the luminal epithelial cells. Like prostate development, initial growth of prostate tumors depends on androgens. Consequently, AR-regulated target genes involved in the proliferation of hormone-dependent prostate tumors might be identical to those needed for normal prostate growth and development.

Treatment of locally confined prostate cancer is based on the surgical removal of the prostate or on radiation therapy. Therapy of metastatic prostate cancer is in general based on the inhibition of androgen-regulated cell growth by blockade of testosterone production (orchiectomy, LHRH analogs) or by inhibition of AR function (antiandrogens). However, although endocrine therapy initially seems effective, local recurrences and metastases develop within a few years. A key question in this regard is whether the AR is still important for the growth of these apparently hormone-refractory prostate tumors, or whether androgen-regulated cell growth has been bypassed by different mechanisms of cell growth stimulation.

In the normal mature prostate, the AR is expressed in both stromal and epithelial cells (Ruizeveld de Winter *et al.*, 1991). In general, immunohistochemical studies demonstrate that in prostate cancer the AR is expressed in the epithelial tumor cells and in the stromal cell compartment, although the stromal expression is variable (Sadi *et al.*, 1991; van der Kwast *et al.*, 1991; Chodak *et al.*, 1992; Sadi and Barrack, 1993; Ruizeveld de Winter *et al.*, 1994; Tilley *et al.*, 1994; Pertschuk *et al.*, 1995). So, androgen-regulated growth of the tumor could be directed by the activated AR in the stromal cell or in the tumor cells. The role of androgen-regulated processes in the stromal cells on prostate tumor cells is not well understood. In model systems, mesenchymal cells derived from the urogenital sinus, or stromal cells from the mature prostate, are able to support growth of human prostate tumors transplanted on male nude mice, whether or not the tumor cells express the AR (Chung *et al.*, 1991). Evidence against a role of stromal cells in androgen-regulated prostate tumor growth is the observation that, in cell culture in the absence of stromal cells, the proliferation of the LNCaP prostate tumor cell line is androgen sensitive

(Schuurmans *et al.*, 1988). In addition, prostate cancer metastases, which are supposed not to be surrounded by prostate-derived stromal cells, respond to AR blockade.

The CAG repeat in the androgen receptor gene and risk of prostate cancer

In epidemiologic studies it has been found that a slightly shorter CAG repeat in exon 1 of the AR gene, which results in a shortened glutamine stretch in the AR protein, correlates with a higher risk of prostate cancer development or progressive tumor growth (Irvine *et al.*, 1995; Giovannucci *et al.*, 1997; Stanford *et al.*, 1997), although more recently this correlation has been disputed (Correa-Cerro *et al.*, 1999). *In vitro* experiments indicate that a short glutamine stretch leads to a somewhat more active AR (Chamberlain *et al.*, 1994; Jenster *et al.*, 1994; Kazemi-Esfarjani *et al.*, 1995). However, the differences in activity are modest, and have been clearly observed only for ARs with a very short glutamine stretch or in AR mutants that completely lack the glutamine repeat. Obviously, small differences in repeat lengths, which can lead to a slow but relevant biologic effect during lifetime, are almost impossible to study in short-term experiments. In one report the AR glutamine repeat in the tumor was substantially shortened compared with nontumor tissue (18 versus 24 residues) (Schoenberg *et al.*, 1994). Again, it is unknown whether this deletion will have a stimulating effect on tumor growth in the long term.

Androgen receptor expression and amplification in prostate cancer

In hormone-refractory prostate cancer, the tumor can grow in the absence of testicular testosterone or, more precisely, its metabolite DHT. Obvious explanations of hormone-resistant tumor growth include bypassing of the AR-regulated pathway or an outgrowth of a tumor cell subpopulation that already was androgen independent at the time of diagnosis and start of therapy. However, several points of evidence also point to the AR as an important component in a proportion of hormone therapy-resistant tumors.

Although more heterogeneous than in the normal prostate, the vast majority of hormone therapy-resistant, locally progressive and metastatic tumors show high levels of AR expression (Hobisch *et al.*, 1995; Kleinerman D.I. *et al.*, unpublished results). The predominant nuclear localization of the AR in these tumors argues in favor of an active conformation. In a proportion of the local recurrences, AR mRNA expression is higher than that in the tumor before endocrine therapy (e.g. orchiectomy) (Koivisto *et al.*, 1997). Interestingly, in bone metastases, which represent a late stage of hormone-refractory prostate cancer, AR expression seems to be identical to or even higher and more homogeneous than that in the corresponding local recurrent tumor (Kleinerman D.I., unpublished results).

AR overexpression may be the result of amplification of the AR gene. *In situ* hybridization studies showed AR gene amplification in approximately 30% of recurrent hormone-refractory tumors following orchiectomy (Visakorpi *et al.*, 1995; Koivisto *et al.*, 1997). Importantly, this amplification was not detected in tumors before therapy. Furthermore, amplification of the gene was found predominantly in tumors that relapsed relatively late after the start of hormone therapy. These findings suggest that tumors with AR amplification responded primarily to hormone therapy, followed later by an escape mechanism including amplification of the AR gene and an overexpressed active AR. It remains to be established whether, in tumors without AR gene amplification, presumed AR overexpression results from other mechanisms of upregulation of gene expression.

Androgen receptor mutations

Another mechanism that could explain a functionally active AR in hormone therapy-resistant prostate cancer would be the occurrence of a truncated, constitutively active, receptor or a structurally altered receptor due to a nonsense or missense mutation in the gene. Nonsense mutations in the first part of exon 4 of the gene, which would lead to a constitutively active AR, have never been reported. Unfortunately, the situation concerning AR amino acid substitutions in prostate tumor tissues is rather confusing. Different reports describe apparently conflicting data. Point mutations in the AR gene have been detected in primary and metastatic tumors with varying frequency (for an overview see http://www.mcgill.ca/androgendb, and Table 4.2). The effect of most of these mutations on the transactivating function of the receptor has not been established, so it is impossible to determine the biologic relevance of these observations. In fact, for most mutations, it cannot be determined whether they are the cause or the result of prostate cancer. Some of the missense mutations have also been found in individuals with AIS, and would most likely lead to an AR with diminished activity or a completely inactive receptor. The few nonsense mutations reported are expected to result in an inactive receptor. If inactivating mutations were selected for during cancer progression, it is unclear how this would be favorable for tumor growth. One possibility is that, if the cancer cells have bypassed the AR pathway to become androgen independent, inactivating mutations could abolish the potential differentiating effect of the AR.

Five mutations in the LBD have been studied in more detail. As the classic and best documented example, the AR gene in the LNCaP prostate cancer cell line contains a point mutation at codon 877 (ACT to GCT), which leads to a Thr to Ala substitution (Veldscholte *et al.*, 1990). This substitution has a dramatic effect on the ligand specificity and transactivating function of the AR. It renders the receptor, and as a consequence the growth of LNCaP cells, responsive to most antiandrogens, and to the natural low-affinity ligands estradiol, progesterone and adrenal gland androgens (Schuurmans *et al.*, 1988; Veldscholte *et al.*, 1992; Tan *et al.*, 1997). The Thr877Ala substitution has been found not only in the LNCaP cell line, but also in patients with prostate cancer (see Table 4.2). Preferentially, this mutation has been found in hormone-refractory tumors, underscoring its functional importance. As sole mutation, the Thr to Ala substitution has been observed by various groups and may, as a consequence, be a hotspot of mutation in prostate cancer. In two patients with prostate cancer a Thr877Ser substitution (ACT-AGT) has been detected, which, like Thr877Ala, results in a receptor with diminished ligand specificity (Taplin *et al.*, 1995, 1999; Fenton *et al.*, 1997).

Preference of the Thr to Ala substitution in prostate cancer might be explained by the type of mutation (A → G versus C → G). Recently, evidence has been presented that AR mutations do not play a significant role in the flutamide withdrawal response in which the interruption of flutamide therapy has a temporarily beneficial effect on the patient (Taplin *et al.*, 1999). Importantly, it has been shown that patients who are resistant to flutamide therapy can still respond to the antiandrogen bicalutamide, which probably affects AR function at a different level. Three individual amino acid substitutions in the LBD correspond to a loss of ligand specificity of the transactivating function, similar to that in Thr877Ala. One of these is very close to 877: His874Tyr (Tan *et al.*, 1997). The other two are in a different part of the LBD: Val715Met and Val730Met respectively (Newmark *et al.*, 1992; Culig *et al.*, 1993; Peterziel *et al.*, 1995).

It might be predicted that these amino acid substitutions modify the folding of the LBD of the AR, or might even affect the direct interaction of the AR with its ligand.

Table 4.2 Androgen receptor mutations in prostate cancer tissue samples

Type of mutation

Nucleotide	Amino acid	Domain	Reference	Comment
163T > C	L54S	A/B	Tilley et al. (1996)	Also F891L
172T > A	L57Q	A/B	Tilley et al. (1996)	
193A > G	Q64R	A/B	Tilley et al. (1996)	Also L830P
338G > T	Q112H	A/B	Tilley et al. (1996)	Also W796X
541A > G	K180R	A/B	Tilley et al. (1996)	
799T > C	M266T	A/B	Tilley et al. (1996)	Also L572P
807C > T	P269S	A/B	Tilley et al. (1996)	
1021C > T	P340L	A/B	Castagnaro et al. (1993)	
1589A > G	D528G	A/B	Tilley et al. (1996)	
1643delG	Frameshift	C	Takahashi et al. (1995)	
1664delC	Frameshift	C	Takahashi et al. (1995)	Found twice
1948G > A	S647N	D	Taplin et al. (1995)	
2011A > G	Q670R	E	Tilley et al. (1996)	Also S791P
2017T > C	I672T	E	Tilley et al. (1996)	
2050G > C	G683A	E	Koivisto et al. (1997)	
2114T > A	L701H	E	Suzuki et al. (1993)	
2114T > A	L701H	E	Watanabe et al. (1997)	
2145A > G	V715M	E	Culig et al. (1993)	Modifies ligand specificity
2160A > G	K720E	E	Kleinerman D.I., unpublished	
2163G > A	A721T	E	Taplin et al. (1995)	
2179G > T	R726L	E	Elo et al. (1995)	Germline
2190G > A	V730M	E	Newmark et al. (1992)	Modifies ligand specificity
2224G > A	W741X	E	Takahashi et al. (1995)	
2231delG	Frameshift	E	Takahashi et al. (1995)	
2245C > T	A748V	E	Takahashi et al. (1995)	
2249G > A	M749F	E	Takahashi et al. (1995)	
2250G > A	G750S	E	Takahashi et al. (1995)	

Nucleotide change	Amino acid change	Domain	Reference	Notes
2254G > A	W751X	E	Takahashi et al. (1995)	Found twice
2255G > A	W751X	E	Takahashi et al. (1995)	
2262T > C	F754L	E	Takahashi et al. (1995)	
2265A > G	T755A	E	Takahashi et al. (1995)	
2278C > T	S759P	E	Takahashi et al. (1995)	
2290A > G	Y763C	E	Takahashi et al. (1995)	
2347G > A	S782N	E	Tilley et al. (1996)	
2390G > A	W796X	E	Tilley et al. (1996)	
2394C > G	Q798E	E	Evans et al. (1996)	
2394C > G	Q798E	E	Castagnaro et al. (1993)	
2598G > A	V866M	E	Takahashi et al. (1995)	
2522C > T	H874Y	E	Taplin et al. (1995)	Modifies ligand specificity
2631A > G	T877A	E	Veldscholte et al. (1990)	Modifies ligand specificity, LNCaP
2631A > G	T877A	E	Suzuki et al. (1993)	
2631A > G	T877A	E	Suzuki et al. (1996)	Found three times
2631A > G	T877A	E	Gaddipati et al. (1994)	Found six times
2631A > G	T877A	E	Kleinerman D.I., unpublished	
2631A > G	T877A	E	Taplin et al. (1999)	
2632C > G	T877S	E	Taplin et al. (1995)	Found five times
2632C > G	T877S	E	Taplin et al. (1999)	
2670G > A	D890N	E	Taplin et al. (1999)	
2707A > G	Q902R	E	Taplin et al. (1995)	
2738A > G	K910R	E	Watanabe et al. (1997)	

Nucleotide substitutions and small nucleotide insertions or deletions are shown. Stop codons are designated by X. The domains are designated as follows: exon 1: A/B = N-terminal domain; exons 2 and 3: C = DNA-binding domain; first part of exon 4: D = hinge region; rest of exon 4 and exons 5–8: E = ligand-binding domain.

Although the three-dimensional structure of the LBD of the AR is unknown, comparison with the progesterone receptor indicates a direct effect on ligand binding of Thr877Ala/Ser (Williams and Sigler, 1998). In conclusion, the authors propose that the most likely explanation for AR mutations in late-stage prostate cancer represent a mixture of functional mutations and random mutations, which might result from genomic instability.

Alternative activation of androgen receptor function

As has been described above, the AR is not activated only by steroid ligands but also in a steroid hormone-independent fashion by signaling through kinase signal transduction pathways. Manipulation of kinase signaling might ultimately result in the phosphorylation of the AR at serine, threonine and/or tyrosine residues. It is unknown which sites in the AR might be most relevant in this respect. Neither is it known whether or not phosphorylation of other components in AR-regulated gene expression, like coactivators, is important for ligand-independent activation. Most relevant for prostate cancer is that, as shown in the LNCaP prostate cancer cell line, androgen-regulated gene expression can be mimicked by growth factors and cytokines such as IGF-1, KGF, EGF and IL-6.

Recently, it has been found that, in the presence or absence of the synthetic androgen R1881, overexpression of the *HER2/neu* oncogene can lead to an increased expression of androgen-regulated prostate-specific antigen (PSA) (Riegman *et al.*, 1991; Cleutjens *et al.*, 1997; Craft *et al.*, 1999; Yeh *et al.*, 1999). Upregulation of PSA expression is, at least partially, due to increased PSA promoter activity. This activation of PSA promoter activity via *HER2/neu* can be blocked by bicalutamide, indicating crosstalk between steroid- and growth factor-activated signal transduction. Overexpressed *HER2/neu* is also able to stimulate the *in vitro* growth and *in vivo* tumorigenicity of LNCaP cells in intact and castrated mice. Although it is tempting to speculate, it remains to be investigated whether this involves AR crosstalk. Furthermore, it has to be realized that *HER2/neu* amplification and overexpression has not been reported convincingly in prostate cancer tissues.

Overexpression of coactivators or downregulation of corepressors could potentially affect AR function. So far, no evidence exists that this mechanism is involved in hormone-resistant prostate cancer. The observation that the coactivator androgen receptor-associated protein 70 kDa (ARA70) can increase the transactivating function of the AR in the presence of estradiol and antiandrogens (Miyamoto *et al.*, 1998; Yeh *et al.*, 1998) deserves further attention. It should be investigated whether candidate coactivators are overexpressed or candidate corepressors downregulated in prostate cancer.

SUMMARY AND CONCLUSIONS

The cloning and characterization of the AR made it possible to understand the molecular basis of androgen action in normal and disease states. Numerous studies have contributed to a detailed knowledge of the working mechanism of this steroid receptor, its functional domains, posttranslational modifications, and target genes. However, despite these successful efforts, there are a multitude of unresolved questions. At least three different areas are the focus of intense research. Undoubtedly, the isolation and characterization of AR-binding proteins, including coactivators and corepressors, is the most extensively studied topic. Since the cloning of the first nuclear receptor coactivators and corepressors in 1995, a large number of nuclear receptor and AR-binding proteins has been identified. Although the role of the different interacting proteins in receptor-mediated transcription has not always been clearly defined, our understanding of receptor-driven transcription initiation has been tremendously enhanced by the molecular characterization of coactivators and corepressors. The ongoing efforts to identify novel receptor-binding proteins will only intensify the challenge of sorting out the components of the receptor-bound complex, their mutual interactions, and their role in receptor function.

Another field of interest is the novel finding that the N-terminal domain of steroid receptors, including the AR, binds its own LBD. It is not known whether this interaction is intramolecular or intermolecular when the AR occupies its ARE as a homodimer. Moreover, the functional importance of this interaction has not yet been established.

The third area of research that will greatly impact on our understanding of AR action is the subcellular distribution of the AR and AR mutants. Although it has been clearly established that the liganded wild-type AR is a nuclear protein, absent in nucleoli and distributed in a granular fashion, certain AR mutants reveal a clustered or aggregate pattern. Besides AR mutants that lack DNA binding, ARs with an extended glutamine repeat from patients with SBMA, will also form large aggregates in the cytoplasm and nucleus. The importance of aggregate formation with respect to symptoms of SBMA is still debated, but molecular characterization of these clusters will probably enhance our knowledge of AR cellular trafficking, matrix binding, protein degradation and stress responses.

A research area in which the AR lags behind is the three-dimensional elucidation of the different functional domains by nuclear magnetic resonance (NMR) or X-ray crystallography. None of the AR domains has been analyzed in this way and structural analysis of the DBD and LBD is predicted based on the known three-dimensional organization of other steroid receptors. The elucidation of the exact structure of the AR domains will be invaluable for our understanding of antagonist action, interaction with binding proteins, and the ultimate three-dimensional organization of the full-length protein.

The AR is a unique protein as a result of its direct involvement in three unrelated diseases. Germline mutations that inhibit or abolish AR function result in partial or complete androgen insensitivity (Fig. 4.3). Those AR mutations are highly variable in type (deletion, substitution, insertion) and location (in introns or the different exons), but all affect AR protein expression or protein activity.

The extension of the CAG repeat in the first exon of the AR gene results in SBMA or Kennedy disease. The subsequent increased length of a glutamine repeat in the AR N-terminal domain might affect AR protein activity and stability, explaining the mild symptoms of androgen insensitivity in these patients. However, the reasons for the loss of motor neurons and aggregation of the AR are not fully understood. The third disease associated with AR aberrations is prostate cancer. The wide spectrum of mutations identified in the AR show that different cancer cells adapt diversely to their environmental and cellular signals. AR amplification, amino acid substitutions that change ligand specificity, and AR protein and gene inactivations are all selective ways to affect the odds of surviving and proliferating in the challenging environment faced by the cell during metastasis or androgen ablation and antiandrogen treatments. Although many mutations have been identified, only a few have been characterized to explain the potential advantage for the cancer cell. This work needs to be continued for

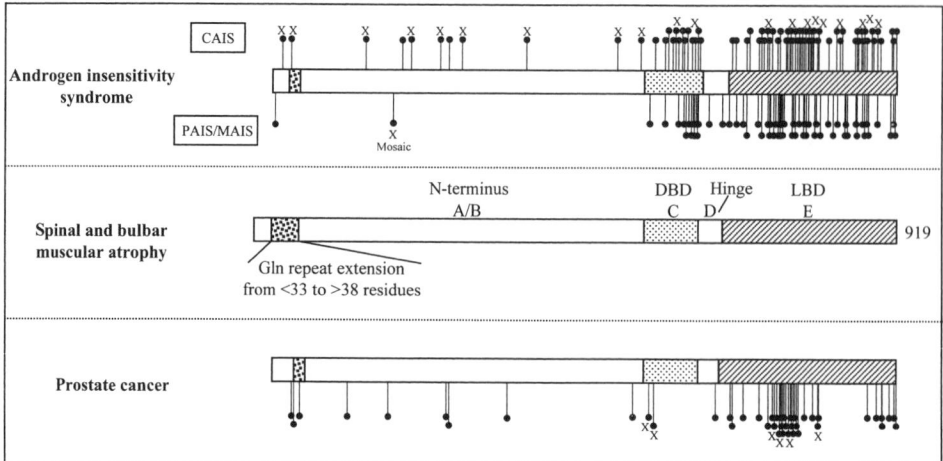

Fig. 4.3 Location of amino acid substitutions, small deletions and insertions identified in the androgen receptor in individuals with androgen insensitivity, spinal and bulbar muscular atrophy, and prostate cancer. Mutations in subjects with androgen insensitivity syndrome (AIS) are subdivided into the complete syndrome (CAIS) and the partial or mild form (PAIS/MAIS). 'X' indicates that the mutation results in a stop codon or in an incorrect reading frame eventually leading to translation termination. In many instances, the same amino acid has been substituted in different residues; only one line will mark that site. DBD, DNA-binding domain; LBD, ligand-binding domain.

a better understanding of the role of the AR in the development and progression of prostate cancer.

REFERENCES

Aarnisalo, P., Palvimo, J.J. and Janne, O.A. (1998) CREB-binding protein in androgen receptor-mediated signaling. *Proc. Natl. Acad. Sci. USA* **95**: 2122–2127.

Adeyemo, O., Kallio, P.J., Palvimo, J.J., Kontula, K. and Janne, O.A. (1993) A single-base substitution in exon 6 of the androgen receptor gene causing complete androgen insensitivity: the mutated receptor fails to transactivate but binds to DNA *in vitro. Hum. Mol. Genet.* **2**: 1809–1812.

Adler, A.J., Scheller, A. and Robins, D.M. (1993) The stringency and magnitude of androgen-specific gene activation are combinatorial functions of receptor and nonreceptor binding site sequences. *Mol. Cell. Biol.* **13**: 6326–6335.

Akin, J.W., Behzadian, A., Tho, S.P. and McDonough, P.G. (1991) Evidence for a partial deletion in the androgen receptor gene in a phenotypic male with azoospermia. *Am. J. Obstet. Gynecol.* **165**: 1891–1894.

Alland, L., Muhle, R., Hou, H. *et al.* (1997) Role for n-cor and histone deacetylase in sin3-mediated transcriptional repression. *Nature* **387**, 49–55.

Allera, A., Herbst, M.A., Griffin, J.E., Wilson, J.D., Schweikert, H.U. and McPhaul, M.J. (1995) Mutations of the androgen receptor coding sequence are infrequent in patients with isolated hypospadias. *J. Clin. Endocrinol. Metab.* **80**: 2697–2699.

Antonarakis, S.E. (1998) Recommendations for a nomenclature system for human gene mutations. Nomenclature Working Group. *Hum. Mutat.* **11**: 1–3.

Arbizu, T., Santamaria, J., Gomez, J.M., Quilez, A. and Serra, J.P. (1983) A family with adult spinal and bulbar muscular atrophy, X-linked inheritance and associated testicular failure. *J. Neurol. Sci.* **59**: 371–382.

Archer, T.K., Lefebvre, P., Wolford, R.G. and Hager, G.L. (1992) Transcription factor loading on the MMTV promoter: a bimodal mechanism for promoter activation. *Science* **255**: 1573–1576.

Baldazzi, L., Baroncini, C., Pirazzoli, P. *et al.* (1994) Two mutations causing complete androgen insensitivity: a frameshift in the steroid binding domain and a Cys → Phe substitution in the second zinc finger of the androgen receptor. *Hum. Mol. Genet.* **3**: 1169–1170.

Batch, J.A., Williams, D.M., Davies, H.R. *et al.* (1992) Androgen receptor gene mutations identified by SSCP in fourteen subjects with androgen insensitivity syndrome. *Hum. Mol. Genet.* **1**: 497–503.

Batch, J.A., Davies, H.R., Evans, B.A., Hughes, I.A. and Patterson, M.N. (1993) Phenotypic variation and detection of carrier status in the partial androgen insensitivity syndrome. *Arch. Dis. Child.* **68**: 453–457.

Beitel, L.K., Kazemi-Esfarjani, P., Kaufman, M. *et al.* (1994a) Substitution of arginine-839 by cysteine or histidine in the androgen receptor causes different receptor phenotypes in cultured cells and coordinate degrees of clinical androgen resistance. *J. Clin. Invest.* **94**: 546–554.

Beitel, L.K., Prior, L., Vasiliou, D.M. *et al.* (1994b) Complete androgen insensitivity due to mutations in the probable alpha-helical segments of the DNA-binding domain in the human androgen receptor. *Hum. Mol. Genet.* **3**: 21–27.

Belsham, D.D., Pereira, F., Greenberg, C.R., Liao, S. and Wrogemann, K. (1995) Leu-676-Pro mutation of the androgen receptor causes complete androgen insensitivity syndrome in a large Hutterite kindred. *Hum. Mutat.* **5**: 28–33.

Berrevoets, C.A., Doesburg, P., Steketee, K., Trapman, J. and Brinkmann, A.O. (1998) Functional interactions of the AF-2 activation domain core region of the human androgen receptor with the amino-terminal domain and with the transcriptional coactivator TIF2 (transcriptional intermediary factor2). *Mol. Endocrinol.* **12**: 1172–1183.

Bevan, C.L., Brown, B.B., Davies, H.R., Evans, B.A., Hughes, I.A. and Patterson, M.N. (1996) Functional analysis of six androgen receptor mutations identified in patients with partial androgen insensitivity syndrome. *Hum. Mol. Genet.* **5**: 265–273.

Bevan, C.L., Hughes, I.A. and Patterson, M.N. (1997) Wide variation in androgen receptor dysfunction in complete androgen insensitivity syndrome. *J. Steroid Biochem. Mol. Biol.* **61**: 19–26.

Blok, L.J., de Ruiter, P.E. and Brinkmann, A.O. (1996) Androgen receptor phosphorylation. *Endocr. Res.* **22**: 197–219.

Blok, L.J., de Ruiter, P.E. and Brinkmann, A.O. (1998) Forskolin-induced dephosphorylation of the androgen receptor impairs ligand binding. *Biochemistry* **37**: 3850–3857.

Boehmer, A.L., Brinkmann, A.O., Niermeijer, M.F., Bakker, L., Halley, D.J. and Drop, S.L. (1997) Germ-line and somatic mosaicism in the androgen insensitivity syndrome: implications for genetic counseling. *Am. J. Hum. Genet.* **60**: 1003–1006.

Bourguet, W., Ruff, M., Chambon, P., Gronemeyer, H. and Moras, D. (1995) Crystal structure of the ligand-binding domain of the human nuclear receptor RXR-alpha. *Nature* **375**: 377–382.

Brinkmann, A.O., Jenster, G., Ris-Stalpers, C. *et al.* (1995) Androgen receptor mutations. *J. Steroid Biochem. Mol. Biol.* **53**: 443–448.

Brinkmann, A.O., Faber, P.W., Jenster, G. *et al.* (1996) Structural and functional aspects of androgen receptors in normal and pathological situations. In: J.R. Pasquilini and B.S. Katzenellenbogen (eds), *Hormone-Dependent Cancer*, pp. 323–342. Marcel Dekker, New York.

Brown, T.R., Lubahn, D.B., Wilson, E.M., Joseph, D.R., French, F.S. and Migeon, C.J. (1988) Deletion of the steroid-binding domain of the human androgen receptor gene in one family with complete androgen insensitivity syndrome: evidence for further genetic heterogeneity in this syndrome. *Proc. Natl. Acad. Sci. USA* **85**: 8151–8155.

Brown, C.J., Goss, S.J., Lubahn, D.B. *et al.* (1989) Androgen receptor locus on the human X chromosome: regional localization to Xq11-12 and description of a DNA polymorphism. *Am. J. Hum. Genet.* **44**: 264–269.

Brown, T.R., Lubahn, D.B., Wilson, E.M., French, F.S., Migeon, C.J. and Corden, J.L. (1990) Functional characterization of naturally occurring mutant androgen receptors from subjects with complete androgen insensitivity. *Mol. Endocrinol.* **4**: 1759–1772.

Brown, T.R., Scherer, P.A., Chang, Y.T. *et al.* (1993) Molecular genetics of human androgen insensitivity. *Eur. J. Pediatr.* **152**: S62–S69.

Brüggenwirth, H.T., Boehmer, A.L., Verleun-Mooijman, M.C. *et al.* (1996) Molecular basis of androgen insensitivity. *J. Steroid Biochem. Mol. Biol.* **58**: 569–575.

Brüggenwirth, H.T., Boehmer, A.L., Ramnarain, S. *et al.* (1997) Molecular analysis of the androgen-receptor gene in a family with receptor-positive partial androgen insensitivity: an unusual type of intronic mutation. *Am. J. Hum. Genet.* **61**: 1067–1077.

Brüggenwirth, H.T., Boehmer, A.L., Lobaccaro, J.M. *et al.* (1998) Substitution of Ala564 in the first zinc cluster of the deoxyribonucleic acid (DNA)-binding domain of the androgen receptor by Asp, Asn, or Leu exerts differential effects on DNA binding. *Endocrinology* **139**: 103–110.

Brzozowski, A.M., Pike, A.C., Dauter, Z. *et al.* (1997) Molecular basis of agonism and antagonism in the oestrogen receptor. *Nature* **389**: 753–758.

Butler, R., Leigh, P.N., McPhaul, M.J. and Gallo, J.M. (1998) Truncated forms of the androgen receptor are associated with polyglutamine expansion in X-linked spinal and bulbar muscular atrophy. *Hum. Mol. Genet.* **7**: 121–127.

Castagnaro, M., Yandell, D.W., Dockhorn-Dworniczal, B., Wolfe, H.J. and Poremba, C. (1993) Androgenrezeptor-gen mutationen und p53-gen analyse in fortgeschritten prostatakarzinomen. *Verh. Dtsch. Ges. Pathol.* **77**: 119–123.

Chamberlain, N.L., Driver, E.D. and Miesfeld, R.L. (1994) The length and location of CAG trinucleotide repeats in the androgen receptor N-terminal domain affect transactivation function. *Nucleic Acids Res.* **22**: 3181–3186.

Chamberlain, N.L., Whitacre, D.C. and Miesfeld, R.L. (1996) Delineation of two distinct type 1 activation functions in the androgen receptor amino-terminal domain. *J. Biol. Chem.* **271**: 26 772–26 778.

Chang, C., Kokontis, J. and Liao, S. (1988) Molecular cloning of human and rat complementary DNA encoding androgen receptors. *Science* **240**: 324–326.

Chang, C., Saltzman, A., Yeh, S. *et al.* (1995) Androgen receptor: an overview. *Crit. Rev. Eukaryot. Gene Expr.* **5**: 97–125.

Chen, J.D. and Li, H. (1998) Coactivation and corepression in transcriptional regulation by steroid/nuclear hormone receptors. *Crit Rev. Eukaryot. Gene Expr.* **8**: 169–190.

Chodak, G.W., Kranc, D.M., Puy, L.A., Takeda, H., Johnson, K. and Chang, C. (1992)

Nuclear localization of androgen receptor in heterogeneous samples of normal, hyperplastic and neoplastic human prostate. *J. Urol.* **147**: 798–803.

Choong, C.S. and Wilson, E.M. (1998) Trinucleotide repeats in the human androgen receptor: a molecular basis for disease. *J. Mol. Endocrinol.* **21**: 235–257.

Choong, C.S., Kemppainen, J.A., Zhou, Z.X. and Wilson, E.M. (1996a) Reduced androgen receptor gene expression with first exon CAG repeat expansion. *Mol. Endocrinol.* **10**: 1527–1535.

Choong, C.S., Quigley, C.A., French, F.S. and Wilson, E.M. (1996b) A novel missense mutation in the amino-terminal domain of the human androgen receptor gene in a family with partial androgen insensitivity syndrome causes reduced efficiency of protein translation. *J. Clin. Invest.* **98**: 1423–1431.

Choong, C.S., Sturm, M.J., Strophair, J.A. *et al.* (1996c) Partial androgen insensitivity caused by an androgen receptor mutation at amino acid 907 (Gly → Arg) that results in decreased ligand binding affinity and reduced androgen receptor messenger ribonucleic acid levels. *J. Clin. Endocrinol. Metab.* **81**: 236–243.

Chung, L.W., Gleave, M.E., Hsieh, J.T., Hong, S.J. and Zhau, H.E. (1991) Reciprocal mesenchymal–epithelial interaction affecting prostate tumour growth and hormonal responsiveness. *Cancer Surv.* **11**: 91–121.

Claessens, F., Alen, P., Devos, A., Peeters, B., Verhoeven, G. and Rombauts, W. (1996) The androgen-specific probasin response element 2 interacts differentially with androgen and glucocorticoid receptors. *J. Biol. Chem.* **271**: 19 013–19 016.

Cleutjens, K.B., van der Korput, H.A., van Eekelen, C.C., van Faber, P.W. and Trapman, J. (1997) An androgen response element in a far upstream enhancer region is essential for high, androgen-regulated activity of the prostate-specific antigen promoter. *Mol. Endocrinol.* **11**: 148–161.

Correa-Cerro, L., Wohr, G., Haussler, J. *et al.* (1999) (CAG)nCAA and GGN repeats in the human androgen receptor gene are not associated with prostate cancer in a French–German population. *Eur. J. Hum. Genet.* **7**: 357–362.

Craft, N., Shostak, Y., Carey, M. and Sawyers, C.L. (1999) A mechanism for hormone-independent prostate cancer through modulation of androgen receptor signaling by the *HER2/neu* tyrosine kinase. *Nat. Med.* **5**: 280–285.

Culig, Z., Hobisch, A., Cronauer, M.V. *et al.* (1993) Mutant androgen receptor detected in an advanced-stage prostatic carcinoma is activated by adrenal androgens and progesterone. *Mol. Endocrinol.* **7**: 1541–1550.

Culig, Z., Hobisch, A., Cronauer, M.V. *et al.* (1994) Androgen receptor activation in prostatic tumor cell lines by insulin-like growth factor-I, keratinocyte growth factor and epidermal growth factor. *Cancer Res.* **54**: 5474–5478.

Culig, Z., Hobisch, A., Hittmair, A. *et al.* (1997) Synergistic activation of androgen receptor by androgen and luteinizing hormone-releasing hormone in prostatic carcinoma cells. *Prostate* **32**, 106–114.

Cunha, G.R., Donjacour, A.A., Cooke, P.S. *et al.* (1987) The endocrinology and developmental biology of the prostate. *Endocr. Rev.* **8**: 338–362.

Danek, A., Witt, T.N., Mann, K. *et al.* (1994) Decrease in androgen binding and effect of androgen treatment in a case of X-linked bulbospinal neuronopathy. *Clin. Invest.* **72**: 892–897.

Danielsen, M., Hinck, L. and Ringold, G.M. (1989) Two amino acids within the knuckle of the first zinc finger specify DNA response element activation by the glucocorticoid receptor. *Cell* **57**: 1131–1138.

Davies, H.R., Hughes, I.A. and Patterson, M.N. (1995) Genetic counselling in complete androgen insensitivity syndrome: trinucleotide repeat polymorphisms, single-strand conformation polymorphism and direct detection of two novel mutations in the androgen receptor gene. *Clin. Endocrinol. (Oxf.)* **43**: 69–77.

De Bellis, A., Quigley, C.A., Cariello, N.F. *et al.* (1992) Single base mutations in the human androgen receptor gene causing complete androgen insensitivity: rapid detection by a modified denaturing gradient gel electrophoresis technique. *Mol. Endocrinol.* **6**: 1909–1920.

De Bellis, A., Quigley, C.A., Marschke, K.B. *et al.* (1994) Characterization of mutant androgen receptors causing partial androgen insensitivity syndrome. *J. Clin. Endocrinol. Metab.* **78**: 513–522.

de Ruiter, P.E., Teuwen, R., Trapman, J., Dijkema, R. and Brinkmann, A.O. (1995) Synergism between androgens and protein kinase-C on androgen-regulated gene expression. *Mol. Cell. Endocrinol.* **110**: 1–6.

Ding, X.F., Anderson, C.M., Ma, H. *et al.* (1998) Nuclear receptor-binding sites of coactivators glucocorticoid receptor interacting protein 1 (GRIP1) and steroid receptor coactivator 1 (SRC-1): multiple motifs with different binding specificities. *Mol. Endocrinol.* **12**: 302–313.

Doesburg, P., Kuil, C.W., Berrevoets, C.A. *et al.* (1997) Functional *in vivo* interaction between the amino-terminal, transactivation domain and the ligand binding domain of the androgen receptor. *Biochemistry* **36**: 1052–1064.

Donjacour, A.A. and Cunha, G.R. (1993) Assessment of prostatic protein secretion in tissue recombinants made of urogenital sinus mesenchyme and urothelium from normal or androgen-insensitive mice. *Endocrinology* **132**: 2342–2350.

Dork, T., Schnieders, F., Jakubiczka, S., Wieacker, P., Schroeder-Kurth, T. and Schmidtke, J. (1998) A new missense substitution at a mutational hot spot of the androgen receptor in siblings with complete androgen insensitivity syndrome. *Hum. Mutat.* **11**: 337–339.

Doyu, M., Sobue, G., Mitsuma, T., Uchida, M., Iwase, T. and Takahashi, A. (1993) Very late onset X-linked recessive bulbospinal neuronopathy: mild clinical features and a mild increase in the size of tandem CAG repeat in androgen receptor gene. *J. Neurol. Neurosurg. Psychiatry* **56**: 832–833.

Edwards, A., Hammond, H.A., Jin, L., Caskey, C.T. and Chakraborty, R. (1992) Genetic variation at five trimeric and tetrameric tandem repeat loci in four human population groups. *Genomics* **12**: 241–253.

Ellerby, L.M., Hackam, A.S., Propp, S.S. *et al.* (1999) Kennedy's disease: caspase cleavage of the androgen receptor is a crucial event in cytotoxicity. *J. Neurochem.* **72**: 185–195.

Elo, J.P., Kvist, L., Leinonen, K. *et al.* (1995) Mutated human androgen receptor gene detected in a prostatic cancer patient is also activated by estradiol. *J. Clin. Endocrinol. Metab.* **80**: 3494–3500.

Enmark, E. and Gustafsson, J.A. (1996) Orphan nuclear receptors – the first eight years. *Mol. Endocrinol.* **10**: 1293–1307.

Evans, B.A.J. (1992) Detection of a point mutation within the androgen receptor gene in a family with complete androgen insensitivity syndrome and subsequent prenatal diagnosis. *J. Endocrinol.* **135**: 26.

Evans, B.A.J., Ismail, R.A., France, T. and Hughes, I.A. (1991) Analysis of the androgen receptor gene structure in a patient with complete androgen insensitivity syndrome. *J. Endocrinol.* **129**: Abstract 65.

Evans, B.A., Harper, M.E., Daniells, C.E. *et al.* (1996) Low incidence of androgen receptor gene mutations in human prostatic tumors using single strand conformation polymorphism analysis. *Prostate* **28**: 162–171.

Evans, B.A., Hughes, I.A., Bevan, C.L., Patterson, M.N. and Gregory, J.W. (1997) Phenotypic diversity in siblings with partial androgen insensitivity syndrome. *Arch. Dis. Child.* **76**: 529–531.

Faber, P.W., Kuiper, G.G.J.M., van Rooij, H.C.J., van der Korput, J.A.G.M., Brinkmann, A.O. and Trapman, J. (1989) The N-terminal domain of the human androgen receptor is encoded by one large exon. *Mol. Cell. Endocrinol.* **61**: 257–262.

Fenton, M.A., Shuster, T.D., Fertig, A.M. *et al.* (1997) Functional characterization of mutant androgen receptors from androgen-independent prostate cancer. *Clin. Cancer Res.* **3**: 1383–1388.

Gaddipati, J.P., McLeod, D.G., Heidenberg, H.B. *et al.* (1994) Frequent detection of codon 877 mutation in the androgen receptor gene in advanced prostate cancers. *Cancer Res.* **54**: 2861–2864.

Gast, A., Neuschmid-Kaspar, F., Klocker, H. and Cato, A.C. (1995) A single amino acid exchange abolishes dimerization of the androgen receptor and causes Reifenstein syndrome. *Mol. Cell. Endocrinol.* **111**: 93–98.

Geissler, W.M., Davis, D.L., Wu, L. *et al.* (1994) Male pseudohermaphroditism caused by mutations of testicular 17 beta-hydroxysteroid dehydrogenase 3. *Nat. Genet.* **7**: 34–39.

Ghirri, P. and Brown, T.R. (1993) Improved detection of point mutations in the human androgen receptor gene by denaturing gradient gel electrophoresis of DNA heteroduplexes under stringent denaturing conditions. *Pediatr. Res.* **33**: S19, Abstract 95.

Giovannucci, E., Stampfer, M.J., Krithivas, K. *et al.* (1997) The CAG repeat within the androgen receptor gene and its relationship to prostate cancer. *Proc. Natl. Acad. Sci. USA* **94**: 3320–3323.

Glass, C.K., Rose, D.W. and Rosenfeld, M.G. (1997) Nuclear receptor coactivators. *Curr. Opin. Cell Biol.* **9**: 222–232.

Gottlieb, B., Lehvaslaiho, H., Beitel, L.K., Lumbroso, R., Pinsky, L. and Trifiro, M. (1998) The Androgen Receptor Gene Mutations Database. *Nucleic Acids Res.* **26**: 234–238.

Green, H. (1993) Human genetic diseases due to codon reiteration: relationship to an evolutionary mechanism. *Cell* **74**: 955–956.

Hard, T., Kellenbach, E., Boelens, R. *et al.* (1990) Solution structure of the glucocorticoid receptor DNA-binding domain. *Science* **249**: 157–160.

Harding, A.E., Thomas, P.K., Baraitser, M., Bradbury, P.G., Morgan-Hughes, J.A. and Ponsford, J.R. (1982) X-linked recessive bulbospinal neuronopathy: a report of ten cases. *J. Neurol. Neurosurg. Psychiatry* **45**: 1012–1019.

Heinzel, T., Lavinsky, R.M., Mullen, T.M. *et al.* (1997) A complex containing N-CoR, mSIN3 and histone deacetylase mediates transcriptional repression. *Nature* **387**: 43–48.

Hiort, O., Huang, Q., Sinnecker, G.H. *et al.* (1993) Single strand conformation polymorphism analysis of androgen receptor gene mutations in patients with androgen insensitivity syndromes: application for diagnosis, genetic counseling and therapy. *J. Clin. Endocrinol. Metab.* **77**: 262–266.

Hiort, O., Wodtke, A., Struve, D., Zollner, A. and Sinnecker, G.H. (1994a) Detection of point mutations in the androgen receptor gene using non-isotopic single strand conformation polymorphism analysis. German Collaborative Intersex Study Group. *Hum. Mol. Genet.* **3**: 1163–1166.

Hiort, O., Klauber, G., Cendron, M. *et al.* (1994b) Molecular characterization of the androgen receptor gene in boys with hypospadias. *Eur. J. Pediatr.* **153**: 317–321.

Hiort, O., Sinnecker, G.H., Holterhus, P.M., Nitsche, E.M. and Kruse, K. (1996) The clinical and molecular spectrum of androgen insensitivity syndromes. *Am. J. Med. Genet.* **63**: 218–222.

Hiort, O., Sinnecker, G.H., Holterhus, P.M., Nitsche, E.M. and Kruse, K. (1998) Inherited and *de novo* androgen receptor gene mutations: investigation of single-case families. *J. Pediatr.* **132**: 939–943.

Hobisch, A., Culig, Z., Radmayr, C., Bartsch, G., Klocker, H. and Hittmair, A. (1995) Distant metastases from prostatic carcinoma express androgen receptor protein. *Cancer Res.* **55**: 3068–3072.

Hobisch, A., Eder, I.E., Putz, T. *et al.* (1998) Interleukin-6 regulates prostate-specific protein expression in prostate carcinoma cells by activation of the androgen receptor. *Cancer Res.* **58**: 4640–4645.

Holterhus, P.M., Bruggenwirth, H.T., Hiort, O. *et al.* (1997) Mosaicism due to a somatic mutation of the androgen receptor gene determines phenotype in androgen insensitivity syndrome. *J. Clin. Endocrinol. Metab.* **82**: 3584–3589.

Horwitz, K.B., Jackson, T.A., Bain, D.L., Richer, J.K., Takimoto, G.S. and Tung, L. (1996) Nuclear receptor coactivators and corepressors. *Mol. Endocrinol.* **10**: 1167–1177.

Igarashi, S., Tanno, Y., Onodera, O. *et al.* (1992) Strong correlation between the number of CAG repeats in androgen receptor genes and the clinical onset of features of spinal and bulbar muscular atrophy. *Neurology* **42**: 2300–2302.

Ikonen, T., Palvimo, J.J., Kallio, P.J., Reinikainen, P. and Janne, O.A. (1994) Stimulation of androgen-regulated transactivation by modulators of protein phosphorylation. *Endocrinology* **135**: 1359–1366.

Ikonen, T., Palvimo, J.J. and Janne, O.A. (1997) Interaction between the amino- and carboxyl-terminal regions of the rat androgen receptor modulates transcriptional activity and is influenced by nuclear receptor coactivators. *J. Biol. Chem.* **272**: 29 821–29 828.

Imai, A., Ohno, T., Iida, K., Ohsuye, K., Okano, Y. and Tamaya, T. (1995) A frame-shift mutation of the androgen receptor gene in a patient with receptor-negative complete testicular feminization: comparison with a single base substitution in a receptor-reduced incomplete form. *Ann. Clin. Biochem.* **32**: 482–486.

Imasaki, K., Hasegawa, T., Okabe, T. *et al.* (1994) Single amino acid substitution (840Arg → His) in the hormone-binding domain of the androgen receptor leads to incomplete androgen insensitivity syndrome associated with a thermolabile androgen receptor. *Eur. J. Endocrinol.* **130**: 569–574.

Imasaki, K., Okabe, T., Murakami, H., Fujita, K., Takayanagi, R. and Nawata, H. (1995) Premature termination mutation (772Glu → stop) in the hormone-binding domain of the androgen receptor in a patient with the receptor-negative form of complete androgen insensitivity syndrome. *Endocr. J.* **42**: 643–648.

Imasaki, K., Okabe, T., Murakami, H. *et al.* (1996) Androgen insensitivity syndrome due to new mutations in the DNA-binding domain of the androgen receptor. *Mol. Cell. Endocrinol.* **120**: 15–24.

Irvine, R.A., Yu, M.C., Ross, R.K. and Coetzee, G.A. (1995) The CAG and GGC microsatellites of the androgen receptor gene are in linkage disequilibrium in men with prostate cancer. *Cancer Res.* **55**: 1937–1940.

Jackson, T.A., Richer, J.K., Bain, D.L., Takimoto, G.S., Tung, L. and Horwitz, K.B. (1997) The partial agonist activity of antagonist-occupied steroid receptors is controlled by a novel hinge domain-binding coactivator L7/SPA and the corepressors N-CoR or SMRT. *Mol. Endocrinol.* **11**: 693–705.

Jakubiczka, S., Nedel, S., Werder, E.A. *et al.* (1997) Mutations of the androgen receptor gene in patients with complete androgen insensitivity. *Hum. Mutat.* **9**: 57–61.

Jenster, G. (1998) Coactivators and corepressors as mediators of nuclear receptor function: an update. *Mol. Cell. Endocrinol.* **143**: 1–7.

Jenster, G., Trapman, J. and Brinkmann, A.O. (1993) Nuclear import of the human androgen receptor. *Biochem. J.* **293**: 761–768.

Jenster, G., de Ruiter, P.E., van der Korput, H.A., Kuiper, G.G., Trapman, J. and Brinkmann, A.O. (1994) Changes in the abundance of androgen receptor isotypes: effects of ligand treatment, glutamine-stretch variation and mutation of putative phosphorylation sites. *Biochemistry* **33**: 14 064–14 072.

Jenster, G., van der Korput, H.A., Trapman, J. and Brinkmann, A.O. (1995) Identification of two transcription activation units in the N-terminal domain of the human androgen receptor. *J. Biol. Chem.* **270**: 7341–7346.

Jenster, G., Spencer, T.E., Burcin, M.M., Tsai, S.Y., Tsai, M.J. and O'Malley, B.W. (1997) Steroid receptor induction of gene transcription: a two-step model. *Proc. Natl. Acad. Sci. USA* **94**: 7879–7884.

Kahlem, P., Terre, C., Green, H. and Djian, P. (1996) Peptides containing glutamine repeats as substrates for transglutaminase-catalyzed cross-linking: relevance to diseases of the nervous system. *Proc. Natl. Acad. Sci. USA* **93**: 14 580–14 585.

Kantoff, P., Giovannucci, E. and Brown, M. (1998) The androgen receptor CAG repeat polymorphism and its relationship to prostate cancer. *Biochim. Biophys. Acta.* **1378**: 1–5.

Kazemi-Esfarjani, P., Trifiro, M.A. and Pinsky, L. (1995) Evidence for a repressive function of the long polyglutamine tract in the human androgen receptor: possible pathogenetic relevance for the (CAG)n-expanded neuronopathies. *Hum. Mol. Genet.* **4**: 523–527.

Kingston, R.E. (1999) A shared but complex bridge. *Nature* **399**: 199–200.

Knoke, I., Jakubiczka, S., Ottersen, T., Goppinger, A. and Wieacker, P. (1997) A(870)E mutation of the androgen receptor gene in a patient with complete androgen insensitivity syndrome and Sertoli cell tumor. *Cancer Genet. Cytogenet.* **98**: 139–141.

Ko, T.M., Yang, Y.S., Wu, M.Y. *et al.* (1997) Complete androgen insensitivity syndrome. Molecular characterization in two Chinese women. *J. Reprod. Med.* **42**: 424–428.

Kobayashi, Y., Miwa, S., Merry, D.E. *et al.* (1998) Caspase-3 cleaves the expanded androgen receptor protein of spinal and bulbar muscular atrophy in a polyglutamine repeat length-dependent manner. *Biochem. Biophys. Res. Commun.* **252**: 145–150.

Koivisto, P., Kononen, J., Palmberg, C. *et al.* (1997) Androgen receptor gene amplification: a possible molecular mechanism for androgen deprivation therapy failure in prostate cancer. *Cancer Res.* **57**: 314–319.

Komori, S., Sakata, K., Tanaka, H., Shima, H. and Koyama, K. (1997) DNA analysis of the androgen receptor gene in two cases with complete androgen insensitivity syndrome. *J. Obstet. Gynaecol. Res.* **23**: 277–281.

Komori, S., Kasumi, H., Sakata, K., Tanaka, H., Hamada, K. and Koyama, K. (1998) Molecular analysis of the androgen receptor gene in 4 patients with complete androgen insensitivity. *Arch. Gynecol. Obstet.* **261**: 95–100.

Kraus, W.L., McInerney, E.M. and Katzenellenbogen, B.S. (1995) Ligand-dependent, transcriptionally productive association of the amino- and carboxyl-terminal regions of a steroid hormone nuclear receptor. *Proc. Natl. Acad. Sci. USA* **92**: 12 314–12 318.

Kuiper, G.G., Faber, P.W., van Rooij, H.C. *et al.* (1989) Structural organization of the human androgen receptor gene. *J. Mol. Endocrinol.* **2**: R1–R4.

Kuiper, G.G.J.M., de Ruiter, P.E., Grootegoed, J.A. and Brinkmann, A.O. (1991) Synthesis and post-translational modification of the androgen receptor in LNCaP cells. *Mol. Cell. Endocrinol.* **80**: 65–73.

Kuiper, G.G.J.M., de Ruiter, P.E., Trapman, J., Boersma, W.J.A., Grootegoed, J.A. and Brinkmann, A.O. (1993) Localization and hormonal stimulation of phosphorylation sites in the LNCaP-cell androgen receptor. *Biochem. J.* **291**: 95–101.

Langley, E., Zhou, Z.X. and Wilson, E.M. (1995) Evidence for an anti-parallel orientation of the ligand-activated human androgen receptor dimer. *J. Biol. Chem.* **270**: 29983–29990.

La Spada, A.R., Wilson, E.M., Lubahn, D.B., Harding, A.E. and Fischbeck, K.H. (1991) Androgen receptor gene mutations in X-linked spinal and bulbar muscular atrophy. *Nature* **352**: 77–79.

La Spada, A.R., Roling, D.B., Harding, A.E. *et al.* (1992) Meiotic stability and genotype–phenotype correlation of the trinucleotide repeat in X-linked spinal and bulbar muscular atrophy. *Nat. Genet.* **2**: 301–304.

Laudet, V., Hänni, C., Coll, J., Catzeflis, F. and Stéhelin, D. (1992) Evolution of the nuclear receptor gene superfamily. *EMBO J.* **11**: 1003–1013.

Lavinsky, R.M., Jepsen, K., Heinzel, T. *et al.* (1998) Diverse signaling pathways modulate nuclear receptor recruitment of N-CoR and SMRT complexes. *Proc. Natl. Acad. Sci. USA* **95**: 2920–2925.

Li, M., Miwa, S., Kobayashi, Y. *et al.* (1998) Nuclear inclusions of the androgen receptor protein in spinal and bulbar muscular atrophy. *Ann. Neurol.* **44**: 249–254.

Lobaccaro, J.M., Lumbroso, S., Belon, C. *et al.* (1993a) Androgen receptor gene mutation in male breast cancer. *Hum. Mol. Genet.* **2**: 1799–1802.

Lobaccaro, J.M., Lumbroso, S., Ktari, R., Dumas, R. and Sultan, C. (1993b) An exonic point mutation creates a *Mae*III site in the androgen receptor gene of a family with complete androgen insensitivity syndrome. *Hum. Mol. Genet.* **2**: 1041–1043.

Lobaccaro, J.M., Lumbroso, S., Belon, C. *et al.* (1993c) Androgen receptor (AR) gene mutations in 6 families with androgen insensitivity syndrome. *Pediatr. Res.* **33**: S22, Abstract 114.

Lobaccaro, J.M., Lumbroso, S., Berta, P., Chaussain, J.L. and Sultan, C. (1993d) Complete androgen insensitivity syndrome associated with a *de novo* mutation of the androgen receptor gene detected by single strand conformation polymorphism. *J. Steroid Biochem. Mol. Biol.* **44**: 211–216.

Lobaccaro, J.M., Belon, C., Lumbroso, S. *et al.* (1994) Molecular prenatal diagnosis of partial androgen insensitivity syndrome based on the *Hind*III polymorphism of the androgen receptor gene. *Clin. Endocrinol. (Oxf.)* **40**: 297–302.

Lobaccaro, J.M., Poujol, N., Chiche, L., Lumbroso, S., Brown, T.R. and Sultan, C. (1996) Molecular modeling and *in vitro* investigations of the human androgen receptor DNA-binding domain: application for the study of two mutations. *Mol. Cell. Endocrinol.* **116**: 137–147.

Lu, J. and Danielsen, M. (1996) A *Stu*I polymorphism in the human androgen receptor gene (AR). *Clin. Genet.* **49**: 323–324.

Lubahn, D.B., Joseph, D.R., Sullivan, P.M., Willard, H.F., French, F.S. and Wilson, E.M. (1988) Cloning of human androgen receptor complementary DNA and localization to the X chromosome. *Science* **240**: 327–330.

Lubahn, D.B., Brown, T.R., Simental, J.A. *et al.* (1989) Sequence of the intron/exon junctions of the coding region of the human androgen receptor gene and identification of a point mutation in a family with complete androgen insensitivity. *Proc. Natl. Acad. Sci. USA* **86**: 9534–9538.

Luisi, B.F., Xu, W.X., Otwinowski, Z., Freedman, L.P., Yamamoto, K.R. and Sigler, P.B. (1991) Crystallographic analysis of the interaction of the glucocorticoid receptor with DNA. *Nature* **352**: 497–505.

Lumbroso, S., Lobaccaro, J.M., Belon, C., Martin, D., Chaussain, J.L. and Sultan, C. (1993) A new mutation within the deoxyribonucleic acid-binding domain of the androgen receptor gene in a family with complete androgen insensitivity syndrome. *Fertil. Steril.* **60**: 814–819.

Lumbroso, S., Lobaccaro, J.M., Georget, V. *et al.* (1996) A novel substitution (Leu707Arg) in exon 4 of the androgen receptor gene causes complete androgen resistance. *J. Clin. Endocrinol. Metab.* **81**: 1984–1988.

Lumbroso, S., Lobaccaro, J.M., Vial, C. *et al.* (1997) Molecular analysis of the androgen receptor gene in Kennedy's disease. Report of two families and review of the literature. *Horm. Res.* **47**: 23–29.

Lyon, M.F. and Glenister, P.H. (1980) Reduced reproductive performance in androgen-resistant *Tfm/Tfm* female mice. *Proc. R. Soc. Lond. [Biol.]* **208**: 1–12.

MacLean, H.E., Chu, S., Warne, G.L. and Zajac, J.D. (1993) Related individuals with different androgen receptor gene deletions. *J. Clin. Invest.* **91**: 1123–1128.

MacLean, H.E., Choi, W.T., Rekaris, G., Warne, G.L. and Zajac, J.D. (1995) Abnormal androgen receptor binding affinity in subjects with Kennedy's disease (spinal and bulbar muscular atrophy). *J. Clin. Endocrinol. Metab.* **80**: 508–516.

Mader, S., Kumar, V., de Verneuil, H. and Chambon, P. (1989) Three amino acids of the oestrogen receptor are essential to its ability to distinguish an oestrogen from a glucocorticoid-responsive element. *Nature* **338**: 271–274.

Mahtani, M.M., Lafreniere, R.G., Kruse, T.A. and Willard, H.F. (1991) An 18-locus linkage map of the pericentromeric region of the human X chromosome: genetic framework for mapping X-linked disorders. *Genomics* **10**: 849–857.

Mangelsdorf, D.J., Thummel, C., Beato, M. *et al.* (1995) The nuclear receptor superfamily: the second decade. *Cell* **83**: 835–839.

Marcelli, M., Tilley, W.D., Wilson, C.M., Griffin, J.E., Wilson, J.D. and McPhaul, M.J. (1990a) Definition of the human androgen receptor gene structure permits the identification of mutants that cause androgen resistance: premature termination of the receptor protein at amino acid residue 588 causes complete androgen resistance. *Mol. Endocrinol.* **4**: 1105–1116.

Marcelli, M., Tilley, W.D., Wilson, C.M., Wilson, J.D., Griffin, J.E. and McPhaul, M.J. (1990b) A single nucleotide substitution introduces a premature termination codon into the androgen receptor gene of a patient with receptor-negative androgen resistance. *J. Clin. Invest.* **85**: 1522–1528.

Marcelli, M., Zoppi, S., Grino, P.B., Griffin, J.E., Wilson, J.D. and McPhaul, M.J. (1991) A mutation in the DNA-binding domain of the androgen receptor gene causes complete testicular feminization in a patient with receptor-positive androgen resistance. *J. Clin. Invest.* **87**: 1123–1126.

Marcelli, M., Zoppi, S., Wilson, C.M., Griffin, J.E. and McPhaul, M.J. (1994) Amino acid substitutions in the hormone-binding domain of the human androgen receptor alter the stability of the hormone receptor complex. *J. Clin. Invest.* **94**: 1642–1650.

McKenna, N.J., Lanz, R.B. and O'Malley, B.W. (1999) Nuclear receptor coregulators: cellular and molecular biology. *Endocr. Rev.* **20**: 321–344.

McLaughlin, B.A., Spencer, C. and Eberwine, J. (1996) CAG trinucleotide RNA repeats interact with RNA-binding proteins. *Am. J. Hum. Genet.* **59**: 561–569.

McPhaul, M.J., Marcelli, M., Tilley, W.D., Griffin, J.E., Isidro-Gutierrez, R.F. and Wilson, J.D. (1991a) Molecular basis of androgen resistance in a family with a qualitative abnormality of the androgen receptor and responsive to high-dose androgen therapy. *J. Clin. Invest.* **87**: 1413–1421.

McPhaul, M.J., Marcelli, M., Tilley, W.D., Griffin, J.E. and Wilson, J.D. (1991b) Androgen resistance caused by mutations in the androgen receptor gene. *FASEB J.* **5**: 2910–2915.

McPhaul, M.J., Marcelli, M., Zoppi, S., Wilson, C.M., Griffin, J.E. and Wilson, J.D. (1992) Mutations in the ligand-binding domain of the androgen receptor gene cluster in two regions of the gene. *J. Clin. Invest.* **90**: 2097–2101.

Merry, D.E., Kobayashi, Y., Bailey, C.K., Taye, A.A. and Fischbeck, K.H. (1998) Cleavage, aggregation and toxicity of the expanded androgen receptor in spinal and bulbar muscular atrophy. *Hum. Mol. Genet.* **7**: 693–701.

Mhatre, A.N., Trifiro, M.A., Kaufman, M. *et al.* (1993) Reduced transcriptional regulatory competence of the androgen receptor in X-linked spinal and bulbar muscular atrophy. *Nat. Genet.* **5**: 184–188.

Miyamoto, H., Yeh, S., Wilding, G. and Chang, C. (1998) Promotion of agonist activity of antiandrogens by the androgen receptor coactivator, ARA70, in human prostate cancer DU145 cells. *Proc. Natl. Acad. Sci. USA* **95**, 7379–7384.

Murono, K., Mendonca, B.B., Arnhold, I.J., Rigon, A.C., Migeon, C.J. and Brown, T.R. (1995) Human androgen insensitivity due to point mutations encoding amino acid substitutions in the androgen receptor steroid-binding domain. *Hum. Mutat.* **6**: 152–162.

Nagy, L., Kao, H.Y., Chakravarti, D. *et al.* (1997) Nuclear receptor repression mediated by a complex containing SMRT, mSin3A and histone deacetylase. *Cell* **89**: 373–380.

Nakamura, M., Mita, S., Murakami, T. *et al.* (1994) Exonic trinucleotide repeats and expression of androgen receptor gene in spinal cord from X-linked spinal and bulbar muscular atrophy. *J. Neurol. Sci.* **122**: 74–79.

Nakao, R., Haji, M., Yanase, T. *et al.* (1992) A single amino acid substitution (Met786 → Val) in the steroid-binding domain of human androgen receptor leads to complete androgen insensitivity syndrome. *J. Clin. Endocrinol. Metab.* **74**: 1152–1157.

Nakao, R., Yanase, T., Sakai, Y., Haji, M. and Nawata, H. (1993) A single amino acid substitution (gly743 → val) in the steroid-binding domain of the human androgen receptor leads to Reifenstein syndrome. *J. Clin. Endocrinol. Metab.* **77**: 103–107.

Nance, M.A. (1997) Clinical aspects of CAG repeat diseases. *Brain Pathol.* **7**: 881–900.

Nazareth, L.V. and Weigel, N.L. (1996) Activation of the human androgen receptor through a protein kinase A signaling pathway. *J. Biol. Chem.* **271**: 19 900–19 007.

Newmark, J.R., Hardy, D.O., Tonb, D.C. *et al.* (1992) Androgen receptor gene mutations in human prostate cancer. *Proc. Natl. Acad. Sci. USA* **89**: 6319–6323.

Onate, S.A., Boonyaratanakornkit, V., Spencer, T.E. *et al.* (1998) The steroid receptor coactivator-1 contains multiple receptor interacting and activation domains that cooperatively enhance the activation function 1 (AF1) and AF2 domains of steroid receptors. *J. Biol. Chem.* **273**: 12 101–12 108.

Pertschuk, L.P., Schaeffer, H., Feldman, J.G. *et al.* (1995) Immunostaining for prostate cancer androgen receptor in paraffin identifies a subset of men with a poor prognosis. *Lab. Invest.* **73**: 302–305.

Peterziel, H., Culig, Z., Stober, J. *et al.* (1995) Mutant androgen receptors in prostatic tumors distinguish between amino-acid-sequence requirements for transactivation and ligand binding. *Int. J. Cancer* **63**: 544–550.

Pinsky, L., Trifiro, M., Kaufman, M. *et al.* (1992) Androgen resistance due to mutation of the androgen receptor. *Clin. Invest. Med.* **15**: 456–472.

Prior, L., Bordet, S., Trifiro, M.A. *et al.* (1992) Replacement of arginine 773 by cysteine or histidine in the human androgen receptor causes complete androgen insensitivity with different receptor phenotypes. *Am. J. Hum. Genet.* **51**: 143–155.

Quigley, C.A., Evans, B.A., Simental, J.A. *et al.* (1992) Complete androgen insensitivity due to deletion of exon c of the androgen receptor gene highlights the functional importance of the second zinc finger of the androgen receptor *in vivo*. *Mol. Endocrinol.* **6**: 1103–1112.

Quigley, C.A., De Bellis, A., Marschke, K.B., el-Awady, M.K., Wilson, E.M. and French, F.S. (1995) Androgen receptor defects: historical, clinical and molecular perspectives. *Endocr. Rev.* **16**: 271–321.

Radmayr, C., Culig, Z., Glatzl, J., Neuschmid-Kaspar, F., Bartsch, G. and Klocker, H. (1997) Androgen receptor point mutations as the underlying molecular defect in 2 patients with androgen insensitivity syndrome. *J. Urol.* **158**: 1553–1556.

Reichardt, H.M., Kaestner, K.H., Tuckermann, J. *et al.* (1998) DNA binding of the glucocorticoid receptor is not essential for survival. *Cell* **93**: 531–541.

Reifenstein, E.C. (1947) Hereditary female hypogonadism. *Proc. Am. Fed. Clin. Res.* **3**: 86.

Reinikainen, P., Palvimo, J.J. and Janne, O.A. (1996) Effects of mitogens on androgen receptor-mediated transactivation. *Endocrinology* **137**: 4351–4357.

Renaud, J.P., Rochel, N., Ruff, M. *et al.* (1995) Crystal structure of the RAR-gamma ligand-binding domain bound to all-*trans* retinoic acid. *Nature* **378**: 681–689.

Riegman, P.H., Vlietstra, R.J., van der Korput, J.A., Brinkmann, A.O. and Trapman, J. (1991) The promoter of the prostate-specific antigen gene contains a functional androgen responsive element. *Mol. Endocrinol.* **5**: 1921–1930.

Ris-Stalpers, C., Kuiper, G.G.J.M., Faber, P.W. *et al.* (1990) Aberrant splicing of androgen receptor mRNA results in synthesis of a nonfunctional receptor protein in a patient with androgen insensitivity. *Proc. Natl. Acad. Sci. USA* **87**: 7866–7870.

Ris-Stalpers, C., Trifiro, M.A., Kuiper, G.G. *et al.* (1991) Substitution of aspartic acid-686 by histidine or asparagine in the human androgen receptor leads to a functionally inactive protein with altered hormone-binding characteristics. *Mol. Endocrinol.* **5**: 1562–1569.

Ris-Stalpers, C., Hoogenboezem, T., Sleddens, H.F. *et al.* (1994a) A practical approach to the detection of androgen receptor gene mutations and pedigree analysis in families with x-linked androgen insensitivity. *Pediatr. Res.* **36**: 227–234.

Ris-Stalpers, C., Verleun-Mooijman, M.C., de Blaeij, T.J., Degenhart, H.J., Trapman, J. and Brinkmann, A.O. (1994b) Differential splicing of human androgen receptor pre-mRNA in X-linked Reifenstein syndrome, because of a deletion involving a putative branch site. *Am. J. Hum. Genet.* **54**: 609–617.

Robbins, J., Dilworth, S.M., Laskey, R.A. and Dingwall, C. (1991) Two interdependent basic domains in nucleoplasmin nuclear targeting sequence: identification of a class of bipartite nuclear targeting sequence. *Cell* **64**: 615–623.

Roche, P.J., Hoare, S.A. and Parker, M.G. (1992) A consensus DNA-binding site for the androgen receptor. *Mol. Endocrinol.* **6**: 2229–2235.

Rodien, P., Mebarki, F., Mowszowicz, I. *et al.* (1996) Different phenotypes in a family with androgen insensitivity caused by the same M780I point mutation in the androgen receptor gene. *J. Clin. Endocrinol. Metab.* **81**: 2994–2998.

Ruizeveld de Winter, J.A., Trapman, J., Vermey, M., Mulder, E., Zegers, N.D. and van der Kwast, T.H. (1991) Androgen receptor expression in human tissues: an immunohisto-chemical study. *J. Histochem. Cytochem.* **39**: 927–936.

Ruizeveld de Winter, J.A., Janssen, P.J., Sleddens, H.M. *et al.* (1994) Androgen receptor status in localized and locally progressive hormone refractory human prostate cancer. *Am. J. Pathol.* **144**: 735–746.

Rundlett, S.E. and Miesfeld, R.L. (1995) Quantitative differences in androgen and glucocorticoid receptor DNA binding properties contribute to receptor-selective transcriptional regulation. *Mol. Cell. Endocrinol.* **109**: 1–10.

Russell, D.W. and Wilson, J.D. (1994) Steroid 5alpha-reductase: two genes/two enzymes. *Annu. Rev. Biochem.* **63**: 25–61.

Sadar, M.D. (1999) Androgen-independent induction of prostate-specific antigen gene expression via cross-talk between the androgen receptor and protein kinase A signal transduction pathways. *J. Biol. Chem.* **274**: 7777–7783.

Sadi, M.V. and Barrack, E.R. (1993) Image analysis of androgen receptor immunostaining in metastatic prostate cancer. Heterogeneity as a predictor of response to hormonal therapy. *Cancer* **71**: 2574–2580.

Sadi, M.V., Walsh, P.C. and Barrack, E.R. (1991) Immunohistochemical study of androgen receptors in metastatic prostate cancer. Comparison of receptor content and response to hormonal therapy. *Cancer* **67**: 3057–3064.

Sai, T.J., Seino, S., Chang, C.S. *et al.* (1990) An exonic point mutation of the androgen receptor gene in a family with complete androgen insensitivity. *Am. J. Hum. Genet.* **46**: 1095–1100.

Saunders, P.T., Padayachi, T., Tincello, D.G., Shalet, S.M. and Wu, F.C. (1992) Point mutations detected in the androgen receptor gene of three men with partial androgen insensitivity syndrome. *Clin. Endocrinol. (Oxf.)* **37**: 214–220.

Schoenberg, M.P., Hakimi, J.M., Wang, S. *et al.* (1994) Microsatellite mutation (CAG24 → 18) in the androgen receptor gene in human prostate cancer. *Biochem. Biophys. Res. Commun.* **198**: 74–80.

Schuurmans, A.L., Bolt, J., Voorhorst, M.M., Blankenstein, R.A. and Mulder, E. (1988) Regulation of growth and epidermal growth factor receptor levels of LNCaP prostate tumor cells by different steroids. *Int. J. Cancer* **42**: 917–922.

Schwabe, J.W.R., Neuhaus, D. and Rhodes, D. (1990) Solution structure of the DNA-binding domain of the oestrogen receptor. *Nature* **348**: 458–461.

Schwartz, M., Skovby, F., Muller, J., Nielsen, O. and Skakkebaek, N.E. (1994) Partial androgen insensitivity (PAIS) in a large eskimo kindred caused by a delta D690 mutation in the androgen receptor (AR) gene. *Horm. Res.* **41**: 117, Abstract 244.

Shibata, H., Spencer, T.E., Onate, S.A. *et al.* (1997) Role of co-activators and co-repressors in the mechanism of steroid/thyroid receptor action. *Recent Prog. Horm. Res.* **52**: 141–165.

Sleddens, H.F., Oostra, B.A., Brinkmann, A.O. and Trapman, J. (1992) Trinucleotide repeat polymorphism in the androgen receptor gene (AR). *Nucleic Acids Res.* **20**: 1427.

Sleddens, H.F., Oostra, B.A., Brinkmann, A.O. and Trapman, J. (1993) Trinucleotide (GGN) repeat polymorphism in the human androgen receptor (AR) gene. *Hum. Mol. Genet.* **2**: 493.

Smith, C.L., Nawaz, Z. and O'Malley, B.W. (1997) Coactivator and corepressor regulation of the agonist/antagonist activity of the mixed antiestrogen, 4-hydroxytamoxifen. *Mol. Endocrinol.* **11**: 657–666.

Smith, D.F. and Toft, D.O. (1993) Steroid receptors and their associated proteins. *Mol. Endocrinol.* **7**: 4–11.

Stanford, J.L., Just, J.J., Gibbs, M. *et al.* (1997) Polymorphic repeats in the androgen receptor gene: molecular markers of prostate cancer risk. *Cancer Res.* **57**: 1194–1198.

Stenoien, D.L., Cummings, C.J., Adams, H.P. *et al.* (1999) Polyglutamine-expanded androgen receptors form aggregates that sequester heat shock proteins, proteasome components and SRC-1 and are suppressed by the HDJ-2 chaperone. *Hum. Mol. Genet.* **8**: 731–741.

Sui, X., Bramlett, K.S., Jorge, M.C., Swanson, D.A., von Eschenbach, A.C. and Jenster, G. (1999) Specific androgen receptor activation by an artificial coactivator. *J. Biol. Chem.* **274**: 9449–9454.

Sultan, C., Lumbroso, S., Poujol, N., Belon, C., Boudon, C. and Lobaccaro, J.M. (1993) Mutations of androgen receptor gene in androgen receptor insensitivity syndromes. *J. Steroid Biochem. Mol. Biol.* **46**: 519–530.

Sutherland, R.W., Wiener, J.S., Hicks, J.P. *et al.* (1996) Androgen receptor gene mutations are rarely associated with isolated penile hypospadias. *J. Urol.* **156**: 828–831.

Suzuki, H., Sato, N., Watabe, Y., Masai, M., Seino, S. and Shimazaki, J. (1993) Androgen receptor gene mutations in human prostate cancer. *J. Steroid Biochem. Mol. Biol.* **46**: 759–765.

Suzuki, H., Akakura, K., Komiya, A., Aida, S., Akimoto, S. and Shimazaki, J. (1996) Codon 877 mutation in the androgen receptor gene in advanced prostate cancer: relation to antiandrogen withdrawal syndrome. *Prostate* **29**: 153–158.

Takahashi, H., Furusato, M., Allsbrook, W.C. *et al.* (1995) Prevalence of androgen receptor gene mutations in latent prostatic carcinomas from Japanese men. *Cancer Res.* **55**: 1621–1624.

Tan, J., Sharief, Y., Hamil, K.G. *et al.* (1997) Dehydroepiandrosterone activates mutant androgen receptors expressed in the androgen-dependent human prostate cancer xenograft CWR22 and LNCaP cells. *Mol. Endocrinol.* **11**: 450–459.

Tanaka, H., Komori, S., Sakata, K., Shima, H. and Koyama, K. (1998) One additional mutation at exon A amplifies thermolability of androgen receptor in a case with complete androgen insensitivity syndrome. *Gynecol. Endocrinol.* **12**: 75–82.

Taplin, M.E., Bubley, G.J., Shuster, T.D. *et al.* (1995) Mutation of the androgen-receptor gene in metastatic androgen-independent prostate cancer. *N. Engl. J. Med.* **332**: 1393–1398.

Taplin, M.E., Bubley, G.J., Ko, Y.J. *et al.* (1999) Selection for androgen receptor mutations in prostate cancers treated with androgen antagonist. *Cancer Res.* **59**: 2511–2515.

Tetel, M.J., Jung, S., Carbajo, P., Ladtkow, T., Skafar, D.F. and Edwards, D.P. (1997) Hinge and amino-terminal sequences contribute to solution dimerization of human progesterone receptor. *Mol. Endocrinol.* **11**: 1114–1128.

Tetel, M.J., Giangrande, P.H., Leonhardt, S.A., McDonnell, D.P. and Edwards, D.P. (1999) Hormone-dependent interaction between the amino- and carboxyl-terminal domains of progesterone receptor *in vitro* and *in vivo*. *Mol. Endocrinol.* **13**: 910–924.

Thigpen, A.E., Davis, D.L., Milatovich, A. *et al.* (1992) Molecular genetics of steroid 5 alpha-reductase 2 deficiency. *J Clin Invest.* **90**: 799–809.

Tilley, W.D., Marcelli, M., Wilson, J.D. and McPhaul, M.J. (1989) Characterization and expression of a cDNA encoding the human androgen receptor. *Proc. Natl. Acad. Sci. USA* **86**: 327–331.

Tilley, W.D., Lim-Tio, S.S., Horsfall, D.J., Aspinall, J.O., Marshall, V.R. and Skinner, J.M. (1994) Detection of discrete androgen receptor epitopes in prostate cancer by immunostaining: measurement by color video image analysis. *Cancer Res.* **54**: 4096–4102.

Tilley, W.D., Buchanan, G., Hickey, T.E. and Bentel, J.M. (1996) Mutations in the androgen receptor gene are associated with progression of human prostate cancer to androgen independence. *Clin. Cancer Res.* **2**: 277–285.

Tincello, D.G., Saunders, P.T., Hodgins, M.B. *et al.* (1997) Correlation of clinical, endocrine and molecular abnormalities with *in vivo* responses to high-dose testosterone in patients with partial androgen insensitivity syndrome. *Clin. Endocrinol. (Oxf.)* **46**: 497–506.

Trapman, J., Klaassen, P., Kuiper, G.G.J.M. *et al.* (1988) Cloning, structure and expression of a cDNA encoding the human androgen receptor. *Biochem. Biophys. Res. Commun.* **153**: 241–248.

Trifiro, M., Prior, R.L., Sabbaghian, N. *et al.* (1991) Amber mutation creates a diagnostic MaeI site in the androgen receptor gene of a family with complete androgen insensitivity. *Am. J. Med. Genet.* **40**: 493–499.

Trifiro, M.A., Lumbroso, R., Beitel, L.K. *et al.* (1997) Altered mRNA expression due to insertion or substitution of thymine at position $+3$ of two splice-donor sites in the androgen receptor gene. *Eur. J. Hum. Genet.* **5**: 50–58.

Tsai, M.J. and O'Malley, B.W. (1994) Molecular mechanisms of action of steroid/thyroid receptor superfamily members. *Annu. Rev. Biochem.* **63**: 451–486.

Tsukada, T., Inoue, M., Tachibana, S., Nakai, Y. and Takebe, H. (1994) An androgen receptor mutation causing androgen resistance in undervirilized male syndrome. *J. Clin. Endocrinol. Metab.* **79**: 1202–1207.

Umesono, K. and Evans, R.M. (1989) Determinants of target gene specificity for steroid/thyroid hormone receptors. *Cell* **57**: 1139–1146.

van der Kwast, T.H., Schalken, J., Ruizeveld de Winter, J.A. *et al.* (1991) Androgen receptors in endocrine-therapy-resistant human prostate cancer. *Int. J. Cancer* **48**: 189–193.

Veldscholte, J., Ris-Stalpers, C., Kuiper, G.G.J.M. *et al.* (1990) A mutation in the ligand binding domain of the androgen receptor of human LNCaP cells affects steroid binding characteristics and response to anti-androgens. *Biochem. Biophys. Res. Commun.* **173**: 534–540.

Veldscholte, J., Berrevoets, C.A., Brinkmann, A.O., Grootegoed, J.A. and Mulder, E. (1992) Anti-androgens and the mutated androgen receptor of LNCaP cells: differential effects on binding affinity, heat shock protein interaction and transcription activation. *Biochemistry* **31**: 2393–2399.

Visakorpi, T., Hyytinen, E., Koivisto, P. *et al.* (1995) *In vivo* amplification of the androgen receptor gene and progression of human prostate cancer. *Nat. Genet.* **9**: 401–406.

Wagner, B.L., Norris, J.D., Knotts, T.A., Weigel, N.L. and McDonnell, D.P. (1998) The nuclear corepressors NCoR and SMRT are key regulators of both ligand- and 8-bromo-cyclic AMP-dependent transcriptional activity of the human progesterone receptor. *Mol. Cell. Biol.* **18**: 1369–1378.

Wagner, R.L., Apriletti, J.W., McGrath, M.E., West, B.L., Baxter, J.D. and Fletterick, R.J. (1995) A structural role for hormone in the thyroid hormone receptor. *Nature* **378**: 690–697.

Warner, C.L., Griffin, J.E., Wilson, J.D. *et al.* (1992) X-linked spinomuscular atrophy: a kindred with associated abnormal androgen receptor binding. *Neurology* **42**: 2181–2184.

Watanabe, M., Ushijima, T., Shiraishi, T. *et al.* (1997) Genetic alterations of androgen receptor gene in Japanese human prostate cancer. *Jpn. J. Clin. Oncol.* **27**: 389–393.

Weidemann, W., Linck, B., Haupt, H. *et al.* (1996) Clinical and biochemical investigations and molecular analysis of subjects with mutations in the androgen receptor gene. *Clin. Endocrinol. (Oxf.)* **45**: 733–739.

Weigel, N.L. and Zhang, Y. (1998) Ligand-independent activation of steroid hormone receptors. *J. Mol. Med.* **76**: 469–479.

Wilbert, D.M., Griffin, J.E. and Wilson, J.D. (1983) Characterization of the cytosol androgen receptor of the human prostate. *J. Clin. Endocrinol. Metab.* **56**: 113–120.

Williams, S.P. and Sigler, P.B. (1998) Atomic structure of progesterone complexed with its receptor. *Nature* **393**: 392–396.

Wilmot, G.R. and Warren, S.T. (1998) A new mutational basis for disease. In: S.T. Warren and R.D. Wells (eds) *Genetic Instability and Hereditary Neurological Diseases*, pp. 3–12. Academic Press, San Diego.

Wilson, C.M., Griffin, J.E., Wilson, J.D., Marcelli, M., Zoppi, S. and McPhaul, M.J. (1992) Immunoreactive androgen receptor expression in subjects with androgen resistance. *J. Clin. Endocrinol. Metab.* **75**: 1474–1478.

Wilson, J.D., Harrod, M.J., Goldstein, J.L., Hemsell, D.L. and MacDonald, P.C. (1974) Familial incomplete male pseudohermaphroditism, type 1. Evidence for androgen resistance and variable clinical manifestations in a family with the Reifenstein syndrome. *N. Engl. J. Med.* **290**: 1097–1103.

Wilson, J.D., Griffin, J.E. and Russell, D.W. (1993) Steroid 5α-reductase 2 deficiency. *Endocr. Rev.* **14**: 577–593.

Wooster, R., Mangion, J., Eeles, R. *et al.* (1992) A germline mutation in the androgen receptor gene in two brothers with breast cancer and Reifenstein syndrome. *Nat. Genet.* **2**: 132–134.

Yeh, S., Miyamoto, H., Shima, H. and Chang, C. (1998) From estrogen to androgen receptor: a new pathway for sex hormones in prostate. *Proc. Natl. Acad. Sci. USA* **95**: 5527–5532.

Yeh, S., Lin, H.K., Kang, H.Y., Thin, T.H., Lin, M.F. and Chang, C. (1999) From *HER2/Neu* signal cascade to androgen receptor and its coactivators: a novel pathway by induction of androgen target genes through MAP kinase in prostate cancer cells. *Proc. Natl. Acad. Sci. USA* **96**: 5458–5463.

Yong, E.L., Chua, K.L., Yang, M., Roy, A. and Ratnam, S. (1994a) Complete androgen insensitivity due to a splice-site mutation in the androgen receptor gene and genetic screening with single-stranded conformation polymorphism. *Fertil. Steril.* **61**: 856–862.

Yong, E.L., Ng, S.C., Roy, A.C., Yun, G. and Ratnam, S.S. (1994b) Pregnancy after hormonal correction of severe spermatogenic defect due to mutation in androgen receptor gene. *Lancet* **344**: 826–827.

Yong, E.L., Tut, T.G., Ghadessy, F.J., Prins, G. and Ratnam, S.S. (1998) Partial androgen insensitivity and correlations with the predicted three dimensional structure of the androgen receptor ligand-binding domain. *Mol. Cell. Endocrinol.* **137**: 41–50.

Zhang, J.S., Guenther, M.G., Carthew, R.W. and Lazar, M.A. (1998) Proteasomal regulation of nuclear receptor corepressor-mediated repression. *Genes Dev.* **12**: 1775–1780.

Zhou, Z.X., Sar, M., Simental, J.A., Lane, M.V. and Wilson, E.M. (1994) A ligand-dependent bipartite nuclear targeting signal in the human androgen receptor. Requirement for the DNA-binding domain and modulation by NH_2-terminal and carboxyl-terminal sequences. *J. Biol. Chem.* **269**: 13 115–13 123.

Zhou, Z.X., Kemppainen, J.A. and Wilson, E.M. (1995) Identification of three proline-directed phosphorylation sites in the human androgen receptor. *Mol. Endocrinol.* **9**: 605–615.

Zhu, X. and Liu, J.P. (1997) Steroid-independent activation of androgen receptor in androgen-independent prostate cancer: a possible role for the MAP kinase signal transduction pathway? *Mol. Cell. Endocrinol.* **134**: 9–14.

Zilliacus, J., Dahlman-Wright, K., Wright, A., Gustafsson, J.A. and Carlstedt-Duke, J. (1991) DNA binding specificity of mutant glucocorticoid receptor DNA-binding domains. *J. Biol. Chem.* **266**: 3101–3106.

Zoppi, S., Marcelli, M., Deslypere, J.P., Griffin, J.E., Wilson, J.D. and McPhaul, M.J. (1992) Amino acid substitutions in the DNA-binding domain of the human androgen receptor are a frequent cause of receptor-binding positive androgen resistance. *Mol. Endocrinol.* **6**: 409–415.

Zoppi, S., Wilson, C.M., Harbison, M.D. *et al.* (1993) Complete testicular feminization caused by an amino-terminal truncation of the androgen receptor with downstream initiation. *J. Clin. Invest.* **91**: 1105–1112.

Chapter 5

DAX1 and Related Orphan Receptors

Eric Vilain and Edward R.B. McCabe

INTRODUCTION AND HISTORICAL PERSPECTIVE

DAX1 adrenal hypoplasia congenita and identification of the *DAX1* gene

In adrenal hypoplasia congenita (AHC), the affected individual's adrenal cortices are small or absent. The hypoplastic or aplastic adrenal cortex is unable to produce normal amounts of adrenal corticosteroids and, therefore, untreated patients risk death from adrenal cortical insufficiency. The estimated incidence of AHC is 1 in 12 500 live births (Kelch *et al.*, 1984).

Two distinct types of AHC have been described, based on their histologic appearance (McCabe, 1999b). The adrenal cortex from a patient with the miniature adult form appears to be primarily permanent adult cortex with the normal layers of the adult tissue small but present, and the fetal cortex is absent or a minimal remnant. The miniature adult form of AHC may be sporadic or inherited in an autosomal recessive fashion, and is frequently associated with developmental abnormalities of the central nervous system (Favara *et al.*, 1972; Online Mendelian Inheritance in Man (OMIM), 1999). The adrenal cortex in patients with the cytomegalic form of AHC is structurally disorganized, with an adult zone that is absent or nearly absent. The residual cortex contains eosinophilic cells in irregular nodular formations, as well as large vacuolated cells. The cytomegalic form of AHC is typically inherited in an X-linked fashion (OMIM *300200), but an autosomal recessive form (OMIM 202155) may occur as well (McCabe, 1999b; OMIM, 1999).

The focus of this chapter is the X-linked cytomegalic form of AHC, which is associated with mutations involving the gene encoding the orphan nuclear hormone receptor protein, *DAX1* (*D*osage-sensitive sex reversal, *A*drenal hypoplasia congenita, *X* chromosome, gene *1*).

AHC and identification of the *DAX1* gene

Patients' deletions delineate the AHC critical region and facilitate the cloning of *DAX1*.   Inheritance of AHC, consistent with X-linked disease, has been recognized for decades (Brochner-Mortensen, 1956; McCabe, 1999b). Cryptorchidism, either unilateral or bilateral, is common in boys with X-linked AHC (Zachmann *et al.*, 1980; Renier *et al.*, 1983; Wise *et al.*, 1987). X-linked AHC is also associated with hypogonadotropic hypogonadism (HH) (McCabe, 1999b) and has mixed hypothalamic and pituitary origins (Habiby *et al.*, 1996). Interestingly, however, affected boys have a normal 'minipuberty' of infancy suggesting a change between infancy and adolescence in the mechanisms for regulation of the hypothalamic–pituitary–gonadal axis or loss of the functional integrity of this axis over time (Takahashi *et al.*, 1997; Kaiserman *et al.*, 1998; Peter *et al.*, 1998).

Some boys, including the two brothers who were originally described with AHC, glycerol kinase deficiency (GKD) and Duchenne muscular dystrophy (DMD) (McCabe *et al.*, 1977; Guggenheim *et al.*, 1980), have a more complex phenotype that is now recognized to be due to a contiguous gene syndrome involving additional loci in Xp21 (McCabe, 1999a,b). Careful clinical characterization of these patients permitted the mapping of the *DAX1* gene to Xp21.3. The deletions in affected patients permitted the fine mapping of DNA markers within this region and facilitated the building of genomic contigs (Love *et al.*, 1990; Walker *et al.*, 1992; Worley *et al.*, 1992, 1993).

Correlating the clinical features of patients with contiguous gene syndromes involving deletions in Xp21.3 (i.e. the presence or absence of AHC) with the physical mapping of their breakpoints, permitted the identification of an AHC 'critical region' in which all or part of the gene responsible for AHC was presumed to be located. Using genomic strategies and focusing on the AHC critical region, the *DAX1* gene was identified (Zanaria *et al.*, 1994; Guo *et al.*, 1995b). Intragenic mutations in the *DAX1* gene among patients without deletions, confirmed that *DAX1* mutations were responsible for the X-linked cytomegalic form of AHC (Muscatelli *et al.*, 1994; Zanaria *et al.*, 1994; Guo *et al.*, 1995b).

Intragenic mutations identify a role for DAX1 in the development of the hypothalamic–pituitary–adrenal/gonadal axis. With the cloning and initial characterization of the *DAX1* gene, intragenic mutations were identified in patients with AHC and HH (Muscatelli *et al.*, 1994; Zanaria *et al.*, 1994). These observations clearly showed that HH is caused by mutations in this gene, and not by a distinct locus as had been suggested previously (Matsumoto *et al.*, 1988; Goonewardena *et al.*, 1989).

Initial data were contradictory regarding the level of interference with normal function caused by *DAX1* mutations, i.e. whether the HH was hypothalamic or pituitary in origin (Partsch and Sippell, 1989; Kletter *et al.*, 1991). The reason for the contradictory information was resolved when it was shown that the HH caused by *DAX1* mutations resulted from combined hypothalamic and pituitary defects in gonadotropin production (Habiby *et al.*, 1996; McCabe, 1996). Independent observations on the DAX1 expression pattern were consistent with the mixed origins of HH in these patients, since DAX1 is expressed in the hypothalamus and pituitary (Guo *et al.*, 1995a; Swain *et al.*, 1996; Ikeda *et al.*, 1996). *DAX1* mutations are associated with impaired spermatogenesis (Tabarin *et al.*, 2000).

Although X-linked AHC is usually associated with HH, a family has been reported in which affected males had androgenic precocity of varying degrees (Wittenberg, 1981). Although this report predated the identification of the *DAX1* gene, the pedigree was consistent with X linkage, since the four affected male cousins were related through their maternal grandmothers who were sisters. One of the males exhibited nonprogressive virilization at birth and the other three had advanced growth and/or skeletal maturation and raised circulating testosterone concentrations.

Dosage-sensitive sex reversal and the possible role of *DAX1*

Patients' tandem duplications define the critical region.   In 1978, the report of a family with 46XY gonadal dysgenesis that appeared to be inherited in an X-linked fashion, first suggested the presence on the X chromosome of a gene involved in sex determination (German *et al.*, 1978). Additional families were described subsequently with inheritance of this disorder in either an X-linked recessive or sex-limited autosomal dominant pattern of inheritance (Fechner *et al.*, 1993). Partial duplication of Xp has been observed in individuals who otherwise have a normal 46XY karyotype and female or ambiguous external genitalia (Narahara *et al.*, 1979; Bernstein *et al.*, 1980; Nielsen and Langkjaer, 1982; Scherer *et al.*, 1989; Stern *et al.*, 1990; May *et al.*, 1991; Ogata *et al.*, 1992; Bardoni *et al.*, 1993; Arn *et al.*, 1994; Bardoni *et al.*, 1994; Ogata and Matsuo, 1994). Cytogenetic analyses had localized the dosage-sensitive sex reversal (DSS) locus to Xp22.11-p21.2 (Arn *et al.*, 1994). Molecular genetic analyses further narrowed the DSS critical region contained in all individuals with dosage-sensitive sex reversal to a 160-kilobase (kb) interval in Xp21 (Bardoni *et al.*, 1994).

DSS critical region contains *DAX1*.   Since the molecular markers that were used to define the DSS critical region (Bardoni *et al.*, 1994) included those that defined the AHC critical region and facilitated the cloning of *DAX1* (Zanaria *et al.*, 1994; Guo *et al.*, 1995b), it became immediately evident that the *DAX1* gene was contained within the DSS critical region. As *DAX1* codes for a putative transcription factor that is expressed in the hypothalamic–pituitary–adrenal/gonadal axis, and because it is contained within the DSS critical region, it is a candidate gene for DSS (Bardoni *et al.*, 1994; Zanaria *et al.*, 1994; Guo *et al.*, 1995a,b; Swain *et al.*, 1996).

Transgenic murine overexpression of DAX1 is consistent with a role in DSS.   Since tandem duplication of a region containing the *DAX1* gene in the human is presumed to result in overexpression of the DAX1 protein, a murine model to test the hypothesis that DSS is caused by *DAX1* would involve a mouse with transgenic overexpression of the murine homolog, *Dax1* (also known as *Ahch*). This experiment has been performed and transgenic overexpression of *Dax1* in *Mus musculus musculus* carrying the Y chromosome from the *Poschiavinus* strain of *Mus musculus domesticus* results in XY mice that are phenotypically female (Swain *et al.*, 1998). The *Poschiavinus* strain of mouse has a low level of *SRY* expression. Interestingly, sex reversal is not observed, but reduced fertility is seen, in *Mus musculus musculus* made transgenic for *Dax1*, and this strain has a higher level of *SRY* expression since it does not carry the 'weak' *SRY*. It is unclear whether the requirement for decreased *SRY* expression in this mouse model is due simply to differences between the mouse and the human, or whether there may be a requirement for one or more additional genes within the DSS locus for sex reversal in humans (Ogata and Matsuo, 1996; Vilain and McCabe, 1998; McCabe, 1999b). These results do suggest that DAX1 may be a negative effector of male development that is antagonized by SRY, findings consistent with a previously proposed model for mammalian sex determination (McElreavey *et al.*, 1993).

DAX1-related orphan receptors

Because of the similarity of the deduced amino acid sequence of the DAX1 with other members of the nuclear hormone receptor superfamily in its C-terminal half, and its unusual sequence in its N-terminal half with very little similarity to other members of the superfamily, relationships were sought between DAX1 and the other members of the superfamily.

SHP is the closest structural relative to DAX1

At the time that *DAX1* was cloned, no other nuclear hormone receptor superfamily member had been identified with a putative DNA-binding domain (DBD) similar to DAX1 (Zanaria *et al.*, 1994; Guo *et al.*, 1995b; Burris *et al.*, 1996). In 1996, a new orphan receptor, SHP (small heterodimer partner), was identified by its specific interaction with another orphan receptor in a yeast two-hybrid system (Seol *et al.*, 1996). Like DAX1, SHP does not contain a conventional DBD, but SHP does have one of the 3.5 N-terminal amino acid repeat sequences that are found in the DAX1 protein (McCabe, 1999b). Northern blot analysis indicated that SHP was expressed most highly in the liver, with lower levels of expression in the pancreas and heart (Seol *et al.*, 1996). SHP interacted with the retinoid receptors, retinoic acid receptor (RAR) and retinoid X receptor (RXR), the thyroid hormone receptor (TR) and the orphan receptor MB67, and inhibited transactivation by these conventional and orphan receptors. These investigators concluded that SHP, which has the closest sequence similarity to DAX1, plays the role of a negative regulator in receptor-dependent signaling pathways in the tissues in which it is expressed.

Subsequently the tissue expression of SHP was expanded to include adult small intestine, as well as liver and adrenal gland from a 12-week fetus, and ovary and testis from individuals of unspecified ages (Lee *et al.*, 1998). These investigators noted the overlapping expression between DAX1 and SHP. They also reported the presence of steroidogenic factor 1 response elements (SF1RE) in the promoter regions for human and mouse SHP (Lee *et al.*, 1998), another similarity with the *DAX1* gene (Burris *et al.*, 1995).

This group has also shown that SHP inhibits the action of the estrogen receptor through a direct interaction (Seol *et al.*, 1998).

SF1 interacts functionally with DAX1

The DAX1 promoter contains a SF1RE. Both the human and murine DAX1 promoters contain consensus sequences for SF1REs in similar positions (Burris *et al.*, 1995; Bae *et al.*, 1996; Guo *et al.*, 1996a). *In vitro* transcribed and translated SF1 specifically bound to the putative SF1RE in the DAX1 promoter by electrophoretic mobility shift assay (Burris *et al.*, 1995). The importance of the SF1RE in DAX1 expression is suggested by transfection experiments in cells that normally express both of these orphan receptors (McCabe, 1999b). Expression of the reporter gene required the presence of the intact SF1RE in the DAX1 promoter and SF1 cotransfection stimulated expression of the reporter gene in the presence of the SF1RE in NCI H295 cells, which normally express both DAX1 and SF1 (Vilain *et al.*, 1997). However, in adrenal cortical cells that do not express DAX1, and in Leydig-derived cells, transfection experiments suggest that the SF1RE is not required for DAX1 expression (Ikeda *et al.*, 1996). Whether these differing observations reflect the different experimental systems, or whether they indicate tissue-dependent differences in the regulation of DAX1 expression, remains to be determined (McCabe, 1999b; Vilain *et al.*, 1999).

SF1 and DAX1 may interact in a concerted fashion during normal development. Fushi tarazu factor 1 (Ftz-F1), the murine homolog of SF1, shows a similarity in tissue profile of expression but a different temporal profile of expression when compared to Dax1 (Ikeda *et al.*, 1994, 1996). Ftz-F1 expression is first noted at 9 days' postconception in the urogenital ridge, which is 3.5 days earlier than Dax1. This pattern of expression, with the murine homolog of Dax1 appearing later than that of SF1, would be compatible with the presence of the SF1RE in the DAX1 promoter and the possibility that SF1 might function upstream of DAX1 in a transcription cascade (McCabe, 1999b). SF1 is essential for the development of the ventromedial hypothalamic nucleus, the adrenal glands and the gonads, as shown by targeted disruption of the *Ftz-F1* gene (Luo *et al.*, 1994; Ikeda *et al.*, 1995). Based on these observations, Parker and Schimmer (1996) suggested that normal development of the hypothalamic–pituitary–gonadal axis requires SF1 and DAX1 acting in a concerted fashion.

SF1 and DAX1 heterodimerize. Direct protein interaction occurs between SF1 and residues in the N-terminal portion of DAX1 (Ito *et al.*, 1997). However, this heterodimerization does not alter SF1 binding to its response element. Similarly, transactivation mediated by SF1 is inhibited by the coexpression of DAX1 and SF1, but this inhibition does not appear to require heterodimerization of SF1 and DAX1 (Ito *et al.*, 1997). A transcriptional silencing domain is contained within the C-terminal portion of DAX1 and mutations in this region reduce transcriptional silencing (Ito *et al.*, 1997; Lalli *et al.*, 1997; Crawford *et al.*, 1998). The nuclear receptor corepressor, NCoR, is recruited to SF1 by DAX1 (Crawford *et al.*, 1998). DAX1 binds to DNA hairpin structures that are found in the promoters for the murine *Dax1* and human steroidogenic acute regulatory protein (*StAR*) genes (Zazopoulos *et al.*, 1997). DAX1 binding to these hairpin structures results in transcriptional repression. Since the DAX1 and SF1REs are present in close proximity in both the *Dax1* and *StAR* promoters, it has been suggested that SF1 binding to its RE is altered by DAX1 in an allosteric fashion (Zazopoulos *et al.*, 1997).

SF1 may also interact directly with DAX1 to antagonize the synergistic action of SF1 on Wilms tumor 1 (WT1) action, resulting in the increased expression of müllerian inhibiting substance (MIS) (Nachtigal *et al.*, 1998). These investigators suggested that mutations in the human *WT1* gene that result in hypogonadism, sexual ambiguity or male pseudohermaphroditism in 46XY individuals cause a decrease in the WT1/DAX1 ratio, favoring the SF1–DAX1 interaction and inhibiting the SF1–WT1 synergy and, therefore, antagonizing male-specific gene expression.

PHYSIOLOGY

DAX1

DAX1 is required for the normal development and function of the adrenal cortex
Mechanisms of adrenal development are poorly understood. Adrenal cortex and medulla have different embryologic origins. The adrenal medulla derives from neural crest cells whereas the cortex develops from mesodermal cells. Mesenchymal cells from the posterior abdominal wall, near the cephalic portion of the mesonephros, proliferate at day 33 postfertilization and migrate to the adrenal area to form the adrenal cortex primordium. Two zones can be distinguished during adrenal development. The inner fetal zone, initially larger than the outer definitive zone, progressively regresses after the second trimester of gestation. It represents 25% of the adrenal cortex at 2 months of age and disappears by the end of the first year (Orth *et al.*, 1992). It is probable that the fetal cortex regresses by apoptosis, and the residual cells give rise to the fasciculata and reticularis zones of the definitive cortex (McCabe, 1999b).

In the human, no data on fetal expression are available. In mouse, *Dax1* is expressed in the adrenals at E12.5, 1 day after the development of the adrenal primordium (Swain *et al.*, 1996). A Cre recombinase-mediated disruption of *Dax1* exon 2 does not result in any abnormal development of fetal and adult cortical zones until puberty (Yu *et al.*, 1998), which differs from the observations in patients with *DAX1* mutations. However, the fetal zone in targeted mice fails to regress after sexual maturation, which is similar to the human AHC phenotype, but differs in that the adult zone persists in mice and not in humans. These results suggest a role for Ahch in fetal adrenal cortical regression.

DAX1 is developmentally expressed in the gonads

In mouse fetal gonads, Dax1 is expressed at E11.5 in both sexes (Swain *et al.*, 1996). This corresponds to the onset of sex-determining region, Y chromosome gene (Sry) expression and the differentiation of the seminiferous tubules in males (Koopman *et al.*, 1990). Dax1 expression is rapidly turned off in male gonads, whereas it persists in ovaries (Swain *et al.*, 1996). These results are consistent with the hypothesis that DAX1 may be involved in ovarian development, and may inhibit testicular development (Bardoni *et al.*, 1994; Swain *et al.*, 1996). These findings are also in accord with a previous hypothesis, based on observations of recessive inheritance of XX maleness, proposing that SRY may antagonize an inhibitor of male development, which could be DAX1 (McElreavey *et al.*, 1993; Vilain and McCabe, 1998). In addition, overexpression of Dax1 antagonizes the function of Sry, therefore interfering with testicular development and enhancing the differentiation of the genital ridge into ovarian tissue (Swain *et al.*, 1998). However, the influence of DAX1 on gonadal development may not be purely ovarian. *DAX1* mutations and deletions do not result in sexual ambiguity or reversal, but are associated with an increased incidence of micropenis and cryptorchidism (Wise *et al.*, 1987; McCabe, 1999a,b). In addition, the hypothesis that DAX1 is the inhibitor of male development that is antagonized by SRY, and, therefore, might be a positive effector of ovarian development (McElreavey *et al.*, 1993; Vilain and McCabe, 1998), would suggest that a 46XX female with deletions or mutations in both copies of her *DAX1* genes ($46X^{-DAX1}/X^{-DAX1}$) might have defective ovarian development and be infertile. Studies in a *Dax1* exon 2 knockout mouse show degeneration of the testicular epithelium, suggesting involvement in spermatogenesis, and indicate normal ovarian development (Yu *et al.*, 1998). These results are consistent with incomplete *Dax1* knockout in this murine model, differences between murine and human development and/or a role for DAX1 as a negative effector of male development, but the requirement for other genes in normal ovarian development.

There is a complex interaction of genes involved in testicular development that leads to normal male sexual differentiation. WT1 and SF1 act synergistically to promote the expression of the MIS gene. DAX1 acts as an antagonist to this synergistic cooperation, possibly by interacting directly with SF1 (Nachtigal *et al.*, 1998). DAX1 can inhibit SF1-mediated transcriptional activation, and mutations causing AHC impair this inhibition (Ito *et al.*, 1997). DAX1 may also inhibit the expression of genes normally enhanced by SF1 by binding (and possibly blocking access of SF1) to hairpin loops present in the promoters of SF1-responsive genes (Zazopoulos *et al.*, 1997) and/or by recruiting NCoR (Crawford *et al.*, 1998). Other studies showed that SF1 and DAX1 also may cooperate functionally, essentially by upregulation of DAX1 expression by SF1 (Vilain *et al.*, 1997; Yu *et al.*, 1998). The functional significance of these gene interactions demonstrated *in vitro* is still unclear, and probably varies with the tissue and with the time of embryonic development.

Dax1-related orphan receptors

SHP inhibits transactivation by the nuclear hormone receptors with which it interacts

The closest relative to DAX1 is SHP. Like DAX1, SHP is an unusual orphan member of the nuclear hormone receptor superfamily, containing a typical ligand-binding domain (LBD), but lacking the conserved DBD (Seol *et al.*, 1996). SHP interacts with various nuclear hormone receptors including retinoid receptors, TR, constitutively active receptor (CAR) and peroxisome proliferator-activated receptor (PPAR). It acts as a negative regulator of receptor-dependent signaling pathways.

SF1 has a broad role in steroidogenic tissues

SF1 is an orphan nuclear hormone receptor that was initially shown to be a regulator of the expression of cytochrome P450 steroid hydroxylases (Morohashi *et al.*, 1992). SF1 has the capacity to bind to a consensus response element AGGTCA which resembles a nuclear receptor-binding half-site (Mangelsdorf *et al.*, 1995). This response element has been found in a variety of genes having a role in steroidogenesis and in the reproductive axis. They include aromatase (Lynch *et al.*, 1993), MIS (Shen *et al.*, 1994), StAR (Sugawara *et al.*, 1996), the α subunit of the pituitary glycoprotein hormones (Barnhart and Mellon, 1994), and the luteinizing hormone (LH) β subunit (Keri and Nilson, 1996). SF1 is expressed at all levels of a functional reproductive axis, including the ventro-medial nucleus of the hypothalamus, the anterior pituitary, the adrenals, the testes and the ovary (Luo *et al.*, 1994). SF1 expression starts at E9 in the urogenital ridge, 3.5 days earlier than DAX1. Mice with targeted disruption of the SF-1 gene lack adrenal tissue and show an abnormal development of the ventromedial nucleus of the hypothalamus as well as an abnormal secretion of LH and follicle-stimulating hormone (FSH) (Ikeda *et al.*, 1995).

Physiologic integration

DAX1 and SF1 are expressed and play a major developmental role at all levels of the hypothalamic–pituitary–adrenal/gonad axis. As reviewed above, there is evidence of DAX1–SF1 protein–protein and functional interactions. In addition, SF1 is involved in the function of this axis, by regulating the expression of adrenal steroid hydroxylases and the expression of LH and FSH. It is therefore possible to hypothesize that sexual development represents an example of functional and developmental integration, in which developmental mechanisms are linked with ongoing physiologic function.

STRUCTURE AND MECHANISM OF ACTION

DAX1

DAX1 structure may provide insight into mechanism of action. The C-terminal portion of DAX1 shows the conventional features for a nuclear hormone receptor superfamily member (Burris *et al.*, 1996). This region typically contains the LBD, and two portions of the LBD, termed regions II and III, have high levels of conservation among superfamily members (Wang *et al.*, 1989). The human receptor with the highest level of similarity in region II is RXRα at 36% and in region III TR2–11 at 57% (Burris *et al.*, 1996). Overall, SHP is the closest relative to DAX1 and has 41% amino acid similarity in the C-terminal portion of the protein. The N-terminal portion of DAX1 is quite unusual.

It is composed of 3.5 amino acid repeats. These tandem repeats consist of a 65–67-amino-acid motif that contains cysteines in conserved positions (Burris *et al.*, 1996). Cysteines within the N-terminal portion are positioned in such a manner that they could form two zinc finger structures, although these would be quite novel compared with those in other nuclear hormone receptor superfamily members (Guo *et al.*, 1995b; Burris *et al.*, 1996). The cysteines in these putative zinc finger structures are absolutely conserved in the mouse Dax1 sequence (Guo *et al.*, 1996b). Thus the DAX1 structure has features consistent with an N-terminal DBD and C-terminal LBD and dimerization domains.

Patients' missense mutations appear to be concentrated in a specific domain. The sequence similarity between DAX1 and other members of the nuclear hormone receptor superfamily in the putative LBD permitted the development of a three-dimensional homology model for DAX1 (Zhang *et al.*, 1998). Specifically this model was based on the known structures of rat rTRα$_1$ (Wagner *et al.*, 1995) and human hRXRα (Bourguet *et al.*, 1995). While most DAX1 intragenic mutations are frameshift or nonsense mutations, the authors recently compiled a group of seven naturally occurring single amino acid changes, including six missense and one in-frame deletion (Zhang *et al.*, 1998). All seven of these single amino acid changes clustered in the hydrophobic core of the LBD when they were mapped to the three-dimensional DAX1 homology model (Fig. 5.1, see color plate section). The seven mutations mapped to the structural subdomain of the LBD and altered amino acids that were identical in the human and murine DAX1 homologs. Since these results represent a relatively small number of single amino acid changes, the structural clustering could be pure chance. However, their conservation and positions suggest that these mutations would disrupt receptor folding, dimerization and overall function in a significant manner. Two of these single amino acid changes have been investigated functionally and reduce transcriptional silencing (Ito *et al.*, 1997; Lalli *et al.*, 1997).

DAX1-related orphan receptors

SF-1
SF1 has structural similarity to the *Drosophila* orphan nuclear receptor FTZ-F1. SF1 was initially cloned as a transcription activator of steroid hydroxylases (Ikeda *et al.*, 1993). It was homologous to Ftz-F1, a regulating factor of the *Drosophila* fushi tarazu (*ftz*) gene (Lavorgna *et al.*, 1991). During *Drosophila* early embryogenesis, *ftz*, a homeobox segmentation gene, is expressed in a seven-stripe pattern, specified by the zebra element upstream of its initiation start site. Ftz-F1, a member of the nuclear hormone receptor superfamily, activates *ftz* transcription (Lavorgna *et al.*, 1991).

SF1 and ELP (embryonal long terminal repeat-binding protein) are alternative products from the same gene. Ftz-F1 encodes two proteins by alternative splicing. One is SF1. The other transcription product, ELP, is expressed very early during embryogenesis and was identified in an embryonic carcinoma cell line as a repressor binding to the negative regulatory region of the Moloney murine leukemia virus long terminal repeat (Tsukiyama *et al.*, 1989). ELP is a member of the nuclear hormone receptor superfamily and is capable of repressing the transcriptional activity of a number of other nuclear receptors, including the RAR (Kotomura *et al.*, 1997). ELP and SF1 have the same expression pattern in steroidogenic tissues. In particular, they are both expressed in the three zones of the adrenal cortex. They also share the same recognition sequence.

However, unlike SF1, ELP does not have the ability to enhance the transcription of steroid hydroxylases. This may be due to a difference in the binding affinity to its target, weaker for ELP than for SF1 (Morohashi *et al.*, 1994).

SHP contains novel domains required for receptor interaction and repressor function

The structures of human and mouse *SHP* genes were recently identified, and are remarkably similar to that of DAX1 (Lee *et al.*, 1998). *SHP* contains two exons, and its unique intron is inserted between the first and second nucleotides of the codon for aspartic acid 181, which is identical to the position of the DAX1 intron. SHP is expressed in the liver, small intestine, spleen, ovary, adrenal gland and testis. Expression in gonads and adrenals is also reminiscent of the DAX1 profile of expression. The *SHP* basal promoter contains a TATA box, and several consensus SF-1 binding sites are found in both the human and mouse promoter regions (Lee *et al.*, 1998), similar to the *DAX1* promoter (Burris *et al.*, 1995). However, the SF1REs are located outside the minimal promoter region necessary for expression in adrenal-derived Y1 cells (Lee *et al.*, 1998).

GENETICS

DAX1

DAX1 mutations are inherited in an X-linked recessive fashion

The clinical characterization of patients with AHC associated with a contiguous gene syndrome that involved GKD with or without Duchenne/Becker muscular dystrophy (DMD/BMD) permitted the mapping of AHC to Xp21 (McCabe, 1999a,b). AHC is an X-linked recessive disorder, in which males are affected and die from adrenal insufficiency and addisonian crisis, whereas, most typically, carrier females have normal adrenal function. However, as with other X-linked recessive diseases, carrier females may occasionally be affected (Bartley *et al.*, 1982). In one family with a deletion resulting in AHC and GKD a 7.5-month-old girl died and autopsy revealed hypoplastic adrenals. Therefore, the authors recommend that prenatal counseling for a family with a female carrier fetus should include discussion that there is a low but definite risk for this girl to develop AHC. We also suggest that a plasma adrenocorticotropic hormone level be considered in the neonatal period in a girl for whom a DNA diagnosis has not yet been obtained and for known carrier female neonates.

Germline mosaicism influences genetic counseling for families with new *DAX1* mutations

The authors have observed an example of germline mosaicism in a family with AHC caused by a 23-base-pair (bp) deletion in *DAX1* (Zhang *et al.*, 1998). The matriarch of this family had a son who was affected with AHC and HH, and a daughter who, in turn, had a son with AHC and a carrier daughter. However, the matriarch did not show evidence of the 23-bp deletion in two analyses of her circulating leukocytes and two different collections of buccal brushings. The matrilineal relationships within this family were confirmed by sequencing of the mitochondrial D loop. Therefore, it is assumed that a mutation occurred early in the embryonic development of the matriarch with involvement of at least a portion of the cells that would give rise to her ova. Based on this experience, if a mutation is identified in a son, but not observed in his mother, the mother should be counseled that there is a relatively small but definite risk of having a carrier daughter or affected son as the product of a future pregnancy.

Not all AHC appears to be due to mutant DAX1

A report from Germany of 18 patients with AHC diagnosed clinically in 16 families included one family with isolated AHC in which no *DAX1* mutation could be detected and another in which mutational analysis was not performed (Peter *et al.*, 1998). Among the remaining families, the mutations were evenly distributed with seven including complete deletion of the *DAX1* gene, with or without GKD and DMD, and seven with intragenic mutations, including five frameshift and two nonsense mutations. In the authors' experience, and that of others (A. Monaco, personal communication), a larger proportion of families referred with the clinical diagnosis of AHC have not had detectable *DAX1* mutations. We have found that, if there is a clear X-linked family history of AHC and HH, a *DAX1* mutation is nearly always found. However, in the absence of this family history, we are frequently unable to detect a *DAX1* mutation. At least six disorders in addition to the X-linked form of AHC associated with *DAX1* mutations have been cataloged in Online Mendelian Inheritance in Man (McCabe, 1999b; OMIM, 1999). These include one family with the cytomegalic form of AHC in two sisters and, therefore, presumed to be inherited in an autosomal recessive manner (Kruger *et al.*, 1993).

DAX1-related orphan receptors

SHP has been mapped to 1p36.1

SHP was initially mapped to human chromosome 1 using a monochromosomal somatic cell hybrid blot (Lee *et al.*, 1998). These investigators showed that *SHP* mapped to a single locus at 1p36.1 by fluorescence *in situ* hybridization (FISH).

SF1 has been mapped to 9q33

The *SF1* gene was mapped to 9q33 by FISH, in a region syntenic with the portion of mouse chromosome 2 to which its murine homolog had been mapped by interspecific backcross analysis (Taketo *et al.*, 1995).

PATHOPHYSIOLOGY

DAX1

DAX1 loss of function results in AHC and HH

Point mutations in the *DAX1* gene are responsible not only for AHC but also for HH (Muscatelli *et al.*, 1994; Zanaria *et al.*, 1994). Most endocrinologic investigations on patients with *DAX1* mutations have been inconclusive or contradictory regarding whether HH was hypothalamic or pituitary in origin (Partsch and Sippell, 1989; Kletter *et al.*, 1991). Recent investigations showed that the HH observed in these patients is of mixed (pituitary and hypothalamic) origin (Habiby *et al.*, 1996). HH in patients with mutated *DAX1* is consistent with the expression of this gene in both pituitary and hypothalamus. However, the 'minipuberty' observed during infancy is normal in patients mutated in *DAX1* (Takahashi *et al.*, 1997; Kaiserman *et al.*, 1998). This suggests that DAX1 may be important for the maintenance of functional integrity of the reproductive axis. It is also possible that the regulation of the hypothalamic–pituitary–adrenal/gonad axis occurs by different mechanisms in infancy and later in adolescence.

DAX1 may have multiple roles in testicular function

46XY patients with a tandem duplication of DSS, a 160-kb region in Xp21 (including *DAX1*), present with either female or ambiguous genitalia (Bardoni *et al.*, 1994). Since *DAX1* is localized within this small region critical for sex reversal, it is a candidate for a role in sex determination. This hypothesis was reinforced by the observation of XY sex reversal in female mice carrying *DAX1* as a transgene (Swain *et al.*, 1998). However, sex reversal occurred only when a specific strain of mouse carrying a 'weak' Sry of *Poschiavinus* origin was used. The absence of sex reversal in mice carrying a 'regular' Y chromosome may be caused by the fact that sex determination mechanisms are evolving rapidly at the molecular level (Whitfield *et al.*, 1993), or by the presence of one or more additional genes within the DSS locus acting together with *DAX1* (Ogata and Matsuo, 1996; Vilain and McCabe, 1998; McCabe, 1999b).

In testes, DAX1 is expressed in Leydig (Ikeda *et al.*, 1996) and Sertoli cells (Tamai *et al.*, 1996). Its expression in Leydig cells and in adrenocortical cells suggests a role in steroidogenesis. Its expression in Sertoli cells, where SRY is also expressed, is compatible with a role in sex determination. Moreover, DAX1 expression peaks during the androgen-sensitive phase of the rat spermatogenic cycle, and during the first spermatogenic wave between postnatal days 20 and 30 (Tamai *et al.*, 1996), suggesting a role for DAX1 in spermatogenesis. Two additional lines of evidence support this hypothesis: the abnormal spermatogenesis observed in the mice targeted for *Dax1* knockout (Yu *et al.*, 1998), and the absence of germ cells together with immature Sertoli cells in the testicular histology of AHC patients with *DAX1* mutations (Fujieda *et al.*, 1998).

DAX1-related orphan receptors

SF1 is required for normal development of the hypothalamus, adrenal cortex and gonads

Mice with targeted disruption of the *SF1* gene have an embryonic lethal phenotype with abnormal development of gonads, adrenals and ventromedial hypothalamic nuclei (Luo *et al.*, 1994; Ikeda *et al.*, 1995), suggesting a critical role for SF1 in the development of the reproductive axis. In humans, a point mutation in *SF1* has been described in a 46XY female patient (Achermann *et al.*, 1999). She presented with adrenal failure within the first 2 weeks of life, showed an endocrine profile similar to that which is usually observed in AHC, and exhibited normal müllerian structures and streak gonads as observed in 46XY females with pure gonadal dysgenesis. The mutation was a heterozygous missense mutation and occurred in the proximal box (P box) of the first zinc finger of SF1, which is involved in the recognition of target DNA-binding sites. It is possible that SF1 acts via a dosage-sensitive mechanism, as already observed for other sex-determining genes such as *SRY*, *DAX1*, *SOX9* and *WT1* (Vilain and McCabe, 1998).

SHP

No human disorder has been shown to be associated with SHP.

DIAGNOSIS

DAX1

***DAX1* gene structure facilitates sequence-based diagnosis of intragenic mutations**
As the two different strategies used for the cloning of the *DAX1* gene involved genomic positional cloning, the gene was immediately recognized to be made up of two exons (Zanaria *et al.*, 1994; Guo *et al.*, 1995b). The 1168-bp exon 1 is separated from the 245-bp exon 2 by an intron that is 3385 bp in length (Guo *et al.*, 1996a). The 1413 combined open reading frame from exons 1 and 2 codes for a protein 470 amino acids in length plus the stop codon. The simple genomic structure of the *DAX1* gene permits direct sequencing from 10 overlapping polymerase chain reaction (PCR) amplification products that include the entire coding sequence from both exons as well as the intron–exon boundaries (Guo *et al.*, 1996a). The authors use this approach routinely for identifying new mutations in DNA samples referred from patients with AHC for investigation (Zhang *et al.*, 1998). *DAX1* deletions may be diagnosed by FISH. For FISH analysis, we use the cosmid, 8E10, in which we initially identified the *DAX1* gene (Guo *et al.*, 1995b). This was utilized initially to show that *DAX1* was appropriately deleted in a family in which two brothers had been previously shown by PCR markers to have a deletion of the AHC locus (Fig. 5.2, see color plate section).

DAX1-related orphan receptors

A human mutation has been identified in *SF1*, but not in *SHP*
A heterozygous G35E missense mutation was reported in the first zinc finger of the SF1 DBD (Achermann *et al.*, 1999). The patient was an XY phenotypic female who presented with adrenal insufficiency and addisonian crisis in the neonatal period. She had normal müllerian structures and small intraabdominal gonads. The gonads were removed and histological analysis showed them to contain immature tubules. Since the G35E mutation was in the first zinc finger of the SF1 DBD, electrophoretic mobility shift assays were carried out with the SF1RE that is conserved in the mouse Dax1 promoter. The protein containing the G35E allele bound weakly to the SF1RE compared with the wild-type *SF1* allele. Similarly the G35E mutant allele stimulated transcriptional activity only weakly compared with the wild-type allele in transient expression studies using reporter gene constructs. The authors concluded that SF1 might act in a dosage-sensitive fashion, since the heterozygous mutation was sufficient for XY sex reversal like other autosomal sex-determining genes in the human, *WT1* and *SOX9*. However, they added that other mechanisms such as a dominant negative effect by the G35E mutant protein could not be excluded.

No human mutation has yet been identified in *SHP*.

PHARMACOLOGY

No bona fide ligands have been identified for DAX1 or related orphan receptors; however, oxysterols were recently shown to activate SF1 (Lala *et al.*, 1997). Oxysterols, many of which are ligands for another nuclear receptor, liver X receptor (LXR) (Peet *et al.*, 1998), enhanced SF1-dependent transcription in transfected CU-1 cells (Lala *et al.*,

1997). These results remain controversial as oxysterols do not increase expression of several SF1-dependent DNA sequences in Leydig MA-10 cells (Mellon and Bair, 1998).

REFERENCES

Achermann, J.C., Ito, M., Hindmarsh, P.C. and Jameson, J.L. (1999) A mutation in the gene encoding steroidogenic factor-1 causes XY sex reversal and adrenal failure in humans (letter). *Nat. Genet.* **22**: 125–126.

Arn, P., Chen, H., Tuck Muller, C.M. *et al.* (1994) A sex reversing locus in Xp21.2 to p22.11. *Hum. Genet.* **93**: 389–393.

Bae, D.S., Schaefer, M.L., Partan, B.W. and Muglia, L. (1996) Characterization of the mouse *DAX1* gene reveals evolutionary conservation of a unique amino-terminal motif and widespread expression in mouse tissue. *Endocrinology* **137**: 3921–3927.

Bardoni, B., Floridia, G., Guioli, S. *et al.* (1993) Functional disomy of Xp22-pter in three males carrying a portion of Xp translocated to Yq. *Hum. Genet.* **91**: 333–338.

Bardoni, B., Zanaria, E., Guioli, S. *et al.* (1994) A dosage sensitive locus at chromosome Xp21 is involved in male to female sex reversal. *Nat. Genet.* **7**: 497–501.

Barnhart, K.M. and Mellon, P.L. (1994) The orphan nuclear receptor, steroidogenic factor-1, regulates the glycoprotein hormone α-subunit gene in pituitary gonadotropes. *Mol. Endocrinol.* **8**: 878–885.

Bartley, J.A., Miller, D.K., Hayford, J.T. and McCabe, E.R.B. (1982) The concordance of X-linked glycerol kinase deficiency with X-linked adrenal hypoplasia in two families. *Lancet* **ii**: 733–736.

Bernstein, R., Jenkins, T., Dawson, B. *et al.* (1980) Female phenotype and multiple abnormalities in sibs with a Y chromosome and partial X chromosome duplication: H-Y antigen and Xg blood group findings. *J. Med. Genet.* **17**: 291–300.

Bourguet, W., Ruff, M., Chambon, P., Gronemeyer, H. and Moras, D. (1995) Crystal structure of the ligand-binding domain of the human nuclear receptor RXR-α. *Nature* **375**: 377–382.

Brochner-Mortensen, K. (1956) Familial occurrence of Addison's disease. *Acta Med. Scand.* **156**: 205–209.

Burris, T.P., Guo, W., Le, T. and McCabe, E.R.B. (1995) Identification of a putative steroidogenic factor-1 response element in the DAX-1 promoter. *Biochem. Biophys. Res. Commun.* **214**: 576–581.

Burris, T.P., Guo, W. and McCabe, E.R.B. (1996) The gene responsible for adrenal hypoplasia congenita, *DAX1*, encodes a nuclear hormone receptor that defines a new class within the superfamily. In: P.M. Conn (ed.) *Recent Progress in Hormone Research*, pp. 241–260. The Endocrine Society, Bethesda, MD.

Crawford, P.A., Dorn, C., Sadovsky, Y. and Milbrandt, J. (1998) Nuclear receptor DAX-1 recruits nuclear receptor corepressor N-CoR to steroidogenic factor 1. *Mol. Cell. Biol.* **18**: 2949–2956.

Favara, B.E., Franciosi, R.A. and Miles, V. (1972) Idiopathic adrenal hypoplasia in children. *Am. J. Clin. Pathol.* **57**: 287–296.

Fechner, P.Y., Marcantonio, S.M., Ogata, T. *et al.* (1993) Report of a kindred with X-linked (or autosomal dominant sex-limited) 46,XY partial gonadal dysgenesis. *J. Clin. Endocrinol. Metab.* **76**: 1248–1252.

Fujieda, K., Nakae, J., Abe, S., Shinohara, N., Murashita, M. and Sato, K. (1998) Analysis of hypothalamic–pituitary–testicular axis in a patient with X-linked adrenal hypoplasia congenita. Presented to the Endocrine Society, June 24–27, 1998, New Orleans, Louisiana.

German, J., Simpson, J.L., Chaganti, R.S., Summitt, R.L., Reid, L.B. and Merkatz, I.R. (1978) Genetically determined sex-reversal in 46,XY humans. *Science* **202**: 53–56.

Goonewardena, P., Dahl, N., Ritzen, M., van Ommen, G.J. and Pettersson, U. (1989) Molecular Xp deletion in a male: suggestion of a locus for hypogonadotropic hypogonadism distal to the glycerol kinase and adrenal hypoplasia loci. *Clin. Genet.* **35**: 5–12.

Guggenheim, M.A., McCabe, E.R.B., Roig, M. *et al.* (1980) Glycerol kinase deficiency with neuromuscular, skeletal and adrenal abnormalities. *Ann. Neurol.* **7**: 441–449.

Guo, W., Burris, T.P. and McCabe, E.R.B. (1995a) Expression of *DAX-1*, the gene responsible for X-linked adrenal hypoplasia congenita and hypogonadotropic hypogonadism, in the hypothalmic–pituitary–adrenal/gonadal axis. *Biochem. Mol. Med.* **56**: 8–13.

Guo, W., Mason, J.S., Stone, C.G. *et al.* (1995b) Diagnosis of X-linked adrenal hypoplasia congenita by mutation analysis of the *DAX-1* gene. *JAMA* **274**: 324–330.

Guo, W., Burris, T.P., Zhang, Y.-H. *et al.* (1996a) Genomic sequence of the *DAX1* gene: an orphan nuclear receptor responsible for X-linked adrenal hypoplasia congenita and hypogonadotropic hypogonadism. *J. Clin. Endocrinol. Metab.* **81**: 2481–2486.

Guo, W., Lovell, R.S., Zhang, Y.-H. *et al.* (1996b) *Ahch*, the mouse homologue of *DAX1*: cloning, characterization and synteny with *GyK*, the glycerol kinase locus. *Gene* **178**: 31–34.

Habiby, R.L., Boepple, P., Nachtigall, L., Sluss, P.M., Crowley. W.F., Jr. and Jameson, J.L. (1996) Adrenal hypoplasia congenita with hypogonadotropic hypogonadism: evidence that *DAX1* mutations lead to combined hypothalamic and pituitary defects in gonadotropin production. *J. Clin. Invest.* **98**: 1055–1062.

Ikeda, Y., Lala, D.S., Luo, X., Kim, E., Moisan, M.P. and Parker, K.L. (1993) Characterization of the mouse *FTZ-F1* gene, which encodes a key regulator of steroid hydroxylase gene expression. *Mol. Endocrinol.* **7**: 852–860.

Ikeda, Y., Shen, W.-H., Ingraham, H.A. and Parker, K.L. (1994) Developmental expression of mouse steroidogenic factor-1, an essential regulator of the steroid hydroxylases. *Mol. Endocrinol.* **8**: 654–662.

Ikeda, Y., Luo, X., Abbud, R., Nilson, J.H. and Parker, K.L. (1995) The nuclear receptor steroidogenic factor 1 is essential for the formation of the ventromedial hypothalamic nucleus. *Mol. Endocrinol.* **9**: 478–486.

Ikeda, Y., Swain, A., Weber, T.J. *et al.* (1996) Steroidogenic factor 1 and Dax-1 colocalize in multiple cell lineages: potential links in endocrine development. *Mol. Endocrinol.* **10**: 1261–1272.

Ito, M., Yu, R. and Jameson, J.L. (1997) DAX-1 inhibits SF-1-mediated transactivation via a carboxy-terminal domain that is deleted in adrenal hypoplasia congenita. *Mol. Cell. Biol.* **17**: 1476–1483.

Kaiserman, K.B., Nakamoto, J.M., Geffner, M.E. and McCabe, E.R.B. (1998) Minipuberty of infancy and adolescent pubertal function in adrenal hypoplasia congenita. *J. Pediatr.* **133**: 300–302.

Kelch, R.P., Virdis, R., Rapaport, R., Greig, F., Levine, L.S. and New, M.I. (1984) Congenital adrenal hypoplasia. *Pediatr. Adolesc. Endocrinol.* **13**: 156–161.

Keri, R.A. and Nilson, J.H. (1996) A steroidogenic factor-1 binding site is required for activity of the luteinizing hormone beta subunit promoter in gonadotropes of transgenic mice. *J. Biol. Chem.* **271**: 10 782–10 785.

Kletter, G.B., Gorski, J.L. and Kelch, R.P. (1991) Congenital adrenal hypoplasia and isolated gonadotropin deficiency. *Trends Endocrinol. Metab.* **2**: 123–128.

Koopman, P., Munserberg, A., Capel, B., Vivian, N. and Lovell-Badge, R. (1990) Expression of a candidate sex-determining gene during mouse testis differentiation. *Nature* **348**: 450–452.

Kotomura, N., Ninomiya, Y., Umesono, K. and Niwa, O. (1997) Transcriptional regulation by competition between ELP isoforms and nuclear receptors. *Biochem. Biophys. Res. Commun.* **230**: 407–412.

Kruger, G., Mix, M., Pelz, L. and Dunker, H. (1993) Cytomegalic type of congenital adrenal hypoplasia due to autosomal recessive inheritance. *Am. J. Med. Genet.* **46**: 475.

Lala, D.S., Syka, P.M., Lazarchik, S.B., Magelsdorf, D.J., Parker, K.L. and Heyman, R.A. (1997) Activation of the orphan receptor steroidogenic factory 1 by oxysterols. *Proc. Natl. Acad. Sci. USA* **94**: 4895–4900.

Lalli, E., Bardoni, B., Zazopoulos, E. *et al.* (1997) A transcriptional silencing domain in *DAX-1* whose mutation causes adrenal hypoplasia congenita. *Mol. Endocrinol.* **11**: 1950–1960.

Lavorgna, G., Ueda, H., Clos, J. and Wu, C. (1991) FTZ-F1, a steroid hormone receptor-like protein implicated in the activation of fushi tarazu. *Science* **252**: 848–851.

Lee, H.-K., Lee, Y.-K., Park, S.-H. *et al.* (1998) Structure and expression of the orphan nuclear receptor *SHP* gene. *J. Biol. Chem.* **273**: 14 398–14 402.

Love, D.R., Bloomfield, J.F., Kenwrick, S.J., Yates, J.R.W. and Davies, K.E. (1990) Physical mapping distal to the DMD locus. *Genomics* **8**: 106–112.

Luo, X., Ikeda, Y. and Parker, K.L. (1994) A cell-specific nuclear receptor is essential for adrenal and gonadal development and sexual differentiation. *Cell* **77**: 481–490.

Lynch, J.P., Lala, D.S., Peluso, J.J., Luo, W., Parker, K.L. and White, B.A. (1993) Steroidogenic factor 1, an orphan nuclear receptor, regulates the expression of the rat aromatase gene in gonadal tissues. *Mol. Endocrinol.* **7**: 776–786.

Mangelsdorf, D.J., Thummel, C., Beato, M. *et al.* (1995) The nuclear receptor superfamily: the second decade. *Cell* **83**: 835–839.

Matsumoto, T., Kondoh, T., Yoshimoto, M. *et al.* (1988) Complex glycerol kinase deficiency: molecular genetic, cytogenetic, and clinical studies of five Japanese patients. *Am. J. Med. Genet.* **31**: 603–616.

May, K.M., Grinzaid, K.A. and Blackston, R.D. (1991) Sex reversible and multiple abnormalities due to abnormal segregation of t(X;16)(p11.4;p13.3). *Am. J. Hum. Genet.* **49**: 19.

McCabe, E.R.B. (1996) Sex and the single DAX1: too little is bad, but can we have too much? *J. Clin. Invest.* **98**: 881–882.

McCabe, E.R.B. (1999a) Disorders of glycerol metabolism. In: C.R. Scriver, A.L. Beaudet, W.S. Sly, D. Valle, B. Childs and B. Vogelstein (eds) *The Metabolic and Molecular Bases of Inherited Disease*, 8th edn. McGraw-Hill, New York (in press).

McCabe, E.R.B. (1999b) Adrenal hypoplasias and aplasias. In: C.R. Scriver, A.L. Beaudet, W.S. Sly, D. Valle, B. Childs and B. Vogelstein (eds) *The Metabolic and Molecular Bases of Inherited Disease*, 8th edn. McGraw-Hill, New York (in press).

McCabe, E.R.B., Fennessey, P.V., Guggenheim, M.A. *et al.* (1977) Human glycerol kinase deficiency with hyperglycerolemia and glyceroluria. *Biochem. Biophys. Res. Commun.* **78**: 1327–1333.

McElreavey, K., Vilain, E., Abbas, N., Herskowitz, I. and Fellous, M. (1993) A regulatory cascade hypothesis for mammalian sex determination: SRY represses a negative regulator of male development. *Proc. Natl. Acad. Sci. USA* **90**: 3368–3372.

Mellon, S.H. and Bair, S.R. (1998) 25-Hydroxycholesterol is not a ligand for the orphan nuclear receptor steroidogenic factor-1 (SF-1). *Endocrinology* **139**: 3026–3029.

Morohashi, K., Honda, S., Inomata, Y., Handa, H. and Omura, T. (1992) A common *trans-acting* factor, Ad4-binding protein, to the promoters of steroidogenic P-450s. *J. Biol. Chem.* **267**: 17 913–17 919.

Morohashi, K., Iida, H., Nomura, M. *et al.* (1994) Functional difference between Ad4BP and ELP, and their distributions in steroidogenic tissues. *Mol. Endocrinol.* **8**: 643–653.

Muscatelli, F., Strom, T.M., Walker, A.P. *et al.* (1994) Mutations in the *DAX-1* gene give rise to both X-linked adrenal hypoplasia congenita and hypogonadotropic hypogonadism. *Nature* 372: 672–676.

Nachtigal, M.W., Hirokawa, Y., Enyeart-VanHouten, D.L., Flanagan, J.N., Hammer, G.D. and Ingraham, H.A. (1998) Wilms' tumor 1 and Dax-1 modulate the orphan nuclear receptor SF-1 in sex-specific gene expression. *Cell* **93**: 445–454.

Narahara, K., Kodama, Y., Kimura, S. and Kimoto, H. (1979) Probable inverted tandem duplicaton of Xp in a 46,Xp + Y boy. *Jpn. J. Hum. Genet.* **24**: 105–110.

Nielsen, K.B. and Langkjaer, F. (1982) Inherited partial X chromosome duplication in a mentally retarded male. *J. Med. Genet.* **19**: 222–224.

Ogata, T. and Matsuo, N. (1994) Testis determining gene(s) on the X chromosome short arm: chromosomal localisation and possible role in testis determination (letter). *J. Med. Genet.* **31**: 349.

Ogata, T. and Matsuo, N. (1996) Sex determining gene on the X chromosome short arm: dosage sensitive sex reversal. *Acta Paediatr. Jpn.* **38**: 390–398.

Ogata, T., Taylor, A., Hawkins, J.R., Matsuo, N., Hata, J. and Goodfellow, P.N. (1992) Sex reversal in a child with a 46,X,Yp+ karyotype: support for the existence of a gene(s), located in distal Xp, involved in testis formation. *J. Med. Genet.* **29**: 226–230.

Online Mendelian Inheritance in Man (OMIM) (1999) Center for Medical Genetics, Johns Hopkins University (Baltimore, MD) and National Center for Biotechnology Information, National Library of Medicine (Bethesda, MD). World Wide Web URL: http:// www.ncbi.nim.nih.gov/Omim

Orth, D.N., Kovacs, W.J. and DeBold, C.R. (1992) The adrenal cortex. In: J.D. Wilson and D.W. Foster (eds) *Williams' Textbook of Endocrinology*, 8th edn, pp. 489–619. Saunders, Philadelphia.

Parker, K.L. and Schimmer, B.P. (1996) The roles of the nuclear hormone receptor steroidogenic factor 1 in endocrine differentiation and development. *Trends Endocrinol. Metab.* **7**: 203–207.

Partsch, C.-J. and Sippell, W.G. (1989) Hypothalamic hypogonadism in congenital adrenal hypoplasia. *Horm. Metab. Res.* **21**: 623–625.

Peet, D.J., Janaowski, B.A. and Magelsdorf, D.J. (1998) The LXRs: a new class of oxysterol receptors. *Curr. Opin. Genet. Dev.* **8**: 571–575.

Peter, M., Viemann, M., Partsch, C.-J. and Sippell, W.G. (1998) Congenital adrenal hypoplasia: clinical spectrum, experience with hormonal diagnosis, and report on new point mutations of the *DAX-1* gene. *J. Clin. Endocrinol. Metab.* **83**: 2666–2674.

Renier, W.O., Nabbe, F.A.E., Hustinx, T.W.J. *et al.* (1983) Congenital adrenal hypoplasia, progressive muscular dystrophy, and severe mental retardation, in association with glycerol kinase deficiency, in male sibs. *Clin. Genet.* **24**: 243.

Scherer, G., Schempp, W., Baccichetti, C. *et al.* (1989) Duplication of an Xp segment that includes the *ZFX* locus causes sex inversion in man. *Hum. Genet.* **81**: 291–294.

Seol, W., Choi, H.-S. and Moore, D.D. (1996) An orphan nuclear hormone receptor that lacks a DNA binding domain and heterodimerizes with other receptors. *Science* **272**: 1336–1339.

Seol, W., Hanstein, B., Brown, M. and Moore, D.D. (1998) Inhibition of estrogen receptor action by the orphan receptor SHP (short heterodimer partner). *Mol. Endocrinol.* **12**: 1551–1557.

Shen, W.H., Moore, C.C., Ikeda, Y., Parker, K.L. and Ingraham, H.A. (1994) Nuclear receptor steroidogenic factor 1 regulates the mullerian inhibiting substance gene: a link to the sex determination cascade. *Cell* **77**: 651–661.

Stern, H.J., Garrity, A.M., Saal, H.M., Wangsa, D. and Disteche, C.M. (1990) Duplication Xp21 and sex reversal: insight into the mechanism of sex determination. *Am. J. Hum. Genet.* **47**: A41.

Sugawara, T., Holt, J.A., Kiriakidou, M. and Strauss, J.F., III (1996) Steroidogenic factor 1-dependent promoter activity of the human steroidogenic acute regulatory protein (*StAR*) gene. *Biochemistry* **35**: 9052–9059.

Swain, A., Zanaria, E., Hacker, A., Lovell-Badge, R. and Camerino, G. (1996) Mouse Dax1 expression is consistent with a role in sex determination as well as in adrenal and hypothalamus function. *Nat. Genet.* **12**: 404–409.

Swain, A., Narvaez, V., Burgoyne, P., Camerino, G. and Lovell-Badge, R. (1998) Dax1 antagonizes Sry action in mammalian sex determination. *Nature* **391**: 761–767.

Tabarin, A., Achermann, J.C., Recan, D. *et al.* (2000) A novel mutation in *DAX1* caused delayed onset adrenal insufficiency and incomplete hypogonadotropic hypogonadism. *J. Clin. Invest.* **105**: 321–328.

Takahashi, T., Shoji, Y., Haraguchi, N., Takahashi, I. and Takada, G. (1997) Active hypothalamic–pituitary–gonadal axis in an infant with X-linked adrenal hypoplasia congenita. *J. Pediatr.* **130**: 485–488.

Taketo, M., Parker, K.L., Howard, T.A. *et al.* (1995) Homologs of *Drosophila* Fushi-Tarazu factor 1 map to mouse chromosome 2 and human chromosome 9q33. *Genomics* **25**: 565–567.

Tamai, K.T., Monaco, L., Alastalo, T.P., Lalli, E., Parvinen, M. and Sassone-Corsi, P. (1996) Hormonal and developmental regulation of DAX-1 expression in Sertoli cells. *Mol. Endocrinol.* **10**: 1561–1569.

Tsukiyama, T., Niwa, O. and Yokoro, K. (1989) Mechanism of suppression of the long terminal repeat of Moloney leukemia virus in mouse embryonal carcinoma cells. *Mol. Cell. Biol.* **9**: 4670–4676.

Vilain, E. and McCabe, E.R.B. (1998) Mammalian sex determination: from gonads to brain. *Mol. Genet. Metab.* **65**: 74–84.

Vilain, E., Guo, W., Zhang, Y.-H. and McCabe, E.R.B. (1997) *DAX1* gene expression upregulated by steroidogenic factor 1 in an adrenocortical carcinoma cell line. *Biochem. Mol. Med.* **61**: 1–8.

Vilain, E., Guo, W., Patel, M. and McCabe, E.R.B. (1999) X-linked adrenal hypoplasia congenita (AHC): a DAX1 deficiency disorder. In: G. Chrousos, J. Olefsky and E. Samols (eds) *Hormone Resistance*. Lippincott-Raven, Philadelphia (in press).

Wagner, R.L., Apriletti, J.W., McGrath, M.E., West, B.L., Baxter, J.D. and Fletterick, R.J. (1995) A structural role for hormone in the thyroid hormone receptor. *Nature* **378**: 690–697.

Walker, A.P., Chelly, J., Love, D.R. *et al.* (1992) A *YAC* contig in Xp21 containing the adrenal hypoplasia congenita and glycerol kinase deficiency genes. *Hum. Mol. Genet.* **1**: 579–585.

Wang, L.H., Tsai, S.Y., Cook, R.G., Beattie, W.G., Tsai, M.J. and O'Malley, B.W. (1989) COUP transcription factor is a member of the steroid receptor superfamily. *Nature* **340**: 163–166.

Whitfield, L.S., Lovell-Badge, R. and Goodfellow, P.N. (1993) Rapid sequence evolution of the mammalian sex-determining gene SRY. *Nature* **364**: 713–715.

Wise, J.E., Matalon, R., Morgan, A.M. and McCabe, E.R.B. (1987) Phenotypic features of patients with congenital adrenal hypoplasia and glycerol kinase deficiency. *Am. J. Dis. Child.* **141**: 744–747.

Wittenberg, D.F. (1981) Familial X-linked adrenocortical hypoplasia association with androgenic precocity. *Arch. Dis. Child.* **56**: 633–636.

Worley, K.C., Towbin, J.A., Zhu, X.M. *et al.* (1992) Identification of three new markers in Xp21 between DXS28 (C7) and DMD. *Genomics* **13**: 957–961.

Worley, K.C., Ellison, K.A., Zhang, Y.-H. *et al.* (1993) Yeast artificial chromosome cloning in the glycerol kinase and adrenal hypoplasia congenita region of Xp21. *Genomics* **16**: 407–416.

Yu, R.N., Ito, M., Saunders, T.L., Camper, S.A. and Jameson, J.L. (1998) Role of Ahch in gonadal development and gametogenesis. *Nat. Genet.* **20**: 353–357.

Zachmann, M., Illig, R. and Prader, A. (1980) Gonadotropin deficiency and cryptorchidism in three prepubertal brothers with congenital adrenal hypoplasia. *J. Pediatr.* **97**: 255–257.

Zanaria, E., Muscatelli, F., Bardoni, B. *et al.* (1994) A novel and unusual member of the nuclear hormone receptor superfamily is responsible for X-linked adrenal hypoplasia congenita. *Nature* **372**: 635–641.

Zazopoulos, E., Lalli, E., Stocco, D.M. and Sassone-Corsi, P. (1997) DNA binding and transcriptional repression by DAX-1 blocks steroidogenesis. *Nature* **390**: 311–315.

Zhang, Y.-H., Guo, W., Wagner, R.L. *et al.* (1998) *DAX1* mutations map to putative structural domains in a deduced three-dimensional model. *Am. J. Hum. Genet.* **62**: 855–864.

Chapter 6

The Vitamin D Receptor

*Paul N. MacDonald, Dennis M. Kraichely and
Alex J. Brown*

INTRODUCTION AND HISTORICAL PERSPECTIVE

Vitamin D plays a central role in calcium and phosphate homeostasis and is essential for the proper development and maintenance of bone. The observation by Sir Edward Mellanby in 1919 that rickets could be caused by a nutritional deficiency led to the isolation of a fat-soluble antirachitic substance in fish liver oil and other foods that was identified as vitamin D_2. At the same time, Huldschinsky (1919) and Hess and Unger (1921) discovered that children with rickets could be cured by exposing them to ultraviolet light. Antirachitic activity could also be induced in various foods by ultraviolet irradiation. Subsequent studies of these antirachitic substances led to the structural identification of vitamin D_2 (ergocalciferol) (Askew *et al.*, 1931) and vitamin D_3 (cholecalciferol) (Windaus *et al.*, 1936) as secosterols, derived from the photolytic cleavage of the B rings of ergosterol and 7-dehydrocholesterol respectively.

These two sterols were considered the biologically active forms of vitamin D until the mid-1960s when the availability of radiolabeled vitamin D_3 (Norman and DeLuca, 1963) permitted the identification of metabolites with greater antirachitic activity. 25-Hydroxyvitamin D_3 (25-$(OH)D_3$) was found to be the major circulating metabolite of vitamin D_3 (Blunt *et al.*, 1968) and subsequently was shown to be produced primarily in the liver. Haussler *et al.* (1968) found a metabolite more polar than 25$(OH)D_3$ in a nuclear fraction isolated from chicken intestine. This molecule, synthesized mainly in the kidney, was identified as 1,25-dihydroxyvitamin D_3 (1,25$(OH)_2D_3$) (Fraser and Kodicek, 1970; Holick *et al.*, 1971; Lawson *et al.*, 1971; Norman *et al.*, 1971) and is now known to be the most active metabolite of vitamin D.

The next major breakthrough in vitamin D research was the discovery of a high-affinity receptor for 1,25$(OH)_2D$ (Haussler *et al.*, 1968). This 50–70-kDa protein, associated with nuclear chromatin, displayed saturable binding of 1,25$(OH)_2D_3$, and had a specificity for other vitamin D metabolites that closely matched their *in vivo* biopotency. The development of monoclonal antibodies against the vitamin D receptor (VDR) allowed the isolation of complementary DNAs (cDNAs) coding for the avian, human, mouse and rat VDRs (McDonnell *et al.*, 1987; Baker *et al.*, 1988; Burmester *et al.*, 1988; Kamei *et al.*, 1995). The sequence of the VDR revealed considerable similarity

Nuclear Receptors and Genetic Disease
ISBN 0-12-146160-2

to other members of the steroid receptor superfamily including the characteristic two zinc finger motifs in a DNA-binding domain (DBD). This suggested that the VDR was also a ligand-activated transcription factor.

The VDR was detected initially in the classical vitamin D target organs involved in mineral homeostasis: the intestine, bone, kidney and parathyroid glands. The VDR has now been demonstrated to be present in many other tissues and cells types as well. These nonclassic vitamin D target organs respond to $1,25(OH)_2D_3$ with a diverse range of biologic actions including immunomodulation, the control of other hormonal systems, inhibition of cell growth and induction of cell differentiation. These actions have suggested a number of new therapeutic applications of $1,25(OH)_2D_3$ in immune dysfunction (autoimmune disease, transplantation), endocrine disorders (hyperparathyroidism) and hyperproliferative disorders (leukemia, cancer, psoriasis). However, the potent calcemic and phosphatemic activities of $1,25(OH)_2D_3$ have precluded its use in most cases. This limitation has been overcome by the development of vitamin D analogs with less calcemic activity. Two of these compounds are currently in use for the treatment of psoriasis and uremic secondary hyperparathyroidism, and many others are in clinical trials.

PHYSIOLOGY

Vitamin D metabolism

Sources of vitamin D

Vitamin D can be obtained from the diet and by the action of sunlight on the skin. Only a few food sources such as fish oils, egg yolks and liver contain significant amounts of vitamins D_2 and D_3. However, many foods are now fortified with the vitamin and minimum daily requirements are easily met. Vitamin D_3 is produced in the skin by an ultraviolet light-induced photolytic conversion of 7-dehydrocholesterol to previtamin D_3 (Holick *et al.*, 1977; Okano *et al.*, 1977) followed by thermal isomerization to vitamin D_3 (Hanewald *et al.*, 1961). The serum vitamin D-binding protein (DBP), which preferentially binds vitamin D_3 over its precursors, mediates the translocation of the vitamin from the epidermis into the circulation (Haddad *et al.*, 1988).

Activation of vitamin D to $1,25(OH)_2D$

25-Hydroxylation. The first step in the metabolic activation of vitamin D is hydroxylation of carbon 25 (Fig. 6.1). This reaction occurs primarily in the liver, although other tissues including skin, intestine and kidney have been reported to catalyze 25-hydroxylation of vitamin D. The contribution of the extrahepatic sources to the circulating levels of 25(OH)D is uncertain. The hepatic 25-hydroxylation involves cytochrome P450 monooxygenase(s). At least two enzymes have been reported, one mitochondrial and the other microsomal, but the roles of these two cytochrome P450s remains controversial (Hayashi *et al.*, 1988; Saarem *et al.*, 1984; Guo *et al.*, 1993). The 25-hydroxylation of vitamin D is poorly regulated. The levels of 25(OH)D increase in proportion to vitamin D intake and, for this reason, plasma 25(OH)D levels are commonly used as an indicator of vitamin D status (Holick, 1981). After it is synthesized in the liver, 25(OH)D appears in the circulation bound primarily to DBP and albumin.

1α-Hydroxylation. The final step in vitamin D bioactivation, the conversion of 25(OH)D to $1,25(OH)_2D$, occurs (Fig. 6.1), under physiologic conditions, mainly in the

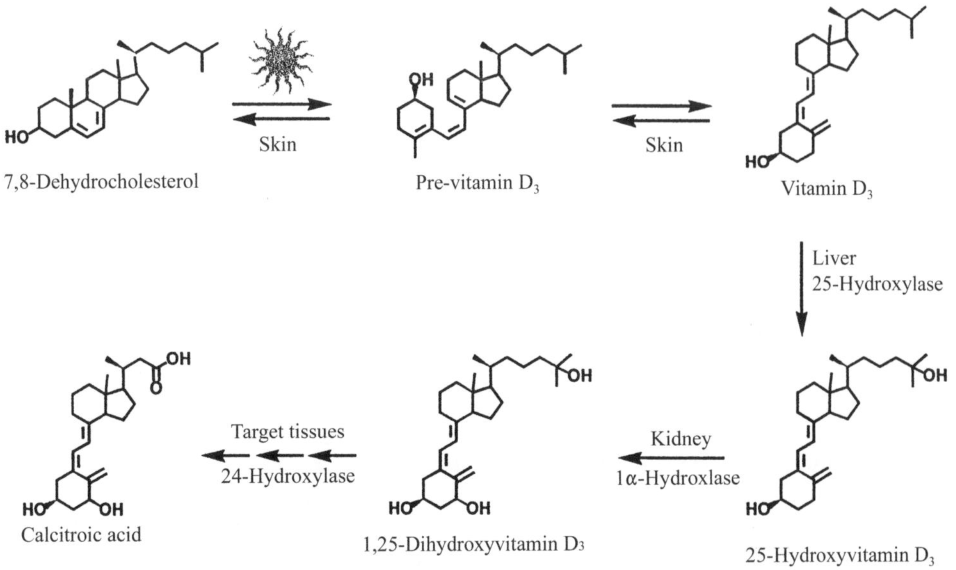

Fig. 6.1 Metabolism of vitamin D to 1,25(OH)$_2$D$_3$. Vitamin D$_3$ is produced in the skin by ultraviolet photolysis of 7,8-dehydrocholesterol followed by thermal isomerization. The vitamin D$_3$ (or vitamin D$_2$ obtained from the diet) is hydroxylated at carbon 25 mainly in the liver. The final activation step, 1α-hydroxylation, occurs almost exclusively in the kidney and is highly regulated. A series of oxidation reactions catalyzed by the 24-hydroxylase converts 1,25(OH)$_2$D to the inactive, secreted metabolite, calcitroic acid.

kidney (Fraser and Kodicek, 1970). The renal enzyme responsible for producing 1,25(OH)$_2$D, 25-(OH)D-1α-hydroxylase, is located in the inner mitochondrial membrane and is a cytochrome P450 monooxygenase requiring molecular oxygen and reduced ferredoxin (Ghazarian *et al.*, 1974). In recent years, many reports have demonstrated that the kidney is not unique in its ability to convert 25(OH)D to 1,25(OH)$_2$D. Numerous cells and tissues express 1α-hydroxylase *in vitro*; however, in humans, these extrarenal sources of 1,25(OH)$_2$D contribute significantly to circulating 1,25(OH)$_2$D levels only during pregnancy, in chronic renal failure and in pathologic conditions such as sarcoidosis, tuberculosis, granulomatous disorders and rheumatoid arthritis.

The 1α-hydroxylase cDNA has been cloned from mouse (Takeyama *et al.*, 1997), rat (Shinki *et al.*, 1997; St-Arnaud *et al.*, 1997) and human (Monkawa *et al.*, 1997) kidney, and human keratinocytes (Fu *et al.*, 1997). Expression of the protein in cultured cells promotes 1α-hydroxylation of 25(OH)D$_3$. Further evidence for the identity of the human 1α-hydroxylase cDNA came from chromosomal mapping and mutational analysis. The cDNA hybridizes solely to chromosomal locus 12q13.1-q13.3, the site to which the defect in patients unable to produce 1,25(OH)$_2$D$_3$ (vitamin D-dependent rickets type I) has been mapped (Labuda *et al.*, 1990). Moreover, mutations in the coding regions of the 1α-hydroxylase gene have been identified in patients with the disease (Fu *et al.*, 1997; Kitanaka *et al.*, 1998; Yoshida *et al.*, 1998).

The circulating 1,25(OH)$_2$D is carried primarily on DBP, with a smaller amount bound to albumin and lipoproteins (Bikle *et al.*, 1985). Although more than 99% of the $1,25(OH)_2D$ is bound *in vivo*, current evidence suggests that the unbound fraction of the hormone (free $1,25(OH)_2D_3$), with greater accessibility to target cells, is the biologically active form (Bikle *et al.*, 1984; Vanham *et al.*, 1988; Bikle and Gee, 1989).

Vitamin D catabolism

The high potency of $1,25(OH)_2D_3$ in raising serum calcium and phosphate levels requires its circulating levels to be tightly regulated. Control of serum $1,25(OH)_2D_3$ usually involves joint reciprocal changes in the rates of synthesis and degradation. Vitamin D compounds are catabolized primarily by oxidation of the side-chain. The major catabolic enzyme is the vitamin D-24-hydroxylase, another mitochondrial cytochrome P450 requiring molecular oxygen and reduced ferredoxin (Knutson and DeLuca, 1974; Burgos-Trinidad *et al.*, 1986). The oxidation of the side-chain of $25(OH)D_3$ and $1,25(OH)_2D_3$ is initiated at carbon 24. This is followed by further oxidation of carbon 24 to a ketone, oxidation of carbon 23 and subsequent oxidative cleavage of the side-chain (Makin *et al.*, 1989; Reddy and Tserng, 1989). Each oxidation step leads to progressive loss of biologic activity. The final cleavage product of $1,25(OH)_2D_3$, calcitroic acid, is biologically inert. The 24-hydroxylase cDNA (Ohyama and Okuda, 1991) and gene (Ohyama *et al.*, 1993) have been cloned. In contrast to the limited tissue distribution of 1α-hydroxylase, 24-hydroxylase is present in all vitamin D target tissues. The 24-hydroxylase is highly inducible by $1,25(OH)_2D_3$, providing a mechanism for attentuating the response to the vitamin D hormone, and reducing $1,25(OH)_2D_3$ levels when they are abnormally high. Mice lacking a functional 24-hydroxylase gene (St-Arnaud *et al.*, 1996) have high serum $1,25(OH)_2D_3$ levels owing to the decreased capacity to degrade it.

Classical actions of $1,25(OH)_2D_3$

Vitamin D is part of the stringent regulatory system that governs calcium and phosphate homeostasis through specific actions in the classic mineral-regulating organs: the intestine, bone, parathyroid glands and kidneys. These actions are summarized in Table 6.1 and depicted schematically in Fig. 6.2.

Intestine

The most crucial function of the active form of vitamin D in mineral homeostasis is to enhance the efficiency of the small intestine to absorb dietary calcium and phosphate. Although the absorption of calcium and phosphate occurs along the entire length of the small intestine, vitamin D primarily stimulates calcium transport in the duodenum and phosphate absorption in the jejunum and ileum (Chen *et al.*, 1974).

Intestinal absorption of calcium and phosphate correlates positively with the need for these minerals (Kletzien *et al.*, 1932). Both hypocalcemia and hypophosphatemia increase the production of $1,25(OH)_2D_3$, which increases the efficiency of the intestine to absorb calcium and phosphate, while vitamin D deficiency causes calcium malabsorption and negative calcium and phosphate balance.

$1,25(OH)_2D_3$ is the only hormone known to stimulate intestinal calcium and phosphate transport directly. Evidence that these actions of $1,25(OH)_2D_3$ require the VDR comes from several sources. Newborn rats do not express VDR in the intestine and are insensitive to the effects of $1,25(OH)_2D_3$ on calcium absorption. The response appears later, concomitant with the expression of intestinal VDR (Halloran and

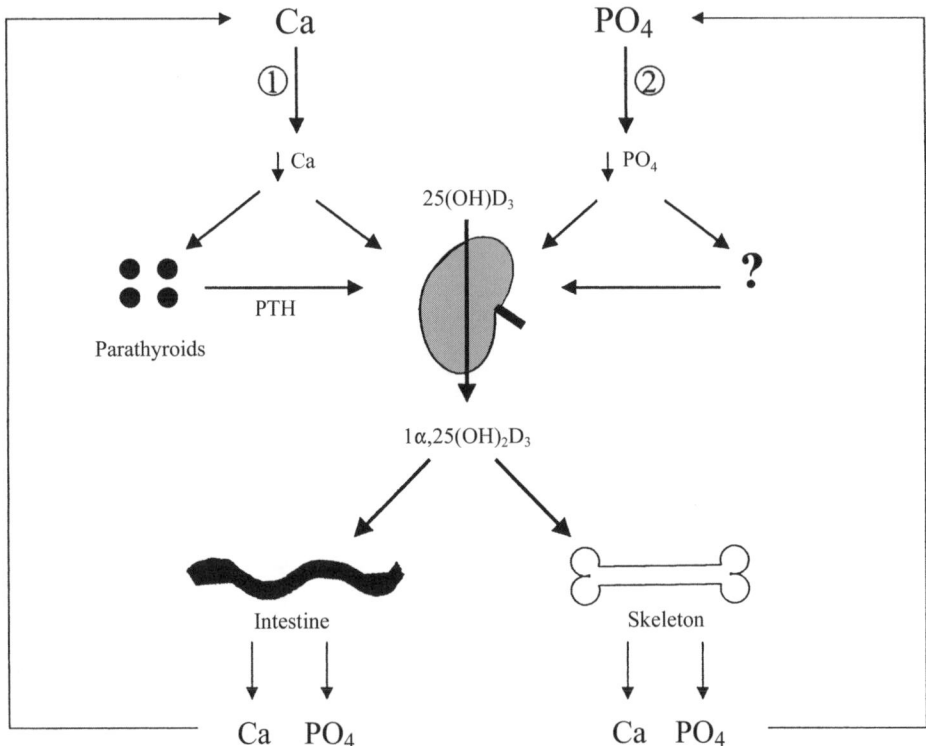

Fig. 6.2 Role of the vitamin D endocrine system in calcium and phosphate homeostasis. A fall in serum calcium concentration (depicted as ①) will enhance renal production of 1,25(OH)$_2$D$_3$ both directly and indirectly via parathyroid hormone (PTH). Similarly, a fall in serum phosphate concentration (depicted as ②) will increase 1,25(OH)$_2$D$_3$ synthesis by unknown mechanisms. 1,25(OH)$_2$D$_3$ stimulates intestinal absorption of calcium and phosphate and the release of calcium and phosphate from bone, normalizing the serum levels of these minerals.

DeLuca, 1981). Hydrocortisone injections cause the precocious appearance of VDR and vitamin D-stimulated calcium absorption, while adrenalectomy delays both (Massaro *et al.*, 1983). Loss of VDR gene expression or activity leads to malabsorption of both calcium and phosphate despite high levels of circulating 1,25(OH)$_2$D$_3$.

The mechanism(s) by which 1,25(OH)$_2$D$_3$ enhances the absorption of calcium and phosphate are not completely understood. There is evidence for at least three pathways for calcium uptake from the intestinal lumen to the blood. The first pathway, the transcellular movement of calcium, involves three steps: (1) entry of calcium through the plasma membrane into the intestinal absorptive cell, (2) the movement of calcium through the cytoplasm, and (3) the transfer of calcium across the basolateral membrane into the circulation. Calcium entry from the intestinal lumen across the brush border membrane into the enterocyte is a diffusional process with a positive calcium gradient which does not require vitamin D (Fullmer, 1992). However, in vitamin D-deficient animals, the calcium entering the enterocyte will be sequestered in the apical region. Movement across the intestinal cell is thought to be facilitated by a calcium-binding protein, calbindin, which is 1,25(OH)$_2$D$_3$ inducible (Wasserman and Taylor, 1968).

Table 6.1 1,25(OH)$_2$D$_3$ actions in classic and nonclassic target tissues

Tissue	Cell type	Action
Classic		
Intestine	Epithelial	Enhancement of calcium and phosphate absorption
Bone	Osteoblast	Enhancement of bone matrix protein synthesis, bone mineralization, and synthesis of mediators of osteoclastogenic and osteoclastic activity
	Osteoclast	Enhancement of bone resorption
Kidney	Epithelial (proximal and distal)	Inhibition of 1,25(OH)$_2$D$_3$ synthesis and induction of 24-hydroxylase Enhancement of calcium and phosphate reabsorption
Parathyroid gland	Chief	Inhibition of cell growth and parathyroid hormone synthesis
Nonclassic		
Hematopoietic tissues	Myeloid cell precursors	Antiproliferative, prodifferentiating
	Colony-forming units	Prodifferentiating
Immune system	Monocyte/macrophages	Enhancement of immune function to control viral and bacterial infections and tumor growth
	Lymphocyte	Immunosuppression
Skin	Keratinocytes, fibroblasts, hair follicle, Langerhans cells, melanocytes	Antiproliferative, prodifferentiating
Muscle	Smooth muscle cell, myoblast	Antiproliferative, prodifferentiating
Heart	Cardiac muscle cell	Antiproliferative, prodifferentiating
	Atrial myocytes	Inhibition of atrial natriuretic factor synthesis
Pancreas	β cells	Enhancement of insulin synthesis and secretion

Cancer cells	Melanoma, breast carcinoma, leukemia, osteosarcoma, fibrosarcoma, pituitary medullary thyroid carcinoma, adenoma, neuroblastoma, pancreatic adenocarcinoma, bladder, cervical, prostate and colonic carcinomas	Antiproliferative, prodifferentiating
Adrenal gland	Medullary cells	Control of catecholamine metabolism
Brain	Hippocampus/selected neurons	Neuronal regeneration, enhancement of nerve growth factor and neurotrophin synthesis, control of sphingomyelin cycle
Cartilage	Chondrocyte	Antiproliferative, prodifferentiating
Female reproductive organs	Myometrial and endometrial cells	Antiproliferative, control of folliculogenesis
Liver	Parenchymal cell (fetal, adult)	Enhancement of liver regeneration, control of glycogen and transferrin synthesis
Lung	Pneumocytes	
	fetal	Enhancement of maturation, phospholipid synthesis and surfactant release
	adult	Cell growth
Male reproductive organs	Sertoli/seminiferous tubule	Enhancement of Sertoli cell function and spermatogenesis
Pituitary	Somatomammotroph	Control of T_3-induced growth hormone, prolactin and tyrotrophyn production
Thyroid	Follicular cells (C cells)	Inhibition of cell function and calcitonin synthesis

Transfer of calcium through the basolateral membrane is accomplished primarily by a calcium-dependent adenosine triphosphatase (calcium pump), which is induced by $1,25(OH)_2D_3$ (Ghijsen and Van Os, 1982; Zelinski et al., 1991; Wasserman et al., 1992; Cai et al., 1993).

A vesicular pathway for intestinal calcium absorption has also been suggested, involving internalization of calcium in endocytic vesicles, fusion of the vesicles with lysosomes, and movement of the lysosomes (along microtubules) to the basolateral membrane for exocytosis (Nemere et al., 1984). $1,25(OH)_2D_3$ augments this vesicular calcium transport through a rapid, nongenomic mechanism termed transcaltachia (Nemere et al., 1991). The contribution of transcaltachia to intestinal calcium absorption is unclear.

A third route for calcium absorption is by a paracellular pathway. Although still controversial, there is evidence of stimulatory effects of vitamin D on this paracellular component of intestinal calcium transport (Nellans and Kimberg, 1978; Karbach, 1992).

$1,25(OH)_2D_3$ also increases dietary phosphate absorption (Harrison and Harrison, 1961). Intestinal phosphate transport appears to be independent of calcium transport and less dependent on $1,25(OH)_2D_3$. In the absence of $1,25(OH)_2D_3$, phosphate absorption proceeds at 50% of the normal rate (Walling, 1978). Sodium is required at the brush border surface (Taylor, 1974), and a Na/P cotransporter moves luminal phosphate uphill into the cell, against a thermodynamic gradient, using the energy from the downhill sodium gradient. $1,25(OH)_2D_3$ appears to stimulate phosphate transport by inducing the synthesis of the cotransporter (Yagci et al., 1992) and by decreasing the permeability of the brush border membrane to sodium, helping to maintain the gradient required for the uphill movement of phosphorus. $1,25(OH)_2D_3$ decreases membrane permeability through a reduction of both the voltage sensitive Na flux across the brush border and the activity of the brush border Na/H antiporter (Fuchs et al., 1985; Cross et al., 1990). $1,25(OH)_2D_3$ also exerts rapid effects on the membrane composition, increasing fluidity and phosphate uptake (Fuchs and Peterlik, 1979; Kurnik and Hruska, 1985). At present, little is known about the extrusion of phosphate across the basolateral membrane into the circulation.

Skeleton

Vitamin D is essential for the development and maintenance of a mineralized skeleton. Vitamin D deficiency results in rickets in young growing animals and osteomalacia in the adult. In both cases, osteoblasts produce normal collagen fibrils which fail to mineralize, resulting in short stature and bony deformities in children. In the adult, new bone is formed only in areas of bone remodeling, so with vitamin D deficiency the new bone fails to calcify properly resulting in weak, easily fractured bones (Holick, 1996). At the same time, in keeping with its primary physiologic function to maintain normal calcium levels, $1,25(OH)_2D_3$ stimulates the mobilization of calcium stores from bone by inducing the dissolution of bone mineral and matrix (Holick, 1996). Thus, $1,25(OH)_2D_3$ can enhance both formation and resorption of bone.

These actions of $1,25(OH)_2D_3$ on bone are mediated by the VDR. The primary target cells are the osteoblast, the bone-forming cells, and the osteoblast precursors (stromal cells and preosteoblasts). The VDR messenger RNA (mRNA) has been detected in osteoclasts, the bone-resorbing cells, by the very sensitive method of in situ polymerase chain reaction (Mee et al., 1996). However, VDR protein has not been detected in mature osteoclast, and therefore the significance of this finding is unclear.

The regulation of osteoblast gene expression by $1,25(OH)_2D_3$ has received considerable attention. $1,25(OH)_2D_3$ controls the expression of several genes associated with osteoblast proliferation and differentiation, including alkaline phosphatase, osteopontin, osteocalcin, osteonectin and matrix Gla protein (Spiess et al., 1986; Fraser et al., 1988; Fraser and Price, 1990), and is capable of stimulating mineralization in cultures of clonal osteoblast-like cells (Matsumoto et al., 1991).

The role of $1,25(OH)_2D_3$ in bone resorption involves enhancement of both osteoclastogenesis and osteoclast activity. Osteoclast precursors express the VDR, but recent studies with VDR-ablated mice indicate that it is the action of $1,25(OH)_2D_3$ on the osteoblast that leads to osteoclast maturation and stimulation of osteoclastic bone resorption (Takeda et al., 1999). When spleen cells (osteoclast precursors) are cocultured with stromal cells (osteoblast precursors), the presence of $1,25(OH)_2D_3$ leads to the formation of mature osteoclasts. This process requires direct interaction between osteoblasts and osteoclasts (Jimi et al., 1996; Yoshizawa et al., 1997). Furthermore, these mixed cultures, containing both osteoblasts and mature osteoclasts, can be plated on to bone slices and stimulated to resorb bone by treatment with $1,25(OH)_2D_3$. If normal stromal cells are cocultured with spleen cells from VDR-ablated mice, there is normal osteoclastogenesis and stimulation of bone resorption in response to $1,25(OH)_2D_3$. However, if the stromal cells are obtained from the VDR-deficient mice, normal spleen cells do not form mature osteoclasts with $1,25(OH)_2D_3$ treatment (Takeda et al., 1999).

The mechanism for the osteoblast-mediated control of osteoclast formation and activity has been the subject of considerable investigation for many years. This osteoclast differentiating factor (ODF) has recently been identified as the osteoblast cell-surface protein osteoprotegerin ligand (OPGL) (Lacey et al., 1998). The receptor for OPGL, RANK, has been shown to be expressed on the surface of both osteoclast precursors and mature osteoclasts (Nakagawa et al., 1998). Osteoblasts also produce a soluble osteoclast inhibitory factor (OIF), which was subsequently identified as osteoprotegerin (OPG) (Simonet et al., 1997). OPG acts as a pseudoreceptor for OPGL and prevents OPGL from interacting with RANK. $1,25(OH)_2D_3$ has been shown to increase OPGL expression and reduce OPG production, consistent with its positive effects on osteoclast formation and activity (Horwood et al., 1998). Thus, the current model for $1,25(OH)_2D_3$-mediated bone resorption involves reciprocal regulation of OPGL and OPG.

Additional effects of $1,25(OH)_2D_3$ on osteoblasts may play a modulating role in osteoclastogenesis. $1,25(OH)_2D_3$ induces the production of interleukin 11 (IL-11) by osteoblasts (Romas et al., 1996), which has been shown to be essential for $1,25(OH)_2D_3$-mediated osteoclastogenesis (Girasole et al., 1994). In the mouse, C3, the third component of the complement system, is produced by osteoblasts in response to $1,25(OH)_2D_3$ and appears to be involved in modulating the differentiation of bone marrow cells into osteoclasts (Sato et al., 1991, 1993).

Despite the extensive studies showing the role of $1,25(OH)_2D_3$ in both bone formation and resorption, it is important to stress that there is considerable redundancy in the hormonal control of these processes. Thus, while VDR-ablated mice and patients with functionally defective VDR develop severe rickets, all bone abnormalities appear to be corrected by providing a diet high in calcium and phosphate (Balsan et al., 1986; Li et al., 1998). As will be discussed elsewhere in this review, most, and perhaps all, abnormalities associated with loss of VDR function may be due to the malabsorption of calcium and phosphate from the diet. Further study is required to test this hypothesis.

Parathyroid glands

The parathyroid glands play a central role in calcium homeostasis by sensing the levels of circulating calcium and secreting the proper amount of parathyroid hormone (PTH), a calciotropic hormone that stimulates release of calcium from bone. Calcium is the dominant regulator of PTH synthesis and secretion, but it is well established that $1,25(OH)_2D_3$ can suppress PTH secretion by inhibiting transcription of the PTH gene (Cantley *et al.*, 1985; Chan *et al.*, 1986; Russell *et al.*, 1986; Silver *et al.*, 1985, 1986). PTH can stimulate $1,25(OH)_2D_3$ synthesis by the kidney (Garabedian *et al.*, 1972; Rasmussen *et al.*, 1972), and therefore suppression of PTH by $1,25(OH)_2D_3$ represents an important feedback loop in calcium homeostasis.

The $1,25(OH)_2D_3$-responsive region of the human PTH gene was defined initially by Okazaki *et al.* (1988). Further analysis revealed a negative response element that bound the VDR, but not its usual heterodimer partner, retinoid X receptor (RXR) (Demay *et al.*, 1992; Mackey *et al.*, 1996). In contrast, analysis of the chicken PTH gene indicated that both VDR and RXR bind to the negative VDRE in this promoter (Liu *et al.*, 1996b). More details of the mechanism for transcriptional repression are discussed below.

The dominant regulatory role of calcium in the parathyroid glands is evident by the blunted response to $1,25(OH)_2D_3$ in the presence of hypocalcemia. Both hypocalcemia and the ensuing increased PTH secretion can stimulate renal $1,25(OH)_2D_3$ synthesis. However, feedback suppression of $1,25(OH)_2D_3$ on PTH synthesis would be counter-productive to the efforts of the parathyroid glands to normalize serum calcium concentration. Therefore, the parathyroid glands reduce the expression of the VDR in response to hypocalcemia (Russell *et al.*, 1993; Brown *et al.*, 1995) to ensure sustained PTH secretion until serum calcium levels return to normal.

Kidney

Vitamin D performs several major functions in the kidney: regulation of calcium and phosphate reabsorption, and control of its own synthesis and degradation. The exact role of vitamin D in the renal handling of calcium and phosphorus continues to be controversial.

Receptors for $1,25-(OH)_2D_3$ have been reported in the distal tubule (Stumpf *et al.*, 1980) and in the proximal tubule (Kawashima *et al.*, 1981). The seco-sterol enhances the tubular reabsorption of calcium (Puschett and Beck, 1975) and also induces the vitamin D-dependent calcium binding protein, calbindin, in the region of the distal nephrons (Roth *et al.*, 1981, 1982) in which calcium reabsorption is known to occur. Calbindin is localized in the cytosol and the nuclear chromatin, but not in the plasma membrane. Therefore, renal calbindin may not be involved directly in transmembrane calcium transport; instead it might play a role in intracellular translocation of calcium ions. While the calcium pump of the intestinal epithelium is induced by $1,25(OH)_2D_3$ treatment, it is not clear whether this induction occurs in the kidney.

The effect of $1,25(OH)_2D_3$ on renal phosphate handling is controversial. While there is evidence that $1,25(OH)_2D_3$ enhances renal reabsorption of phosphate (Puschett *et al.*, 1972; Costanzo *et al.*, 1974; Puschett and Beck, 1975), it is unclear whether this effect is mediated by suppression of PTH, which inhibits phosphate transport (Bonjour *et al.*, 1977).

Perhaps the most important effect of $1,25(OH)_2D_3$ in the kidney is the suppression of 1α-hydroxylase activity and the stimulation of 24-hydroxylase activity. This homeostatic feedback loop ensures that the proper amount of $1,25(OH)_2D_3$ will be released by the proximal tubules. It is important to note that, under conditions of prolonged hypocalcemia or hypophosphatemia, renal VDR expression is decreased to prevent feedback until the mineral levels are normalized (Iida et al., 1995).

Nonclassic actions of 1,25(OH)$_2$D$_3$

The discovery of VDRs in tissues unrelated to mineral homeostasis raised numerous questions about the physiologic role of vitamin D in nonclassic targets. It is beyond the scope of this treatise to catalog all actions of 1,25(OH)$_2$D$_3$ in various tissues and cell types. This section will focus on the actions of vitamin D in hematopoietic tissue, the immune system, skin and pancreas because of their physiologic and potential therapeutic significance. Table 6.1 presents a complete list of the reported actions of 1,25(OH)$_2$D$_3$ in various target tissues and cell types.

Hematopoietic tissues

1,25(OH)$_2$D$_3$ can inhibit growth and promotes differentiation of myeloid leukemic cells to a macrophage phenotype (Abe et al., 1981; Dodd et al., 1983; Goldman, 1984). The mechanism appears to involve decreased oncogene expression (Studzinski et al., 1985) and induction of the cyclin-dependent kinase inhibitor p21 (Liu et al., 1996a). Transient overexpression of p21, independent of 1,25(OH)$_2$D$_3$, results in cell surface expression of monocyte–macrophage markers, suggesting that the 1,25(OH)$_2$D$_3$ transcriptional induction of the p21 gene directly induces the differentiation of this monoblastic cell line.

The therapeutic potential of 1,25-(OH)$_2$D$_3$ to reduce clonal proliferation of myeloid precursors and to induce differentiation has been tested in vivo. The survival of mice inoculated with the M1 leukemia cell line was prolonged significantly after 1,25-(OH)$_2$D$_3$ or 1α-hydroxyvitamin D$_3$ administration (Honma et al., 1983). However, a study of patients with myelodysplasia demonstrated only a transient improvement in peripheral blood counts with 1,25(OH)$_2$D$_3$ administration, and the beneficial effects were offset by hypercalcemia (Koeffler, 1985). Similar limitations were observed in treatment of patients with idiopathic myelofibrosis (McKinley et al., 1987). These findings indicate a need for less calcemic vitamin D compounds.

The immune system

The role of the vitamin D endocrine system on the immune response is complex. In general, 1,25(OH)$_2$D$_3$ enhances the activity of monocytes/macrophages, but suppresses the proliferation of lymphocytes. Vitamin D deficiency is associated with recurrent infections (Stroder and Kasal, 1970), tuberculosis (Davies, 1985), depressed natural cell killer activity (Asaka et al., 1988), defective neutrophil motility (Lorente et al., 1976) and impaired monocyte phagocytosis (Bar-Shavit et al., 1981). In these studies it was unclear whether the abnormalities were due to the vitamin D deficiency per se or to the coincident hypocalcemia. However, the subsequent discovery of VDR in cells of the immune system prompted an evaluation of the immunoregulatory role of 1,25(OH)$_2$D$_3$.

Monocytes and activated macrophages express the VDR. 1,25(OH)$_2$D$_3$ enhances antigen processing by macrophages, thus stimulating their defense against bacterial infection and tumor cell growth (Manolagas et al., 1994). Induction of heat shock protein synthesis by 1,25(OH)$_2$D$_3$ appears to promote macrophage survival and function at the increased temperatures associated with tissue inflammation (Polla et al., 1987). In addition, 1,25(OH)$_2$D$_3$ promotes fusion of macrophages and increases expression of Fc receptors, which are instrumental in recognizing antibody-coated bacteria (Abe et al., 1984).

In contrast to the stimulatory effects of the hormone on monocytes and macrophages, the principal action of $1,25(OH)_2D_3$ in lymphocytes is to decrease the rate of proliferation and the activity of T cells and B cells (Lemire *et al.*, 1984; Tsoukas *et al.*, 1984; Manolagas *et al.*, 1994). $1,25(OH)_2D_3$ also directs the type of T-cell response by suppressing cytokine (interleukin 12 and interferon γ) production (Lemire *et al.*, 1995). Furthermore, $1,25(OH)_2D_3$ increases the availability of suppressor T cells (Meehan *et al.*, 1992). These immunosuppressive actions of $1,25(OH)_2D_3$ have suggested a potential application in the treatment of autoimmune diseases such as rheumatoid arthritis, systemic lupus erythematosus and juvenile (type I) diabetes.

Skin

The skin is not only the site of vitamin D_3 production: keratinocytes also have the capacity to produce $1,25(OH)_2D_3$. Dermal production of $1,25(OH)_2D_3$ appears to play an important autocrine/paracrine function in controlling the proliferation and differentiation of various skin cells. The VDR is expressed by most of cell types found in the dermal layers, including keratinocytes, Langerhans cells, melanocytes, fibroblasts and endothelial cells.

The effects of $1,25(OH)_2D_3$ on epidermal keratinocytes varies with their state of differentiation. $1,25(OH)_2D_3$ stimulates proliferation of keratinocytes committed to terminal differentiation, but inhibits the growth of undifferentiated keratinocytes (Gniadecki, 1996). A similar paradoxical response is observed *in vivo* (Lutzow-Holm *et al.*, 1993). Topical application of $1,25(OH)_2D_3$ to normal mouse skin increases keratinocyte proliferation and causes epidermal hyperplasia. In contrast, petrolatum-induced epidermal hyperplasia is inhibited by $1,25(OH)_2D_3$ (Kato *et al.*, 1987).

These actions of $1,25(OH)_2D_3$ have been exploited in the treatment of various skin disorders. In psoriasis, normal maturation and proliferation of epidermal cells are disturbed by lymphokines released from antigen-presenting cells in the skin lesions, resulting in a hyperproliferative state. $1,25(OH)_2D_3$, administered systemically or topically, resulted in an obvious improvement in the skin condition of psoriatic patients (Holick, 1988). In addition to its effect on keratinocyte proliferation, $1,25(OH)_2D_3$ reduces the number of Langerhans cells, the antigen-presenting cells of the epidermis, as well as their accessory cell function (Bagot *et al.*, 1994). However, administration of $1,25(OH)_2D_3$, even topically, is limited by its potent calcemic activity. New vitamin D analogs may offer a safer alternative for the treatment of psoriasis (Kragballe, 1989; Morimoto *et al.*, 1989) as well as other hyperproliferative skin diseases including scleroderma, pityriasis rubra pilaris, disseminated superficial actinic parakeratosis, inflammatory linear verrucous epidermal naevus and Grover's disease (van de Kerkhof and de Jong, 1991; Harrison and Stollery, 1994; Bottomley *et al.*, 1995; Keohane and Cork, 1995).

Hair follicles may also be a target for $1,25(OH)_2D_3$. The VDR is highly expressed in the hair follicle, and patients with hereditary vitamin D-resistant rickets often present with complete alopecia. Unlike other phenotypes associated with loss of the VDR, however, alopecia is not corrected by normalization of calcium and phosphate levels (Li *et al.*, 1998).

Pancreas

The finding of the VDR in the insulin-secreting β cells of the pancreas suggested a potential role of $1,25(OH)_2D_3$ in controlling insulin secretion. Vitamin D deficiency in rats was shown to impair glucose-mediated insulin secretion (Chertow *et al.*, 1983), and more recent studies have shown that this effect is independent of change in calcium or caloric intake (Cade and Norman, 1986). Vitamin D supplementation enhanced the insulin response to an oral glucose test in vitamin D-deficient women (Gedik and Akalin, 1986), and calcitriol therapy significantly increased serum insulin concentrations in uremic patients (Quesada *et al.*, 1990).

The mechanism for the effect of $1,25(OH)_2D_3$ on insulin secretion is unknown. Expression of the vitamin D-inducible calcium-binding protein, calbindin, in β cells (Christakos *et al.*, 1979) suggested that $1,25(OH)_2D_3$, through modulation of the calbindin concentration, may control intracellular calcium flux (Billaudel *et al.*, 1988), which in turn affects insulin release.

STRUCTURE AND MECHANISM OF ACTION OF THE VDR

The VDR belongs to the superfamily of nuclear receptors, the members of which all share on overall structural relatedness. Five subdomains within the nuclear receptors express varying degrees of sequence similarity and are given the following general designations: A/B, C, D, E and F. A brief summary of these domains for the VDR is presented below, and are depicted schematically in Fig. 6.3.

The N-terminal A/B domain

The A/B domain is a hypervariable region that extends from the N-terminus to the DBD of the receptors. It varies in sequence and in length between receptors, ranging in size from approximately 20 residues in the VDR to over 200 residues in the glucocorticoid receptor (GR). Although a precise functional role for the A/B domain is not well understood, its lack of sequence conservation between receptors suggests that this domain may be important for hormone- or receptor-selective functions. The A/B domain of most receptors contains a ligand-independent transactivation function (AF-1) that is required for the full transcriptional activity of the receptor (Giguere *et al.*, 1986; Gronemeyer *et al.*, 1987; Kumar *et al.*, 1987; Webster *et al.*, 1988; Nagpal *et al.*, 1993; Hadzic *et al.*, 1995). In addition, the human progesterone receptor contains a unique third activation function (AF-3) which lies at the extreme N-terminus of the

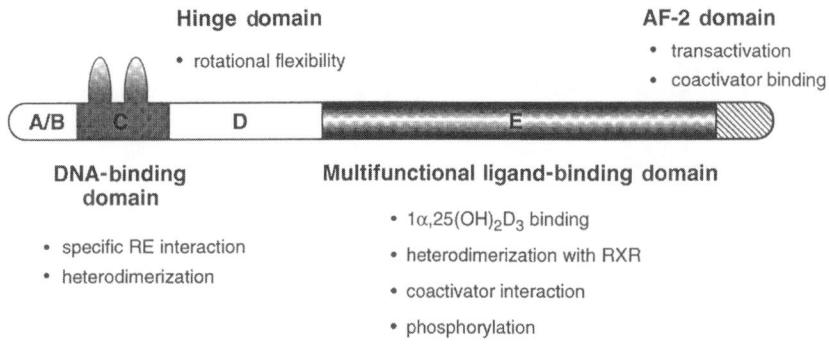

Fig. 6.3 Functional domains of the vitamin D receptor. AF, activation function; RE, response element; RXR, retinoid X receptor.

receptor (Sartorius *et al.*, 1994). The VDR is somewhat atypical among the nuclear receptors in that its A/B domain is small, consisting of only 20 amino acid residues. Deletion of these 20 residues does not affect VDR-mediated transcription in transient reporter gene assays (Sone *et al.*, 1991), suggesting that in this system VDR function is independent of the A/B domain and any intrinsic activation functions such as AF-1 that may reside there.

The DNA-binding domain

Region C or the DNA-binding domain (DBD) is the most highly conserved domain among the nuclear receptors. There are nine cysteine residues that are conserved throughout the members of the nuclear receptor superfamily. The first eight of these cysteines (counting from the N-terminus) tetrahedrally coordinate two zinc atoms to form two zinc-binding modules that together function as a DNA-binding motif (Fig. 6.4). Mutagenesis studies of the VDR DBD provide support for this zinc coordination scheme. Mutation of the first eight of the nine cysteine residues (Cys to Ser) eliminated VDR binding to both nonspecific and specific DNA sequences and eliminated VDR-mediated transactivation (Sone *et al.*, 1991). Mutation of the ninth cysteine residue (C84S) had little effect on VDR function, suggesting that this residue is not functionally analogous to the first eight cysteines.

Much of what is known of the nature of VDR–DNA interactions is modeled on the structural data of other related nuclear receptors. Three important subdomains within the DBDs of nuclear receptors are required to recognize and bind specific nucleotide sequences of DNA (see Fig. 6.4). The first region is an α-helical domain referred to as the proximal or P box, which confers target sequence selectivity for many receptors including the GR and the estrogen receptor (ER) (Danielsen *et al.*, 1989; Mader *et al.*, 1989; Umesono and Evans, 1989). A second region is known as the distal or D box, and is important for homodimerization of the type I subfamily of receptors (Umesono and

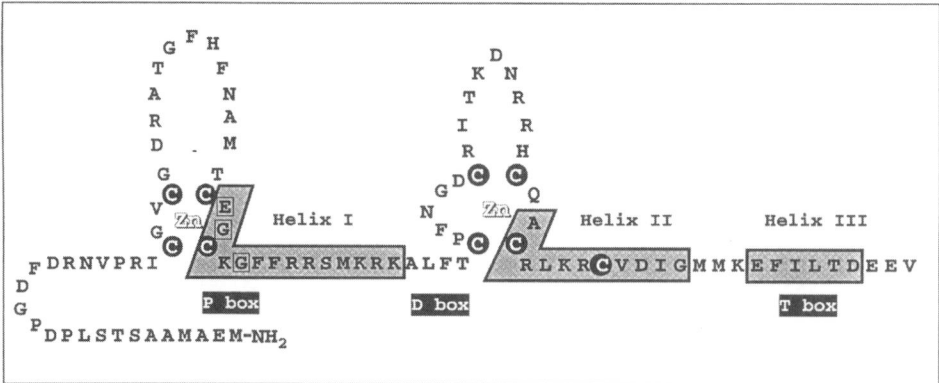

Fig. 6.4   The DNA-binding domain of the vitamin D receptor (VDR). The amino acid sequence of the human VDR is presented. The nine conserved cysteine residues are highlighted by solid circles. The P box residues putatively involved in sequence-specific nucleotide recognition are boxed. The D box is located between the fifth and sixth conserved cysteines. Helices I–III are boxed and shaded, and are based on the crystal structures of related receptors.

Evans, 1989). A third α-helical region, referred to as the T box, resides just C-terminal to the second zinc finger and mediates homodimer and monomer interactions with DNA (Wilson *et al.*, 1992; Lee *et al.*, 1993). Structural predictions of the VDR–RXR heterodimer were generated from the crystal structure of the TR–RXR heterodimer bound to DNA (Rastinejad *et al.*, 1995). An important role was suggested for the T box residues of the VDR in making direct contacts with the D box residues of RXR. Indeed, mutations in the T box of the VDR show a dramatic reduction in both VDR binding to DNA and in transactivation, indicating an important role for this third α-helical domain in VDR–DNA interactions (Hsieh *et al.*, 1995). However, altering the P or D box residues of the VDR to those of the GR did not confer GR target gene selectivity to the VDR (Hsieh *et al.*, 1995). Moreover, these mutations did not significantly affect VDR interaction with DNA or VDR-activated transcription. Thus, it is apparent that the specificity determinants for VDR are more complex than previously thought; perhaps owing to the heterodimeric nature in which VDR binds DNA compared with homodimeric interactions of ER or GR with their response elements.

Three-dimensional structural analysis of the purified DBDs for many of the nuclear receptors has provided detailed insights into the mechanism of receptor–response element interaction (Hard *et al.*, 1990; Schwabe *et al.*, 1990; Luisi *et al.*, 1991; Katahira *et al.*, 1992; Lee *et al.*, 1993). A common structural feature of the DBDs for all these receptors is the folding of two α-helices in the C-terminal portion of the each zinc finger into a single DBD. The first α-helix in the N-terminal finger (denoted helix H1 in Fig. 6.5) lies across the major groove of DNA, making specific contacts with the DNA-binding site; as mentioned above, it is this region that contains the residues that determine response element specificity. The second α-helix (denoted helix H2 in Fig. 6.5) folds across the first in a perpendicular arrangement. The DBD is rich in the positively charged amino acids, lysine and arginine, several of which form favorable electrostatic interactions with the negatively charged phosphate backbone of the DNA helix. Mutations that alter the charge distribution near the tips of each zinc-binding module have been identified in patients with hereditary vitamin D-resistant rickets (HVDRR) (see Genetics below), resulting in mutant VDRs that did not bind effectively to DNA, thus illustrating the importance of net charge in VDR–DNA interaction *in vivo* (Hughes *et al.*, 1988).

The hinge region

The hinge domain or region D lies between the DBD and the ligand-binding domain (LBD). The amino acid sequence of the VDR hinge domain is not homologous to other nuclear receptor hinge domains, and is approximately 50 residues longer than that found in the other receptors. Moreover, this region also exhibits little, if any, sequence homology between VDRs in other species. This raises intriguing questions about the specific functional role of the hinge domain in nuclear receptor action. One apparent role of the hinge domain is to mediate interactions of nuclear receptors with several corepressor proteins such as NCoR (Horlein *et al.*, 1995). However, a more general role was proposed recently in which the hinge domain may serve as a highly flexible link between regions C and E, imparting a high degree of rotational freedom between these two domains to allow nuclear receptors to bind a variety of different DNA response elements (McDonnell *et al.*, 1989; Mangelsdorf and Evans, 1995). Such a model suggests that the precise sequence of the hinge domain is not as important as the overall fluidity or flexibility of the domain.

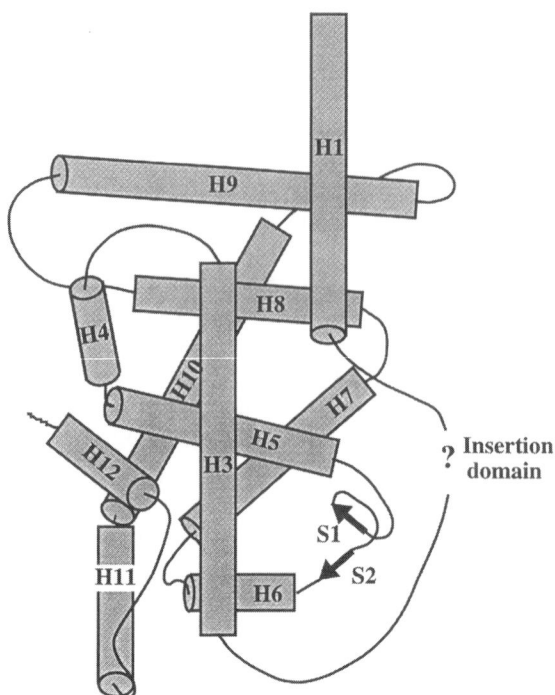

Fig. 6.5 Model for the helical arrangement of the vitamin D receptor ligand-binding domain (VDR LBD). The VDR LBD is proposed to contain 12 α helices (denoted as shaded cylinders) and several β strands (denoted as solid arrows S1 and S2) arranged as an antiparallel α-helical sandwich (Wurtz *et al.*, 1996). The region from beginning of helix H1 to the beginning of helix H3 is not predicted owing to a rather unique insertion region of VDR that is poorly conserved among other nuclear receptors and between the VDRs of other species (denoted as ?).

The multifunctional LBD

The E domain of the nuclear receptors is a multifunctional domain which is generally referred to as the ligand-binding domain, or LBD. From a comparison of crystallographic data obtained from several nuclear receptors, it is clear that the LBDs are structurally similar. In general, the LBDs consist of three layers comprised of 12 α helices and several β strands that are organized around a lipophilic ligand-binding pocket (Bourguet *et al.*, 1995; Renaud *et al.*, 1995; Wagner *et al.*, 1995; Wurtz *et al.*, 1996; Brzozowski *et al.*, 1997). This arrangement has been termed an antiparallel α-helical sandwich (Bourguet *et al.*, 1995). In addition to serving as a binding site for the $1,25(OH)_2D_3$ ligand, it also fulfills several other critical roles. A prominent role of this region of VDR is to mediate interaction with RXR, the heterodimeric partner that is required for high-order binding of the VDR to DNA. Key serine residues in this domain serve as sites of phosphorylation that may be important in regulating the transcriptional activity of the VDR. Finally, this region of the VDR also plays a central role in forming part of a protein–protein interaction surface through which the VDR contacts other proteins that are important for VDR-mediated transcription, such as transcription factor (TFIIB) and transcriptional coactivators. The predicted structure of the VDR

LBD is presented in Fig. 6.5 and various aspects of VDR LBD function are discussed below.

Ligand binding

As its name implies, one of the important roles of this domain in the VDR is to bind the small, lipophilic $1,25(OH)_2D_3$ ligand. This binding event is complicated by the inherent conformational flexibility of the $1,25(OH)_2D_3$ seco-steroid and by the ability of the apo- and holo-VDR to assume different conformations. Thus, major goals of vitamin D research are to obtain a detailed understanding of the ligand–receptor interaction at the molecular level and to determine the three-dimensional structure of $1,25(OH)_2D_3$–VDR complex. Predictions of residues in the VDR that are important in forming the ligand-binding pocket for $1,25(OH)_2D_3$ have been made based on the crystal structures of the liganded retinoic acid receptor γ (RARγ) LBD and the ER LBD bound by either 17β-estradiol or a selective antagonist (Renaud et al., 1995; Wurtz et al., 1996; Brzozowski et al., 1997). Residues involved in direct ligand-binding contacts in the VDR are proposed to reside helices H3, H5, H11 and H12, as well as portions of helices H6 and H7 along with their intervening loop.

RXR heterodimerization

In addition to hormone binding, the LBD has a central role in mediating heterodimerization of VDR with receptor auxiliary factors (RAFs) such as the RXR. VDR–RXR heterodimer formation is required for high-affinity interaction of the receptor with vitamin D response elements (VDREs) (Liao et al., 1990; Yu et al., 1991; Kliewer et al., 1992; MacDonald et al., 1993) and at least three putative regions in the LBD of VDR mediate protein–protein contacts with RXRs and perhaps other RAFs (Nakajima et al., 1994; Whitfield et al., 1995). A predominant, C-terminal heterodimerization domain resides between residues 382 and 403 in the human VDR (hVDR) sequence (Nakajima et al., 1994). Mutagenesis of several specific residues in this domain (Lys382, Met383 and Glu385) disrupted VDR–RAF and VDR–RXR interaction in vitro and eliminated transcriptional activation by the VDR. A second putative interaction domain was identified between residues 318 and 339 (Nakajima et al., 1994). These two regions correspond to helices H10 and H11 and helices H7 and H8 respectively. A third putative heterodimerization surface was defined in the N-terminal segment of the LBD between amino acids 244 and 263 (Whitfield et al., 1995). Selected point mutations within this region do not interfere with ligand binding, but they affect the ability of the VDR to heterodimerize with RAFs or RXRs and disable transcription from vitamin D-responsive constructs. One important outcome of these studies is that, in all the receptor mutants examined, heterodimerization of VDR with RXR was required for VDRE interaction and for $1,25(OH)_2D_3$–VDR-mediated transcriptional activation, suggesting that heterodimerization between VDR and RXR is a requisite step in this mechanism.

The crystal structure of the RXR–RXR homodimeric complex provides additional insight into the putative heterodimerization surface of the VDR–RXR complex (Bourguet et al., 1995). The RXR dimer is symmetrically arranged with the interaction surface being formed mainly by helix H10 and, to a lesser extent, by helix H9. Helix H10 of RXR corresponds to the C-terminal region of VDR identified by Nakajima et al. (1994) as being crucial for heterodimer formation (amino acids 382–403). Interestingly, a natural mutation in helix H10 of the VDR LBD (R391C) was described recently in patients with HVDRR (Whitfield et al., 1996). This mutation abrogates VDR–RXR heterodimer formation in these patients, further suggesting that helix H10 may directly contact helix H10 of RXR to comprise the major interaction surface yielding a structurally symmetric VDR–RXR heterodimeric complex.

VDR phosphorylation

Phosphorylation is widely regarded as a key means of regulating cellular processes. Most of the steroid/thyroid hormone receptors, including the VDR, are phosphoproteins, and key residues that serve as substrates for phosphorylation reside in the E domain. The VDR present in mouse 3T6 cells is hyperphosphorylated in response to physiologic concentrations of $1,25(OH)_2D_3$ (Pike and Sleator, 1985). Ligand-dependent phosphorylation has also been demonstrated in ROS 17/2.8 osteoblasts and in chick duodenal organ culture, two relevant target systems for vitamin D action (Brown and DeLuca, 1990; Jones et al., 1991). In this last system, the effect is observed within 15 min following the addition of $1,25(OH)_2D_3$. The rapid onset of this response to $1,25(OH)_2D_3$ indicates that phosphorylation of the VDR may play an initiating event in the transcriptional processes mediated by the VDR.

The major phosphorylated residues of the VDR have been determined. Using domain-specific antibodies, Brown and DeLuca (1991) mapped the major phosphorylation site(s) to the N-terminal region of the LBD in porcine VDR. Studies in ROS 17/2.8 cells revealed that the main phosphorylated domain of hVDR resided between Met197 and Val234 (Jones et al., 1991). Within this domain is a cluster of serine residues, many of which resemble consensus sites for casein kinase II. Indeed, hVDR is an effective substrate for in vitro phosphorylation by purified casein kinase II, and site-directed mutagenesis defined Ser208 as the site phosphorylated by casein kinase II in vitro and in vivo when the VDR was transiently expressed in COS-7 cells (Jurutka et al., 1993). Furthermore, coexpression of casein kinase II in this system augmented VDR phosphorylation of Ser208 and VDR-activated transcription of a reporter gene construct, showing that this kinase phosphorylates Ser208 in the cell and upregulates the transactivation activity of the VDR (Jurutka et al., 1996). Hilliard et al. (1994) systematically identified this same Ser208 residue as the main phosphorylated residue of VDR using phosphopeptide mapping studies. In this study, phosphorylation at Ser208 was augmented eightfold when the cells were treated for 4 hours with $1,25(OH)_2D_3$. Thus, Ser208 is the major phosphorylated residue of the VDR and likely represents the hormone-dependent phosphorylation site observed in earlier studies.

A second major phosphorylated site is Ser51, which resides between the two zinc-binding modules in the DBD of the VDR (Hsieh et al., 1991). Ser51 is a consensus site for protein kinase C (PKC) and it is selectively phosphorylated by the PKCβ isoform in vitro and in vivo. Phosphorylation at Ser51 interferes with VDR binding to DNA and suggests an intriguing negative regulatory loop of VDR-activated transcription mediated by PKC phosphorylation of Ser51.

Although the global phosphorylated state of the cell clearly affects VDR-mediated transcriptional activity (Desai et al., 1995; Matkovits and Christakos, 1995), a demonstration that VDR phosphorylation per se is important for vitamin D-mediated transcription has remained elusive. For example, conservative (Ser to Ala) mutations that disrupt phosphorylation at Ser208 and Ser51 do not dramatically affect VDR-activated transcription (Hsieh et al., 1993; Jurutka et al., 1996). One caveat here is that mutations in one serine residue may actually promote phosphorylation of an adjacent serine residue as a compensatory mechanism (Hilliard et al., 1994). At present, the steps beyond these phosphorylation events are not well defined. One intriguing possibility is that phosphorylation may alter the interaction of the VDR with nuclear receptor comodulatory proteins, which may be critical for the transcriptional response.

The AF-2 helix, a ligand-dependent transactivation domain

The activation function-2 (AF-2) domain of the nuclear receptors generally is found at the C-terminal end of region E. It is a highly conserved, ligand-dependent transactivation domain that is essential for receptor-mediated transcription (Hollenberg *et al.*, 1989; Lees *et al.*, 1989; Thompson and Evans, 1989; Danielian *et al.*, 1992). Deletions of or select mutations within the AF-2 domains of the nuclear receptors selectively abolish ligand-activated transcription without disrupting other receptor functions such as ligand binding, response element interaction, dimerization or nuclear localization. The transactivation activity of the AF-2 domain is transferable. Fusion of the AF-2 domain to a heterologous DBD results in a fusion protein that activates transcription autonomously (Webster *et al.*, 1988; Hollenberg *et al.*, 1989; Thompson and Evans, 1989). The VDR AF-2 domain resides at the extreme C-terminus of the VDR between residues 416 and 423. Based on structural determinations of related nuclear receptors (Bourguet *et al.*, 1995; Renaud *et al.*, 1995; Wagner *et al.*, 1995; Brzozowski *et al.*, 1997), the VDR AF-2 domain is an α helix (H12). The core activation function consists of a centrally conserved glutamic acid residue (E420 in the hVDR sequence) flanked on either side by hydrophobic residues (Leu417/Val418/Leu419 and Val421/Phe422 respectively). Mutation of key residues in both the hydrophobic and hydrophilic faces selectively abolish $1,25(OH)_2D_3$-activated transcription mediated by the VDR (Jurutka *et al.*, 1997; Masuyama *et al.*, 1997a).

One of the most striking features gleaned from the structural studies of the nuclear receptors is the dramatic repositioning of the AF-2 activation helix (H12) when ligand occupies the receptor (Bourguet *et al.*, 1995; Renaud *et al.*, 1995; Wagner *et al.*, 1995; Wurtz *et al.*, 1996; Brzozowski *et al.*, 1997). By comparing the crystal structures of the unliganded RXRα with that of the liganded RARγ, it is evident that in the unliganded state the AF-2 domain (H12) projects out away from the globular core of the LBD, and in the liganded state the AF-2 domain is folded back on helix H11 and interfaces with the surface of the LBD globular core domain (Bourguet *et al.*, 1995; Renaud *et al.*, 1995). Consequently, helix H12 and its resident AF-2 domain moves from a relatively solvent exposed position to a position in which it is folded or packed on to the LBD in the ligand-activated receptor. This repositioning of helix H12 likely plays several important roles in mediating ligand-activated transcription by the receptor. First, the folding down of helix H12 appears to act as a 'hinged door' sealing off the channel through which the lipophilic ligand enters the ligand-binding pocket. Second, this ligand-induced repositioning of helix H12 may lock the AF-2 domain into a stable conformation, with the hydrophobic residues of the helix facing toward the ligand-binding cavity and the charged residues exposed to the solvent. Finally, folding of helix H12 on to the LBD core creates a platform or protein interaction surface through which nuclear receptor coactivator proteins effectively dock with the VDR (Feng *et al.*, 1998).

Mechanisms involved in VDR-mediated transactivation

VDR interaction with VDREs

The specific DNA promoter sequences that respond to $1,25(OH)_2D_3$ and the VDR are termed vitamin D-responsive elements or VDREs. VDREs serve as binding sites for the liganded VDR and they are the focal point from which the receptor elicits its effects on gene transcription. Positive VDREs are generally described as imperfect direct repeats of a core hexanucleotide sequence, G/A G G T G/C A, with a spacer region of three nucleotides separating each half-element (also termed DR3 for direct repeat with a three-nucleotide spacer) (Morrison *et al.*, 1989; Demay *et al.*, 1990; Markose *et al.*, 1990; Noda *et al.*, 1990; Ozono *et al.*, 1990; Terpening *et al.*, 1991; Cao *et al.*, 1993; Gill and Christakos, 1993; Ohyama *et al.*, 1994). This direct repeat motif is analogous to DNA elements that mediate retinoic acid and thyroid hormone (TR) responsiveness (RAREs and TREs) and it contrasts with responsive elements that mediate glucocorticoid or estrogen (GREs and EREs) responsive genes which are generally palindromic or inverted repeat sequences. Variations on the DR3 motif for VDREs have been identified in the promoters of several vitamin D responsiveness genes (Darwish and DeLuca, 1992) as well as in several synthetic elements (Carlberg *et al.*, 1993). This plasticity may be a reflection of the affinity of VDR binding to a particular element, thus affecting the degree to which a particular promoter or gene responds to vitamin D in a particular cell type. Alternatively, different VDREs may induce different conformational changes in the VDR (Staal *et al.*, 1996) which may alter its interaction with various coactivator proteins. Regardless, the DR3 element seems to be the primary motif through which VDR exerts its positive effects on gene transcription.

VDR interaction with the RXR
The mechanism of VDR binding to VDREs is reflected in the direct repeat nature of the element. For example, the class I members of the nuclear receptor superfamily (e.g. GR or progesterone receptor (PR)) bind to palindromic response elements as symmetric homodimers (for a review see Glass, 1994). In contrast, the class II receptors (of which the VDR is a member) generally bind to direct repeat elements as asymmetric heterodimers. For the VDR, this was originally documented in experiments that showed that purified VDR alone does not bind to a DR3 VDRE with high affinity (Liao *et al.*, 1990). Rather, an unidentified nuclear factor was required for high-affinity binding of the VDR to VDREs (Liao *et al.*, 1990), indicating that the VDR bound as a heteromeric complex to a DR3 VDRE.

It is generally accepted that this accessory factor is RXR. This is based on several observations: (1) RXRs substitute effectively for RAF in VDR/VDRE binding assays (Yu *et al.*, 1991; Kliewer *et al.*, 1992; MacDonald *et al.*, 1993), (2) highly purified RAF contains RXR immunoreactivity (MacDonald *et al.*, 1993; Munder *et al.*, 1995), (3) RXRs augment vitamin D-mediated transcription (Yu *et al.*, 1991; Carlberg *et al.*, 1993; MacDonald *et al.*, 1993), (4) numerous VDR mutants that do not interact with RXR also fail to activate transcription *in vivo* (Nakajima *et al.*, 1994), and (5) in a yeast system that lacks proteins that are analogous to the mammalian nuclear receptors, both the VDR and the RXR are required to elicit activated transcription from a VDRE-driven reporter gene construct (Jin and Pike, 1996). These data support the concept of the VDR–RXR heterodimer as the functional enhancer in vitamin D-activated transcription.

The $1,25(OH)_2D_3$ ligand dramatically enhances VDR–RXR heterodimerization, both the direct interaction of the VDR with the RXR in solution (Sone *et al.*, 1991; MacDonald *et al.*, 1995) and interaction of the VDR–RXR heterodimer with the VDRE (Liao *et al.*, 1990; Sone *et al.*, 1991; MacDonald *et al.*, 1993). Surface plasmon resonance quantitated the binding constants for these interactions and showed a clear $1,25(OH)_2D_3$-dependent decrease in VDR interaction with itself (i.e. VDR homodimers) and a concomitant increase in VDR heterodimerization with RXR (Cheskis and

Freedman, 1996). These ligand-induced changes in VDR–RXR interactions are likely due to altered conformations of the VDR in the absence and presence of $1,25(OH)_2D_3$ (Peleg *et al.*, 1995). Thus, one important role for the $1,25(OH)_2D_3$ ligand in VDR-mediated transcription is to induce a conformational change in VDR that disrupts weak homodimers of unliganded VDR and promotes liganded VDR heterodimerization with RXR. The interaction of VDR and RXR generates a heterodimeric complex that is highly competent to bind DR3 like VDREs and subsequently affect the transcriptional process.

The DR3 direct repeat motif is asymmetric and the VDR–RXR heterodimer binds to the VDRE with a defined polarity. The RXR occupies the 5′ half-site and the VDR occupies the 3′ half-site (Schrader *et al.*, 1995; Jin *et al.*, 1996). Modeling studies based on the crystal structure of the TR–RXR heterodimer suggest that the DBDs of a VDR–RXR heterodimer on a DR3 element are arranged in an asymmetric, head-to-tail assembly with the VDR T box forming a direct contact with the RXR D box (Rastinejad *et al.*, 1995). In contrast, the predominant heterodimerization interface in the receptor C-termini (helices 10 and 9) would place the LBDs of VDR and RXR in a symmetric arrangement. To rationalize this discrepancy, the DBDs of the nuclear receptors are proposed to have an inherent rotational flexibility, presumably a function of the hinge region, which permits a high degree of rotational freedom in the DBD (up to 180°) and would accommodate such a binding scheme (Mangelsdorf and Evans, 1995).

VDR interaction with general transcription factors

The VDR binds the $1,25(OH)_2D_3$ ligand, it then heterodimerizes with RXRs and binds as a VDR–RXR complex to VDREs in promoters of genes that are regulated by vitamin D. Following these initial events, the mechanistic details and communication pathways that link the VDR–RXR heterodimer to the transcription preinitiation complex (PIC) are less clear. Recent approaches have focused on defining various protein–protein contacts that may occur between the VDR and other nuclear factors that may establish a physical link between the VDR–RXR heterodimer and the PIC (Fig. 6.6).

Many transcriptional regulatory factors including the nuclear receptors interact directly with the general transcription factors (GTFs) that comprise the PIC. Of these factors, TFIIB is considered a central target. TFIIB is a bridging protein between TATA-binding protein and RNA polymerase II, and its entry into the PIC is may be a rate-limiting event. Transactivator interaction with TFIIB recruits this factor into the PIC and increases PIC assembly. The VDR interacts directly with TFIIB (Blanco *et al.*, 1995; MacDonald *et al.*, 1995). This interaction is functionally important since TFIIB expression augments vitamin D-activated transcription in transient gene expression studies and a dominant-negative inhibitor of TFIIB–VDR complexes selectively impairs VDR-activated transcription (Blanco *et al.*, 1995; Masuyama *et al.*, 1997b). Thus, formation of the VDR–TFIIB complex is an important step in the mechanism of VDR-activated transcription. VDR targets a second GTF. Lemon *et al.* (1997) demonstrated that the VDR forms a direct protein–protein contact with TFIIA, and $1,25(OH)_2D_3$ binding to the VDR promotes the formation and recruitment of TBP:TFIIA into higher-mobility complexes on a VDRE-linked promoter.

A third general target in the GTF class are the TBP-associated factors (TAFs). The VDR has been shown to interact with TAFII28 (May *et al.*, 1996), TAFII135 (Mengus *et al.*, 1997) and TAFII55 (Lavigne *et al.*, 1999). The expression of human TAFII28 in COS cells significantly augments hormone-dependent transcriptional activation by the estrogen, vitamin D and RXRs (May *et al.*, 1996). Receptor selectivity was observed with

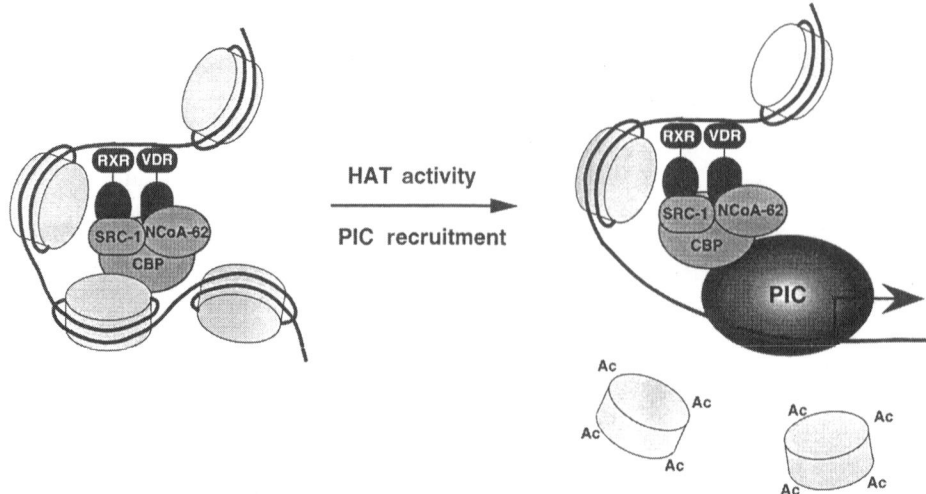

Fig. 6.6 Model of vitamin D receptor (VDR)-activated transcription. The initial event in this model is high-affinity binding of the 1,25(OH)$_2$D$_3$ ligand to the VDR. Ligand binding induces VDR–retinoid X receptor (RXR) heterodimerization and the heterodimer specifically binds vitamin D response elements in the promoter regions of vitamin D-responsive genes. The VDR portion of the heterodimer contacts transcription factor (TF) IIB, TBP-associated factors (TAFs) and coactivator proteins to form protein–protein contacts between the VDR and the transcription initiation complex (PIC). It is the interaction with and the communication between VDR, RXR, TFIIB and other ligand-dependent coactivator proteins such as steroid receptor coactivator 1 (SRC-1) and nuclear receptor coactivator NCoA-62 that may determine the overall transcriptional activity of a vitamin D-responsive gene. Several coactivators possess histone acetyltransferase (HAT) activity, which acetylates histones resulting in a loosening of chromatin structure and greater accessibility of the promoter to the transcription machinery. CBP, cyclic adenosine monophosphate response element-binding protein.

hTAFII135 which was shown to potentiate RAR, TR and VDR-activated transcription without affecting ligand-activated transcription by ER and RXR (Mengus *et al.*, 1997). These data indicate that TAFIIs possess the ability to interact selectively and directly with nuclear receptors including the VDR, and that their interaction with VDR promotes 1,25(OH)$_2$D$_3$-mediated transcriptional activation. Again, the mechanism may involve the facilitated recruitment or stabilization of the PIC through a VDR–TAFII bridging interaction, but the details remain to be elucidated.

VDR interaction with nuclear receptor coactivators
The steroid receptor coactivator (SRC) family of nuclear receptor coactivators (recently reviewed by McKenna *et al.*, 1999) includes three members at present: SRC-1 (NCoA-1), SRC-2 (glucocorticoid receptor-interacting protein 1 (GRIP-1), TIF-2, NCoA-2) and SRC-3 (pCIP, RAC-3, ACTR, AIB-1, TRAM-1). These proteins interact with the receptor in a ligand-dependent manner and augment hormone-activated RNA polymerase II-directed transcription. SRC-1 was the first of this coactivator family to be cloned, and was identified in a yeast two-hybrid screen (Onate *et al.*, 1995). SRC-1

interacts with the PR in an agonist-dependent manner and augments PR-dependent transcription when expressed transiently in mammalian cell lines. SRC-1 also enhances ER-, GR-, TR-, RXR- and RAR-mediated transcription, suggesting a general role for this coactivator in nuclear receptor-mediated transcription.

Since the initial identification of SRC-1, a number of SRC family members and novel proteins with nuclear receptor coactivator properties have been described that interact with and whose expression affects VDR-mediated transcription. Many of these proteins were identified by screening cDNA libraries in the yeast two-hybrid system or by *in vitro* biochemical approaches. Putative coactivators for VDR include: SRC-1 (Masuyama *et al.*, 1997a), GRIP-1/TIF-2 (Hong *et al.*, 1996, 1997; Voegel *et al.*, 1996), RIP-140 (Masuyama *et al.*, 1997a), the DRIP–TRAP complex (Fondell *et al.*, 1996; Rachez *et al.*, 1998, 1999), cyclic adenosine monophosphate response element-binding (CREB) protein (CBP)/P300 (Kamei *et al.*, 1996), SMAD-3 (Yanagisawa *et al.*, 1999) and NCoA-62 (Baudino *et al.*, 1998). Many of these coactivators interact with the AF-2 transactivation domain of the receptors. In the VDR, mutation of two key residues in the AF-2 domain, L417S and E420Q, did not affect the binding of $1,25(OH)_2D_3$, heterodimerization with RXR, or binding to a VDRE, but they did disrupt VDR interaction with SRC-1 and transcriptional activation (Jurutka *et al.*, 1997; Masuyama *et al.*, 1997a). Thus, there is a correlation between the transcriptional activity of VDR mutants and the ability of these mutants to interact with coactivator proteins. Another coactivator protein that contacts VDR is GRIP-1 (or SRC-2), an 86-kDa protein named for its interaction with the GR (Hong *et al.*, 1996, 1997). As with SRC-1, GRIP-1 is rather promiscuous in that it interacts with a variety of nuclear receptors and augments their transcriptional pathways. The VDR contacts GRIP-1 in a $1,25(OH)_2D_3$-dependent fashion, and again this interaction is mediated, in part, through the AF-2 domain (Hong *et al.*, 1996, 1997).

While the AF-2 domain is centrally involved in mediating ligand-dependent interactions of nuclear receptors with some coactivators, in all likelihood the coactivator-binding surface of the VDR is comprised of the AF-2 domain and surrounding residues. Experimental data from several laboratories show that the conserved AF-2 core domain is important, but not sufficient, for full transactivation when assayed outside the context of the full-length receptor (Barettino *et al.*, 1994; Masuyama *et al.*, 1997a). From crystal structural determination of related nuclear receptors, candidate regions for these other areas include those immediately surrounding the AF-2 core, such as exposed residues on the surfaces of H3, H4, the loop between H11 and H12, and the region between H1 and H3 comprising the omega loop (25–29). These structural predictions were confirmed by scanning mutagenesis of the TR which defined a coactivator interaction surface consisting of helices H12, H3, H5 and H6 (Feng *et al.*, 1998). This surface is composed predominantly of helix H12 and also by surrounding residues from helices H3, H5 and H6. A similar surface is likely to exist on the VDR since select mutations within helices H12 and H3 are also involved in coactivator interaction and in the transactivation process (Masuyama *et al.*, 1997a; Jimenez-Lara and Aranda, 1999; Kraichely *et al.*, 1999).

Recently, the authors' laboratory identified a novel coactivator protein termed NCoA-62, which contacts the VDR and augments VDR-activated transcription (Baudino *et al.*, 1998). NCoA-62 is a 62 000-Da protein that forms protein–protein contacts with the VDR LBD both in yeast and *in vitro* in GST–VDR pulldown assays. Interestingly, NCoA-62 interaction with VDR does not require the AF-2 domain, as deletion of the VDR AF-2 domain does not affect NCoA-62 binding, indicating that this novel coactivator may contact other important transactivation domains in the VDR

LBD. NCoA-62 belongs to a growing class of AF-2-independent coactivators. The likelihood exists that distinct classes of coactivator proteins may function through different mechanisms and cooperatively enhance VDR-activated transcription. Indeed, NCoA-62 synergizes with CBP/P300 or with SRC coactivators cooperatively to enhance VDR-dependent transactivation. NCoA-62 is highly related to a nuclear protein in *Drosophila melanogaster* that is putatively involved in ecdysone-mediated transcription (Wieland *et al.*, 1992), suggesting a high functional conservation of NCoA-62 between lower forms and mammals.

The mechanisms through which coactivators function are diverse. Some coactivators may function as macromolecular bridges between the nuclear receptor and the PIC. Their interaction with the PIC (either direct or indirect) may promote PIC assembly or enhance the stability of the PIC, thereby leading to activated transcription. An emerging property of several coactivator proteins including CREB-binding protein (CBP) and SRC-1 is that these coactivators possess intrinsic histone acetyltransferase (HAT) activity (Bannister and Kouzarides, 1996; Orgryzko *et al.*, 1996; Chen *et al.*, 1997; Spencer *et al.*, 1997). The acetylated state of histones is highly correlated with promoter activity. Histone acetylation results in a disruption or loosening of the chromatin structure making promoters more accessible to the transcription machinery and ultimately leading to an increase in the rate of transcription (Roth and Allis, 1996; Grunstein, 1997; Wade and Wolffe, 1997). The ability of nuclear receptor coactivators to express HAT activity or to interact with and recruit HAT activities provides an attractive model for nuclear receptor-activated transactivation. Specifically, nuclear receptors interact in a ligand-dependent manner to recruit enzymes that modify chromatin structure at a particular promoter. Acetylation of histones around a promoter results in a disordered structure, increasing the accessibility of the transcriptional machinery to the promoter ultimately leading to activated transcription (Jenster *et al.*, 1997; Wade and Wolffe, 1997) (see Fig. 6.6).

GENETICS

The *VDR* gene

The location of the *VDR* gene within the human genome was determined using Southern blot analysis of DNA from human–Chinese hamster cell hybrids, and localized the *VDR* gene to human chromosome 12 (Faraco *et al.*, 1989). This was later extended to 12q by means of somatic cell hybrid mapping (Szpirer *et al.*, 1991) and further refined to the 12q13-14 region using *in situ* hybridization and linkage analysis (Labuda *et al.*, 1992). Similar approaches identified the *VDR* sequence on chromosome 7 of the rat genome (Szpirer *et al.*, 1991). In both the human and rat genomes the *VDR* gene is located in regions where distantly related DNA-binding proteins also map, including transcription factor SP-1 (Szpirer *et al.*, 1991) and the γ isoform of the retinoic acid receptor (RARγ) (Ishikawa *et al.*, 1990; Mattei *et al.*, 1991). This clustering of related DNA-binding proteins suggests the possibility that they were derived from some common ancestral gene.

The gene encoding the hVDR was isolated originally from a human liver genomic DNA library (Pike *et al.*, 1988) and later described in detail (Miyamoto *et al.*, 1997). The hVDR gene is contained within approximately 75 kilobases (kb) of genomic DNA. It has a complex structure consisting of 11 coding exon sequences interrupted by intronic sequences ranging in size from 0.2 to 13 kb (Hughes *et al.*, 1988; Pike *et al.*, 1988; Sone *et al.*, 1990). Its overall structural organization is related to other genes encoding nuclear receptor family members. The translation initiation codon is located in exon 2, which also encodes sequences for the first zinc finger of the DBD (see below). Exon 3 contains the sequence for the second zinc finger. The observation that the two motifs of the zinc-finger DBD are encoded by separate exons is a characteristic trait of the steroid receptor superfamily. There is a high degree of complexity in the 5' region of the gene in which multiple start sites and alternate splicing events take place among exons IA, IB and IC. The majority of the C-terminal LBD is encoded by exons 8 and 9.

Hereditary vitamin D-resistant rickets

HVDRR is a rare autosomal recessive disorder characterized by early onset of rickets, hypocalcemia, hyperparathyroidism and, in many cases, total alopecia (lack of body hair). Most of these traits are classic symptoms of vitamin D deficiency and are present in these patients despite raised serum concentrations of $1,25(OH)_2D_3$. The disease is also refractory to supraphysiologic doses of oral or intravenous vitamin D, indicative of a generalized tissue resistance to $1,25(OH)_2D_3$. The disease was first recognized in 1978 (Brooks *et al.*, 1978; Marx *et al.*, 1978) and the genetic basis has been identified as functional mutations of the VDR (Hughes *et al.*, 1988). All of the HVDRR kindreds and the hVDR mutations were reviewed in detail recently (Malloy *et al.*, 1999).

Currently, over 20 specific mutations have been identified in the VDR from over 50 individual patients with HVDRR. These mutations are summarized in Table 6.2 and are illustrated in Fig. 6.7. They may be broadly classified as mutations that: (1) affect DNA binding and (2) impair $1,25(OH)_2D_3$ binding. Hypothetically, *VDR* mutations may also exist that impair other functional parameters of the receptor such as protein–protein contacts, for example mutations that alter key residues involved in heterodimerization or coactivator interaction. At least two candidate mutations have been described that fall into this classification (Whitfield *et al.*, 1996; Cockerill *et al.*, 1997). The general approaches used to define the molecular basis of this disorder were to establish fibroblast cell cultures from human skin biopsy samples to characterize VDR functional defects (Feldman *et al.*, 1980), obtain genomic DNA samples from these cells or from peripheral

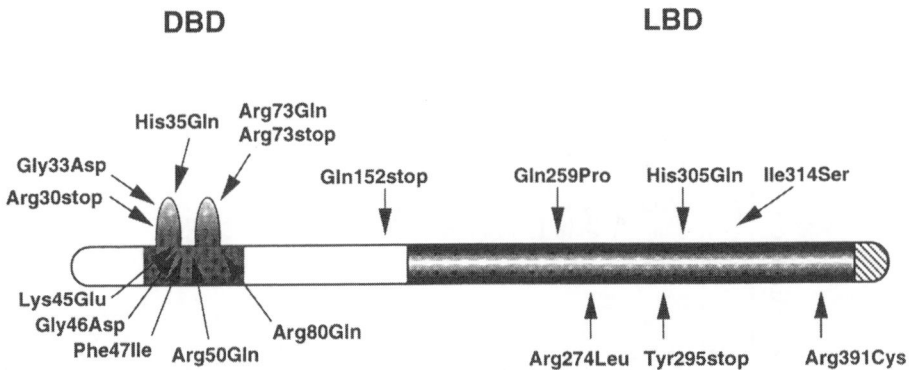

Fig. 6.7 Point mutations identified in patients with hereditary vitamin D-resistant rickets. DBD, DNA-binding domain; LBD, ligand-binding domain.

Table 6.2 Mutations identified in patients with hereditary vitamin D-resistant rickets

VDR mutation	Nucleotide mutation	Domain	Biochemical defect	Reference
Arg30stop	CGA → TGA	C	No detectable VDR	Mechica *et al.* (1997), Zhu *et al.* (1998)
Gly33Asp	GCC → GAC	C	DNA binding	Hughes *et al.* (1988)
His35Gln	CAC → CAG	C	DNA binding	Yagi *et al.* (1993)
Lys45Glu	AAA → GAA	C	DNA binding	Rut *et al.* (1994)
Gly46Asp	GGC → GAC	C	DNA binding	Lin *et al.* (1996)
Phe47Ile	TTC → ATC	C	DNA binding	Rut *et al.* (1994)
Arg50Gln	CGA → CAA	C	DNA binding	Saijo *et al.* (1991)
Arg73Gln	CGA → CAA	C	DNA binding	Hughes *et al.* (1988)
Arg73stop	CGA → TGA	C	No detectable VDR	Wiese *et al.* (1993), Cockerill *et al.* (1997)
Arg80Gln	CGG → CAG	C	DNA binding	Sone *et al.* (1990), Malloy *et al.* (1994)
Gln152stop	CAG → TAG	D	No detectable VDR	Kristjansson *et al.* (1993)
Leu233fs	GTC → GTG	E	Ligand binding	Cockerill *et al.* (1997)
Gln259Pro	CAG → CCG	E	RXR interaction	Cockerill *et al.* (1997)
Arg274Leu	CGC → CTC	E	Ligand binding	Kristjansson *et al.* (1993)
Try295stop	TAC → TAA	E	No detectable VDR	Ritchie *et al.* (1989), Malloy *et al.* (1990), Wiese *et al.* (1993)
His305Gln	CAC → CAG	E	Ligand binding	Malloy *et al.* (1997)
Ile314Ser	ATC → AGC	E	Ligand binding	Whitfield *et al.* (1996)
Arg391Cys	CGC → TGC	E	Ligand–RXR interaction	Whitfield *et al.* (1996)

Adapted from Malloy *et al.* (1999). VDR, vitamin D receptor; RXR, retinoid X receptor.

blood cells, amplify specific regions of interest by polymerase chain reaction (PCR), and sequence the amplified product in order to identify genetic mutations. The mutation is then introduced into the *VDR* cDNA and the mutant VDR is expressed in mammalian cells to confirm that a particular mutation results in the appropriate receptor defect in VDRs from HVDRR patient skin fibroblasts.

Defects in DNA binding

One classification of the disease, based on early biochemical studies of the hVDR in skin fibroblasts extracts obtained from patients with HVDRR, revealed a receptor-positive phenotype that was impaired in binding to DNA. The VDR obtained from these patients displayed normal binding of $1,25(OH)_2D_3$, but the receptors eluted from a DNA–cellulose column with a lower salt concentration than normal VDR. $1,25(OH)_2D_3$ was unable to induce vitamin D 24-hydroxylase activity in fibroblast cultures, which is a useful marker of $1,25(OH)_2D_3$-mediated action. The defect was suspected to reside in the DBD of the VDR. Following the isolation of the hVDR cDNA and genomic DNA in the mid-1980s (Baker *et al.*, 1988; Pike *et al.*, 1988), Hughes *et al.* (1988) used PCR to amplify exon sequences of the *VDR* gene from these patients and identified the first mutations in the *VDR* resulting in the HVDRR phenotype. Point mutations in two separate kindreds were identified in exons 2 and 3, which encode the first and second zinc-binding modules respectively. These mutations led to a single amino acid substitution in each kindred, G33D (a Gly to Asp substitution at residue 33) and R73Q (an Arg to Gln substitution at residue 73); both residues are positioned at the tips of each zinc-binding module. Importantly, when these same mutations were introduced into the VDR cDNA and then expressed in COS1 cells, the mutant receptor bound $1,25(OH)_2D_3$ normally but showed weaker affinity for DNA–cellulose (Hughes *et al.*, 1988) and was unable to activate a vitamin D-responsive reporter gene construct (Sone *et al.*, 1989). Thus, these data conclusively showed that a single mutation yielded the appropriate phenotype and provided the first demonstration of the molecular basis of HVDRR.

Since the initial identification of mutations in the VDR DBD, a number of other mutations have been described in this domain that yield nearly equivalent phenotypes in HVDRR. Yagi *et al.* (1993) described a mutation in a patient with HVDRR whose defect mapped to the 'finger tip' region of the first zinc-binding module. A single base substitution of C to G was identified in exon 2 which resulted in the replacement of histidine with a glutamine at residue 35 (H35Q) of the first zinc finger. Rut *et al.* (1994) described mutations in exon 2 that were localized to the C-terminal base of the first zinc-binding module in two HVDRR kindreds, K45D and F47I. A G46D mutation was also identified in a separate study of another family in this region of the VDR gene (Lin *et al.*, 1996). These last three residues lie within the α-helical region which makes direct, sequence specific, minor groove contacts. Mutations here directly disrupt the ability of VDR to recognize and bind VDREs tightly.

Mutations were identified in three Japanese patients from two families with HVDRR (Saijo *et al.*, 1991). Their genomic VDR sequence was found to have the same unique G to A transition in exon 3, resulting in a substitution of Arg by Gln at residue 50 (R50Q). This Arg residue is located in the region between the two zinc-binding motifs immediately adjacent to Ser51, a key residue phosphorylated by PKC (Hsieh *et al.*, 1991). It is not known whether this precise, single mutation affects the phosphorylation state of the VDR; however, mutagenesis of residues surrounding Ser51 dramatically reduces PKC phosphorylation *in vitro* (Hsieh *et al.*, 1993). Moreover, mutations in this region severely compromise VDRE interaction.

Sone *et al.* (1990) defined a distinct DNA-binding mutation in two unrelated patients with the HVDRR phenotype. This mutation was present in exon 3 but, instead of the fingertip region, it was located at the base of the second zinc finger. The nucleotide mutation caused the substitution of a glutamine residue for an arginine at position 80 (R80Q). Malloy *et al.* (1994) identified the same mutation in two patients from another family. Site-directed mutagenesis was used to introduce this mutation into the normal hVDR cDNA. Following transient expression in mammalian cell lines, the mutated receptor was found to be severely compromised in transcriptional activation and in DNA binding.

Two distinct mutations from four separate families result in the introduction of premature stop codons within the DBD of the VDR. One is a C to T transition that creates a TAG stop codon at Arg 73 (Arg73STOP or R73X) in exon 3 (Wiese *et al.*, 1993; Cockerill *et al.*, 1997) and another found in a French-Canadian patient had a C to T transition in exon 2 that resulted in premature termination of the VDR at Arg 30 (Zhu *et al.*, 1998). The same mutation was also identified in two related Brazilian patients (Mechica *et al.*, 1997). Skin fibroblasts obtained from these patients exhibit no detectable levels of VDR protein by western blot analysis or by $1,25(OH)_2D_3$ binding assays (Wiese *et al.*, 1993; Cockerill *et al.*, 1997; Zhu *et al.*, 1998).

Defects in the ligand binding

A common defect described in early biochemical studies of VDR obtained from several different HVDRR kindreds was undetectable binding of $1,25(OH)_2D_3$, which indicated a potential defect in the LBD of the VDR. Surprisingly, the vast majority of distinct mutations in HVDRR families that have been mapped to the LBD still retain the capacity to bind the $1,25(OH)_2D_3$ ligand, although in some cases there is a reduced affinity for the ligand. This discrepancy is explained by the occurrence of a common LBD truncation in one rather large kindred that also occurs in other unrelated HVDRR family members.

Ritchie *et al.* (1989) were the first to identify the defect in three patients from one family. Biochemical studies demonstrated that the children have no detectable binding of $1,25(OH)_2D_3$, that the 24-hydroxylase activity could not be induced, and that VDR protein could not be detected by immunoblot analysis. All affected children had a single C to A conversion within exon 7 at nucleotide 970 that resulted in the conversion of the normal codon for tyrosine (TAC) into a stop codon (TAA) that results in premature termination of translation at amino acid 295 (Y295X) (Fig. 6.7). This mutation caused a large truncation in the LBD (residues 295–427). Introduction of this mutation in hVDR by site-directed mutagenesis also generated a truncated VDR that did not bind $1,25(OH)_2D_3$ and failed to activate gene transcription in transient expression assays.

A follow-up study examined an extension of this kindred containing seven related family branches and eight affected children with HVDRR (Malloy *et al.*, 1990). All affected individuals from which cells were available ($n = 7$) had the same mutation. Subsequently, this same mutation was also identified in two other patients from an unrelated kindred (Wiese *et al.*, 1993). No truncated receptor species were apparent in extracts of cells from these patients, nor were there any truncated species from cellular extracts derived from their parents that were heterozygous for the trait. There was also no detectable mRNA for the VDR in the patients with HVDRR. The authors suggested that additional mechanisms beyond the mutation itself may play a role in this phenomenon. Little is known about the processes that control VDR mRNA stability or the half-life of the protein. However, it is clear that $1,25(OH)_2D_3$ binding increases the stability of the VDR protein (McCain *et al.*, 1978; Wiese *et al.*, 1992). The inability of the truncated receptor to interact with hormone may leave it more susceptible to proteolytic attack and degradation processes such as those mediated by the proteosome (Masuyama and MacDonald, 1998). Although it is possible, it seems unlikely that a single point mutation would enhance transcript degradation. Perhaps the more intriguing possibility is the VDR protein may play a role in stabilizing its own mRNA transcript.

A second truncation mutation that eliminates all the LBD of the VDR was identified by Kristjansson *et al.* (1993). The mutation occurs within exon 4, which encodes the hinge domain. Here, a C to T transition within the codon for Gln152 produces a premature STOP codon (Q152X) that effectively eliminates over 300 amino acids or virtually all of the LBD. The mutant receptor does not bind $1,25(OH)_2D_3$ or activate a vitamin D-responsive reporter gene construct.

In that same study, Kristjansson *et al.* (1993) identified a missense mutation in exon 7 (CGC to CTC) in a distinct kindred resulting in a substitution of arginine 274 by leucine (R274L) in the LBD. Expression of the mutant VDR in CV-1 cells showed little or no activation of a vitamin D-responsive reporter gene at levels of $1,25(OH)_2D_3$ that activated wild-type VDR. However, transcription was activated with the R274L mutant if 1000-fold higher levels of $1,25(OH)_2D_3$ were added. It was noted in that study that the Arg274 residue is highly conserved among the receptors that heterodimerize, including the VDR, TR and RAR. Indeed, structural analysis of a related receptor reveals that the analogous arginine in RARγ is involved in a direct contact with the all-*trans*-retinoic acid ligand (Renaud *et al.*, 1995). Thus, Arg274 may be involved in high-affinity hormone binding to the VDR, and replacing the basic arginine residue for a nonpolar hydrophobic leucine decreases $1,25(OH)_2D_3$ binding affinity by a factor of 1000.

Two other mutations that resulted in impaired ligand binding were identified in exon 8 of the *VDR* gene from separate individuals with HVDRR (Whitfield *et al.*, 1996; Malloy *et al.*, 1997). One was a His to Gln conversion at residue 305 (H305Q) (Malloy *et al.*, 1997) and the other was an Ile to Ser mutation at amino acid 314 (I314S) (Whitfield *et al.*, 1996) of the hVDR. These particular mutants also reduced the affinity of VDR for the $1,25(OH)_2D_3$ ligand and displayed an attenuated transcriptional response compared with wild-type VDR in transfection experiments. Interestingly, the latter patient showed nearly complete recovery of the defect when treated with high doses of vitamin D (Griffin and Zerwekh, 1983).

Defects in protein–protein interaction

The LBD fulfills a number of vital functions in VDR-mediated transcription in addition to binding the $1,25(OH)_2D_3$ ligand. One key role is as a protein–protein interaction surface through which the VDR contacts other nuclear receptors (e.g. RXR) or nuclear receptor comodulatory proteins. Therefore, the potential exists for VDR mutational events in the LBD that do not significantly impact on ligand binding, but that disrupt heterodimerization or other important protein–protein contacts. Whitfield *et al.* (1996) identified a mutation in exon 9 of the VDR in a young girl with HVDRR. A C to T transition was defined that changed Arg391 to a cysteine residue (R391C). The mutant VDR exhibited a modest decrease in $1,25(OH)_2D_3$ binding; it did not bind effectively to a VDRE as a heterodimer with RXR; and it was virtually inactive in a VDR-responsive reporter gene assay. Addition of high doses of $1,25(OH)_2D_3$ or expression of additional RXR in the system partially rescued the mutation. These data indicate, in part, that a reduced ability of the Arg391Cys mutant to heterodimerize with the RXR may define the molecular basis of the HVDRR phenotype in this individual. Similarly, Cockerill *et al.* (1997) identified a Gln259Pro mutation in two related patients with HVDRR. The mutant receptor was inactive in a 1,25(OH)$_2$D$_3$-mediated transcription assay, yet in bound ligand with an affinity similar to wild-type VDR. Interaction of the mutant VDR with the VDRE differed from the pattern observed with wild-type VDR, suggesting an impaired interaction with the RXR or other unidentified nuclear proteins.

PATHOPHYSIOLOGY AND DIAGNOSIS

The primary physiologic role for 1,25(OH)$_2$D$_3$ is to maintain adequate absorption of calcium and phosphate at the intestine (see Physiology above). These effects are mediated through the VDR and its regulation of gene expression in the gut. It is generally accepted that active transport of calcium across the enterocyte accounts for approximately 50% of the total calcium that is absorbed across the intestine. Mutations that inactivate or disrupt the genomic actions of the VDR therefore lead to a general decrease in calcium and phosphate transport across the intestine. This leads to hypocalcemia and eventually to secondary hyperparathyroidism as the parathyroid glands respond to hypocalcemic challenges by increasing PTH synthesis and secretion to combat the reduction in serum calcium levels. Prolonged hypocalcemia eventually leads to chief cell hyperplasia and gland enlargement. In addition to a direct effect of PTH on the mobilization of calcium reserves in the skeleton, it also stimulates the renal 1α-hydroxylase enzyme leading to increased 1,25(OH)$_2$D$_3$ production. Thus, serum levels of 1,25(OH)$_2$D$_3$ are often raised in individuals with HVDRR. The compensatory increases in serum 1,25(OH)$_2$D$_3$ levels have minimal effect due to the end-organ resistance of the syndrome (VDR defects). In the absence of normal calcium absorption from the diet and secondary increases in serum PTH levels, mineral reserves in the bone are mobilized resulting in a disruption of normal bone mineralization and homeostasis. This results in rickets in children and osteomalacia in adults, the characteristic clinical features of both classical vitamin D deficiency and HVDRR. Rickets is generally evident within months after birth and ranges from mild to severe.

Alopecia is present in most, but not all, patients with HVDRR. In the most severe cases, this is not limited to the scalp, but some patients may also have a complete lack of body hair, including eyebrows and eyelashes. Subjects with HVDRR appear to have normal hair follicles, but lack the hair shaft itself. This curious anomaly in HVDRR highlights the more pleotropic physiologic roles for vitamin D and the VDR in addition to traditional calciotropic actions (see Physiology).

PHARMACOLOGY

Treatment of HVDRR involves administration of calcium supplements and active vitamin D compounds. The effectiveness of vitamin D therapy varies widely and is dependent on the nature of the defect in the VDR. In general, patients with HVDRR without alopecia respond more favorably to treatment with vitamin D compounds (Marx *et al.*, 1986). Vitamin D$_3$ has been used at doses of 5000–40 000 IU per day (Brooks *et al.*, 1978; Marx *et al.*, 1978; Zerwekh *et al.*, 1979). Nonalopecic patients also responded to 20–200 μg per day of 25(OH)D$_3$ and 17–20 μg per day of 1,25(OH)$_2$D$_3$ (Marx *et al.*, 1978). Patients with *VDR* mutations in the LBD that reduce the affinity for

1,25$(OH)_2D_3$ may be treated with high doses of the hormone. For example, a patient with the His305Gln genotype responded to 1,25$(OH)_2D_3$ at a daily dose of 12.5 μg (Van Maldergem *et al.*, 1996) and fibroblasts derived from this patient also responded to high concentrations of 1,25$(OH)_2D_3$ (Malloy *et al.*, 1997). On the other hand, a patient with the Arg274Leu genotype could not be treated successfully and the VDR isolated from his fibroblasts did not bind 1,25$(OH)_2D_3$ (Fraher *et al.*, 1986). While patients without alopecia tend to be more responsive to vitamin D therapy, there are numerous cases of successful treatment of patients with alopecia with vitamin D compounds (Fujita *et al.*, 1980; Tsuchiya *et al.*, 1980; Kudoh *et al.*, 1981; Balsan *et al.*, 1983; Hirst *et al.*, 1985; Castells *et al.*, 1986; Takeda *et al.*, 1987, 1989). Of interest is the case of two siblings with the Glu152stop mutation, lacking most of the LBD, that surprisingly responded to therapy with vitamin D (Kruse and Feldmann, 1995).

Calcium supplementation is necessary in patients who do not respond to vitamin D therapy. Sakati *et al.* (1986) reported successful treatment of a patient that was completely resistant to 1,25$(OH)_2D_3$ administration (48 μg per day) with high doses of oral calcium (3–4 g daily). Calcium infusion has also been used to treat a patient who did not respond to vitamin D compounds or to oral calcium (Balsan *et al.*, 1986). Improvement was noted within 2 weeks of initiation of nocturnal infusion, and weight and height gain was achieved by 7 months. However, the beneficial effects reversed rapidly when the infusions were stopped. Infusion therapy has subsequently been employed successfully in other cases (Weisman *et al.*, 1987; Bliziotes *et al.*, 1988; Hochberg *et al.*, 1992). The current therapeutic regimen for children diagnosed with HVDRR entails calcium infusion until the rachitic symptoms are corrected, and then maintenance with oral calcium (Weisman and Hochberg, 1994).

REFERENCES

Abe, E., Miyaura, C., Sakagami, H. *et al.* (1981) Differentiation of mouse myeloid leukemia cells induced by 1α,25-dihydroxyvitamin D_3. *Proc. Natl. Acad. Sci. USA* **78**: 4990–4994.

Abe, E., Shiina, Y., Miyaura, C. *et al.* (1984) Activation and fusion induced by 1α,25-dihydroxyvitamin D_3 and their relation in alveolar macrophages. *Proc. Natl. Acad. Sci. USA* **81**: 7112–7116.

Asaka, M., Iida, H., Izumino, K. and Sasayama, S. (1988) Depressed natural killer cell activity in uremia. Evidence for immunosuppressive factor in uremic sera. *Nephron* **49**: 291–295.

Askew, F.A., Bourdillon, R.B., Bruce, H.M., Jenkins, R.G.C. and Webster, T.A. (1931) The distillation of vitamin D. *Proc. Roy. Soc. Lond. [Biol]* **107**: 76–90.

Bagot, M., Charue, D., Lescs, M.C., Pamphile, R.P. and Revuz, J. (1994) Immunosuppressive effects of 1,25-dihydroxyvitamin D_3 and its analogue calcipotriol on epidermal cells. *Br. J. Dermatol.* **130**: 424–431.

Baker, A.R., McDonnell, D.P., Hughes, M. *et al.* (1988) Cloning and expression of full-length cDNA encoding human vitamin D receptor. *Proc. Natl. Acad. Sci. USA* **85**: 3294–3298.

Balsan, S., Garabedian, M., Liberman, U.A. *et al.* (1983) Rickets and alopecia with resistance to 1,25-dihydroxyvitamin D: two different clinical courses with two different cellular defects. *J. Clin. Endocrinol. Metab.* **57**: 803–811.

Balsan S., Garabedian M., Larchet, M. *et al.* (1986) Long-term nocturnal calcium infusions can cure rickets and promote normal mineralization in hereditary resistance to 1,25-dihydroxyvitamin D. *J. Clin. Invest.* **77**: 1661–1667.

Bannister, A.J. and Kouzarides, T. (1996) The CBP co-activator is a histone acetyltransferase. *Nature* **384**: 641–643.

Barettino, D., Vivanco Ruiz, M.M. and Stunnenberg, H.G. (1994) Characterization of the ligand-dependent transactivation domain of thyroid hormone receptor. *EMBO J.* **13**: 3039–3049.

Bar-Shavit, Z., Noff, D., Edelstein, S., Meyer, M., Shibolet, S. and Goldman, R. (1981) 1,25-Dihydroxyvitamin D_3 and the regulation of macrophage function. *Calcif. Tissue Int.* **33**: 673–676.

Baudino, T.A., Kraichely, D.M., Jefcoat, S.C., Winchester, S.K., Partridge, N.C. and Macdonald, P.N. (1998) Isolation and characterization of a novel coactivator protein, NCoA-62, involved in vitamin D-mediated transcription. *J. Biol. Chem.* **273**: 16434–16441.

Bikle, D.D. and Gee, E. (1989) Free, and not total, 1,25-dihydroxyvitamin D regulates 25-hydroxyvitamin D metabolism by keratinocytes. *Endocrinology* **124**: 649–654.

Bikle, D.D., Gee, E., Halloran, B. and Haddad, J.G. (1984) Free 1,25-dihydroxyvitamin D levels in serum from normal subjects, pregnant subjects, and subjects with liver disease. *J. Clin. Invest.* **74**: 1966–1971.

Bikle, D.D., Siiteri, P.K., Ryzen, E. and Haddad, J.G. (1985) Serum protein binding of 1,25-dihydroxyvitamin D: a reevaluation by direct measurement of free metabolite levels. *J. Clin. Endocrinol. Metab.* **61**: 969–975.

Billaudel, B., Labriji-Mestaghanmi, H., Sutter, B.C. and Malaisse, W.J. (1988) Vitamin D and pancreatic islet function. II. Dynamics of insulin release and cationic fluxes. *J. Endocrinol. Invest.* **11**: 585–593.

Blanco, J.C., Wang, I.M., Tsai, S.Y. *et al.* (1995) Transcription factor TFIIB and the vitamin D receptor cooperatively activate ligand-dependent transcription. *Proc. Natl. Acad. Sci. USA* **92**: 1535–1539.

Bliziotes, M., Yergey, A.L., Nanes, M.S. *et al.* (1988) Absent intestinal response to calciferols in hereditary resistance to 1,25-dihydroxyvitamin D: documentation and effective therapy with high dose intravenous calcium infusions. *J. Clin. Endocrinol. Metab.* **66**: 294–300.

Blunt, J.W., DeLuca, H.F. and Schnoes, H.K. (1968) 25-Hydroxycholecalciferol. A biologically active metabolite of vitamin D_3. *Biochemistry* **7**: 3317–3322.

Bonjour, J.P., Preston, C. and Fleisch, H. (1977) Effect of 1,25-dihydroxyvitamin D_3 on the renal handling of Pi in thyroparathyroidectomized rats. *J. Clin. Invest.* **60**: 1419–1428.

Bottomley, W.W., Jutley, J., Wood, E.J. and Goodfield, M.D. (1995) The effect of calcipotriol on lesional fibroblasts from patients with active morphoea. *Acta Derm. Venereol. (Stockh.).* **75**: 364–366.

Bourguet, W., Ruff, M., Chambon, P., Gronemeyer, H. and Moras, D. (1995) Crystal structure of the ligand-binding domain of the human nuclear receptor RXR-α. *Nature* **375**: 377–382.

Brooks, M.H., Bell, N.H., Love, L. *et al.* (1978) Vitamin-D-dependent rickets type II. Resistance of target organs to 1,25-dihydroxyvitamin D. *N. Engl. J. Med.* **298**: 996–999.

Brown, A.J., Zhong, M., Finch, J., Ritter, C. and Slatopolsky, E. (1995) The roles of calcium and 1,25-dihydroxyvitamin D_3 in the regulation of vitamin D receptor expression by rat parathyroid glands. *Endocrinology* **136**: 1419–1425.

Brown, T.A. and DeLuca, H.F. (1990) Phosphorylation of the 1,25-dihydroxyvitamin D_3 receptor. A primary event in 1,25-dihydroxyvitamin D_3 action. *J. Biol. Chem.* **265**: 10025–10029.

Brown, T.A. and DeLuca, H.F. (1991) Sites of phosphorylation and photoaffinity labeling of the 1,25-dihydroxyvitamin D_3 receptor. *Arch. Biochem. Biophys.* **286**: 466–472.

Brzozowski, A.M., Pike, A.C., Dauter, Z. *et al.* (1997) Molecular basis of agonism and antagonism in the oestrogen receptor. *Nature* **389**: 753–758.

Burgos-Trinidad, M., Brown, A.J. and DeLuca, H.F. (1986) Solubilization and reconstitution of chick renal mitochondrial 25-hydroxyvitamin D_3 24-hydroxylase. *Biochemistry* **25**: 2692–2696.

Burmester, J.K., Maeda, N. and DeLuca, H.F. (1988) Isolation and expression of rat 1,25-dihydroxyvitamin D_3 receptor cDNA. *Proc. Natl. Acad. Sci. USA* **85**: 1005–1009.

Cade, C. and Norman, A.W. (1986) Vitamin D_3 improves impaired glucose tolerance and insulin secretion in the vitamin D-deficient rat *in vivo*. *Endocrinology* **119**: 84–90.

Cai, Q., Chandler, J.S., Wasserman, R.H., Kumar, R. and Penniston, J.T. (1993) Vitamin D and adaptation to dietary calcium and phosphate deficiencies increase intestinal plasma membrane calcium pump gene expression. *Proc. Natl. Acad. Sci. USA* **90**: 1345–1349.

Cantley, L.K., Russell, J., Lettieri, D. and Sherwood, L.M. (1985) 1,25-Dihydroxyvitamin D_3 suppresses parathyroid hormone secretion from bovine parathyroid cells in tissue culture. *Endocrinology* **117**: 2114–2119.

Cao, X., Ross, F.P., Zhang, L., MacDonald, P.N., Chappel, J. and Teitelbaum, S.L. (1993) Cloning of the promoter for the avian integrin $\beta3$ subunit gene and its regulation by 1,25-dihydroxyvitamin D_3. *J. Biol. Chem.* **268**: 27 371–27 380.

Carlberg, C., Bendik, I., Wyss, A. *et al.* (1993) Two nuclear signalling pathways for vitamin D. *Nature* **361**: 657–660.

Castells, S., Greig, F., Fusi, M.A. *et al.* (1986) Severely deficient binding of 1,25-dihydroxyvitamin D to its receptors in a patient responsive to high doses of this hormone. *J. Clin. Endocrinol. Metab.* **63**: 252–256.

Chan, Y.L., McKay, C., Dye, E. and Slatopolsky, E. (1986) The effect of 1,25 dihydroxycholecalciferol on parathyroid hormone secretion by monolayer cultures of bovine parathyroid cells. *Calcif. Tissue Int.* **38**: 27–32.

Chen, H., Lin, R.J., Schiltz, R.L. *et al.* (1997) Nuclear receptor coactivator ACTR is a novel histone acetyltransferase and forms a multimeric activation complex with P/CAF and CBP/p300. *Cell* **90**: 569–580.

Chen, T.C., Castillo, L., Korycka-Dahl, M. and DeLuca, H.F. (1974) Role of vitamin D metabolites in phosphate transport of rat intestine. *J. Nutr.* **104**: 1056–1060.

Chertow, B.S., Sivitz, W.I., Baranetsky, N.G., Clark, S.A., Waite, A. and Deluca, H.F. (1983) Cellular mechanisms of insulin release: the effects of vitamin D deficiency and repletion on rat insulin secretion. *Endocrinology* **113**: 1511–1518.

Cheskis, B. and Freedman, L.P. (1996) Modulation of nuclear receptor interactions by ligands: kinetic analysis using surface plasmon resonance. *Biochemistry* **35**: 3309–3318.

Christakos, S., Friedlander, E.J., Frandsen, B.R. and Norman, A.W. (1979) Studies on the mode of action of calciferol. XIII. Development of a radioimmunoassay for vitamin D-dependent chick intestinal calcium-binding protein and tissue distribution. *Endocrinology* **104**: 1495–1503.

Cockerill, F.J., Hawa, N.S., Yousaf, N. *et al.* (1997) Mutations in the vitamin D receptor gene in three kindreds associated with hereditary vitamin D resistant rickets. *J. Clin. Endocrinol. Metab.* **82**: 3156–3160.

Costanzo, L.S., Sheehe, P.R. and Weiner, I.M. (1974) Renal actions of vitamin D in D-deficient rats. *Am. J. Physiol.* **226**: 1490–1495.

Cross, H.S., Debiec, H. and Peterlik, M. (1990) Mechanism and regulation of intestinal phosphate absorption. *Min. Electrol. Metab.* **16**: 115–124.

Danielian, P.S., White, R., Lees, J.A. and Parker, M.G. (1992) Identification of a conserved region required for hormone dependent transcriptional activation by steroid hormone receptors. *EMBO J.* **11**: 1025–1033.

Danielsen, M., Hinck, L. and Ringold, G.M. (1989) Two amino acids within the knuckle of the first zinc finger specify DNA response element activation by the glucocorticoid receptor. *Cell* **57**: 1131–1138.

Darwish, H.M. and DeLuca, H.F. (1992) Identification of a 1,25-dihydroxyvitamin D_3-response element in the 5'-flanking region of the rat calbindin *D-9k* gene. *Proc. Natl. Acad. Sci. USA* **89**: 603–607.

Davies, P.D.O. (1985) A possible link between vitamin D deficiency and impaired host defense to mycobacterium tuberculosis. *Tubercle* **66**: 301–306.

Demay, M.B., Gerardi, J.M., DeLuca, H.F. and Kronenberg, H.M. (1990) DNA sequences in the rat osteocalcin gene that bind the 1,25-dihydroxyvitamin D_3 receptor and confer responsiveness to 1,25-dihydroxyvitamin D_3. *Proc. Natl. Acad. Sci. USA* **87**: 369–373.

Demay, M.B., Kiernan, M.S., DeLuca, H.F. and Kronenberg, H.M. (1992) Sequences in the human parathyroid hormone gene that bind the 1,25-dihydroxyvitamin D_3 receptor and mediate transcriptional repression in response to 1,25-dihydroxyvitamin D_3. *Proc. Natl. Acad. Sci. USA* **89**: 8097–8101.

Desai, R.K., van Wijnen, A.J., Stein, J.L., Stein, G.S. and Lian, J.B. (1995) Control of 1,25-dihydroxyvitamin D_3 receptor-mediated enhancement of osteocalcin gene transcription: effects of perturbing phosphorylation pathways by okadaic acid and staurosporine. *Endocrinology* **136**: 5685–5693.

Dodd, R.C., Cohen, M.S., Newman, S.L. and Gray, T.K. (1983) Vitamin D metabolites change the phenotype of monoblastic U937 cells. *Proc. Natl. Acad. Sci. USA* **80**: 7538–7541.

Faraco, J.H., Morrison, N.A., Baker, A., Shine, J. and Frossard, P.M. (1989) ApaI dimorphism at the human vitamin D receptor gene locus. *Nucleic Acids Res.* **17**: 2150.

Feldman, D., Chen, T., Hirst, M., Colston, K., Karasek, M. and Cone, C. (1980) Demonstration of 1,25-dihydroxyvitamin D_3 receptors in human skin biopsies. *J. Clin. Endocrinol. Metab.* **51**: 1463–1465.

Feng, W., Ribeiro, R.C., Wagner, R.L. *et al.* (1998) Hormone-dependent coactivator binding to a hydrophobic cleft on nuclear receptors. *Science* **280**: 1747–1749.

Fondell, J.D., Ge, H. and Roeder, R.G. (1996) Ligand induction of a transcriptionally active thyroid hormone receptor coactivator complex. *Proc. Natl. Acad. Sci. USA* **93**: 8329–8333.

Fraher, L.J., Karmali, R., Hinde, F.R. *et al.* (1986) Vitamin D-dependent rickets type II: extreme end organ resistance to 1,25-dihydroxyvitamin D_3 in a patient without alopecia. *Eur. J. Pediatr.* **145**: 389–395.

Fraser, D.R. and Kodicek, E. (1970) Unique biosynthesis by kidney of a biological active vitamin D metabolite. *Nature* **228**: 764–766.

Fraser, J.D. and Price, P.A. (1990) Induction of matrix Gla protein synthesis during prolonged 1,25-dihydroxyvitamin D_3 treatment of osteosarcoma cells. *Calcif. Tissue Int.* **46**: 270–279.

Fraser, J.D., Otawara, Y. and Price, P.A. (1988) 1,25-Dihydroxyvitamin D_3 stimulates the synthesis of matrix γ-carboxyglutamic acid protein by osteosarcoma cells. Mutually exclusive expression of vitamin K-dependent bone proteins by clonal osteoblastic cell lines. *J. Biol. Chem.* **263**: 911–916.

Fu, G.K., Lin, D., Zhang, M.Y.H., Bikle, D.D. *et al.* (1997) Cloning of human 25-

hydroxyvitamin D-1α-hydroxylase and mutations causing vitamin D-dependent rickets type I. *Mol. Endocrinol.* **11**: 1961–1970.

Fuchs, R. and Peterlik, M. (1979) Pathways of phosphate transport in chick jejunum: influence of vitamin D and extracellular sodium. *Pflugers Arch.* **381**: 217–222.

Fuchs, R., Graf, J. and Peterlik, M. (1985) Effects of 1α,25-dihydroxycholecalciferol on sodium-ion translocation across chick intestinal brush-border membrane. *Biochem. J.* **230**: 441–449.

Fujita, T., Nomura, M., Okajima, S. and Furuya, H. (1980) Adult-onset vitamin D-resistant osteomalacia with the unresponsiveness to parathyroid hormone. *J. Clin. Endocrinol. Metab.* **50**: 927–931.

Fullmer, C.S. (1992) Intestinal calcium absorption: calcium entry. *J. Nutr.* **122**: 644–650.

Garabedian, M., Holick, M.F., Deluca, H.F. and Boyle, I.T. (1972) Control of 25-hydroxycholecalciferol metabolism by parathyroid glands. *Proc. Natl. Acad. Sci. USA* **69**: 1673–1676.

Gedik, O. and Akalin, S. (1986) Effects of vitamin D deficiency and repletion on insulin and glucagon secretion in man. *Diabetologia* **29**: 142–145.

Ghazarian, J.G., Jefcoate, C.R., Knutson, J.C., Orme-Johnson, W.H. and DeLuca, H.F. (1974) Mitochondrial cytochrome P450. A component of chick kidney 25-hydrocholecalciferol-1α-hydroxylase. *J. Biol. Chem.* **249**: 3026–3033.

Ghijsen, W.E. and Van Os, C.H. (1982) 1α,25-Dihydroxy-vitamin D-3 regulates ATP-dependent calcium transport in basolateral plasma membranes of rat enterocytes. *Biochim. Biophys. Acta* **689**: 170–172.

Giguere, V., Hollenberg, S.M., Rosenfeld, M.G. and Evans, R.M. (1986) Functional domains of the human glucocorticoid receptor. *Cell* **46**: 645–652.

Gill, R.K. and Christakos, S. (1993) Identification of sequence elements in mouse calbindin-D28k gene that confer 1,25-dihydroxyvitamin D_3- and butyrate-inducible responses. *Proc. Natl. Acad. Sci. USA* **90**: 2984–2988.

Girasole, G., Passeri, G., Jilka, R.L. and Manolagas, S.C. (1994) Interleukin-11: a new cytokine critical for osteoclast development. *J. Clin. Invest.* **93**: 1516–1524.

Glass, C.K. (1994) Differential recognition of target genes by nuclear receptor monomers, dimers, and heterodimers. *Endocr. Rev.* **15**: 391–407.

Gniadecki, R. (1996) Stimulation versus inhibition of keratinocyte growth by 1,25-dihydroxyvitamin D_3: dependence on cell culture conditions. *J. Invest. Dermatol.* **106**: 510–516.

Goldman, R. (1984) Induction of a high phagocytic capability in P388D1, a macrophage-like tumor cell line, by 1α,25-dihydroxyvitamin D_3. *Cancer Res.* **44**: 11–19.

Griffin, J.E. and Zerwekh, J.E. (1983) Impaired stimulation of 25-hydroxyvitamin D–24-hydroxylase in fibroblasts from a patient with vitamin D-dependent rickets, type II. A form of receptor-positive resistance to 1,25-dihydroxyvitamin D_3. *J. Clin. Invest.* **72**: 1190–1199.

Gronemeyer, H., Turcotte, B., Quirin-Stricker, C. *et al.* (1987) The chicken progesterone receptor: sequence, expression and functional analysis. *EMBO J.* **6**: 3985–3994.

Grunstein, M. (1997) Histone acetylation in chromatin structure and transcription. *Nature* **389**: 349–352.

Guo, Y.D., Strugnell, S., Back, D.W. and Jones, G. (1993) Transfected human liver cytochrome P-450 hydroxylates vitamin D analogs at different side-chain positions. *Proc. Natl. Acad. Sci. USA* **90**: 8668–8672.

Haddad, J.G., Jennings, A.S. and Aw, T.C. (1988) Vitamin D uptake and metabolism by perfused rat liver: influences of carrier proteins. *Endocrinology* **123**: 498–504.

Hadzic, E., Desai-Yajnik, V., Helmer, E. *et al.* (1995) A 10-amino-acid sequence in the N-terminal A/B domain of thyroid hormone receptor α is essential for transcriptional activation and interaction with the general transcription factor TFIIB. *Mol. Cell. Biol.* **15**: 4507–4517.

Halloran, B.P. and DeLuca, H.F. (1981) Appearance of the intestinal cytosolic receptor for 1,25-dihydroxyvitamin D_3 during neonatal development in the rat. *J. Biol. Chem.* **256**: 7338–7342.

Hanewald, K.H., Rappoldt, M.P. and Roborgh, J.R. (1961) The antirachitic activity of previtamin D_3. *Rec. Trav. Chim. Pays-Bas. Belg.* **80**: 1063–1069.

Hard, T., Kellenbach, E., Boelens, R. *et al.* (1990) Solution structure of the glucocorticoid receptor DNA-binding domain. *Science* **249**: 157–160.

Harrison, H.E.M. and Harrison, H.C. (1961) Intestinal transport of phosphate: action of vitamin D, calcium and potassium. *Am. J. Physiol.* **201**: 1007–1012.

Harrison, P.V. and Stollery, N. (1994) Disseminated superficial actinic porokeratosis responding to calcipotriol. *Clin. Exp. Dermatol.* **19**: 95.

Haussler, M.R., Myrtle, J.F. and Norman, A.W. (1968) The association of a metabolite of vitamin D_3 with intestinal mucosa chromatin *in vivo*. *J. Biol. Chem.* **243**: 4055–4064.

Hayashi, S., Usui, E. and Okuda, K. (1988) Sex-related differences in vitamin D_3 25-hydroxylase of rat liver microsomes. *J. Biochem.* **103**: 863–866.

Hess, A.F. and Unger, L.J. (1921) Cure of infantile rickets by sunlight. *JAMA* **77**: 39.

Hilliard, G.M.T., Cook, R.G., Weigel, N.L. and Pike, J.W. (1994) 1,25-Dihydroxyvitamin D_3 modulates phosphorylation of serine 205 in the human vitamin D receptor: site-directed mutagenesis of this residue promotes alternative phosphorylation. *Biochemistry* **33**: 4300–4311.

Hirst, M.A., Hochman, H.I. and Feldman, D. (1985) Vitamin D resistance and alopecia: a kindred with normal 1,25-dihydroxyvitamin D binding, but decreased receptor affinity for deoxyribonucleic acid. *J. Clin. Endocrinol. Metab.* **60**: 490–495.

Hochberg, Z., Tiosano, D. and Even, L. (1992) Calcium therapy for calcitriol-resistant rickets. *J. Pediatr.* **121**: 803–808.

Holick, M.F. (1981) The cutaneous photosynthesis of previtamin D_3: a unique photoendocrine system. *J. Invest. Dermatol.* **77**: 51–58.

Holick, M.F. (1988) Skin: site of the synthesis of vitamin D and a target tissue for the active form, 1,25-dihydroxyvitamin D_3. *Ann. N. Y. Acad. Sci.* **548**: 14–26.

Holick, M.F. (1996) Vitamin D and bone health. *J. Nutr.* **126**: 1159S–1164S.

Holick, M.F., Schnoes, H.K., DeLuca, H.F., Suda, T. and Cousins, R.J. (1971) Isolation and identification of 1,25-dihydroxycholecalciferol. A metabolite of vitamin D active in intestine. *Biochemistry* **10**: 2799–2804.

Holick, M.F., Frommer, J.E., McNeill, S.C., Richtand, N.M., Henley, J.W. and Potts, J.T., Jr. (1977) Photometabolism of 7-dehydrocholesterol to previtamin D_3 in skin. *Biochem. Biophys. Res. Commun.* **76**: 107–114.

Hollenberg, S.M., Giguere, V. and Evans, R.M. (1989) Identification of two regions of the human glucocorticoid receptor hormone binding domain that block activation. Cancer Res. **49**: 15.

Hong, H., Kohli, K., Trivedi, A., Johnson, D.L. and Stallcup, M.R. (1996) GRIP1, a novel mouse protein that serves as a transcriptional coactivator in yeast for the hormone binding domains of steroid receptors. *Proc. Natl. Acad. Sci. USA* **93**: 4948–4952.

Hong, H., Kohli, K., Garabedian, M.J. and Stallcup, M.R. (1997) GRIP1, a transcriptional coactivator for the AF-2 transactivation domain of steroid, thyroid, retinoid, and vitamin D receptors. *Mol. Cell. Biol.* **17**: 2735–2744.

Honma, Y., Hozumi, M., Abe, E. *et al.* (1983) 1α,25-Dihydroxyvitamin D_3 and 1α-hydroxyvitamin D₃ prolong survival time of mice inoculated with myeloid leukemia cells. *Proc. Natl. Acad. Sci. USA* **80**: 201–204.

Horlein, A.J., Naar, A.M., Heinzel, T. *et al.* (1995) Ligand-independent repression by the thyroid hormone receptor mediated by a nuclear receptor co-repressor. *Nature* **377**: 397–404.

Horwood, N.J., Elliott, J., Martin, T.J. and Gillespie, M.T. (1998) Osteotropic agents regulate the expression of osteoclast differentiation factor and osteoprotegerin in osteo-blastic stromal cells. *Endocrinology* **139**: 4743–4746.

Hsieh, J.C., Jurutka, P.W., Galligan, M.A. *et al.* (1991) Human vitamin D receptor is selectively phosphorylated by protein kinase C on serine 51, a residue crucial to its trans-activation function. *Proc. Natl. Acad. Sci. USA* **88**: 9315–9319.

Hsieh, J.C., Jurutka, P.W., Nakajima, S. *et al.* (1993) Phosphorylation of the human vitamin D receptor by protein kinase C. Biochemical and functional evaluation of the serine 51 recognition site. *J. Biol. Chem.* **268**: 15 118–15 126.

Hsieh, J.C., Jurutka, P.W., Selznick, S.H. *et al.* (1995) The T-box near the zinc fingers of the human vitamin D receptor is required for heterodimeric DNA binding and transactiva-tion. *Biochem. Biophys. Res. Commun.* **215**: 1–7.

Hughes, M.R., Malloy, P.J., Kieback, D.G. *et al.* (1988) Point mutations in the human vitamin D receptor gene associated with hypocalcemic rickets. *Science* **242**: 1702–1705.

Huldschinsky, K. (1919) Heilung von Rachitis durch kunstliche Hohensonne. *Dtsch. Med. Wochenschr.* **45**: 712–713.

Iida, K., Shinki, T., Yamaguchi, A., DeLuca, H.F., Kurokawa, K. and Suda, T. (1995) A possible role of vitamin D receptors in regulating vitamin D activation in the kidney. *Proc. Natl. Acad. Sci. USA* **92**: 6112–6116.

Ishikawa, T., Umesono, K., Mangelsdorf, D.J. *et al.* (1990) A functional retinoic acid receptor encoded by the gene on human chromosome 12. *Mol. Endocrinol.* **4**: 837–844.

Jenster, G., Spencer, T.E., Burcin, M.M., Tsai, S.Y., Tsai, M.J. and O'Malley, B.W. (1997) Steroid receptor induction of gene transcription: a two-step model. *Proc. Natl. Acad. Sci. USA* **94**: 7879–7884.

Jimenez-Lara, A.M. and Aranda, A. (1999) Lysine 246 of the vitamin D receptor is crucial for ligand-dependent interaction with coactivators and transcriptional activity. *J. Biol. Chem.* **274**: 13 503–13 510.

Jimi, E., Nakamura, I., Amano, H. *et al.* (1996) Osteoclast function is activated by osteoblastic cells through a mechanism involving cell-to-cell contact. *Endocrinology* **137**: 2187–2190.

Jin, C.H. and Pike, J.W. (1996) Human vitamin D receptor-dependent transactivation in *Saccharomyces cerevisiae* requires retinoid X receptor. *Mol. Endocrinol.* **10**: 196–205.

Jin, C.H., Kerner, S.A., Hong, M.H. and Pike, J.W. (1996) Transcriptional activation and dimerization functions in the human vitamin D receptor. *Mol. Endocrinol.* **10**: 945–957.

Jones, B.B., Jurutka, P.W., Haussler, C.A., Haussler, M.R. and Whitfield, G.K. (1991) Vitamin D receptor phosphorylation in transfected ROS 17/2.8 cells is localized to the N-terminal region of the hormone-binding domain. *Mol. Endocrinol.* **5**: 1137–1146.

Jurutka, P.W., Hsieh, J.C., MacDonald, P.N. *et al.* (1993) Phosphorylation of serine 208 in the human vitamin D receptor. The predominant amino acid phosphorylated by casein kinase II, *in vitro*, and identification as a significant phosphorylation site in intact cells. *J. Biol. Chem.* **268**: 6791–6799.

Jurutka, P.W., Hsieh, J.C., Nakajima, S., Haussler, C.A., Whitfield, G.K. and Haussler, M.R. (1996) Human vitamin D receptor phosphorylation by casein kinase II at Ser-208 potentiates transcriptional activation. *Proc. Natl. Acad. Sci. USA* **93**: 3519–3524.

Jurutka, P.W., Hsieh, J.C., Remus, L.S. *et al.* (1997) Mutations in the 1,25-dihydroxyvitamin D$_3$ receptor identifying C-terminal amino acids required for transcriptional activation that are functionally dissociated from hormone binding, heterodimeric DNA binding, and interaction with basal transcription factor IIB, *in vitro. J. Biol. Chem.* **272**: 14 592–14 599.

Kamei, Y., Kawada, T., Fukuwatari, T., Ono, T., Kato, S. and Sugimoto, E. (1995) Cloning and sequencing of the gene encoding the mouse vitamin D receptor. *Gene* **152**: 281–282.

Kamei, Y., Xu, L., Heinzel, T. *et al.* (1996) A CBP integrator complex mediates transcriptional activation and AP-1 inhibition by nuclear receptors. *Cell* **85**: 403–414.

Karbach, U. (1992) Paracellular calcium transport across the small intestine. *J. Nutr.* **122**: 672–677.

Katahira, M., Knegtel, R.M., Boelens, R. *et al.* (1992) Homo- and heteronuclear NMR studies of the human retinoic acid receptor β DNA-binding domain: sequential assignments and identification of secondary structure elements. *Biochemistry* **31**: 6474–6480.

Kato, T., Terui, T. and Tagami, H. (1987) Topically active vitamin D$_3$ analogue, 1α,24-dihydroxy-cholecalciferol, has an anti-proliferative effect on the epidermis of guinea pig skin. *Br. J. Dermatol.* **117**: 528–530.

Kawashima, H., Torikai, S. and Kurokawa, K. (1981) Localization of 25-hydroxyvitamin D$_3$ 1α-hydroxylase and 24-hydroxylase along the rat nephron. *Proc. Natl. Acad. Sci. USA* **78**: 1199–1203.

Keohane, S.G. and Cork, M.J. (1995) Treatment of Grover's disease with calcipotriol (Dovonex). *Br. J. Dermatol.* **132**: 832–833.

Kitanaka, S., Takeyama, K., Murayama, A. *et al.* (1998) Inactivating mutations in the 25-hydroxyvitamin D$_3$ 1α-hydroxylase gene in patients with pseudovitamin D-deficiency ricket. *N. Engl. J. Med.* **338**: 653–661.

Kletzien, S.W., Templin, V.M., Steenbock, H. and Thomas, B.R. (1932) Vitamin D and the conservation of calcium in the adult. *J. Biol. Chem.* **97**: 265–280.

Kliewer, S.A., Umesono, K., Mangelsdorf, D.J. and Evans R.M. (1992) Retinoid X receptor interacts with nuclear receptors in retinoic acid, thyroid hormone and vitamin D$_3$ signalling. *Nature* **355**: 446–449.

Knutson, J.C. and DeLuca, H.F. (1974) 25-Hydroxyvitamin D$_3$–24-hydroxylase. Subcellular location and properties. *Biochemistry* **13**: 1543–1548.

Koeffler, H.P. (1985) Vitamin D: myeloid differentiation and proliferation. *Hamatologie und Bluttransfusion* **29**: 409–417.

Kragballe, K. (1989) Treatment of psoriasis by the topical application of the novel cholecalciferol analogue calcipotriol (MC 903). *Arch. Dermatol.* **125**: 1647–1652.

Kraichely, D.M., Collins, J.J. III, DeLisle, R.K. and MacDonald, P.N. (1999) The autonomous transactivation domain in helix H3 of the vitamin D receptor is required for transactivation and coactivator interaction. *J. Biol. Chem.* **274**: 14 352–14 358.

Kristjansson, K., Rut, A.R., Hewison, M. *et al.* (1993) Two mutations in the hormone binding domain of the vitamin D receptor cause tissue resistance to 1,25-dihydroxyvitamin D$_3$. *J. Clin. Invest.* **92**: 12–16.

Kruse, K. and Feldmann, E. (1995) Healing of rickets during vitamin D therapy despite defective vitamin D receptors in two siblings with vitamin D-dependent rickets type II. *J. Pediatr.* **126**: 145–148.

Kudoh, T., Kumagai, T., Uetsuji, N. *et al.* (1981) Vitamin D dependent rickets: decreased sensitivity to 1,25-dihydroxyvitamin D. *Eur. J. Pediatr.* **137**: 307–311.

Kumar, V., Green, S., Stack, G., Berry, M., Jin, J.R. and Chambon, P. (1987) Functional domains of the human estrogen receptor. *Cell* **51**: 941–951.

Kurnik, B.R. and Hruska, K.A. (1985) Mechanism of stimulation of renal phosphate transport by 1,25-dihydroxycholecalciferol. *Biochim. Biophys. Acta* **817**: 42–50.

Labuda, M., Morgan, K. and Glorieux, F.H. (1990) Mapping autosomal recessive vitamin D dependency type I to chromosome 12q14 by linkage analysis. *Am. J. Hum. Genet.* **47**: 28–36.

Labuda, M., Fujiwara, T.M., Ross, M.V. *et al.* (1992) Two hereditary defects related to vitamin D metabolism map to the same region of human chromosome 12q13-14. *J. Bone Miner. Res.* **7**: 1447–1453.

Lacey, D.L., Timms, E., Tan, H.L. *et al.* (1998) Osteoprotegerin ligand is a cytokine that regulates osteoclast differentiation and activation. *Cell* **93**: 165–176.

Lavigne, A.C., Mengus, G., Gangloff, Y.G., Wurtz, J.M. and Davidson, I. (1999) Human TAF(II)55 interacts with the vitamin D_3 and thyroid receptors and with derivatives of the retinoid X receptor that have altered transactivation properties. *Mol. Cell. Biol.* **19**: 5486–5494.

Lawson, D.E., Fraser, D.R., Kodicek, E., Morris, H.R. and Williams, D.H. (1971) Identification of 1,25-dihydroxycholecalciferol, a new kidney hormone controlling calcium metabolism. *Nature* **230**: 228–230.

Lee, M.S., Kliewer, S.A., Provencal, J., Wright, P.E. and Evans, R.M. (1993) Structure of the retinoid X receptor α DNA binding domain: a helix required for homodimeric DNA binding. *Science* **260**: 1117–1121.

Lees, J.A., Fawell, S.E. and Parker, M.G. (1989) Identification of two transactivation domains in the mouse oestrogen receptor. *Nucleic Acids Res.* **17**: 5477–5488.

Lemire, J.M., Adams, J.S., Sakai, R. and Jordan, S.C. (1984) 1α,25-Dihydroxyvitamin D_3 suppresses proliferation and immunoglobulin production by normal human peripheral blood mononuclear cells. *J. Clin. Invest.* **74**: 657–661.

Lemire, J.M., Archer, D.C., Beck, L. and Spiegelberg, H.L. (1995) Immunosuppressive actions of 1,25-dihydroxyvitamin D_3: preferential inhibition of Th1 functions. *J. Nutr.* **125**: 1704S–1708S.

Lemon, B.D., Fondell, J.D. and Freedman, L.P. (1997) Retinoid X receptor:vitamin D_3 receptor heterodimers promote stable preinitiation complex formation and direct 1,25-dihydroxyvitamin D_3-dependent cell-free transcription. *Mol. Cell. Biol.* **17**: 1923–1937.

Li, Y.C., Amling, M., Pirro, A.E. *et al.* (1998) Normalization of mineral ion homeostasis by dietary means prevents hyperparathyroidism, rickets, and osteomalacia, but not alopecia in vitamin D receptor-ablated mice. *Endocrinology* **139**: 4391–4396.

Liao, J., Ozono, K., Sone, T., McDonnell, D.P. and Pike, J.W. (1990) Vitamin D receptor interaction with specific DNA requires a nuclear protein and 1,25-dihydroxyvitamin D_3. *Proc. Natl. Acad. Sci. USA* **87**: 9751–9755.

Lin, N.U., Malloy, P.J., Sakati, N., al-Ashwal, A. and Feldman, D. (1996) A novel mutation in the deoxyribonucleic acid-binding domain of the vitamin D receptor causes hereditary 1,25-dihydroxyvitamin D-resistant rickets. *J. Clin. Endocrinol. Metab.* **81**: 2564–2569.

Liu, M., Lee, M.H., Cohen, M., Bommakanti, M. and Freedman, L.P. (1996a) Transcriptional activation of the Cdk inhibitor p21 by vitamin D_3 leads to the induced differentiation of the myelomonocytic cell line U937. *Genes Dev.* **10**: 142–153.

Liu, S.M., Koszewski, N., Lupez, M., Malluche, H.H., Olivera, A. and Russell, J. (1996b) Characterization of a response element in the 5'-flanking region of the avian (chicken) PTH gene that mediates negative regulation of gene transcription by 1,25-dihydroxyvitamin D_3 and binds the vitamin D_3 receptor. *Mol. Endocrinol.* **10**: 206–215.

Lorente, F., Fontan, G., Jara, P., Casas, C., Garcia-Rodriguez, M.C. and Ojeda, J.A. (1976)

Defective neutrophil motility in hypovitaminosis D rickets. *Acta Paediatr. Scand.* **65**: 695–699.

Luisi, B.F., Xu, W.X., Otwinowski, Z., Freedman, L.P., Yamamoto, K.R. and Sigler, P.B. (1991) Crystallographic analysis of the interaction of the glucocorticoid receptor with DNA. *Nature* **352**: 497–505.

Lutzow-Holm, C., De Angelis, P., Grosvik, H. and Clausen, O.P. (1993) 1,25-Dihydroxyvitamin D₃ and the vitamin D analogue KH1060 induce hyperproliferation in normal mouse epidermis. A BrdUrd/DNA flow cytometric study. *Exp. Dermatol.* **2**: 113–120.

MacDonald, P.N., Dowd, D.R., Nakajima, S. *et al.* (1993) Retinoid X receptors stimulate and 9-*cis* retinoic acid inhibits 1,25-dihydroxyvitamin D₃-activated expression of the rat osteocalcin gene. *Mol. Cell. Biol.* **13**: 5907–5917.

MacDonald, P.N., Sherman, D.R., Dowd, D.R., Jefcoat, S.C., Jr. and DeLisle, R.K. (1995) The vitamin D receptor interacts with general transcription factor IIB. *J. Biol. Chem.* **270**: 4748–4752.

Mackey, S.L., Heymont, J.L., Kronenberg, H.M. and Demay, M.B. (1996) Vitamin D receptor binding to the negative human parathyroid hormone vitamin D response element does not require the retinoid X receptor. *Mol. Endocrinol.* **10**: 298–305.

Mader, S., Kumar, V., de Verneuil, H. and Chambon, P. (1989) Three amino acids of the oestrogen receptor are essential to its ability to distinguish an oestrogen from a glucocorticoid-responsive element. *Nature* **338**: 271–274.

Makin, G., Lohnes, D., Byford, V., Ray, R. and Jones, G. (1989) Target cell metabolism of 1,25-dihydroxyvitamin D₃ to calcitroic acid. Evidence for a pathway in kidney and bone involving 24-oxidation. *Biochem. J.* **262**: 173–180.

Malloy, P.J., Hochberg, Z., Tiosano, D., Pike, J.W., Hughes, M.R. and Feldman, D. (1990) The molecular basis of hereditary 1,25-dihydroxyvitamin D₃ resistant rickets in seven related families. *J. Clin. Invest.* **86**: 2071–2079.

Malloy, P.J., Weisman, Y. and Feldman, D. (1994) Hereditary 1α,25-dihydroxyvitamin D-resistant rickets resulting from a mutation in the vitamin D receptor deoxyribonucleic acid-binding domain. *J. Clin. Endocrinol. Metab.* **78**: 313–316.

Malloy, P.J., Eccleshall, T.R., Gross, C., Van Maldergem, L., Bouillon, R. and Feldman, D. (1997) Hereditary vitamin D resistant rickets caused by a novel mutation in the vitamin D receptor that results in decreased affinity for hormone and cellular hyporesponsiveness. *J. Clin. Invest.* **99**: 297–304.

Malloy, P.J., Pike, J.W. and Feldman, D. (1999) The vitamin D receptor and the syndrome of hereditary 1,25-dihydroxyvitamin D-resistant rickets. *Endocr. Rev.* **20**: 156–188.

Mangelsdorf, D.J. and Evans, R.M. (1995) The RXR heterodimers and orphan receptors. *Cell* **83**: 841–850.

Manolagas, S.C., Yu, X.P., Girasole, G. and Bellido, T. (1994) Vitamin D and the hematolymphopoietic tissue: a 1994 update. *Semin. Nephrol.* **14**: 129–143.

Markose, E.R., Stein, J.L., Stein, G.S. and Lian, J.B. (1990) Vitamin D-mediated modifications in protein–DNA interactions at two promoter elements of the osteocalcin gene. *Proc. Natl. Acad. Sci. USA* **87**: 1701–1705.

Marx, S.J., Spiegel, A.M., Brown, E.M. *et al.* (1978) A familial syndrome of decrease in sensitivity to 1,25-dihydroxyvitamin D. *J. Clin. Endocrinol. Metab.* **47**: 1303–1310.

Marx, S.J., Bliziotes, M.M. and Nanes, M. (1986) Analysis of the relation between alopecia and resistance to 1,25-dihydroxyvitamin D. *Clin. Endocrinol.* **25**: 373–381.

Massaro, E.R., Simpson, R.U. and DeLuca, H.F. (1983) Glucocorticoids and appearance of 1,25-dihydroxyvitamin D₃ receptor in rat intestine. *Am. J. Physiol.* **244**: E230–E235.

Masuyama, H. and MacDonald, P.N. (1998) Proteasome-mediated degradation of the vitamin D receptor (VDR) and a putative role for SUG1 interaction with the AF-2 domain of VDR. *J. Cell. Biochem.* **71**: 429–440.

Masuyama, H., Brownfield, C.M., St-Arnaud, R. and MacDonald, P.N. (1997a) Evidence for ligand-dependent intramolecular folding of the AF-2 domain in vitamin D receptor-activated transcription and coactivator interaction. *Mol. Endocrinol.* **11**: 1507–1517.

Masuyama, H., Jefcoat, S.C., Jr. and MacDonald, P.N. (1997b) The N-terminal domain of transcription factor IIB is required for direct interaction with the vitamin D receptor and participates in vitamin D-mediated transcription. *Mol. Endocrinol.* **11**: 218–228.

Matkovits, T. and Christakos, S. (1995) Ligand occupancy is not required for vitamin D receptor and retinoid receptor-mediated transcriptional activation. *Mol. Endocrinol.* **9**: 232–242.

Matsumoto, T., Igarashi, C., Takeuchi, Y. *et al.* (1991) Stimulation by 1,25-dihydroxy-vitamin D_3 of *in vitro* mineralization induced by osteoblast-like MC3T3-E1 cells. *Bone* **12**: 27–32.

Mattei, M.G., Riviere, M., Krust, A. *et al.* (1991) Chromosomal assignment of retinoic acid receptor (RAR) genes in the human, mouse, and rat genomes. *Genomics* **10**: 1061–1069.

May, M., Mengus, G., Lavigne, A.C., Chambon, P. and Davidson, I. (1996) Human TAF(II28) promotes transcriptional stimulation by activation function 2 of the retinoid X receptors. *EMBO J.* **15**: 3093–3104.

McCain, T.A., Haussler, M.R., Okrent, D. and Hughes, M.R. (1978) Partial purification of the chick intestinal receptor for 1,25-dihydroxyvitamin D by ion exchange and blue dextran–Sepharose chromatography. *FEBS Lett.* **86**: 65–70.

McDonnell, D.P., Mangelsdorf, D.J., Pike, J.W., Haussler, M.R. and O'Malley, B.W. (1987) Molecular cloning of complementary DNA encoding the avian receptor for vitamin D. *Science* **235**: 1214–1217.

McDonnell, D.P., Scott, R.A., Kerner, S.A., O'Malley, B.W. and Pike, J.W. (1989) Functional domains of the human vitamin D_3 receptor regulate osteocalcin gene expression. *Mol. Endocrinol.* **3**: 635–644.

McKenna, N.J., Lanz, R.B. and O'Malley, B.W. (1999) Nuclear receptor coregulators: cellular and molecular biology. *Endocr. Rev.* **20**: 321–344.

McKinley, R., Kwan, Y.L., Ford, D., Lam-po-tang, P.R., Mason, R.S. and Manoharan, A. (1987) Clinical and laboratory studies of 1,25-dihydroxycholecalciferol in myelofibrosis. *Br. J. Haematol.* **65**: 252–254.

Mechica, J.B., Leite, M.O., Mendonca, B.B., Frazzatto, E.S., Borelli, A. and Latronico, A.C. (1997) A novel nonsense mutation in the first zinc finger of the vitamin D receptor causing hereditary 1,25-dihydroxyvitamin D_3-resistant rickets. *J. Clin. Endocrinol. Metab.* **82**: 3892–3894.

Mee, A.P., Hoyland, J.A., Braidman, I.P., Freemont, A.J., Davies, M. and Mawer, E.B. (1996) Demonstration of vitamin D receptor transcripts in actively resorbing osteoclasts in bone sections. *Bone* **18**: 295–299.

Meehan, M.A., Kerman, R.H. and Lemire, J.M. (1992) 1,25-Dihydroxyvitamin D_3 enhances the generation of nonspecific suppressor cells while inhibiting the induction of cytotoxic cells in a human MLR. *Cell. Immunol.* **140**: 400–409.

Mellanby, E. (1919) An experimental investigation on rickets. *Lancet* **i**: 407–412.

Mengus, G., May, M., Carre, L., Chambon, P. and Davidson, I. (1997) Human TAF(II)135 potentiates transcriptional activation by the AF-2s of the retinoic acid, vitamin D_3, and thyroid hormone receptors in mammalian cells. *Genes Dev.* **11**: 1381–1395.

Miyamoto, K., Kesterson, R.A., Yamamoto, H. *et al.* (1997) Structural organization of the human vitamin D receptor chromosomal gene and its promoter. *Mol. Endocrinol.* **11**: 1165–1179.

Monkawa, T., Yoshida, T., Wakino, S. *et al.* (1997) Molecular cloning of cDNA and genomic DNA for human 25-hydroxyvitamin D_3 1α-hydroxylase. *Biochem. Biophys. Res. Commun.* **239**: 527–533.

Morimoto, S., Imanaka, S., Koh, E. *et al.* (1989) Comparison of the inhibitions of proliferation of normal and psoriatic fibroblasts by 1α,25-dihydroxyvitamin D_3 and synthetic analogues of vitamin D_3 with an oxygen atom in their side chain. *Biochem. Int.* **19**: 1143–1149.

Morrison, N.A., Shine, J., Fragonas, J.C., Verkest, V., McMenemy, M.L. and Eisman, J.A. (1989) 1,25-Dihydroxyvitamin D-responsive element and glucocorticoid repression in the osteocalcin gene. *Science* **246**: 1158–1161.

Munder, M., Herzberg, I.M., Zierold, C. *et al.* (1995) Identification of the porcine intestinal accessory factor that enables DNA sequence recognition by vitamin D receptor. *Proc. Natl. Acad. Sci. USA* **92**: 2795–2799.

Nagpal, S., Friant, S., Nakshatri, H. and Chambon, P. (1993) RARs and RXRs: evidence for two autonomous transactivation functions (AF-1 and AF-2) and heterodimerization *in vivo*. *EMBO J.* **12**: 2349–2360.

Nakagawa, N., Kinosaki, M., Yamaguchi, K. *et al.* (1998) RANK is the essential signaling receptor for osteoclast differentiation factor in osteoclastogenesis. *Biochem. Biophys. Res. Commun.* **253**: 395–400.

Nakajima, S., Hsieh, J.C., MacDonald, P.N. *et al.* (1994) The C-terminal region of the vitamin D receptor is essential to form a complex with a receptor auxiliary factor required for high affinity binding to the vitamin D-responsive element. *Mol. Endocrinol.* **8**: 159–172.

Nellans, H.N. and Kimberg, D.V. (1978) Cellular and paracellular calcium transport in rat ileum: effects of dietary calcium. *Am. J. Physiol.* **235**: E726–E737.

Nemere, I., Yoshimoto, Y. and Norman, A.W. (1984) Calcium transport in perfused duodena from normal chicks: enhancement within fourteen minutes of exposure to 1,25-dihydroxyvitamin D_3. *Endocrinology* **115**: 1476–1483.

Nemere, I., Feld, C. and Norman, A.W. (1991) 1,25-Dihydroxyvitamin D_3-mediated alterations in microtubule proteins isolated from chick intestinal epithelium: analyses by isoelectric focusing. *J. Cell. Biochem.* **47**: 369–379.

Noda, M., Vogel, R.L., Craig, A.M., Prahl, J., DeLuca, H.F. and Denhardt, D.T. (1990) Identification of a DNA sequence responsible for binding of the 1,25-dihydroxyvitamin D_3 receptor and 1,25-dihydroxyvitamin D_3 enhancement of mouse secreted phosphoprotein 1 (SPP-1 or osteopontin) gene expression. *Proc. Natl. Acad. Sci. USA* **87**: 9995–9999.

Norman, A.W. and DeLuca, H.F. (1963) The preparation of ^{3}H-vitamins D_2 and D_3 and their localization in kidney and intestine. *Biochemistry* **2**: 1160–1168.

Norman, A.W., Myrtle, J.F., Midgett, R.J., Nowicki, H.G., Williams, V. and Popjak, G. (1971) 1,25-Dihydroxycholecalciferol: identification of the proposed active form of vitamin D_3 in the intestine. *Science* **173**: 51–54.

Ohyama, Y. and Okuda, K. (1991) Isolation and characterization of a cytochrome P-450 from rat kidney mitochondria that catalyzes the 24-hydroxylation of 25-hydroxyvitamin D_3. *J. Biol. Chem.* **266**: 8690–8695.

Ohyama, Y., Noshiro, M., Eggertsen, G. *et al.* (1993) Structural characterization of the gene encoding rat 25-hydroxyvitamin D_3 24-hydroxylase. *Biochemistry* **32**: 76–82.

Ohyama, Y., Ozono, K., Uchida, M. *et al.* (1994) Identification of a vitamin D-responsive element in the 5′-flanking region of the rat 25-hydroxyvitamin D_3 24-hydroxylase gene. *J. Biol. Chem.* **269**: 10 545–10 550.

Okano, T., Yasumura, M., Mizuno, K. and Kobayashi, T. (1977) Photochemical conversion of 7-dehydrocholesterol into vitamin D$_3$ in rat skins. *J. Nutr. Sci. Vitaminol.* **23**: 165–168.

Okazaki, T., Igarashi, T. and Kronenberg, H.M. (1988) 5'-Flanking region of the parathyroid hormone gene mediates negative regulation by 1,25-(OH)$_2$ vitamin D$_3$. *J. Biol. Chem.* **263**: 2203–2308.

Onate, S.A., Tsai, S.Y., Tsai, M.J. and O'Malley, B.W. (1995) Sequence and characterization of a coactivator for the steroid hormone receptor superfamily. *Science* **270**: 1354–1357.

Orgryzko, V.V., Schiltz, R.L., Russanova, V., Howard, B.H. and Nakatani, Y. (1996) The transcriptional coactivators p300 and CBP are histone acetyltransferases. *Cell* **87**: 953–959.

Ozono, K., Liao, J., Kerner, S.A., Scott, R.A. and Pike, J.W. (1990) The vitamin D-responsive element in the human osteocalcin gene. Association with a nuclear proto-oncogene enhancer. *J. Biol. Chem.* **265**: 21881–21888.

Peleg, S., Sastry, M., Collins, E.D., Bishop, J.E. and Norman, A.W. (1995) Distinct conformational changes induced by 20-epi analogues of 1α,25-dihydroxyvitamin D$_3$ are associated with enhanced activation of the vitamin D receptor. *J. Biol. Chem.* **270**: 10 551–10 558.

Pike, J.W. and Sleator, N.M. (1985) Hormone-dependent phosphorylation of the 1,25-dihydroxyvitamin D$_3$ receptor in mouse fibroblasts. *Biochem. Biophys. Res. Commun.* **131**: 378–385.

Pike, J.W., Kesterson, R.A., Scott, R.A., Kerner, S.A., McDonnell, D.P. and O'Malley, B.W. (1988) Vitamin D receptors: molecular structure of the protein and its chromosomal gene. In: A.W. Norman and K. Schaefer (eds) *Vitamin D: Molecular, Cellular and Clinical Endocrinology*, pp. 215–224. Walter de Gruyter, Berlin.

Polla, B.S., Healy, A.M., Wojno, W.C. and Krane, S.M. (1987) Hormone 1α,25-dihydroxy-vitamin D$_3$ modulates heat shock response in monocytes. *Am. J. Physiol.* **252**: C640–C649.

Puschett, J.B. and Beck, W.S., Jr. (1975) Parathyroid hormone and 25-hydroxy vitamin D$_3$: synergistic and antagonistic effects on renal phosphate transport. *Science* **190**: 473–475.

Puschett, J.B., Moranz, J. and Kurnick, W.S. (1972) Evidence for a direct action of cholecalciferol and 25-hydroxycholecalciferol on the renal transport of phosphate, sodium, and calcium. *J. Clin. Invest.* **51**: 373–385.

Quesada, J.M., Martin-Malo, A., Santiago, J. *et al.* (1990) Effect of calcitriol on insulin secretion in uraemia. *Nephrol. Dial. Transplant.* **5**: 1013–1017.

Rachez, C., Suldan, Z., Ward, J. *et al.* (1998) A novel protein complex that interacts with the vitamin D-3 receptor in a ligand-dependent manner and enhances VDR transactivation in a cell-free system. *Genes Dev.* **12**: 1787–1800.

Rachez, C., Lemon, B.D., Suldan, Z. *et al.* (1999) Ligand-dependent transcription activation by nuclear receptors requires the DRIP complex. *Nature* **398**: 824–828.

Rasmussen, H., Wong, M., Bikle, D. and Goodman, D.B. (1972) Hormonal control of the renal conversion of 25-hydroxycholecalciferol to 1,25-dihydroxycholecalciferol. *J. Clin. Invest.* **51**: 2502–2504.

Rastinejad, F., Perlmann, T., Evans, R.M. and Sigler, P.B. (1995) Structural determinants of nuclear receptor assembly on DNA direct repeats. *Nature* **375**: 203–211.

Reddy, G.S. and Tserng, K.Y. (1989) Calcitroic acid, end product of renal metabolism of 1,25-dihydroxyvitamin D$_3$ through C-24 oxidation pathway. *Biochemistry* **28**: 1763–1769.

Renaud, J.P., Rochel, N., Ruff, M. *et al.* (1995) Crystal structure of the RAR-γ ligand-binding domain bound to all-*trans* retinoic acid. *Nature* **378**: 681–689.

Ritchie, H.H., Hughes, M.R., Thompson, E.T. *et al.* (1989) An ochre mutation in the vitamin

D receptor gene causes hereditary 1,25-dihydroxyvitamin D$_3$-resistant rickets in three families. *Proc. Natl. Acad. Sci. USA* **86**: 9783–9787.

Romas, E., Udagawa, N., Zhou, H. *et al.* (1996) The role of gp130-mediated signals in osteoclast development: regulation of interleukin 11 production by osteoblasts and distribution of its receptor in bone marrow cultures. *J. Exp. Med.* **183**: 2581–2591.

Roth, J., Thorens, B., Hunziker, W., Norman, A.W. and Orci, L. (1981) Vitamin D-dependent calcium binding protein: immunocytochemical localization in chick kidney. *Science* **214**: 197–200.

Roth, J., Brown, D., Norman, A.W. and Orci, L. (1982) Localization of the vitamin D-dependent calcium-binding protein in mammalian kidney. *Am. J. Physiol.* **243**: F243–F252.

Roth, S.Y. and Allis, C.D. (1996) Histone acetylation and chromatin assembly: a single escort, multiple dances? *Cell* **87**: 5–8.

Russell, J., Lettieri, D. and Sherwood, L.M. (1986) Suppression by 1,25(OH)$_2$D$_3$ of transcription of the pre-proparathyroid hormone gene. *Endocrinology* **119**: 2864–2866.

Russell, J., Bar, A., Sherwood, L.M. and Hurwitz, S. (1993) Interaction between calcium and 1,25-dihydroxyvitamin D$_3$ in the regulation of preproparathyroid hormone and vitamin D receptor messenger ribonucleic acid in avian parathyroids. *Endocrinology* **132**: 2639–2644.

Rut, A.R., Hewison, M., Kristjansson, K. *et al.* (1994) Two mutations causing vitamin D resistant rickets: modelling on the basis of steroid hormone receptor DNA-binding domain crystal structures. *Clin. Endocrinol.* **41**: 581–590.

Saarem, K., Bergseth, S., Oftebro, H. and Pedersen, J.I. (1984) Subcellular localization of vitamin D$_3$ 25-hydroxylase in human liver. *J. Biol. Chem.* **259**: 10 936–10 940.

Saijo, T., Ito, M., Takeda, E. *et al.* (1991) A unique mutation in the vitamin D receptor gene in three Japanese patients with vitamin D-dependent rickets type II: utility of single-strand conformation polymorphism analysis for heterozygous carrier detection. *Am. J. Hum. Genet.* **49**: 668–673.

Sakati, N., Woodhouse, N.J., Niles, N., Harfi, H., de Grange, D.A. and Marx, S. (1986) Hereditary resistance to 1,25-dihydroxyvitamin D: clinical and radiological improvement during high-dose oral calcium therapy. *Horm. Res.* **24**: 280–287.

Sartorius, C.A., Melville, M.Y., Hovland, A.R., Tung, L., Takimoto, G.S. and Horwitz, K.B. (1994) A third transactivation function (AF3) of human progesterone receptors located in the unique N-terminal segment of the B-isoform. *Mol. Endocrinol.* **8**: 1347–1360.

Sato, T., Hong, M.H., Jin, C.H. *et al.* (1991) The specific production of the third component of complement by osteoblastic cells treated with 1α,25-dihydroxyvitamin D$_3$. *FEBS Lett.* **285**: 21–24.

Sato, T., Abe, E., Jin, C.H. *et al.* (1993) The biological roles of the third component of complement in osteoclast formation. *Endocrinology* **133**: 397–404.

Schrader, M., Nayeri, S., Kahlen, J.P., Muller, K.M. and Carlberg, C. (1995) Natural vitamin D$_3$ response elements formed by inverted palindromes: polarity-directed ligand sensitivity of vitamin D$_3$ receptor-retinoid X receptor heterodimer-mediated transactivation. *Mol. Cell. Biol.* **15**: 1154–1161.

Schwabe, J.W., Neuhaus, D. and Rhodes, D. (1990) Solution structure of the DNA-binding domain of the oestrogen receptor. *Nature* **348**: 458–461.

Shinki, T., Shimada, H., Wakino, S. *et al.* (1997) Cloning and expression of rat 25-hydroxyvitamin D$_3$-1α-hydroxylase cDNA. *Proc. Natl. Acad. Sci. USA* **94**: 12 920–12 925.

Silver, J., Russell, J. and Sherwood, L.M. (1985) Regulation by vitamin D metabolites of messenger ribonucleic acid for preproparathyroid hormone in isolated bovine parathyroid cells. *Proc. Natl. Acad. Sci. USA* **82**: 4270–4273.

Silver, J., Naveh-Many, T., Mayer, H., Schmelzer, H.J. and Popovtzer, M.M. (1986) Regulation by vitamin D metabolites of parathyroid hormone gene transcription *in vivo* in the rat. *J. Clin. Invest.* **78**: 1296–1301.

Simonet, W.S., Lacey, D.L., Dunstan, C.R. *et al.* (1997) Osteoprotegerin: a novel secreted protein involved in the regulation of bone density. *Cell* **89**: 309–319.

Sone, T., Scott, R.A., Hughes, M.R. *et al.* (1989) Mutant vitamin D receptors which confer hereditary resistance to 1,25-dihydroxyvitamin D_3 in humans are transcriptionally inactive *in vitro*. *J. Biol. Chem.* **264**: 20230–20234.

Sone, T., Marx, S.J., Liberman, U.A. and Pike, J.W. (1990) A unique point mutation in the human vitamin D receptor chromosomal gene confers hereditary resistance to 1,25-dihydroxyvitamin D_3. *Mol. Endocrinol.* **4**: 623–631.

Sone, T., Kerner, S. and Pike, J.W. (1991) Vitamin D receptor interaction with specific DNA association as a 1,25-dihydroxyvitamin D_3-modulated heterodimer. *J. Biol. Chem.* **266**: 23296–23305.

Spencer, T.E., Jenster, G., Burcin, M.M. *et al.* (1997) Steroid receptor coactivator-1 is a histone acetyltransferase. *Nature* **389**: 194–198.

Spiess, Y.H., Price, P.A., Deftos, J.L. and Manolagas, S.C. (1986) Phenotype-associated changes in the effects of 1,25-dihydroxyvitamin D_3 on alkaline phosphatase and bone GLA-protein of rat osteoblastic cells. *Endocrinology* **118**: 1340–1346.

Staal, A., van Wijnen, A.J., Birkenhager, J.C. *et al.* (1996) Distinct conformations of vitamin D receptor/retinoid X receptor-α heterodimers are specified by dinucleotide differences in the vitamin D-responsive elements of the osteocalcin and osteopontin genes. *Mol. Endocrinol.* **10**: 1444–1456.

St-Arnaud, R., Arabian, A. and Glorieux, F.H. (1996) Abnormal bone development in mice deficient for the vitamin D 24-hydroxylase gene. *J. Bone Miner. Res.* **11**: S126.

St-Arnaud, R., Messerlian, S., Moir, J.M., Omdahl, J.L. and Glorieux, F.H. (1997) The 25-hydroxyvitamin D 1-α-hydroxylase gene maps to the pseudovitamin D-deficiency rickets (PDDR) disease locus. *J. Bone Miner. Res.* **12**: 1552–1559.

Stroder, J. and Kasal, P. (1970) Phagocytosis in vitamin D deficient rickets. *Klin. Wochenschr.* **48**: 383–384.

Studzinski, G.P., Bhandal, A.K. and Brelvi, Z.S. (1985) Cell cycle sensitivity of HL-60 cells to the differentiation-inducing effects of 1-α,25-dihydroxyvitamin D_3. *Cancer Res.* **45**: 3898–3905.

Stumpf, W.E., Sar, M., Narbaitz, R., Reid, F.A., DeLuca, H.F. and Tanaka, Y. (1980) Cellular and subcellular localization of 1,25-$(OH)_2$-vitamin D_3 in rat kidney: comparison with localization of parathyroid hormone and estradiol. *Proc. Natl. Acad. Sci. USA* **77**: 1149–1153.

Szpirer, J., Szpirer, C., Riviere, M. *et al.* (1991) The Sp1 transcription factor gene (SP1) and the 1,25-dihydroxyvitamin D_3 receptor gene (VDR) are colocalized on human chromosome arm 12q and rat chromosome 7. *Genomics* **11**: 168–173.

Takeda, E., Kuroda, Y., Saijo, T. *et al.* (1987) 1α-Hydroxyvitamin D_3 treatment of three patients with 1,25-dihydroxyvitamin D-receptor-defect rickets and alopecia. *Pediatrics* **80**: 97–101.

Takeda, E., Yokota, I., Kawakami, I., Hashimoto, T., Kuroda, Y. and Arase, S. (1989) Two siblings with vitamin-D-dependent rickets type II: no recurrence of rickets for 14 years after cessation of therapy. *Eur. J. Pediatr.* **149**: 54–57.

Takeda, S., Yoshizawa, T., Nagai, Y. *et al.* (1999) Stimulation of osteoclast formation by 1,25-dihydroxyvitamin D requires its binding to vitamin D receptor in osteoblastic cells: studies using VDR knock out mice. *Endocrinology* **140**: 1005–1008.

Takeyama, K., Kitanaka, S., Sato, T., Kobori, M., Yanagisawa, J. and Kato, S. (1997) 25-Hydroxyvitamin D_3 1α-hydroxylase and vitamin D synthesis. *Science* **277**: 1827–1830.

Taylor, A.N. (1974) *In vitro* phosphate transport in chick ileum: effect of cholecalciferol, calcium, sodium and metabolic inhibitors. *J. Nutr.* **104**: 489–494.

Terpening, C.M., Haussler, C.A., Jurutka, P.W., Galligan, M.A., Komm, B.S. and Haussler, M.R. (1991) The vitamin D-responsive element in the rat bone Gla protein gene is an imperfect direct repeat that cooperates with other *cis*-elements in 1,25-dihydroxyvitamin D_3-mediated transcriptional activation. *Mol. Endocrinol.* **5**: 373–385.

Thompson, C.C. and Evans, R.M. (1989) *Trans*-activation by thyroid hormone receptors: functional parallels with steroid hormone receptors. *Proc. Natl. Acad. Sci. USA* **86**: 3494–3498.

Tsoukas, C.D., Provvedini, D.M. and Manolagas, S.C. (1984) 1,25-Dihydroxyvitamin D_3: a novel immunoregulatory hormone. *Science* **224**: 1438–1440.

Tsuchiya, Y., Matsuo, N., Cho, H. *et al.* (1980) An unusual form of vitamin D-dependent rickets in a child: alopecia and marked end-organ hyposensitivity to biologically active vitamin D. *J. Clin. Endocrinol. Metab.* **51**: 685–690.

Umesono, K. and Evans, R.M. (1989) Determinants of target gene specificity for steroid/thyroid hormone receptors. *Cell* **57**: 1139–1146.

van de Kerkhof, P.C. and de Jong, E.M. (1991) Topical treatment with the vitamin D_3 analogue MC903 improves pityriasis rubra pilaris: clinical and immunohistochemical observations. *Br. J. Dermatol.* **125**: 293–294.

Van Maldergem, L., Bachy, A., Feldman, D. *et al.* (1996) Syndrome of lipoatrophic diabetes, vitamin D resistant rickets, and persistent Mullerian ducts in a Turkish boy born to consanguineous parents. *Am. J. Med. Genet.* **64**: 506–513.

Vanham, G., Van Baelen, H., Tan, B.K. and Bouillon, R. (1988) The effect of vitamin D analogs and of vitamin D-binding protein on lymphocyte proliferation. *J. Steroid Bioch.* **29**: 381–386.

Voegel, J.J., Heine, M.J., Zechel, C., Chambon, P. and Gronemeyer, H. (1996) TIF2, a 160 kDa transcriptional mediator for the ligand-dependent activation function AF-2 of nuclear receptors. *EMBO J.* **15**: 3667–3675.

Wade, P.A. and Wolffe, A.P. (1997) Histone acetyltransferases in control. *Curr. Biol.* **7**: 1.

Wagner, R.L., Apriletti, J.W., McGrath, M.E., West, B.L., Baxter, J.D. and Fletterick, R.J. (1995) A structural role for hormone in the thyroid hormone receptor. *Nature* **378**: 690–697.

Walling, M.W. (1978) Intestinal inorganic phosphate transport. *Adv. Exp. Med. Biol.* **103**: 131–147.

Wasserman, R.H. and Taylor, A.N. (1968) Vitamin D-dependent calcium-binding protein. Response to some physiological and nutritional variables. *J. Biol. Chem.* **243**: 3987–3993.

Wasserman, R.H., Smith, C.A., Brindak, M.E. *et al.* (1992) Vitamin D and mineral deficiencies increase the plasma membrane calcium pump of chicken intestine. *Gastroenterology* **102**: 886–894.

Webster, N.J., Green, S., Jin, J.R. and Chambon, P. (1988) The hormone-binding domains of the estrogen and glucocorticoid receptors contain an inducible transcription activation function. *Cell* **54**: 199–207.

Weisman, Y. and Hochberg, Z. (1994) Genetic rickets and osteomalacia. *Curr. Ther. Endocrinol. Metab.* **5**: 492–495.

Weisman, Y., Bab, I., Gazit, D., Spirer, Z., Jaffe, M. and Hochberg, Z. (1987) Long-term intracaval calcium infusion therapy in end-organ resistance to 1,25-dihydroxyvitamin D. *Am. J. Med.* **83**: 984–990.

Whitfield, G.K., Hsieh, J.C., Nakajima, S. *et al.* (1995) A highly conserved region in the hormone-binding domain of the human vitamin D receptor contains residues vital for heterodimerization with retinoid X receptor and for transcriptional activation. *Mol. Endocrinol.* **9**: 1166–1179.

Whitfield, G.K., Selznick, S.H., Haussler, C.A. *et al.* (1996) Vitamin D receptors from patients with resistance to 1,25-dihydroxyvitamin D_3: point mutations confer reduced transactivation in response to ligand and impaired interaction with the retinoid X receptor heterodimeric partner. *Mol. Endocrinol.* **10**: 1617–1631.

Wieland, C., Mann, S., von Besser, H. and Saumweber, H. (1992) The *Drosophila* nuclear protein Bx42, which is found in many puffs on polytene chromosomes, is highly charged. *Chromosoma* **101**: 517–525.

Wiese, R.J., Uhland-Smith, A., Ross, T.K., Prahl, J.M. and DeLuca, H.F. (1992) Up-regulation of the vitamin D receptor in response to 1,25-dihydroxyvitamin D_3 results from ligand-induced stabilization. *J. Biol. Chem.* **267**: 20 082–20 086.

Wiese, R.J., Goto, H., Prahl, J.M. *et al.* (1993) Vitamin D-dependency rickets type II: truncated vitamin D receptor in three kindreds. *Mol. Cell. Endocrinol.* **90**: 197–201.

Wilson, T.E., Paulsen, R.E., Padgett, K.A. and Milbrandt, J. (1992) Participation of non-zinc finger residues in DNA binding by two nuclear orphan receptors. *Science* **256**: 107–110.

Windaus, A., Schenck, F.R. and von Werder, F. (1936) Uber das antirachitisch wirksame Bestrahlungsprodukt aus 7-dehydrocholesterin. *Z. Physiol.* **241**: 100–103.

Wurtz, J.M., Bourguet, W., Renaud, J.P. *et al.* (1996) A canonical structure for the ligand-binding domain of nuclear receptors. *Nat. Struct. Biol.* **3**: 87–94.

Yagci, A., Werner, A., Murer, H. and Biber, J. (1992) Effect of rabbit duodenal mRNA on phosphate transport in *Xenopus laevis* oocytes: dependence on 1,25-dihydroxy-vitamin-D_3. *Pflugers Arch.* **422**: 211–216.

Yagi, H., Ozono, K., Miyake, H., Nagashima, K., Kuroume, T. and Pike, J.W. (1993) A new point mutation in the deoxyribonucleic acid-binding domain of the vitamin D receptor in a kindred with hereditary 1,25-dihydroxyvitamin D-resistant rickets. *J. Clin. Endocrinol. Metab.* **76**: 509–512.

Yanagisawa, J., Yanagi, Y., Masuhiro, Y. *et al.* (1999) Convergence of transforming growth factor-β and vitamin D signaling pathways on SMAD transcriptional coactivators. *Science* **283**: 1317–1321.

Yoshida, T., Monkawa, T., Tenenhouse, H.S. *et al.* (1998) Two novel 1-α-hydroxylase mutations in French-Canadians with vitamin D dependency rickets type I. *Kidney Int.* **54**: 1437–1443.

Yoshizawa, T., Handa, Y., Uematsu, Y. *et al.* (1997) Mice lacking the vitamin D receptor exhibit impaired bone formation, uterine hypoplasia and growth retardation after weaning. *Nat. Genet.* **16**: 391–396.

Yu, V.C., Delsert, C., Andersen, B. *et al.* (1991) RXRβ: a coregulator that enhances binding of retinoic acid, thyroid hormone, and vitamin D receptors to their cognate response elements. *Cell* **67**: 1251–1266.

Zelinski, J.M., Sykes, D.E. and Weiser, M.M. (1991) The effect of vitamin D on rat intestinal plasma membrane Ca-pump mRNA. *Biochem. Biophys. Res. Commun.* **179**: 749–755.

Zerwekh, J.E., Glass, K., Jowsey, J. and Pak, C.Y. (1979) An unique form of osteomalacia associated with end organ refractoriness to 1,25-dihydroxyvitamin D and apparent defective synthesis of 25-hydroxyvitamin D. *J. Clin. Endocrinol. Metab.* **49**: 171–175.

Zhu, W.J., Malloy, P.J., Delvin, E., Chabot, G. and Feldman, D. (1998) Hereditary 1,25-dihydroxyvitamin D-resistant rickets due to an opal mutation causing premature termination of the vitamin D receptor. *J. Bone Min. Res.* **13**: 259–264.

Chapter 7

Retinoid Receptors

Arthur C.-K. Chung and Austin J. Cooney

INTRODUCTION AND HISTORICAL PERSPECTIVE

The retinoid receptors are members of the steroid hormone/nuclear receptor superfamily that mediate the biologic effects of retinoic acid (RA), an active metabolite of vitamin A. RA is essential in many biologic processes, such as embryonic development, the regulation of differentiation, as well as the regulation of proliferation in cell types throughout all stages of life (reviewed in Giguere, 1994; Mangelsdorf *et al.*, 1994; Chambon, 1996; Durston *et al.*, 1998). This chapter focuses on how these receptors interact with RAs, transduce their biologic effects and how mutation of the retinoid receptor leads to pathophysiologic effects.

Early research demonstrated that either a deficiency or an excess of vitamin A significantly affected the differentiation state of normal epithelial cells in the human body (reviewed in Johnson and Scadding, 1991; Valhquist, 1994). Studies on the effects of vitamin A deficiency in the rat showed that vitamin A was required for the normal differentiation of the epithelia (Wolbach and Howe, 1925). Other animal studies showed that vitamin A deficiency may increase the risk of cancer and the susceptibility to chemical carcinogens (reviewed in Sporn *et al.*, 1994). Retinoids, one of the important vitamin A derivatives, can modulate differentiation in several cell types both *in vitro* and *in vivo*, and are required for proper differentiation and maintenance of most epithelial tissues (Kopan *et al.*, 1987).

The investigation of the molecular mechanisms of RA action began with the identification of cellular retinol and RA binding proteins (CRBPs and CRABPs) in the mid-1970s (Ong and Chytil, 1978; Chytil and Ong, 1979). This initial characterization and biochemical analysis of CRABP led many investigators to speculate that CRABPs were the cognate receptors for all-*trans* RA. However, the absence of nuclear localization and any apparent DNA-binding properties did not support this speculation (reviewed in Mangelsdorf *et al.*, 1994). Finally, in late 1987, two independent groups successfully identified the nuclear receptor that was specific for RA (Giguere *et al.* 1987; Petkovitch *et al.*, 1987). This receptor, called retinoic acid receptor (RAR), was activated by RA and it regulated specific gene expression via binding to short DNA sequences, known as hormone response elements (HREs), in target genes. Shortly thereafter, a

Nuclear Receptors and Genetic Disease
ISBN 0-12-146160-2

study showed that the RAR required a partner to perform its functions (Glass *et al.*, 1990). This partner was eventually cloned and its sequence was shown to be related to RAR (Mangelsdorf *et al.*, 1990). It is termed the retinoid X receptor (RXR). Studies of these two types of retinoid receptors permitted a greater understanding of the mechanism of retinoid signaling and provided deeper insight into the diseases that arise from defects in retinoid signaling.

This review focuses on an update of our understanding about these two types of retinoid receptor and their involvement in human diseases and cancers. The involvement of these receptors in other biologic areas, such as retinoid-induced apoptosis, embryonic development, and the development of the central nervous system, is not discussed extensively here, as these topics have been reviewed elsewhere (Rogers, 1997; Durston *et al.*, 1998; Nagy *et al.*, 1998; Niles, 1998; Morriss-Kay and Ward, 1999).

PHYSIOLOGY

Metabolism of the retinoids

Unlike most hormones, vitamin A is a nutrient that is present in the blood and available to all cells most of the time (Napoli *et al.*, 1993). Since vitamin A and its derivatives, retinoids, are hydrophobic and insoluble in water, small retinoid-binding proteins are produced in the vertebrate body to ensure proper uptake, transport and storage of biologically active retinoids (Kanai *et al.*, 1968; Sani and Hill, 1974; Ong and Chytil, 1975; Chytil and Ong, 1979). These cellular retinoid-binding proteins consist of four proteins: two cellular retinol-binding proteins (CRBP-I and CRBP-II) and two cellular RA-binding proteins (CRABP-I and CRABP-II) (Ong and Chytil, 1975; Soprano *et al.*, 1981; Bailey and Siu, 1988; Giguere *et al.*, 1990). In the cytoplasm of cells, the CRBPs and CRABPs bind their ligands (retinol and RA respectively) with high affinity and specificity (Blomhoff *et al.*, 1990; Giguere *et al.*, 1990). The precise function of these proteins is unknown at present, but it has been speculated that they play a role in retinoid metabolism and in the regulation of free retinoid concentration in the cytoplasm (Lohnes *et al.*, 1992; Fiorella and Napoli, 1994; Ong *et al.*, 1994).

After absorption in the intestine, vitamin A, in the form of retinol, is released from the liver and transported in the plasma bound to retinol-binding protein (RBP) (Kanai *et al.*, 1968). Inside the cells, retinol is then converted to different biologically active derivatives, such as retinal, 9-*cis* RA, 13-*cis* RA, all-*trans* RA, and some oxo-retinoid products (Napoli *et al.*, 1993). The CRBPs are actively involved in these metabolic processes. Retinol bound to CRBP-I serves as a substrate for a specific nicotinamide–adenine dinucleotide phosphate (NADP)-dependent dehydrogenase and a cytosolic retinal dehydrogenase in the metabolic cascade leading to all-*trans* RA synthesis (Posch *et al.*, 1991, 1992). In this process, the CRBPs protect retinoids from oxidation by nonspecific microsomal and cytoplasmic dehydrogenases. After synthesis, all-*trans* RA then binds to CRABP, and this complex either delivers RA to the nucleus or acts as a carrier for the catabolism of all-*trans* RA to more polar metabolites by cytochrome P450 (Leo *et al.*, 1989; Fiorella and Napoli, 1991, 1994). The resultant 4-oxo-metabolites are then conjugated with glucuronic acid and excreted in the bile.

Interestingly, 9-*cis* RA, a stereoisomer of all-*trans* RA, has biologic effects similar to all-*trans* RA, but studies to date have shown that both CRABP-I and CRABP-II are unable to bind 9-*cis* RA (Allenby *et al.*, 1993; Fiorella *et al.*, 1993; Fogh *et al.*, 1993).

These findings suggest that 9-*cis* retinoids may have their own distinct metabolic pathway.

Retinoid signaling pathway

In the nucleus, the retinoid signaling pathway is mediated by nuclear retinoid receptors. These receptors are products of two multigene families, the RARs and RXRs. Each family has three receptor subtypes: RARα, β, γ, (NR1B1, 2 and 3 respectively), and RXRα, β, γ (NR2B1, 2 and 3 respectively). Both receptors are ligand-dependent transcription factors, belonging to subfamilies one and two in the nuclear receptor (NR) superfamily respectively (Evans, 1988; Green and Chambon, 1988; Beato, 1989; Laudet, 1997). Three different subtypes of RARs are found in mammals, birds and amphibians. Although the newt RAR is named RARδ, it appears to be the homolog of mammalian RARγ (Kastner *et al.*, 1994). To date, RARs have not been found among invertebrates, suggesting that they may have evolved after the separation of the invertebrates and vertebrates. Like the RARs, three distinct RXR subtypes have been found in mammals, birds, amphibians and fish. Unlike steroid hormone receptors, retinoid receptors do not bind to heat shock protein (hsp) 90 and are located within the nucleus. Current evidence shows that retinoid receptors need to bind cofactors to regulate gene transcription (reviewed in Chen and Li, 1998; Torchia *et al.*, 1998).

General expression of retinoid receptors

In normal adult tissues (human and mouse), RARα is expressed ubiquitously, with very high levels observed in certain regions of the brain (de The *et al.*, 1989). One of its isoforms, RARα1, has been shown to have the expression pattern of a housekeeping gene (Zelent *et al.*, 1989; Leroy *et al.*, 1991). In contrast, RARβ is found to be specifically expressed in the kidney, prostate, pituitary, adrenal glands, spinal cord and the cerebral cortex (Benbrook *et al.*, 1988). Unlike the previous two receptor subtypes, RARγ has the most restricted distribution and is expressed predominantly in the skin and lungs (Noji *et al.*, 1989; Zelent *et al.*, 1989; Kastner *et al.*, 1990). One of its isoforms, RARγ1, is expressed almost exclusively in skin where it is expressed at a high level in both the dermal and epidermal layers, suggesting that this isoform plays an important role in mediating the RA signaling in this tissue (Dolle *et al.*, 1989; Zelent *et al.*, 1989; Kastner *et al.*, 1990, 1994; Elder *et al.*, 1991).

RXRα and β are also widely expressed in normal adult human and mouse tissues (Niles, 1998). In the adult mouse, RXRα is expressed at high levels in the liver, skin and intestine, and at lower levels in a variety of other tissues, although it is undetectable in the brain (Mangelsdorf *et al.*, 1992). RXRβ appears to possess a rather ubiquitous and uniform expression pattern (Mangelsdorf *et al.*, 1992). The expression of RXRγ, however, is more restricted and has been observed in the liver, kidney, adrenal gland, lung and brain (Mangelsdorf *et al.*, 1990, 1992).

Expression of different subtypes of RARs and RXRs has also been demonstrated in embryos of various animals. In mouse embryos, RARα is expressed ubiquitously at the gastrulation stages (6.5–7.5 days postcoitum (dpc)) (Ruberte *et al.*, 1991; Ang and Duester, 1997). At 8 dpc, RARα appears in neural epithelium in brain and then at 8.5 dpc it is widely and abundantly expressed in the lateral neural epithelium of both forebrain and hindbrain, from where neural crest cells migrate.

At gastrulation stages, RARβ is expressed primarily in the presumptive hindbrain ectoderm and the adjacent mesenchyme (Ruberte *et al.*, 1991; Ang and Duester, 1997). During neural fold stages of mouse embryos (four to seven somites stage or 8.25 dpc), RARβ is found in the mesoderm of the caudal hindbrain region and in the hindgut endoderm (Ruberte *et al.*, 1990, 1991). At 8.5 dpc, its expression is in the neural epithelium, and it is restricted in the transition area between the closed and open neural fold. At 13.5 dpc, RARβ is expressed in several mesenchyme- and endoderm-derived tissues and its expression may be related to neurogenesis and programmed cell death. In addition, RARβ is found in dopamine-innervated areas of the brain (Ruberte *et al.*, 1990, 1991).

The expression of RARγ is temporally and spatially restricted and its pattern does not overlap with the expression of RARβ (Ruberte *et al.*, 1991; Ang and Duester, 1997). RARγ is expressed only after 8.5 dpc in the open portion of the neural epithelium of the posterior neuropore. After the neural tube closes, RARγ neural expression disappears. At 13.5 dpc, RARγ expression reappears and is strong in the developing limbs and skin (Dolle *et al.*, 1990).

There is no close relationship between the expression of RARs and RXRs (Mangelsdorf *et al.*, 1990, 1992). In rodent embryos, RXRα is expressed abundantly in liver, kidney, spleen, visceral tissues and skin (Mangelsdorf *et al.*, 1992). The expression of RXRβ is mainly in the central nervous system, while RXRγ is found in the peripheral nervous system and in muscle (Mangelsdorf *et al.*, 1992; Georgiades *et al.*, 1998; Zetterstrom *et al.*, 1999). In both chick and murine embryos, the expression of RXRγ is also found in dopaminergic neurons in which RARβ is expressed (Rowe *et al.*, 1991; Zetterstrom *et al.*, 1999). Recently, expression of RXRs has been demonstrated in the developing spinal cord (Solomin *et al.*, 1998).

The expression of retinoid receptor genes has been correlated to disease states. For example, altered RARα expression is found in patients with acute promyelocytic leukemia. In addition, many tumor cells lack RARβ expression, suggesting that RARβ may act as a tumor suppressor gene (Niles, 1998). Current evidence therefore suggests the involvement of RARs and/or RXRs in some human diseases and cancers (for a more detailed discussion see Pathophysiology below).

STRUCTURE AND MECHANISM OF ACTION OF RETINOID RECEPTORS

General description

Retinoic acid receptor (RAR)

The first RAR was isolated by two independent research groups (Giguere *et al.*, 1987; Petkovitch *et al.*, 1987). This human RAR (the hRARα1 isoform) was identified by low-stringency screening for complementary DNAs (cDNAs) encoding steroid hormone receptor-like molecules (Giguere *et al.*, 1987; Petkovitch *et al.*, 1987). This hRARα contains a DNA-binding domain (DBD) and a ligand-binding domain (LBD) that are structurally and functionally conserved, and is related to other members of the nuclear receptor family.

The three RAR subtype genes have been mapped to different chromosomes in the human (Table 7.1) (Ishikawa *et al.*, 1990; Mattei *et al.*, 1991). The chromosomal locations of the RARα, β and γ genes in humans are 17q21.1, 3q24 and 12q13 respectively, and the three receptor genes have been mapped to the syntenic chromosomal regions in mice and rats (Ishikawa *et al.*, 1990; Mattei *et al.*, 1991). Each RAR

Table 7.1 Characteristics of retinoid receptors

Gene	Major isoforms	Chromosomal location in human	Ligand	Nonretinoid ligand
RARα	α1, α2	17q21.1	All-*trans* RA and 9-*cis* RA	–
RARβ	β1, β2, β3, β4	3p24	All-*trans* RA and 9-*cis* RA	–
RARγ	γ1, γ2	12q13	All-*trans* RA and 9-*cis* RA	–
RXRα	α1, α2	9q34.3	9-*cis* RA	Phytanic and phytenic acids
RXRβ	β1, β2	6p21.3	9-*cis* RA	Phytanic and phytenic acids
RXRγ	γ1, γ2	1q22-q23	9-*cis* RA	Phytanic and phytenic acids
Jellyfish RXR	–	–	9-*cis* RA	–
Insect USP	A, B1, B2	–	–	Juvenile hormone

RAR, retinoic acid receptor; RXR, retinoid X receptor; USP, ultraspiracle.

gene can generate multiple isoforms by using a combination of different promoters, alternative splicing of exons, and initiation of translation at an internal CUG codon (Kastner *et al.*, 1990; Leroy *et al.*, 1991; Zelent *et al.*, 1991; Nagpal *et al.*, 1992a). These RAR isoforms with different N-terminal regions could potentially add to the diversity of the retinoid signal (Kastner *et al.*, 1994). In addition, RARs can efficiently bind to and be activated by either all-*trans* RA or 9-*cis* RA at $50 \, \mathrm{nmol \, l^{-1}}$ concentrations of ligand (Lohnes *et al.*, 1992; Kastner *et al.*, 1994).

Retinoid X receptor (RXR)
Using a cloning approach similar to the RAR cloning strategy, a second retinoid-responsive receptor, human RXRα, was isolated (Mangelsdorf *et al.*, 1990). Sequencing of this receptor gene revealed similarity to the RARs, especially in the DBD. The LBDs of RXR and RAR, however, were divergent (Mangelsdorf *et al.*, 1992). In addition, since RAR is more closely related to the thyroid hormone receptor than to RXR, it has been proposed that the response of RAR and RXR to RA has evolved independently (Kastner *et al.*, 1994; Mangelsdorf *et al.*, 1994; Laudet, 1997).

The three RXR genes have been mapped to different chromosomes in the human and mouse (Hoopes *et al.*, 1992). The positions of the RXRα, β and γ genes on human chromosomes are 9p34.3, 6p21.3 and 1q22-23 respectively (Table 7.1) (Hoopes *et al.*, 1992; Mangelsdorf *et al.*, 1992). In addition to α and γ subtypes, δ and ε subtypes of RXRs are found in zebrafish (Jones *et al.*, 1995). Isoforms exist in most RXR subtypes and they differ from one another in their N-terminal regions due to alternative usage of two different promoters (Liu and Linney, 1993; Jindra *et al.*, 1997; Guo *et al.*, 1998).

Initially, RXRs were shown to be activated by high concentrations of all-*trans* RA; however, it did not bind with high affinity to the receptor (Mangelsdorf *et al.*, 1990). Thus, it was postulated that another retinoid metabolite was the true agonist (retinoid X). Using different biochemical strategies, several groups identified 9-*cis* RA as the ligand for RXR (Heyman *et al.*, 1992; Levin *et al.*, 1992; Mangelsdorf *et al.*, 1992).

Unlike RARs, invertebrate homologs of RXR have been found in flatworms, crustaceans, insects and ticks (Oro *et al.*, 1990; Escriva *et al.*, 1997; Jindra *et al.*, 1997; Chung *et al.*, 1998; Guo *et al.*, 1998). Recently, a jellyfish RXR homolog has been cloned. This result suggests that RXRs might have been present early in the evolution from diploblastic animals, which contain two germ layers, to triploblastic animals, which have three germ layers, such as the vertebrates (Kostrouch *et al.*, 1998). Ultraspiracle (USP), an insect RXR homolog, has been shown to be required for both DNA binding and ligand binding of its dimerization partner, the ecdysteroid receptor (EcR) (Thomas *et al.*, 1993; Yao *et al.*, 1993). However, insect USP is reportedly unable to bind 9-*cis* RA, suggesting that the ability of RXRs to bind RA may have evolved after the separation of vertebrates and invertebrates during evolution (Yao *et al.*, 1993; Escriva *et al.*, 1997).

Functional domains

Like most members in the superfamily, both RAR and RXR are composed of several functional regions. Similar to other nuclear receptors, the N-terminal A/B region of RARs and RXRs has been shown to function as a ligand-independent transcriptional activation domain and to contain an activation function (AF) 1 subdomain (Nagpal *et al.*, 1992b, 1993; Folkers *et al.*, 1993). Unlike many receptors, the B region is well conserved among the different RAR subtypes, but the A region is more divergent (Kastner *et al.*, 1994).

Similar to other nuclear receptors, the DBD (region C) of retinoid receptors comprises 66 amino acids conformed as two zinc finger modules (Giguere *et al.*, 1987; Petkovitch *et al.*, 1987; Mangelsdorf *et al.*, 1992). The DBD is the most highly conserved domain throughout the superfamily of nuclear receptors. There is 94–97% identity between the DBDs of the three RARs (Kastner *et al.*, 1994). Both vertebrate and invertebrate RXRs also show great amino acid identity in this domain (Mangelsdorf *et al.*, 1990; Jindra *et al.*, 1997; Chung *et al.*, 1998). Furthermore, the DBD is responsible for dimerization and for binding of the receptor to HREs (Tsai and O'Malley, 1994; Mangelsdorf and Evans, 1995). DNA-binding specificity is governed by a short amino acid stretch in the DBD, called the P box. The P boxes of both RARs and RXRs are identical and the retinoic acid response elements (RAREs) usually are composed of repeats of the nucleotide sequence PuGGTCA.

Nuclear magnetic resonance (NMR) and X-ray crystallographic structure analyses demonstrate that the DBD of retinoid receptor contains three α helices (Knegtel *et al.*, 1993; Lee *et al.*, 1993a; Rastinejad *et al.*, 1995). Like steroid hormone receptors, each of the first two helices extends from the base of each zinc finger module. Specific residues in the N-terminal-most region of the α helix form hydrogen bonds with specific bases of target DNA in the major groove of HREs (Cooney and Tsai, 1994). The C-terminal-most helix lies above and perpendicular to the previous helix and does not contact the DNA. Unlike steroid hormone receptors, retinoid receptors have a third α helix after the second zinc finger (Lee *et al.*, 1993a). This helix participates in both protein–protein and protein–DNA interactions that are necessary for RXR to bind DNA as a dimer. Although this third helix was not observed in the crystal structure of the heterodimer–DNA complex, two recent high-resolution NMR studies support the existence of this third helix and its importance for dimer formation, as well as DNA binding (Rastinejad *et al.*, 1995; Holmbeck *et al.*, 1998a,b). This third helix has also been suggested to contribute to the dimerization interface of RXR DBD bound at the downstream (3′) half-site (Holmbeck *et al.*, 1998a).

Immediately after the DBD is region D (the hinge region). The hinge regions of RARs are well conserved among RAR subtypes and even more highly conserved within a given receptor subtype among different species (Leid *et al.*, 1992; Kastner *et al.*, 1994). These findings suggest that this region may have an RAR subtype-specific function, which has not yet been determined. The hinge region contains a stretch of several basic residues, which may correspond to a nuclear localization signal (Ylikomi *et al.*, 1992). Recently, it has been shown that the hinge regions of RARs are also the binding sites for some corepressors (see below). In RXRs, the T box, a short sequence of amino acids located at the N-terminal part of region D, is highly conserved in both invertebrate and vertebrate RXRs (Chung *et al.*, 1998). It has been shown that the T box participates in HRE recognition (Schwabe and Rhodes, 1991; Wilson *et al.*, 1992; Lee *et al.*, 1993a).

The LBD (region E) is the largest and most complex domain in the retinoid receptors. It mediates several functions, such as ligand binding, ligand-dependent transcriptional activation, and receptor dimerization (Forman *et al.*, 1989; Baniahmad *et al.*, 1992; Nagpal *et al.*, 1992b). X-ray crystallography studies of the LBD of ligand-free (apo-) hRXRα (Bourguet *et al.*, 1995) and ligand-bound (holo-) hRARγ (Renaud *et al.*, 1995; Rochel *et al.*, 1997) have demonstrated that both structures display the same overall tertiary structure of an antiparallel α-helical sandwich which consists of 12 α-helices and a β-turn region. In the ligand-bound structure, the ligand is bound to the β-turn region and is completely buried in the interior of the LBD, leading to the conclusion that ligand binding causes a conformational change in the LBD (Renaud *et al.*, 1995). Through biochemical methods, this conformational change has been shown to occur in several nuclear receptors (Allan *et al.*, 1992; Leng *et al.*, 1995).

The inability to crystallize the apo-hRARγ LBD has impeded direct comparisons (Rochel *et al.*, 1997). From the comparison of the holo-hRARγ with the apo-hRXRα, it has been proposed that ligand binding has the effect of setting off a 'mousetrap', with helix 12 (H12) flipping over to cover the ligand-binding cavity (Bourguet *et al.*, 1995; Renaud *et al.*, 1995). Within this α-helical sandwich structure of the LBD, the most C-terminal α helix, H12, contains a motif critical for ligand-dependent transactivation (AF-2) activity (known as the AF-2 core) (Wurtz *et al.*, 1996). This motif contains conserved hydrophobic and hydrophilic amino acid residues that form an amphipathic helix in which the conserved acidic amino acid residues in the AF-2 core form a salt bridge with basic residues in helix 4 after ligand binds to the receptor (Renaud *et al.*, 1995). This salt bridge may be important to the mousetrap model. Mutations of the conserved acidic amino acid residues in the AF-2 core in RAR eliminate its transactivation function, but do not affect ligand binding (Durand *et al.*, 1994; Baniahmad *et al.*, 1995; Renaud *et al.*, 1995; LeDouarin *et al.*, 1996).

Crystal structure and functional studies have revealed that the positioning of H12 plays a major role in the AF-2 activity of nuclear receptors. First, agonist binding results in a major structural transition of H12, placing it on the ligand-binding pocket and generating a surface for coactivator interaction (Renaud *et al.*, 1995; Vivat *et al.*, 1997). Second, antagonist action of raloxifene is linked to H12 positioning because the crystal structure of the estrogen receptor (ERα)-raloxifene complex demonstrates that, in the presence of this antagonist, H12 occupies a distinct position that does not allow coactivator interaction to occur (Brzozowski *et al.*, 1997). Finally, H12 appears to play a role in the corepressor release, as its deletion results in constitutive receptor binding to corepressors (Wurtz *et al.*, 1996).

The LBD also contributes to the dimerization activity of nuclear receptors. A series of heptad repeats, which is different from the leucine zipper, has been identified in RARs

(Rosen *et al.*, 1993). Later, the ninth heptad repeat was found to be essential for dimerization (Leid *et al.*, 1995). This heptad repeat is located in helix 10. Together with part of helix 9, helix 10 has been shown to form the main dimerization interface of hRXRα (Bourguet *et al.*, 1995; Wurtz *et al.*, 1996). However, there are some discrepancies between the previously identified crucial residues of the ninth heptad and the results from the X-ray crystallographic structures.

In human ERα, the last 42 amino acids of the receptor constitute an F domain. These amino acids are not required for steroid binding but seem to modulate the transcriptional activity of estrogen and antiestrogens (Montano *et al.*, 1995). hRARα, β and γ have F domains of 42, 35 and 34 amino acids respectively (Brand *et al.*, 1988; Krust *et al.*, 1989). A point mutation (M406A) at the N-terminal part of the F domain of hRARα has been shown to have a greater detrimental effect on the transcriptional activity of the full-length receptor than on the truncated receptor lacking the F domain (Tate *et al.*, 1996). Recently, this F region has been suggested to bind the coactivator CBP/p300 (Niles, 1998). On the other hand, RXRs do not have F domains, although the removal of the C-terminal 29, not the C-terminal 18 or 21, residues caused the loss of ligand binding (Leng *et al.*, 1995).

Ligand-binding properties

As ligand-dependent transcription factors, both RARs and RXRs have endogenous ligands. RARs can bind to and be activated by both all-*trans* and 9-*cis* RA (Cavey *et al.*, 1990; Schwabe and Rhodes, 1991; Torma *et al.*, 1994). The dissociation constants of all-*trans* RA for different RARs are in the low nanomolar range (K_d 0.2, 0.36 and 0.3 nmol L^{-1} for RARα, β and γ respectively) (summarized in Kastner *et al.*, 1994), while 9-*cis* RA appears to have a similar affinity for RARs as all-*trans* RA (Allenby *et al.*, 1993). The natural RAR ligands also include all-*trans*-3,4-didehydroretinoic acid (ddRA), all-*trans*-4-oxo-retinoic acid (4-oxo-RA), all-*trans*-4-oxoretinol (4-oxo-ROL), all-*trans*-4-oxoretinal (4-oxo-RAL) (Thaller and Eichele, 1990; Allenby *et al.*, 1993; Pijnappel *et al.*, 1993; Achkar *et al.*, 1996; Blumberg *et al.*, 1996).

All-*trans* RA does not bind to RXR *in vitro* (Heyman *et al.*, 1992; Levin *et al.*, 1992; Mangelsdorf *et al.*, 1992) nor activate RXRs in yeast (Heery *et al.*, 1993). Thus, in early studies, all-*trans* RA activation of RXRs at high concentrations in transfected animal cells may probably reflect intracellular isomerization to 9-*cis* RA (Mangelsdorf *et al.*, 1990; Levin *et al.*, 1992). The affinities of 9-*cis* RA for RXRα, β and γ are 15, 18 and 14 nmol L^{-1} respectively (Heyman *et al.*, 1992; Allenby *et al.*, 1993). In reality, it seems that 9-*cis* RA binds to RARs with an affinity that is approximately 20-fold higher than its affinity for RXRs. Thus, at low 9-*cis* RA concentrations, RARs may be the favored receptors to be activated by this ligand (Kastner *et al.*, 1994). Endogenous 9-*cis* RA has been detected in mouse liver and kidney at a level that is about threefold lower than that of all-*trans* RA (Heyman *et al.*, 1992). Another identified ligand for RXRs is 9-*cis*-3,4-didehydroretinoic acid (9-*cis*-ddRA) (Allenby *et al.*, 1993).

Unlike most of the vertebrate RXRs, two RXRs in zebrafish (δ and ε), and the *Drosophila* USP do not bind to RAs (Yao *et al.*, 1993; Jones *et al.*, 1995). The failure to bind RA may be due to the presence of an additional stretch of amino acids in the putative ligand-binding pockets of these RXRs (Chung *et al.*, 1998). In contrast, the new jellyfish RXR is able to, bind 9-*cis* RA and, activate the crystallin gene, which contains a RARE (Kostrouch *et al.*, 1998). It is not clear why jellyfish RXR can bind to 9-*cis* RA while *Drosophila* USP cannot, although phylogenetic analysis of the amino sequences of the RXRs implies that jellyfish RXR is more closely related to vertebrate RXR than is insect USP (Kostrouch *et al.*, 1998).

In addition, two nonretinoids, phytanic acid and phytenic acid, have been reported as endogenous ligands for RXRs but not RARs (Kitareewan *et al.*, 1996; Lemotte *et al.*, 1996). These terpenoids can bind to and activate RXR with a binding affinity of about 5 μmol L^{-1}. This binding affinity is about two to three orders of magnitude less than that for 9-*cis* RA. However, these two terpenoids may be physiologically relevant ligands because they are available in the blood at high concentrations (4–6 μmol L^{-1} in serum), which are also similar to their dissociation constants for RXRs. As the main source of these two terpenoids is the diet, phytanic and phytenic acid may represent nutrients that activate the RXR-dependent signaling pathway (reviewed in Forman, 1998). An *in vitro* study has shown that an insect terpenoid, juvenile hormone, binds to the *Drosophila* USP with a moderate binding affinity (Jones and Sharp, 1997). These results suggest that terpenoids may be alternative endogenous signaling molecules for RXRs.

The development of synthetic retinoids, which are either RAR- or RXR-selective, as well as receptor subtype-specific ligands, has become an important tool to explore the physiologic and molecular consequences of retinoid receptor action. In addition, the therapeutic potential of these synthetic retinoids is important in the treatment of cancer (see Pharmacology below). Both RARs and RXRs are capable of binding 9-*cis* RA. Each retinoid receptor has three receptor subtypes. Depending on the ligands, RXRs may mediate different transactivation pathways by forming homodimers or heterodimers with other nuclear receptors. The resulting pleiotropic effects of retinoid signaling create increasing complexity for the investigation of ligand-dependent transactivation by specific receptors or receptor subtypes. The development of synthetic retinoids that are selective for RARs, RXRs or their receptor subtypes has enabled a better elucidation of the roles of individual receptors or receptor subtypes in retinoid-dependent biologic functions, and has facilitated the development of clinically effective retinoids with higher therapeutic indices.

In the past two decades, a variety of synthetic retinoids has been assayed for binding and activation of RARs and RXRs (Apfel *et al.*, 1991; Graupner *et al.*, 1991; Lehmann *et al.*, 1992a). Some of these retinoids appear to display selectivity for given RARs and RXRs, as well as their subtypes. With some exceptions, the data obtained from both binding and activation experiments are often in agreement (Apfel *et al.*, 1991). Three generations of synthetic retinoids have been developed to date (reviewed in Chandraratna, 1998a; Orfanos *et al.*, 1997). The first generation of synthetic retinoids was basically artificially manufactured natural RAs. They included tretinoin (all-*trans* RA), 9-*cis* RA, and the polyene side-chain modified compounds, such as isotretinoin (13-*cis* RA). Second-generation retinoids were created by changing the cyclic end-group to give monoaromatic compounds. They included etretinate, acitretin and isoacitretin (reviewed in Orfanos *et al.*, 1997). Third-generation synthetic retinoids were then developed by cyclizing the polyene side-chain to produce polyaromatic compounds, such as arotinoids, adapalene (CD271) and tazarotene (AGN 190168) (Verschoore *et al.*, 1991; Esgleyes-Ribot *et al.*, 1994).

Crystal structures of the hRARγ LBD bound to all-*trans* RA, 9-*cis* RA and a synthetic retinoid agonist (BMS961) have recently been reported (Klaholz *et al.*, 1998). Although the structures of these three retinoids are different, each of these three molecules is able to fill almost entirely the ligand-binding pocket and produce an identical holo-LBD conformation. These results suggest that there is a conformational alteration of these molecules, allowing them to fit in a common cavity that produces an identical final shape of the holo-LBD (Klaholz et al., 1998). The recent strategy to improve receptor selectivity of new synthetic retinoids is to increase the conformational rigidity (Chandraratna, 1998a). This reduces the chance of conformational changes due to isomerization of retinoids in cells. Thus the retinoids can fit to the specific ligand-binding pocket of retinoid receptor subtypes.

On the basis of their transactivation properties, these synthetic retinoids can also be grouped into three distinct classes: agonists, inverse agonists and neutral antagonists (Chandraratna, 1998b). Agonists bind to retinoid receptors and increase the level of interaction between coactivators and the receptors, as well as reduce the level of interaction between corepressors and the receptors. Therefore, the level of ligand-dependent transactivation increases. Graupner et al. (1991) and Lehmann et al. (1991) were the first two groups to investigate the RAR subtype-selective agonists. They demonstrated that RARα was generally more sensitive to the variation of the retinoid conformation in the 4-substituted benzoic acid and 6-substituted 2-naphthalenecarboxylic acid classes than RARβ and RARγ. The RXR-selective agonists, SR11217 and SR11237, were strong activators of RXR response elements (RXREs) in the CRBP-II gene (Lehmann et al., 1992a). Inverse agonists, such as AGN193109, inhibit the basal level of transcription (Chandraratna, 1998b; Thacher et al., 1999). Unlike inverse agonists, neutral antagonists do not affect the basal levels of transcription because they do not exert any effect on the basal equilibrium between the receptor and coactivators and corepressors. RO41-5253 and SR11335 display selective antagonist activity, and SR11335 has been shown to be a potent inhibitor of 9-cis RA (Apfel et al., 1992; Lee et al., 1994). Examples of therapeutic uses of these compounds will be discussed in Pharmacology below.

Hormone response elements

As dimeric transcription factors, retinoid receptors regulate the transactivation of the target genes by binding to short specific DNA sequences known as HREs, which are generally located in the promoter region of responsive genes (reviewed in Beato, 1989; Tsai and O'Malley, 1994; Glass, 1996). Most of the HREs of steroid hormone receptors are palindromes consisting of inverted repeats (IRs) separated by 0 or 3 intervening base pairs (bp) since the steroid hormone receptors binds to these HREs as symmetric homodimers (Evans, 1988; Green and Chambon, 1988).

In contrast, most RAREs identified in the promoters of natural RA target genes are direct repeats (DRs) of a 6-bp sequence separated by 1, 2 or 5 bp (DR1, DR2 and DR5 respectively) (Naar et al., 1991; Umesono et al., 1991). Genes containing these RAREs include CRBP-I, CRABP-II, RARα, RARβ and RARγ (de The et al., 1990a; Sucov et al., 1990; Leroy et al., 1991; Smith et al., 1991; Durand et al., 1992; Lehmann et al., 1992b). RAREs of either palindromic arrangements (IR0 and IR1) in bovine growth hormone (bGH) and human oxytocin (hOXY), everted palindromes (ER8) in mouse γF-crystallin, or more complex arrangements, such as in rat growth hormone (rGH) and mouse LamB1, have also been identified (Table 7.2) (Richard and Zingg, 1991; Williams et al., 1992; Tini et al., 1993).

Based on the naturally occurring and synthetic RAREs, the consensus sequence of the RARE half-site is 5'-PuG(G/T)TCA-3' (reviewed in Chambon, 1996). Although the spacing, orientation and precise sequences of these half-sites vary in naturally occurring RAREs, the basic direct repeat of half-sites is in the form of PuG(G/T)TCA(X)nPuG(G/T)TCA, where X is any deoxynucleotide and n is the number of deoxynucleotides. The

Table 7.2 Examples of retinoic acid response elements

Type	Sequence	Gene	Reference
Half-site	AGGTCA		
DR1	GA**AGGGCA**GA**GGTCA**CA	mCRBP-II	Durand *et al.* (1992)
DR2	GT**AGGTCA**AA**AGGTCA**GA	mCRBP-I	Smith *et al.* (1991)
DR5	AG**GGTTCA**CCGAA**AGTTCA**CT	hRARβ2	Sucov *et al.* (1990)
IR0	GG**GGGACATGACCC**CA	bGH	Williams *et al.* (1992)
ER8	AG**TGACCC**TTTTAACC**AGGTCA**GT	mγF-crystallin	Tini *et al.* (1993)
Composite	AA**AGGTAA**GATCA**GGGACGTGACCG**CA	rGH	Williams *et al.* (1992)
Composite	GA**AGGTGA**GCTA**GGTTAA**(N₁₃)**GGGTCA**AC	LamB1	Williams *et al.* (1992)

DR, direct repeat; IR, inverted repeat; ER, estrogen receptor; mCRBP, mouse cellular retinoic acid binding protein; hRAR, human retinoic acid receptor; bGH, bovine growth hormone; mGH, mouse growth hormone.

most potent RAREs have been suggested to be direct repeats of the core AGGTCA half-site (Mangelsdorf and Evans, 1995).

RXR is the common dimerization partner of many nuclear receptors, such as RAR, RXR, thyroid hormone receptor (TR), vitamin D_3 receptor (VDR), as well as other receptors, such as farnesoate X receptor (FXR), liver X receptor (LXR), and peroxisome proliferator-activated receptors (PPAR), chicken ovalbumin upstream promoter-transcription factor (COUP-TF), nerve growth factor-induced protein B type (NGF-I-B) and Nur-related factor 1 (Nurr1) (reviewed in Cooney *et al.*, 1992; Mangelsdorf *et al.*, 1994; Chambon, 1996; Glass 1996; Willy and Mangelsdorf, 1998). The spacing between two half-sites of the DR motifs has been shown to be the important factor in determining which heterodimer partner will bind to the HREs with RXR, i.e. RAR, RXR, TR, VDR, etc. The early findings of DNA binding by RAR, TR and VDR were developed into the '3-4-5 rule' model (Umesono *et al.*, 1991). Heterodimers of RXR with VDR, TR and RAR have preferences for HREs composed of DRs spaced by 3, 4 and 5 bp respectively. These relative preferences are related to the formation of the second dimerization interface between the DBD of RXRs and its dimer partners (Perlmann *et al.*, 1993; Zechel *et al.*, 1994a,b). As more dimer partners of RXRs have been demonstrated and more HREs have been investigated, the model has been expanded to the DR 1-to-5 rule (reviewed in Mangelsdorf *et al.*, 1994). RXR homodimers, RAR–RXR and RXR–PPAR bind to DR1 motifs, while DR2 serves as a weaker RARE (Heery *et al.*, 1994).

Among the three RAREs examined, DR1, 2 and 5 motifs have different properties. First, these three DR motifs exhibit different affinities for the receptors (DR5 > DR2 > DR1) (Mader *et al.*, 1993). Second, DR1 motifs can be bound by both RXR homodimers and RAR–RXR heterodimers. As a receptor of 9-*cis* RA, RXR was first shown to bind as a homodimer to DR1 response elements (Mangelsdorf *et al.*, 1991). In the presence of 9-*cis* RA, RXR homodimers are able to transactivate the CRBP-II gene, which contains a DR1 motif (Kliewer *et al.*, 1992a; Lehmann *et al.*, 1992b; Zhang *et al.*, 1992). In addition, the DR1 element is also a HRE for RAR–RXR heterodimers; however, the binding of this heterodimer to DR1 sequences does not lead to transactivation of the target gene, rather it represses it (Mangelsdorf *et al.*, 1991). This repression cannot be relieved by the addition of ligands after the heterodimer binds DNA (Mangelsdorf *et al.*, 1991; Kurokawa *et al.*, 1994).

In the presence of ligands, RXR–RAR heterodimers activate transcription in response to ligands when bound to DR2 and DR5 elements. The DR5 element was the first element identified and is the most potent RARE (de The *et al.*, 1990a; Sucov *et al.*, 1990). To date, most RAREs examined are DR5 elements, including all three RAR genes (reviewed in Mangelsdorf *et al.*, 1994). The DR2 motif is also an effective RARE, although it was shown to have a weaker activation ability (Heery *et al.*, 1994).

Taken together, DR1 is a negative RARE when it is bound by RAR–RXR heterodimers, but when it is bound by RXR homodimers DR1 is a positive RARE responsive to 9-*cis* RA. Both DR2 and DR5 are positive RAREs and are bound by RXR–RAR heterodimers in the presence of ligands. In the presence of ligands, the binding to DR5 RAREs by RXR–RAR heterodimers leads to stronger activation than binding to a DR2 element.

Dimerization

As mentioned in the previous section, retinoid receptors bind to HREs as dimers. RXRs can either form homodimers or act as a partner for other nuclear receptors, including RARs. In this section we describe RXR–RAR heterodimers to illustrate the basic properties of receptor dimerization. RXR homodimers are then described. Finally, we summarize the findings about the relationship of RXR heterodimers and ligand binding.

RXR–RAR heterodimer
An early study of the DNA binding by RARs indicated that RARs needed to have an accessory factor to function (Glass *et al.*, 1990). Several laboratories independently identified the RXR as the necessary partner for functional RARs and other nuclear receptors (Bugge *et al.*, 1992; Kliewer *et al.*, 1992b; Zhang *et al.*, 1992).

The RXR–RAR heterodimer targets and regulates transcription from three different RAREs, such as DR1, DR2 and DR5 (Heery *et al.*, 1994; Zechel *et al.*, 1994a,b; Kurokawa *et al.*, 1995). In the case of the DR2 and DR5 response elements, RXR occupies the 5′ half-site while RAR binds the 3′ half-site (Kurokawa *et al.*, 1993; Perlmann *et al.*, 1993; Zechel *et al.*, 1994a,b). Transactivation occurs in the presence of RAR ligands. When the RAR–RXR heterodimer binds to DR1, the heterodimer represses the transactivation (Mangelsdorf *et al.*, 1991). This repression by RAR–RXR heterodimers is the result of the reversed polarity of the heterodimers because the RAR occupies the 5′ half-site (Kurokawa *et al.*, 1994). This opposite polarity of the RXR–RAR heterodimers binding to DR1 sequences may result in conformational differences, which may in turn regulate the interactions between RARs and corepressors, and promote repression (Glass, 1996).

RXR homodimer
RXRs can bind to a DR1 response element as either a heterodimer with RAR or a homodimer (Zhang *et al.*, 1992; Lehmann *et al.*, 1993). In addition to the zinc fingers, a short stretch of amino acids at the N-terminus of the hinge region of RXRs, called the T box, promotes binding to the DR1 RARE (Lee *et al.*, 1993a; Rastinejad *et al.*, 1995). Most studies have been carried out on the CRBP-II promoter, which has a DR1

RARE. RXR–RXR homodimers are able to transactivate the DR1 response element in the presence of 9-*cis* RA while RXR–RAR heterodimers repress the transcription of genes with DR1 in the absence or presence of a ligand (Mangelsdorf *et al.*, 1991; Kurokawa *et al.*, 1994). RXRs appear to play two different roles. They act as cofactors of RARs for transactivation when binding to DR2 or DR5 response elements and in the presence of all-*trans* RA, or for transrepression when binding to a DR1 response element. RXRs also behave as an independent transcription factor in the presence of 9-*cis* RA. The availability of 9-*cis* RA appears to be the governing factor of the RXR signaling pathway. RXRs, however, have much weaker DNA-binding and transactivation activities than RAR heterodimers. This result may imply that the main physiologic role of RXRs is as a partner for other nuclear receptors (reviewed in Edwards, 1999).

RXR as a partner in other hormone response systems
In addition to RARs, RXRs are also the partners of TR, VDR and other receptors, such as COUP-TFs, FXR, LXR and PPAR (reviewed in Chambon, 1996; Willy and Mangelsdorf, 1998). USP, the insect RXR homolog, has been shown to heterodimerize with the ecdysteroid receptor (EcR) and DHR38, the *Drosophila* homolog of NGF-IB. USP also has the ability to dimerize with some mammalian RXR partners, such as TRα, TRβ, RARα, VDR and PPAR (Christianson *et al.*, 1992; Yao *et al.*, 1992; Sutherland *et al.*, 1995). These results indicate that the function of RXRs as partners of other receptors is conserved evolutionarily.

9-*cis* RA has been identified as the activating ligand for RXR when it forms homodimers (Mangelsdorf *et al.*, 1992). The next question is whether 9-*cis* RA or other RXR ligands participate in the signaling pathways of its dimer partners. At least three mechanisms of action are proposed for the vertebrate RXRs: first, RXRs are unable to bind their ligands and act as silent partners; second, RXR ligands induce RXR homodimerization and downregulate the transactivation of RXR heterodimers; third, RXR ligands participate in the signaling pathway of their dimerization partners.

In the first proposed mechanism of action, RXR fails to bind its ligand and acts as a silent partner when it forms a heterodimer with RAR, TR and VDR. The pairing of RXR in these dimers stabilizes DNA binding of the heterodimer (Forman *et al.*, 1995). However, addition of 9-*cis* RA does not show any effect on RXR–RAR or RXR–TR heterodimer formation either free in solution or bound to HREs (Kurokawa *et al.*, 1994; Forman *et al.*, 1995; Kersten *et al.*, 1996). Introduction of an RXR-specific ligand also fails to transactivate the response genes of all-*trans* RA, thyroid hormone and vitamin D_3 (Boehm *et al.*, 1994). These results suggest that RAR and TR are able physically to block RXR from binding to its own ligand upon dimerization (Forman *et al.*, 1995). Unliganded TR or RAR recruit corepressors and the presence of these corepressors on RXR heterodimers may suppress the RXR-dependent transactivation (Chen and Evans, 1995).

The second proposed mechanism of action occurs when the presence of 9-*cis* RA induces RXR homodimer formation (Pfahl *et al.*, 1994). RXR homodimers regulate RXR-responsive gene activation and may sequester RXR, preventing formation of heterodimers (Zhang *et al.*, 1992; Lehmann *et al.*, 1993). In the presence of 9-*cis* RA and the absence of thyroid hormone or vitamin D_3, many fewer RXR–TR–DNA or RXR–VDR–DNA complexes can be detected and the affinity of this interaction is much lower (MacDonald *et al.*, 1993; Cheskis and Freedman, 1994). This result implies that RXR homodimerization induced by 9-*cis* RA is able to limit RXR–TR or RXR–VDR

complex formation by reducing the concentration of available RXR monomers. Therefore the intracellular concentration of RXR may be a limiting factor for the expression
of TR- or VDR-regulated genes (Zhang *et al.*, 1992; Lehmann *et al.*, 1993; Cheskis and
Freedman, 1994).

The third and final mechanism of action proposes that RXR is able to bind to its
ligand and acts as a transcriptionally active receptor when it binds to permissive
partners. Addition of RXR-specific ligands appears to enhance transcriptional activation as well as physiologic effects of liganded RXR heterodimer partners, such as
RAR, TR or VDR, in the presence of the respective partner's ligand (Li *et al.*, 1997a;
Minucci *et al.*, 1997; Puzianowska-Kuznicka *et al.*, 1997). The RXR ligands can bind
to RXR after RXR's partner binds its ligand (Westin *et al.*, 1998). These results
suggest a model in which RXR functions as a transcriptionally active receptor/partner
alone or with other nuclear receptors in either a ligand-dependent or ligand-
independent manner.

Recent results from the study of RXR–LXR, RXR–NGF-IB and RXR–PPAR
heterodimers further supports this model (Mukherjee *et al.*, 1997). These heterodimers
permit RXR ligands to transactivate reporter genes that contain the cognate HREs of
RXR–LXR, RXR–NGF-IB and RXR–PPAR heterodimers (Bardot *et al.*, 1993;
Perlmann and Jansson, 1995; Willy *et al.*, 1995). For RXR–LXR and RXR–PPAR
heterodimers, ligands of both partners can induce transactivation (Kliewer *et al.*, 1992c;
Willy and Mangelsdorf, 1997). The ligands of LXR and PPAR alone can activate
transcription, as well as ligands of RXR alone. This is the first example of nuclear
receptors where both ligands of the heterodimer partners can independently activate
transcription. Part of the reason that PPAR permits RXR ligands to activate transcription may be due to the interaction with cofactors (see Interaction of retinoid receptors
with cofactors below).

A recent study has shown that, in the RXR–PPAR heterodimer, the binding of ligands
to both receptors can synergistically activate transcription (Schulman *et al.*, 1998). In
addition, the AF-2 domain is required for PPARγ to respond synergistically to
hormone-dependent transactivation but the AF-2 domain of RXR is not required.
After RXR is bound by its ligands, the conformational changes of RXR may be
responsible for this phenomenon.

Interaction of retinoid receptors with cofactors

As mentioned above, the RAR–RXR heterodimer can bind to a DR1 response element
and repress gene expression, but RXR–RAR can activate gene transcription if bound to
a DR2 or DR5 RARE. Both transrepression and transactivation properties have
recently been shown to involve many cofactors (reviewed in Chen and Li, 1998; Torchia
et al., 1998). This recent research suggests that members of the nuclear receptor
superfamily switch, in a ligand-dependent manner, between binding of a multisubunit
corepressor complex containing factors with histone deacetyltransferase activity and
binding of a coactivator complex containing factors with histone acetyltransferase
(HAT) activity (reviewed in Chen and Li, 1998; Torchia *et al.*, 1998; Xu *et al.*, 1999).
These cofactors have a unique motif (LXXLL), called the NR box, which binds to
receptors. Some cofactors have an enzymatic activity, either histone acetylation or
deacetylation. Here we discuss the findings about cofactors related to the retinoid
signaling pathway.

Corepressors

Nuclear receptor corepressor (NCoR) was the first reported corepressor (Horlein *et al.*, 1995). It is a 270-kDa protein. Through its CoR box sequence, NCoR binds to unliganded RARα, either in solution or bound to DNA, but not to RXR (Horlein *et al.*, 1995; Kurokawa *et al.*, 1995). The presence of RAR agonists can inhibit the binding of NCoR to RARα when RXR–RAR heterodimers are bound to the DR5 sequence (Kurokawa *et al.*, 1995).

Another corepressor, silencing mediator of the RAR and TR (SMRT), was first identified with a yeast two-hybrid assay through its functional association with the hRXRα LBD (Chen and Evans, 1995). However, this 169-kDa protein (SMRT) interacts more strongly with hRARα and TR than with hRXRα. As with NCoR, the association of SMRT with RAR is attenuated in the presence of RAR ligands, either free in solution or bound to DNA (Chen and Evans, 1995; Lala *et al.*, 1996).

Unliganded RARs bind to NCoR and SMRT to form a corepressor complex to transrepress specific genes. NCoR and SMRT in turn interact with switch-independent-3 (Sin-3) and reduced potassium dependency-3 (RPD-3), the latter displaying histone deacetylase activity (Chen and Evans, 1995). In addition, the core complex contains other stably associated polypeptides, including the histone deacetylases HDAC-1 and HDAC-2 (Hassig *et al.*, 1997). Ligand binding to RAR results in the displacement of the NCoR–Sin-3–RPD-3 repressor complex, by a coactivator complex with multiple potential HAT proteins, which catalyze histone acetylation (Torchia *et al.*, 1997; Hong *et al.*, 1998).

Additional corepressors have been identified (reviewed in Torchia *et al.*, 1998; Xu *et al.*, 1999). Some of these corepressors also interact with RAR and RXR, such as T$_3$ receptor-associated cofactor 1 (TRAC-1) and thyroid receptor uncoupling protein (TRUP) (Burris *et al.*, 1995; Hollenberg *et al.*, 1996; Sande and Privalsky, 1996). TRAC-1 was shown to function in a fashion similar to NCoR and SMRT, while TRUP is able to inhibit both DNA- and ligand binding by RAR, but not RXR (Burris *et al.*, 1995).

As mentioned in the previous section, PPARγ permits RXR to bind to its ligand and activate the PPARγ signaling pathway. This feature may be due to the weaker binding of the PPARγ–RXR heterodimer to NCoR and SMRT compared with either RAR or TR (Zamir *et al.*, 1997). PPARγ can bind NCoR and SMRT in solution, but when bound to the naturally occurring PPRE in the acetylcoenzyme A oxidase gene, PPARγ cannot bind NCoR and can bind only weakly to SMRT (Zamir *et al.*, 1997). Thus, RXR ligands binding to RXR in the RXR–PPAR heterodimer may activate transcription of responsive genes.

Coactivators

Since the transcriptional activation of nuclear receptors depends on both ligand binding and the existence of the AF-2 core, these findings suggest that a common protein(s) may be present to mediate the transactivation of nuclear receptors (Nagpal *et al.*, 1992b; Tzukerman *et al.*, 1994). Several research groups independently isolated a novel family of AF-2-interacting proteins, called coactivators.

Steroid receptor coactivator 1 (SRC-1)/nuclear receptor coactivator (NCoA) was the first coactivator identified. It is a 160-kDa protein (p160) that interacts directly with nuclear receptors in an agonist and receptor AF-2 domain-dependent manner (Onate *et al.*, 1995). So far, there are three distinct but related p160 family members which are encoded by three different genes: (1) SRC-1/NCoA-1; (2) SRC-2 – transcription intermediary factor-2 (TIF-2)/glucocorticoid receptor interacting protein-1 (GRIP-1)/ NCoA-2; and (3) SRC-3 – p300/CBP interacting protein (p/CIP)/activator of thyroid and retinoic acid receptor (ACTR)/amplified in breast cancer-1 (AIB-1)/receptor associated coactivator-3 (RAC-3)/thyroid receptor activator molecule-1 (TRAM-1) (Onate *et al.*, 1995; Voegel *et al.*, 1996; Anzick *et al.*, 1997; Hong *et al.*, 1997a; Torchia *et al.*, 1997). All three SRC family members have been shown to potentiate the transcriptional activity of several nuclear hormone receptors, including retinoid receptors (reviewed in Chen and Li, 1998; Torchia *et al.*, 1998). Unlike SRC-1 and SRC-2, SRC-3 has a different specificity. It cannot coactivate RAR-mediated transcriptional activation, but coactivates other nuclear receptors and other activators, such as signal transducer and activator of transcription 5 (STAT-5) and the cyclic adenosine monophosphate (cAMP) response element-binding protein (CREB) (Li and Chen, 1998).

The nuclear receptor interaction domain of the coactivators contains multiple copies of a highly conserved motif that shares a consensus amino acid sequence, LXXLL, where X is any amino acid (Heery *et al.*, 1997; Ding *et al.*, 1998). Similar motifs have also been found in other coactivators or coregulators, such as CBP, TIF-1 and receptor interacting protein-140 (RIP-140) (Torchia *et al.*, 1997). Since a comparable LXXLL motif is present in the AF-2 domain of nuclear receptors and mutational analysis has shown it to be functionally important, it is conceivable that the LXXLL motif provides the critical interactive surface to coordinate the entire nuclear receptor–coactivator complex (Takeshita *et al.*, 1996; Chen *et al.*, 1997a; Jeyakumar *et al.*, 1997; Li *et al.*, 1997b).

The results of recent X-ray crystallography analyses of the co-crystal of ligand-bound PPARγ LBD (complexed with a region of SRC-1/NCoA-1), also showed that two LXXLL motifs of SRC-1 bind to the AF-2 region of the PPARγ LBD (Nolte *et al.*, 1998). These results are consistent with biochemical findings of the interactions between SRC-1 and RXR–RAR or RXR–PPARγ heterodimers (Zhu *et al.*, 1996; Torchia *et al.*, 1997).

Conformational changes in the AF-2 domain can control the binding of coactivators and ligands. As mentioned in a previous section, unliganded RAR allosterically restricts RXR in RXR–RAR heterodimers from binding to its ligand *in vitro* (Kurokawa *et al.*, 1994; Forman *et al.*, 1995). However, RXR ligands can potentiate the transcriptional effects of RAR ligands in the cells (Chen *et al.*, 1996). It was suggested that the RXR AF-2 domain binds to the LXXLL-binding site of the unliganded RAR. A recent study has demonstrated that the RXR AF-2 domain contributes to this allosteric inhibition of RXR (Westin *et al.*, 1998). In the absence of a RAR ligand, the RXR AF-2 domain binds to the coactivator-binding site of unliganded RAR and this conformation prevents RXR from binding to its own ligand. The addition of a RAR-specific ligand enhances the recruitment of SRC-1 to RAR and displaces the RXR AF-2 domain. As a result, RXR can bind to a RXR ligand when its AF-2 domain becomes available.

Another nuclear protein, p300/CBP associated factor (p/CAF) is suggested to be a component of the coactivator complex. p/CAF was the first mammalian homolog of the yeast GCN5, a HAT protein (Yang *et al.*, 1996). Structurally, p/CAF contains a unique N-terminal domain and a C-terminal region that contains a HAT domain (Yang *et al.*, 1996). Its C-terminal domain can potentially interact with components of the coactivator complex, including CBP and SRC-1/NCoA-1, as well as the nuclear receptors such as RAR (Korzus *et al.*, 1998). The HAT activities allow coactivator complexes to relieve histone/chromatin-mediated transcriptional repression.

Cointegrators

In addition to coactivators and corepressors, other factors appear to participate in these complexes, such as CREB-binding protein (CBP) and p300. These two proteins are termed cointegrators because they are common components of coactivator complexes for several families of sequence specific transcriptional activators (Mannervik et al., 1999). CBP and p300 are closely related proteins and share similar activities (Lunblad et al., 1995). In a yeast two-hybrid system, CBP associates with the LBDs of both RAR and RXR in a ligand-dependent manner (Kamei et al., 1996). Specific amino acid mutations in the LBD, which cause the receptors to lose transcriptional activity, are able to inhibit the interaction with CBP, leading to the proposal that nuclear receptors activate transcription via complexes with CBP/p300 and other cofactors.

The study of p300 null mutant mice demonstrates that loss of p300 severely affects RA-dependent transcription (Yao et al., 1998). Reduction of either endogenous CBP or p300 messenger RNA (mRNA) by ribozymes alters the expression of RA-inducible genes (Kawasaki et al., 1998). These results suggest that CBP or p300 plays a crucial role in the retinoid signaling pathway.

GENETICS

In order to understand the physiologic functions of both RARs and RXRs in vivo, genetic analysis has been carried out. Most of the genetic analyses on the retinoid receptors have been performed by using transgenic mice and teratocarcinoma cells as model systems. Dominant-negative approaches have also been used. The latter two methodologies have been discussed in other papers (Durston et al., 1998, Morriss-Kay and Ward, 1999).

Gene-targeting strategies via homologous recombination in embryonic stem cells have been used to produce many different RAR and RXR knockout mouse models. Single, double and even triple mutants have been generated. The details of the phenotypes generated by these knockouts are complex and have been described previously (Lohnes et al., 1993, 1995; Sucov et al., 1994; Grononda et al., 1996; Kastner et al., 1996; Krezel et al., 1996). The results from these studies reveal a spectrum of developmental defects with considerable functional redundancy among the different RAR subtypes. A comprehensive review on this subject is beyond the scope of this chapter, but has been published by Kastner et al. (1995). This section focuses on the recent findings concerning this topic.

Most of the single subtype knockouts are viable and have less significant phenotypes (Lohnes et al., 1994; Mendelsohn et al., 1994). For instance, RARγ knockout mice have minor skeletal malformations (Lohnes et al., 1993). The significant aspect of the phenotype of RARγ homozygous mutants is resistance to the teratogenic effects of RA on vertebral column development (Lohnes et al., 1993). RA can exert teratogenic effects on the craniofacial and limb development of these embryos because the loss of function of RARγ may have been compensated by either RARα or RARβ. These results suggest that RARγ plays an important role in normal vertebral development.

Double knockouts, however, show more severe phenotypes (Lohnes et al., 1994; Mendelsohn et al., 1994; Kastner et al., 1997a). The RARαγ double mutants show rhombencephalic defects (Lohnes et al., 1994). The developmental defects generated include axial transformations. Some of the RAR knockout mice have phenotypes resembling those generated by knocking out *Hox* genes, which are suspected targets of retinoid signaling (Durston *et al.*, 1998). Other defects belong to the fetal vitamin A deficiency syndromes (Lohnes *et al.*, 1993, 1994; Lufkin *et al.*, 1993).

Single knockout mice of RXRβ and RXRγ are viable and do not show defects related to the vitamin A deficiency syndrome (Durston *et al.*, 1998). RXRβ was shown to be necessary for spermatogenesis (Kastner *et al.*, 1994, 1995; Sucov *et al.*, 1994). A recent study has shown that RARβ and RXRγ may play a role in hippocampal long-term potentiation (LTP) and long-term depression (LTD), which may be related to spatial learning and memory (Chiang *et al.*, 1998). In RXRγ null mutant mice, the expression of choline acetyltransferase in the cholinergic interneurons has recently been shown to be downregulated (Saga *et al.*, 1999). These results suggest that RXRγ may play an important role in the proper functioning of neurons.

RXRα has been shown to be involved in cardiac and liver organogenesis, and homozygous mutant fetuses die from embryonic days 12.5–16.5 (Kastner *et al.*, 1994; Sucov *et al.*, 1994). The defects in these fetuses resemble those with the fetal vitamin A deficiency syndromes, including myocardial hypoplasia, conotruncal and ocular defects. In addition, the lethality of RXRα null mutant fetuses may be due to cardiac, hepatic and placental defects (Kastner *et al.*, 1997b; Chen *et al.*, 1998; Trans and Sucov, 1998). More indepth investigations have been performed by inactivation of either RXRβ or RXRγ in a RXRα null mutant mouse. RXR$\alpha\gamma$ double mutants do not show any significant abnormalities while RXR$\alpha\beta$ double mutant fetuses die earlier than the RXRα null mutant (Krezel *et al.*, 1996; Kastner *et al.*, 1997a). In addition, a synergistic effect was observed in RXRα–RAR mutants but no synergy was observed between the effects of mutations of either RXRβ or RXRγ mutations and those of RAR mutations (Kastner *et al.*, 1997a). These results suggest that RXRα–RAR heterodimers are the most common functional unit in the RA signaling pathway during embryogenesis.

A recent study has shown that the failure to form the chorioallantoic placenta is responsible for the death of RXR$\alpha\beta$ double mutant fetuses (Wendling *et al.*, 1999). It was also suggested that heterodimerization of RARs and PPARγ with RXRs is involved in normal early embryogenesis and placentogenesis (Wendling *et al.*, 1999). Surprisingly, RXR$\alpha^{+/-}$/RXR$\beta^{-/-}$/RXR$\gamma^{-/-}$ triple mutants are viable although their body size is smaller than wild-type mice (Krezel *et al.*, 1996).

In early studies, ectopic addition of RA has been shown to affect normal embryonic development (Maden, 1982; Durston *et al.*, 1989). A RA gradient has been shown to exist along the anterior–posterior axis of the embryo (Maden *et al.*, 1998). The findings from a recent knockout of the retinaldehyde dehydrogenase 2 (*Raldh2*) gene have demonstrated that this enzyme is essential for mouse normal embryonic survival and early morphogenesis (Niederreither *et al.*, 1999). Raldh2 is an NAD-dependent aldehyde dehydrogenase and may be responsible for embryonic RA synthesis (Wang *et al.*, 1996; Niederreither *et al.*, 1997). The absence of Raldh2 blocks proper RA synthesis in embryos. Subsequently, its loss causes the loss of axial rotation, incomplete neural tube closure, lack of heart looping and death of the embryo.

The underlying mechanism of these phenotypes is still unknown. It may be due to a deficient response to RXR ligands or to a disruption of one or more heterodimer complexes (reviewed in Forman, 1998). Investigation of the physiologic functions of retinoid receptors using knockout techniques seems to be even more complicated because of the interactions between different receptor subtypes and isoforms and their ligands, retinoids.

Genetic analysis has also been used to dissect the function of the AF-2 domain in RXR (Mascrez *et al.*, 1998). Targeted deletion of the AF-2 domain of RXRα

(RXRαAF2o) is embryonic lethal and the null mutant embryos exhibit some abnorm-alities that have been shown in RXRα null mutant embryos (Kastner *et al.*, 1994; Sucov *et al.*, 1994). Double mutants of RXRαAF2o and RARα, β or γ show defects similar to those that are observed in the corresponding RXRα–RAR double mutants. These findings indicate that an intact AF-2 domain plays an important role in mediating retinoid signaling during embryonic development.

PATHOPHYSIOLOGY

Retinoid receptors are the mediators of retinoid signaling pathways. Therefore, it is plausible to assume that changes in their expression and function may cause aberrations in the response of cells to retinoids and thereby may alter the regulation of cell growth, differentiation and expression of RA-responsive genes. Investigation of the expression pattern of retinoid receptors in normal, premalignant and malignant tissues has provided clues as to the roles of these receptors in cancer development and responses of these tissues to retinoid treatments.

The identification of the RARβ gene as the integration site of a hepatitis B virus from a human hepatocellular carcinoma was the first indirect evidence linking a mutation in a RAR gene to a disease state (Dejean *et al.*, 1986; Brand *et al.*, 1988). Furthermore, the RARβ gene is frequently deleted in various cancer cells and RARβ shows tumor suppressive activity in epidermoid cancer cells (Kok *et al.*, 1987). These two findings implicate RARβ in tumor formation. Definitive proof that disruption of the retinoid signaling pathway can lead to carcinogenesis was provided by the discovery that the t(15:17) translocation breakpoint of acute promyelocytic leukemia (APL) lies within the RARα locus (Borrow *et al.*, 1990; Alcalay *et al.*, 1991). Here we discuss several major examples of retinoid receptor involvement in different cancers and diseases.

Acute promyelocytic leukemia (APL)

APL is the most common and potentially curable subtype of acute myeloid leukemia (AML). AML comprises a group of acute malignancies of bone marrow progenitor cells, which give rise to several different types of leukocytes (reviewed in Norum, 1994). All hematopoietic cells are derived from common multipotent stem cells (Clark and Keating, 1995; Lenny *et al.*, 1997). The production of functional cells in the bloodstream is the result of several regulated processes involving cell division and differentiation, under the control of both soluble and adhesive factors. Leukemia occurs when the hematopoietic cells proliferate continuously. These cells do not have the capacity to differentiate into mature blood cells and are unable to undergo apoptosis, as in the normal condition (Olsson *et al.*, 1996; Grimwade and Solomon, 1997). APL is thus characterized by a potentially devastating disorder that can lead to an accumulation of promyelocytes, which are hypergranulated immature granulocytes (Grimwade and Solomon, 1997). All-*trans* and 9-*cis* RAs have been shown to release the block at the promyelocyte stage of normal myeloid differentiation (Castaigne *et al.*, 1990; Elliott *et al.*, 1992; Sakashita *et al.*, 1993). Clinical trials have demonstrated that retinoids can achieve complete remission in most patients with APL (Huang *et al.*, 1988; Chomienne *et al.*, 1990).

At the molecular level, APL involves a chromosomal translocation and subsequent expression of fusion proteins between several proteins and RARα (Grimwade *et al.*,

1996; Hong *et al.*, 1997b; Rowley *et al.*, 1997). These fusion proteins include promyelocytic leukemia protein (PML), PML zinc-finger protein (PLZF), nucleophosmin (NPM) and nuclear mitotic apparatus protein (NuMA) (Table 7.3) (Redner *et al.*, 1996; Wells *et al.*, 1996). Structurally, these four proteins are not related but they all are nuclear proteins with a homodimerization domain (de The *et al.*, 1991; Kakizuka *et al.*, 1991; Chen *et al.*, 1993a,b; Wells *et al.*, 1997). In the majority of cases of APL, a reciprocal translocation has occurred between chromosome 15 and 17, t(15:17)(q22;q21) (Borrow *et al.*, 1990; de The *et al.*, 1990b; Alcalay *et al.*, 1991). The breakpoint on chromosome 17 lies within the RARα locus while the breakpoint on chromosome 15 lies within the PML locus. This translocation produces two fusion proteins, PML–RARα and RARα–PML. In some cases, the translocation occurs between chromosomes 17 and 11, t(11,17) q(23,21) where the PLZF gene is located (Chen *et al.*, 1993a,b). The other two translocations with NPM and NuMA are rare. In most cases, the fusion protein that contains the DBD and LBD sequences of RARα and the N-terminus of its translocation partner is expressed at higher levels in promyelocytes than the product from the reciprocal translocation (de The *et al.*, 1991; Chen *et al.*, 1993a; Redner *et al.*, 1996; Wells *et al.*, 1997).

PML–RARα and PLZF–RARα are found to be responsible for most of the molecular actions of APL (reviewed in Lin *et al.*, 1999; Slack and Gallagher, 1999; Stunnenberg *et al.*, 1999). Comparable to wild-type RARα, they can bind to both RA and RAREs as either homodimers or heterodimers with RXR (Dong *et al.*, 1996; Benedetti *et al.*, 1997). These findings suggest that PML–RARα and PLZF–RARα may interfere with the normal retinoid signaling pathway.

Like natural RARs in the absence of RA, both PML–RARα and PLZF–RARα interact with corepressor complexes and repress transcription of reporter genes in several cell lines (Fig. 7.1) (Dong *et al.*, 1996; Ruthardt *et al.*, 1997). At low or physiologic RA concentrations (10^{-9} to 10^{-8} mol L^{-1}), both fusion proteins continue to function as constitutive transcriptional repressors bound to RAREs. The addition of all-*trans* RA at high doses (10^{-7} to 10^{-6} mol L^{-1}) causes the PML–RARα to dissociate from the corepressor complex and activate the transcription of the reporter genes (Hong *et al.*, 1997b; Grignani *et al.*, 1998; He *et al.*, 1998). By contrast, PLZF–RARα interacts constitutively with the corepressor in the presence of RA because the PLZF has its own corepressor-binding site (Hong *et al.*, 1997b; Grignani *et al.*, 1998; He *et al.*, 1998). Although the corepressor has been released from the RAR moiety of PLZF–RARα in the presence of high doses of RA, the corepressor complex still exists in the PLZF moiety and this complex continues to repress the transactivation of responsive genes. Addition of specific inhibitors of histone deacetylase, such as trichostatin A (TSA), restores RA responsiveness of both PML–RARα and PLZF–RARα and allows leukemic cells to differentiate in response to all-*trans* RA (Grignani *et al.*, 1998; He *et al.*, 1998; Lin *et al.*, 1998). These results could provide a potential mechanistic pathway and explain why patients with PML–RARα APL achieve complete remission following RA treatment, while those with PLZF–RARα APL respond poorly to treatment because the binding of corepressor to the PLZF moiety of PLZF–RARα represses its transactivation ability (Licht *et al.*, 1995).

Upper aerodigestive tract cancer

Upper aerodigestive tract cancer mainly consists of lung cancer and head and neck cancers. Epidemiologic studies have shown that vitamin A deficiency is associated with

Table 7.3   Translocations in acute promyelocytic leukemia (APL)

Gene	Trans-location	Fusion proteins found	Frequency in APL*	Response to all-*trans* RA*	Response to inhibitors of histone deacetyltransferase
PML	t(15;17) (q22;q21)	PML–RARα RARα–PML	95	Yes	Yes
PLZF	t(11;17) (q23;q21)	PLZF–RARα RARα–PLZF	< 5	No	Yes
NuMA	t(11;17) (q35;q21)	NuMA–RARα	< 1	Yes	–
NPM	t(11;17) (q13;q21)	NPM–RARα	< 1	Yes	–

*Lin *et al.* (1999). PML, promyelocytic leukemia protein; PLZF, PML zinc-finger protein; RAR, retinoic acid receptor; NuMA, nuclear mitotic apparatus protein; NPM, nucleophosmin.

increased incidence of lung cancer (Hong and Itri, 1994). In addition, vitamin A deficiency induces squamous metaplasia in the mucosa of the upper aerodigestive tract (Wolbach and Howe, 1925). Vitamin A supplementation can reverse squamous metaplasia in the trachea of vitamin A-deficient animals *in vivo* (Wolbach, 1956), and various retinoids exhibit similar activities *in vitro* (Sporn and Newton, 1979). In humans, retinoids prevent the development of new primary tumors in patients with head and neck cancer and those with lung cancer (Pastorino *et al.*, 1993; Lippman *et al.*, 1994b). The association between vitamin A status and cancer development in the upper airway and the ability of retinoids to suppress premalignant lesions suggests that physiologic levels of retinoids are required for natural cell growth and differentiation, and that the retinoid-dependent signaling pathway may play a role in suppression of carcinogenesis (Lotan, 1997).

Several human lung cancer cell lines, head and neck squamous cell carcinoma (HNSCC) cell lines, and normal buccal mucosa specimens express RARα, RARγ and RXRβ (Gebert *et al.*, 1991; Geradts *et al.*, 1993). The suppression of both gene expression and inducibility by RA of RARβ has also been shown in these cells. These results suggest that the selective suppression of RARβ expression may correlate with the malignant transformation of epithelial cells (Gebert *et al.*, 1991; Houle *et al.*, 1991; Geradts *et al.*, 1993).

RARs and RXRs can act as ligand-dependent transrepressors of activation protein-1 (AP-1) activity and, reciprocally, AP-1 can inhibit transactivation by RARs and RXRs (reviewed in Resche-Rigon and Gronemeyer, 1998). The reduction of RARβ may relieve antagonism of AP-1 transcriptional activity and thus increase the expression of c-*jun* and c-*fos* genes, which contain AP-1 sites in their promoter regions (reviewed in Pfahl *et al.*, 1994; Karin *et al.*, 1997). The increase of c-Fos and c-Jun proteins, as well as the AP-1 transcriptional activity, may enhance cell proliferation, which may, in turn, cause tumors.

In vivo studies have shown that the RARβ gene, which is located on a region of the short arm of the chromosome 3p24, is frequently deleted in lung cancer cells that show resistance to RA (Gebert *et al.*, 1991; Geradts *et al.*, 1993). However, rearrangements or deletions of the RARβ gene have not been detected in surgical specimens of lung cancer

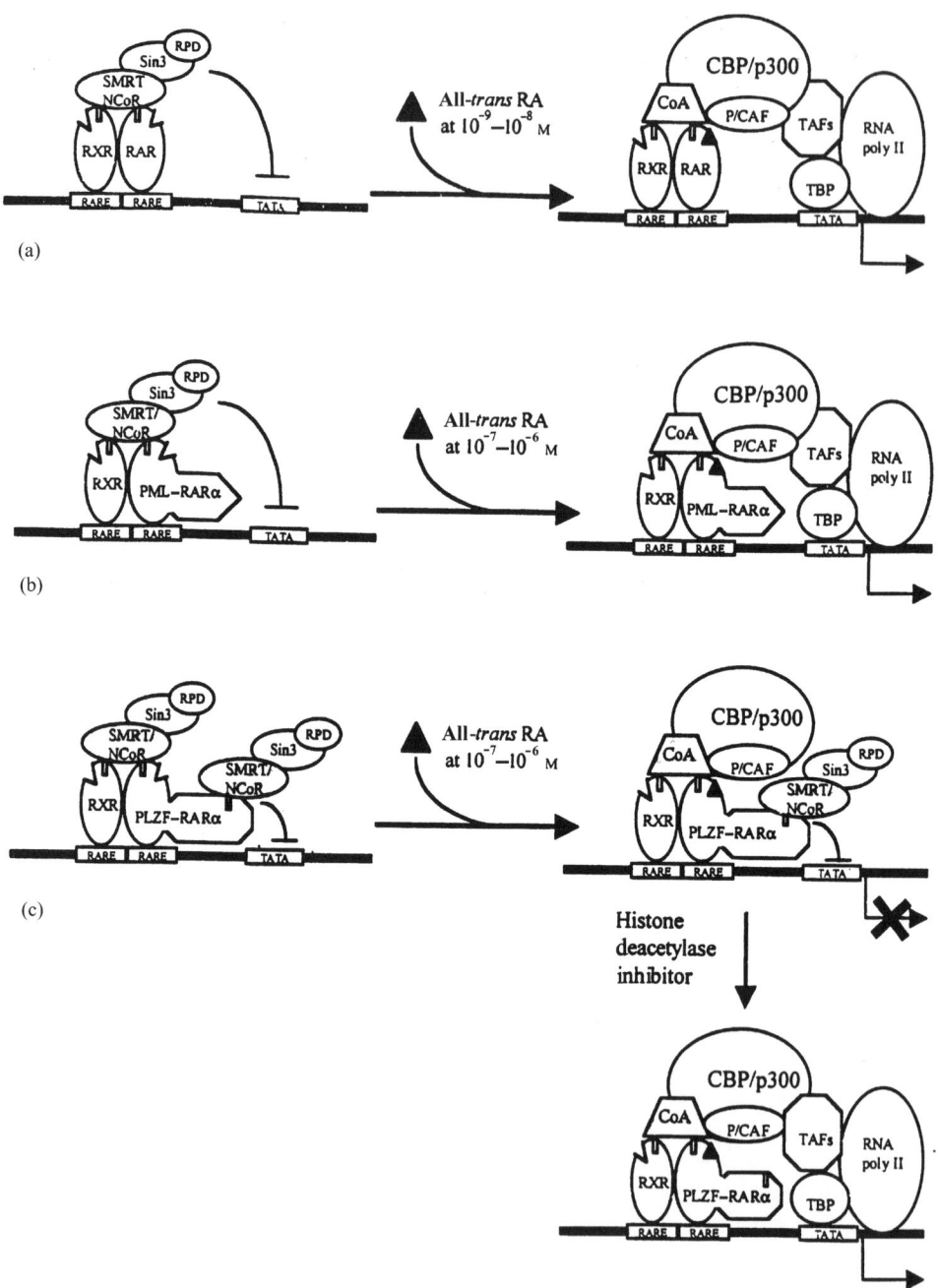

Fig. 7.1

(Gebert *et al.*, 1991). Thus, deletion of RARβ may be related to the upper aerodigestive cancers, although the underlying mechanism is still unknown.

Skin

RA controls epidermal growth and differentiation (Fisher and Voorhees, 1996). A deficiency or excess of RA can affect the keratinization process of the epidermis. Early studies showed that a vitamin A deficiency is associated with hyperkeratinization of the skin (reviewed in Vahlquist, 1994). Currently, *in vivo* studies support the idea that RA promotes keratinocyte proliferation while *in vitro*, depending on the cell culture conditions, RA has been shown to either stimulate or decrease epidermal keratinization (Fisher and Voorhees, 1996).

The expression of retinoid receptors in the human epidermis is specific. RARγ and RXRα are expressed abundantly in the human epidermis while the transcript level of RARα is low and no RARβ is detected in normal human epidermis (Elder *et al.*, 1992; Fisher *et al.*, 1994). Unlike other cell types, all-*trans* RA is unable to induce RARβ expression in keratinocytes (Elder *et al.*, 1991). In transient transfections, however, the isolated RARβ2 RARE confers induction by all-*trans* RA (Xiao *et al.*, 1995).

DR2 and DR5 are the principal RAREs used for the transactivation of RA-responsive genes in keratinocytes. In gel shift studies, RAREs with DR1, DR2 and DR5 orientations have been shown to bind the RARs and RXRs in nuclear extracts from both human skin and cultured keratinocytes (Kurokawa *et al.*, 1994; Fisher *et al.*, 1995; Xiao *et al.*, 1995). By transient transfection in keratinocytes, all-*trans* RA, 9-*cis* RA and an RAR-specific ligand, CD367, can activate reporters with RAREs of either DR2 or DR5 configuration. The RXR-specific ligand SR11237, however, is unable to activate a reporter with the DR1 RARE (Xiao *et al.*, 1995). Thus, in human skin and keratinocytes, liganded RXR–RAR heterodimers may be the functional unit to activate the RAREs through DR2 and DR5 elements, and the unliganded RXR–RAR hetero-dimer may repress the DR1 RARE. Target genes in the skin are CRBPs and CRABPs, since these genes are induced by all-*trans* RA (Astrom *et al.*, 1991; Fisher *et al.*, 1995).

Fig. 7.1 Hormonal response of wild-type RAR and fusion proteins in acute promyelocytic leukemia. (a) In the absence of ligand, the RXR–RAR heterodimer associates with Sin3/RPD histone deacetyltransferase complex through interactions with SMRT or NCoR. In the physiologic concentration of the ligand, the binding of RA releases the corepressor complex and allows RXR–RAR to recruit the coactivator, p/CAF and CBP/p300 histone acetyltransferases, resulting in transcriptional activation. (b) In the absence of ligand, the PML–RARα fusion protein represses the transcription through the corepressor complex. Administration of pharmacologic doses of RA is sufficient to release the repressor complex. The recruitment of coactivator complex allows the transcription to occur. (c) The PLZF–RARα fusion protein binds to two repressor complexes, one of which is RA sensitive. Addition of pharmacologic doses of RA is able only to release the corepressor complex which binds to RAR moiety and has no effect to the one bound to PLZF. Coadministration of RA and TSA, a histone deacetyltransferase inhibitor, is sufficient to induce transactivation. This induction may involve the release of repressor complex from PLZF moiety.

The human CRABP-II gene contains a DR5 RARE, while the rat CRBP has a DR2 type RARE (Astrom *et al.*, 1994; Fisher *et al.*, 1995).

Several reports correlate the alteration of RARα and RARγ expression with skin tumorigenesis (Kumar *et al.*, 1994; Darwiche *et al.*, 1995, 1996). A reduction of RARα and RARγ transcripts occurs in benign skin tumors in mice (Darwiche *et al.*, 1995, 1996). In human undifferentiated squamous cell carcinomas, these two transcripts were absent (Lotan *et al.*, 1995). A recent report showed that ultraviolet (UV) irradiation substantially decreased both RARγ and RXRα transcripts and proteins in human skin *in vivo* (Wang *et al.*, 1999). In addition, UV irradiation diminished almost completely the transactivation of two RAR-responsive genes, CRBP-II and RA 4-hydroxylase in human skin *in vivo*. Chronic UV irradiation is known to promote skin cancer in humans (Campbell *et al.*, 1993). Upregulation of AP-1 activity and transactivation of AP-1-responsive genes by UV radiation may be the result of RARγ and RXRα gene expression in skin (Fisher *et al.*, 1997, 1998).

Breast cancer

Breast cancer is currently the second leading cause of death in women. RARs and RXRs are present in normal and malignant breast tissues (Pasquali *et al.*, 1997). Several lines of evidence suggest that there is mutual regulation between steroids and retinoids in breast cancer cells and tissues. RARα is found to have higher expression in ERα-positive cell lines, while RARβ is expressed more in cancer tissues with low levels of estrogen and progesterone receptors (PRs), especially in ERα-negative cell lines (Roman *et al.*, 1992, 1993). RARα expression can be induced by estradiol in human breast cells but its expression is independent of the levels of ER and PR (Roman *et al.*, 1992, 1993).

In addition, low RARα expression has been suggested to be responsible for regulating the signaling switch from RAR- to RXR-mediated growth inhibition in breast cancer cells (Wu *et al.*, 1997a). On the other hand, RARβ may play an essential role in RA induction of growth and apoptosis in human breast cancer cells through antagonizing AP-1 effects on cytostatasis, G_0/G_1 arrest and apoptosis (Nagpal *et al.*, 1995; Shao *et al.*, 1995; Fanjul *et al.*, 1996).

Parkinson's disease

RA signaling plays a major role in the development of the fetal central nervous system (reviewed in Durston *et al.*, 1998). Class I aldehyde dehydrogenase is found to be expressed at high levels in dopaminergic neurons in the forebrain (McCaffery and Drager, 1994). This enzyme is important in retinoid metabolism and is responsible for oxidation of retinal to RA. In addition, RARα is uniformly expressed in the brain while RARβ and RXRγ occur specifically in the striated regions of the brain that express dopaminergic neurons (Ruberte *et al.*, 1993).

Studies of dopamine receptor 2 (*D2R*) knockout mice suggest that the retinoid signaling may be involved in Parkinson's disease (Saiardi *et al.*, 1997). These knockout mice show a parkinsonian-like locomotor defect and suffer from pituitary tumors. A RARE has been found in the enhancer region of the *D2R* gene (Valdenaire *et al.*, 1994). Both all-*trans* RA and 9-*cis* RA can induce the expression of D2R *in vitro* and increase the activities of a D2R reporter (Valdenaire *et al.*, 1994). Gel shift assays show that the RXR–RAR heterodimer can bind specifically to the *D2R* RARE (Samad *et al.*, 1997).

Using the RAR and/or RXR single or double knockout mice, it was demonstrated that the expression of the *D1R* and *D2R* genes are reduced in the RARβ and RXRγ single knockout mice, as well as in the RXRβ/RXRγ, RARα/RXRγ, RARβ/RXRβ double knockout mice (Samad *et al.*, 1997; Krezel *et al.*, 1998). Taken together, liganded retinoid receptors may be necessary for the normal expression of the *D2R* gene (Wolf, 1998).

A study of the orphan receptor, Nurr1, also suggests that retinoids may play a role in Parkinson's disease. Nurr1 is expressed in the mesencephalic flexure of the fetal brain where the dopamine-producing neurons finally develop (Zetterstrom *et al.*, 1997). These neurons contribute to movement coordination and their loss causes Parkinson's disease. In Nurr1 knockout mice, dopamine-producing neurons are missing only in the midbrain region (Zetterstrom *et al.*, 1997). The absence of these neurons suggests that Nurr1 is required for specification and/or differentiation of dopamine-producing neurons in the midbrain. The reduction of dopamine production may contribute to Parkinson's disease. Functionally Nurr1 receptors heterodimerize with RXRs, and the RXR–Nurr1 heterodimer binds 9-*cis* RA (Forman *et al.*, 1995; Perlmann and Jansson, 1995). These observations suggest that RXR-specific ligands may be capable of adjusting dopamine production, which may be beneficial to parkinsonian patients (Eichele, 1997).

DIAGNOSIS

Techniques involved in the research of retinoid receptors have been used in clinical diagnosis. They include Northern blot analysis, *in situ* hybridization, and reverse transcriptase–polymerase chain reaction (RT-PCR). Northern blot analysis can determine the receptor transcript levels in tissues or cell lines. This method has been used to determine the transcript concentrations of different subtypes or isoforms of retinoid receptors in different cancer cells (Jing *et al.*, 1996; Lotan, 1997). *In situ* hybridization is widely used in diagnosis and serves to visualize nucleic acids, both DNA and RNA, in particular mRNA coding for specific proteins. The technique of *in situ* hybridization can detect the expression of specific genes in cells. Compared with other hybridization methods, such as Northern blot analysis, the most important advantage of *in situ* hybridization is the accurate visualization of specific gene transcripts within specific tissues and individual cells. *In situ* hybridization is used widely to examine the presence of retinoid receptors in surgically removed tumor tissues (Lotan, 1997). The RT-PCR method is a powerful, sensitive and discriminating analytic tool for the detection and study of gene expression in clinical diagnosis, particularly with low-abundance mRNAs. This method has been used to determine the transcripts of retinoid receptors in many cell lines and tumor tissues (Alcalay *et al.*, 1996; Pasquali *et al.*, 1997).

In the diagnosis of APL and other cancers, conventional cytogenetics is routinely used to identify the translocations involved (Grimwade *et al.*, 1998). Unfortunately, this method sometimes fails to detect the presence of t(15;17), which causes the PML–RARα rearrangement. Fluorescence *in situ* hybridization (FISH) has been used to improve the diagnosis of APL (Chen *et al.*, 1994). The identification of structural abnormalities by using FISH cytogenetic studies has become an important tool in the diagnosis of APL and other diseases.

RT-PCR has also become a common and invaluable method to detect the presence of PML–RARα rearrangements in cases of APL (Borrow *et al.*, 1992; Grimwade *et al.*, 1996). Both FISH and RT-PCR are able to detect the presence of small translocations that are below the resolution of conventional cytogenetic methods. RT-PCR has the additional advantage of detecting a single APL cell among tens to hundreds of thousands of normal blood cells (Borrow *et al.*, 1992; Alcalay *et al.*, 1996). In clinical use, RT-PCR allows earlier identification of patients at risk of relapse and the evaluation of the relative efficacy of different treatments for different cancers. This method is also capable of determining the possible relationship between PML breakpoint patterns and prognostic information that may be applicable to future clinical trials (Grimwade *et al.*, 1998).

PHARMACOLOGY

As retinoids are able to interrupt proliferative processes and induce differentiation and apoptosis, they have been shown to be effective inhibitors of chemical carcinogenesis in epithelial tumors such as those of the skin, lung, breast, prostate, cervix, bladder and esophagus (Darwiche *et al.*, 1994; De Vos *et al.*, 1997). RA can also be utilized without any chronic storage in the liver (Huang *et al.*, 1988; Lee *et al.*, 1993b). The potential of retinoids for chemotherapeutic use has prompted several pharmaceutical and academic groups to generate RAR isotype-selective ligands or RXR-selective agonists and antagonists (Resche-Rigon and Gronemeyer, 1998).

Each RAR subtype has been implicated in the regulation of some sort of cancer development and the mediation of retinoid anticancer activities. Translocation of the RARα gene is responsible for the development of APL (Warrell *et al.*, 1993). RARγ may play a role in mediating growth inhibition and apoptosis by certain retinoids (Fanjul *et al.*, 1996). RARβ may play a role in the regulation of cancer cell growth. The RARβ gene is either deleted or not expressed in a variety of cancer cell lines (Kok *et al.*, 1987; Houle *et al.*, 1991; Liu *et al.*, 1996), and re-expression of RARβ in RARβ-negative cancer cells restores the ability of RA to induce growth inhibition and apoptosis (Liu *et al.*, 1996).

Retinoid receptors are involved in retinoid-induced apoptosis in normal and neoplastic tissues (reviewed in Nagy *et al.*, 1998). RARβ expression has been shown to correlate to the apoptosis of mesenchyme of the interdigital regions during mouse limb development. It is also necessary for RA-induced apoptosis of breast cancer and lung cancer cells (Liu *et al.*, 1996; Li *et al.*, 1998). Activation of RXR is required for RA-induced HL-60 cell apoptosis (Nagy *et al.*, 1995). Both RARs and RXRs are required for 9-*cis* RA-induced T-cell apoptosis (Szondy *et al.*, 1998). Two categories of retinoid-induced apoptosis have been proposed (Rosati *et al.*, 1998). The first category is the differentiation-dependent apoptosis in malignancies, with an immature phenotype, after exposure to natural retinoids. The second category is apoptosis without differentiation, induced primarily by synthetic retinoids (Kalemkerian and Ramnath, 1996). These findings support the idea that receptor-specific ligands could be used as therapeutic choices in cancer treatment.

Treatments with natural retinoids

Natural retinoids, such as all-*trans* RA, 9-*cis* RA and 13-*cis* RA, have been used for cancer treatment, particularly for APL, early lesions of head and neck cancer, squamous cell carcinoma of the cervix, and skin cancer (Warrell *et al.*, 1991; Lippman *et al.*,

1992a,b). Listed below are some examples of the uses of natural retinoids in the treatment of various cancers or cancer cell lines.

Uses of all-*trans* RA in APL

All-*trans* RA has been used as a therapeutic agent in APL because it leads to a higher complete remission rate (95%) than therapies using chemotherapy alone (60–70%) (Tallman *et al.*, 1997; Asou *et al.*, 1998). However, most patients who are maintained on all-*trans* RA treatment alone often relapse and have a RA-resistant syndrome within 1 year (Huang *et al.*, 1988; Warrell *et al.*, 1994).

Induction chemotherapy of APL with all-*trans* RA improves the disease-free interval compared with chemotherapy alone (Avvisati *et al.*, 1996; Tallman, 1998). APL has been shown to be highly sensitive to induction chemotherapy using all-*trans* RA with anthracyclines, daunorubicin and idarubicin (Avvisati *et al.*, 1996). The underlying mechanism is still unknown and this combined therapy is unable to resolve the problem of RA-resistant relapse.

To date, four approaches have been proposed to cope with RA resistance in patients with APL. First, combination treatments with all-*trans* RA and interferons (IFNs) may be able to control cell growth and differentiation of hematopoietic cells in leukemia (Gaboli *et al.*, 1998). Similar therapeutic approaches have been utilized in many neoplastic diseases (Zheng *et al.*, 1996; Lippman *et al.*, 1997). Promising results have been reported when giving a combination of 13-*cis* RA and IFNα to patients with advanced solid tumors (Lippman *et al.*, 1993; Moore *et al.*, 1994). Recently similar results have been demonstrated in breast cancer cell lines by using RARγ-selective retinoids and IFNγ (Widschwendter *et al.*, 1997). Type I and II IFNs can induce PML expression in many cell lines (Stadler *et al.*, 1995; Nason-Burchenal *et al.*, 1996). Since the promoter region of PML is maintained in PML–RARα, IFNs can stimulate PML–RARα expression (Peruzzi *et al.*, 1996; Seale *et al.*, 1996). This induction may contribute to the restoration of RA sensitivity of the patients (Koller *et al.*, 1991; Lazzarino *et al.*, 1995). RT-PCR can then help to detect minimal residual disease (MRD) in patients with APL (Peruzzi *et al.*, 1996; Seale *et al.*, 1996).

Second, treatment of patients with RA-resistant APL with arsenic trioxide has been shown to be effective (Chen *et al.*, 1997b; Shao *et al.*, 1998). PML is localized in nuclear bodies inside the nucleus, and the presence of PML–RARα disrupts these nuclear bodies. Both RA and arsenic trioxide treatment can restore the nuclear body reorganization, and arsenic trioxide can also induce apoptosis in NB4 cells (Chen *et al.*, 1997b; Shao *et al.*, 1998). It has been suggested that apoptosis may lead to the degradation of the PML–RARα (Chen *et al.*, 1997b).

Third, a combination therapy of 1,25-dihydroxyvitamin D_3 (1,25$(OH)_2D_3$) with RA exhibits cooperative effects on differentiation in established leukemia cell lines (James *et al.*, 1999; Makishima *et al.*, 1999). 1,25$(OH)_2D_3$ is a differentiation agent which directs monocytic maturation of normal and malignant hematopoietic cells and can potentiate apoptosis induced by RAs (Hewison *et al.*, 1996). 1,25$(OH)_2D_3$ has been shown to induce differentiation in RA-resistant APL cell lines, suggesting that this synergistic property can be applied to RA resistance in patients with APL (Muto *et al.*, 1999).

Finally, inhibitors of histone deacetylase may be new therapeutic agents for APL. The presence of specific inhibitors of histone deacetyltransferases, such as trichostatin A (TSA), restores RA responsiveness of both PML–RARα and PLZF–RARα, and allows leukemic cells to differentiate in response to all-*trans* RA (Grignani *et al.*, 1998; He *et al.*,

1998; Lin *et al.*, 1998). Recently, clinical treatment with sodium phenylbutyrate, another inhibitor of histone deacetyltransferases, caused hyperacetylation of chromatin in the blood cells and restored RA sensitivity in a patient with APL (Warrell *et al.*, 1998). This approach may be applicable to APL and other cancers in which nuclear receptors are involved.

Uses of 13-*cis* RA to restore RARβ transcript levels

The RXR–RAR heterodimer can mediate ligand-dependent transcriptional activation of the RARβ gene, which contains a DR5 response element in its enhancer region (Hong and Itri, 1994; Oridate *et al.*, 1994).

The expression of RARβ is suppressed selectively at early stages of carcinogenesis in the oral cavity and bronchial epithelium. About 50–60% of oral premalignant lesions in patients with leukoplakia, and dysplastic lesions adjacent to head and neck squamous cell carcinomas, fail to express RARβ mRNA (Xu *et al.*, 1994). The decrease in RARβ in early premalignant lesions may enhance the development of malignancies. Treatment of leukoplakia patients with 13-*cis* RA causes a marked restoration of RARβ expression (Lotan *et al.*, 1995).

Uses of 9-*cis* RA

The antiproliferative and differentiating activities of 9-*cis* RA have been demonstrated in various *in vitro* models of diseases, such as breast cancer, head and neck cancers, lung cancer, leukemia and lymphoma (Rubin *et al.*, 1994; Gottardis *et al.*, 1996; Giannini *et al.*, 1997a). Additional studies in rats have reported that 9-*cis* RA possesses significant anticarcinogenic activity in the mammary gland, colon and prostate gland (Anzano *et al.*, 1994, 1996; Zheng *et al.*, 1997). From the results of a phase I clinical trial, 9-*cis* RA has been found to have much less cutaneous and mucous membrane toxicity than other natural retinoids, and is suggested for chronic treatment (Rizvi *et al.*, 1998). Presently, 9-*cis* RA is being studied in phase II clinical trials.

Treatments with synthetic retinoids

Natural RAs have been shown to be efficacious chemopreventive and therapeutic agents in both preclinical models and human cancer clinical trials (Lippman *et al.*, 1994b; Moon *et al.*, 1994). Various side-effects, such as hyperglycemia, mucocutaneous toxicity and teratogenicity, however, limit the uses of these retinoids for chronic treatment. A large panel of synthetic ligands of retinoid receptors has been generated that exhibit a great variety of RAR isotype or RXR selectivity and functional specificity (reviewed in Resche-Rigon and Gronemeyer, 1998). Some of these synthetic ligands have distinct biologic activities and therapeutic potential in both oncology- and nononcology-related diseases, such as type II diabetes (McCormick *et al.*, 1998). Some of them do not manifest the side-effects that are normally associated with retinoid treatment. Many synthesized agonists can selectively induce transcriptional activation of specific receptor subtypes, while some of them are pan-agonists, activating all retinoid receptors. Many antagonistic synthetic retinoids can also selectively inhibit receptor subtype-specific transactivation and they may show antiproliferative effects via anti-AP-1 activity (Fanjul *et al.*, 1994; Chen *et al.*, 1995).

RAR agonists and antagonists

Example 1 ALRT1550, an RAR-selective retinoid, has been shown to have potent antitumor activity against human oral squamous carcinoma (Shalinsky *et al.*, 1997). This activity was mediated through growth inhibition and not through suppression of differentiation. In this model, treatment with ALRT1550 did not cause upregulation of RARβ, suggesting that RARβ did not mediate the antiproliferative effects.

Example 2   AHPN (6-[3-(1-admantyl)-4-hydroxyphenyl]-2-napthalene carboxylic acid or CD437) was first identified as a RARγ-selective retinoid by receptor binding and transactivation assays and has been shown effectively to inhibit growth and induce apoptosis of a variety of cancer cells, including breast cancer, cervical cancer, melanoma, leukemia and lung cancer cells (Shao *et al.*, 1995; Schadendorf *et al.*, 1996; Chao *et al.*, 1997; Hsu *et al.*, 1997; Oridate *et al.*, 1997; Widschwendter *et al.*, 1997). AHPN-induced apoptosis may involve the induction of c-Jun and nur77 (Li *et al.*, 1998). Nur77 is an orphan nuclear receptor which can heterodimerize with RXR to modulate the binding activities of some RAREs and the sensitivity of RXR to all-*trans* RA (Perlmann and Jansson, 1995; Wu *et al.*, 1997b).

RXR agonists and antagonists

Example 1   LGD1069 (Targretin), one of a class of new synthetic RXR-specific ligands, is currently undergoing study in phase I and II clinical trials. It inhibits the growth of tumor cell lines of both hematopoietic and squamous epithelial origin and has the ability to induce apoptosis (Boehm *et al.*, 1994). LGD1069 has been reported to be as efficacious as tamoxifen, an antiestrogen used in breast cancer treatment, as a chemopreventive agent in rat mammary carcinoma (Bischoff *et al.*, 1998). LGD1069 can successfully cause complete regression in most treated tumors, and a higher efficacy is found when a combined treatment of LGD1069 and tamoxifen is applied. The underlying mechanism is still unclear. The effect of LGD1069 may be mediated by one of the RXR heterodimers since the RXR activation of the RXR–PPAR heterodimer can inhibit ER-mediated transactivation (Nunez *et al.*, 1997).

Example 2   Using *in vivo* mouse models for tumor promoter-induced papilloma formation, it has been shown that the antitumor effect of SR11302, an AP-1 inhibition-specific retinoid, is due to AP-1 inhibition but not to RARE activation (Huang *et al.*, 1997). HL60 cells differentiate and cease proliferation in response to RAR–RXR agonists but not to a retinoid with anti-AP-1 activity that does not transactivate target genes (Kizaki *et al.*, 1996). Similar AP-1 antagonism-independent antitumor effects of RXR-selective retinoids, such as SR11383 and SR11246, have been reported for neuroblastoma and prostate cancer cells respectively (De Vos *et al.*, 1997; Giannini *et al.*, 1997b).

The results discussed above suggest that it may be possible to target the antitumor effect of retinoids to specific cell types or tissues by designing synthetic retinoids with a defined pattern of RAR and RXR selectivity, and transactivation or transrepression ability (Resche-Rigon and Gronemeyer, 1998). Their uses in future long-term prevention trials and their eventual application in chemoprevention regimens will require strategies to improve efficacy, by decreasing side-effects of existing retinoids or through the identification of retinoids with few or no side-effects (Bollag *et al.*, 1997).

SUMMARY

The investigation of cancers and some tumors has demonstrated that retinoids homeo-statically regulate the normal processes of cellular growth and differentiation. Disturbance of these processes, by alteration of expression or gene products of the retinoid receptor genes, may result in certain cancers and tumors in humans. Currently, treatment strategies for cancers and tumors, such as for patients with APL, have been improved because of significant advances made in the past decade in our understanding of retinoid receptor function. Based on the knowledge of the molecular mechanism of retinoid receptor's action is the clinical application of synthetic retinoids, which are receptor selective. Structural analyses from X-ray crystallography or NMR studies yielded greater insight into the structure–function relationship of both the DBD and LBD of retinoid receptors, and confirmed what had been found in many biochemical studies. This structural information yielded clues about how to design receptor-selective ligands that have distinct biologic activities. For example, these studies have shown that increasing the rigidity of the molecule appears to improve the receptor selectivity of synthetic retinoids and the therapeutic potential in the treatment of cancers and other diseases.

In addition to their therapeutic functions, synthetic retinoids have become an important tool in examining the molecular mechanism of action of retinoid receptors. Different agonists and antagonists of both RARs and RXRs permit investigation of the functions of each receptor subfamily. Genetic studies have become another essential tool to understand the functions of retinoid receptors *in vivo*. Although the interpretation of these studies may be complicated by functional redundancy of retinoid receptors, a recent report examining the role of the RXR AF-2 subdomain during embryogenesis supports the role of knockout mouse models to study the function of retinoid receptors. Together with structural and biochemical analyses of the receptors, synthetic retinoids and genetic studies will enable greater understanding of the retinoid signal pathway. Such studies will lead to the development of more efficacious and selective retinoid treatments for retinoid-dependent diseases.

REFERENCES

Achkar, C.C., Derguini, F., Blumberg, B. *et al.* (1996) 4-Oxoretinol, a new natural ligand and transactivator of the retinoic acid receptors. *Proc. Natl. Acad. Sci. USA* **93**: 4879–4884.

Alcalay, M., Zangrilli, D., Pandolfi, P.P. *et al.* (1991) Translocation breakpoint of acute promyelocytic leukemia lies within the retinoic acid receptor α locus. *Proc. Natl. Acad. Sci. USA* **88**: 1977–1981.

Alcalay, M., Balitrand, N., Barraga, E. *et al.* (1996) RT-PCR in acute pormyelocytic leukemia: second workshop of the European Retinoic Group. *Leukemia* **10**: 368–371.

Allan, G.F., Leng, X., Tsai, S.F. *et al.* (1992) Hormone and antihormone induce distinct conformational changes which are central to steroid receptor activation. *J. Biol. Chem.* **267**: 19 513–19 520.

Allenby, G., Bocquel, M.-T., Saunders, M. *et al.* (1993) Retinoic acid receptors and retinoid X receptors: interactions with endogenous retinoic acids. *Proc. Natl. Acad. Sci. USA* **90**: 30–34.

Ang, H.L. and Duester, G. (1997) Initiation of retinoid signaling in primitive streak mouse embryos: spatiotemporal expression patterns of receptors and metabolic enzymes for ligand synthesis. *Dev. Dyn.* **108**: 536–543.

Anzano, M.A., Byers, S.W., Smith, J.M. *et al.* (1994) Prevention of breast cancer in the rat with 9-*cis*-retinoic acid as a single agent and in combination with tamoxifen. *Cancer Res.* **54**: 4614–4617.

Anzano, M.A., Peer, C.W., Smith, J.M. *et al.* (1996) Chemoprevention of mammary carcinogenesis in the rat: combined use of raloxifene and 9-*cis*-retinoic acid. *Natl. Cancer Inst.* **88**: 123–125.

Anzick, S.L., Kononen, J., Walker, R.L. *et al.* (1997) AIB1, a steroid receptor coactivator amplified in breast and ovarian cancer. *Science* **277**: 965–968.

Apfel, C., Crettaz, M., Siegenthaler, G. and Hunziker, M. (1991) Synthetic retinoids: differential binding to retinoic acid receptors. In: J.H. Saurat (ed.) *Retinoids: 10 Years On*, pp. 110–120. Karger, Basel.

Apfel, C., Bauer, F.M., Crettaz, M. *et al.* (1992) A retinoic acid receptor α antagonist selectively counteracts retinoic acid effects. *Proc. Natl. Acad. Sci. USA* **89**: 7129–7133.

Asou, N., Adachi, K., Tamura, J. *et al.* (1998) Analysis of prognostic factors in newly diagnosed acute promyelocytic leukemia treated with all-*trans* retinoic acid and chemotherapy. Japanese Adult Leukemia Study Group. *Clin. Oncol.* **16**: 78–85.

Astrom, A., Tavakkol, A., Pettersson, U., Cromie, M., Elder, J.T. and Voorhees, J.J. (1991) Molecular cloning of two human cellular retinoic acid binding proteins (CRABP) Retinoic acid-induced expression of CRABP-II but not CRABP-I in adult human skin *in vivo* and in skin fibroblast *in vitro*. *J. Biol. Chem.* **226**: 17 662–17 666.

Astrom, A., Petterson, U., Chambon, P. and Voorhees, J.J. (1994) Retinoic acid induction of human cellular retinoic acid binding protein-II gene transcription is mediated by retinoic acid receptor-retinoid X receptor heterodimers bound to one far upstream retinoic acid response element with 5-base pair spacing. *J. Biol. Chem.* **269**: 22 334–22 339.

Avvisati, G., Lo Coco, F., Diverio, D. *et al.* (1996) AIDA (all-*trans* retinoic acid + idarubicin) in newly diagnosed acute promyelocytic leukemia. *Blood* **88**: 1390–1398.

Bailey, J.S. and Siu, C.-H. (1988) Purification and partial characterization of a novel binding protein for retinoic acid from neonatal rat. *J. Biol. Chem.* **263**: 9326–9332.

Baniahmad, A., Kohne, A.C. and Renkawitz, R. (1992) A transferable silencing domain is present in the thyroid hormone receptor, in the v-*erbA* oncogene product and in the retinoic acid receptor. *EMBO J.* **11**: 1015–1023.

Baniahmad, A., Leng, X., Burris, T.P., Tsai, S.Y., Tsai, M.-J. and O'Malley, B.W. (1995) The τ4 activation domain of the thyroid hormone receptor is required for release of a putative corepressor(s) necessary for transcriptional silencing. *Mol. Cell. Biol.* **15**: 76–86.

Bardot, O., Aldridge, T.C., Latruffe, N. and Green, S. (1993) PPAR–RXR heterodimer activates a peroxisome proliferator response element upstream of the bifunctional enzyme gene. *Biochem Biophys. Res. Commun.* **192**: 37–45.

Beato, M. (1989) Gene regulation by steroid hormones. *Cell* **56**: 335–344.

Benbrook, D., Lernharadt, E. and Pfahl, M. (1988) A new retinoic acid receptor identified from a hepatocellular carcinoima. *Nature* **333**: 669–672.

Benedetti, L., Levin, A.A., Scicchitano, B.M. *et al.* (1997) Characterization of the retinoid binding properties of the major fusion products present in acute promyelocytic leukemia cells. *Blood* **90**: 1175–1185.

Bischoff, E.D., Gottardis, M.M., Moon, T.E., Heyman, R.A. and Lamph, W.W. (1998) Beyond tamoxifen: the retinoid X receptor-selected ligand LGD1069 (Targretin) causes complete regression of mammary carcinoma. *Cancer Res.* **58**: 479–484.

Blomhoff, R., Green, M.H., Berg, T. and Norum, K.R. (1990) Transport and storage of vitamin A. *Science* **250**: 399–404.

Blumberg, B., Balando, J., Derguini, F. *et al.* (1996) Novel retinoic acid receptor ligands in *Xenopus* embryos. *Proc. Natl. Acad. Sci. USA* **93**: 4873–4878.

Boehm, M.F., Zhang, L., Badea, B.A. *et al.* (1994) Synthesis and structure–activity relationships of novel retinoid X receptor-selective retinoids. *Med. Chem.* **37**: 2930–2941.

Bollag, W., Isnardi, L., Jablonska, S. *et al.* (1997) Links between pharmacological properties of retinoids and nuclear retinoid receptor. *Int. J. Cancer* **70**: 470–472.

Borrow, J., Goddard, A.D., Sheer, D. and Solomon, E. (1990) Molecular analysis of acute promyelocytic leukemia breakpoint cluster region on chromosome 17. *Science* **249**: 1577–1580.

Borrow, J., Goddard, A.D., Fagioli, M. *et al.* (1992) Diagnosis of acute promyelocyte leukemia by RT-PCR: detection of PML–RARα and RARα–PML fusion transcript. *Proc. Natl. Acad. Sci. USA* **89**: 4840–4844.

Bourguet, W., Ruff, M., Chambon, P., Gronemeyer, H. and Moras, D. (1995) Crystal structure of the ligand-binding domain of the retinoid X receptor RXR-α. *Nature* **375**: 377–382.

Brand, N., Petkovich, M., Krust, A. *et al.* (1988) Identification of a second human retinoic acid receptor. *Nature* **332**: 850–853.

Brzozowski, A.M., Pike, A.C.W., Dauter, Z. *et al.* (1997) Molecular basis of agonism and antagonism in the estrogen receptor. *Nature* **389**: 753–758.

Bugge, T.H., Pohl, J., Lonnoy, O. and Stunnenberg, H.G. (1992) RXR α, a promiscuous partner of retinoic acid and thyroid hormone receptors. *EMBO J.* **11**: 1409–1418.

Burris, T.P., Nawaz, Z., Tsai, M.-J. and O'Malley, B.W. (1995) Nuclear hormone receptor-associated protein that inhibits transactivation by the thyroid hormone and retinoic acid receptors. *Proc. Natl. Acad. Sci. USA* **92**: 9525–9529.

Campbell, C., Quinn, A.G., Angus, B., Farr, P.M. and Rees, J.L. (1993) Wavelength specific patterns of p53 induction in human skin following exposure to UV radiation. *Cancer Res.* **53**: 2697–2699.

Castaigne, S., Chomienne, C., Daniel, M.T. *et al.* (1990) All-*trans* retinoic acid as a differentiation therapy for acute myelocytic leukemia. I. Clinical results. *Blood* **76**: 1704–1709.

Cavey, M.T., Martin, B., Carlavan, I. and Shroot, B. (1990) *In vitro* binding of retinoids to the nuclear retinoic acid receptor α. *Anat. Biochem.* **186**: 19–25.

Chambon, P. (1996) A decade of molecular biology of retinoic acid receptors. *FASEB J.* **10**: 940–954.

Chandraratna, R.A.S. (1998a) Rational design of receptor-selective retinoids. *J. Am. Acad. Dermatol.* **39**: S124–128.

Chandraratna, R.A.S. (1998b) Future trends: a new generation of retinoids. *J. Am. Acad. Dermatol.* **39**: S149–152.

Chao, W.R., Hobbs, P.D., Jong, L. *et al.* (1997) Effects of receptor class- and subtype-selective retinoids and an apoptosis-inducing retinoid on the adherent growth of the NIH:OVCAR-3 ovarian cancer cell line in culture. *Cancer Lett.* **15**: 1–7.

Chen, H., Lin, R.J., Schiltz, R.L. *et al.* (1997a) Nuclear receptor coactivator ACTR is a novel histone acetyltransferase and forms a multimeric activation complex with p/CAF and CBP/p300. *Cell* **90**: 569–580.

Chen, G.Q., Shi, X.G., Tang, W. *et al.* (1997b) Use of arsenic trioxide (As_2O_3) in the treatment of acute promyelocytic leukemia (APL) I: As_2O_3 exerts dose-dependent dual effects on APL cells. *Blood* **89**: 3345–3353.

Chen, J., Kubalak, S.W. and Chien, K.R. (1998) Ventricular muscle-restricted targeting of the RXR (α) gene reveals a non-cell-autonomous requirement in cardiac chamber morphogenesis. *Development* **125**: 1943–1949.

Chen, J.D. and Evan, R.M. (1995) A transcriptional corepressor that interacts with nuclear hormone receptors. *Nature* **377**: 454–457.

Chen, J.D. and Li, H. (1998) Coactivation and corepression in transcriptional regulation by steroid/nuclear hormone receptors. *Crit. Rev. Eukaryo. Gene Expr.* **8**: 169–190.

Chen, J.Y., Clifford, J., Zusi, C. *et al.* (1995) RAR-specific agonist/antagonist which dissociate transactivation and AP-1 transrepression inhibit anchorage-independent cell proliferation. *EMBO J.* **14**: 1187–1197.

Chen, J.Y., Clifford, J., Zusi, C. *et al.* (1996) Two distinct actions of retinoid receptor ligands. *Nature* **382**: 819–822.

Chen, S.J., Zelent, A., Tong, J.H. *et al.* (1993a) Rearrangements of the retinoic acid receptor α and promyelocytic leukemia zinc finger genes resulting from t(11;17) (q23;21) in a patient with acute promyelocytic leukemia. *J. Clin. Invest.* **91**: 2260–2267.

Chen, Z., Brand, N.J., Chen, A. *et al.* (1993b) Fusion between a novel Kruppel-like zinc finger gene and the retinoic acid receptor α locus due to a variant t(11;17) translocation associated with acute promyelocytic leukemia. *EMBO J.* **12**: 1161–1167.

Chen, Z., Morgan, R., Stone, J.F. and Sandberg, A.A. (1994) Identification of complex t(15;17) in APL by FISH. *Cancer Genet. Cytogen.* **72**: 73–74.

Cheskis, B. and Freedman, L.P. (1994) Ligand modulates the conversion of DNA-bound vitamin D_3 receptor (VDR) homodimers into VDR-retinoid X receptor heterodimers. *Mol. Cell. Biol.* **14**: 3329–3338.

Chiang, M.Y., Misner, D., Kempermann, G. *et al.* (1998) An essential role for retinoid receptors RARbeta and RXRgamma in long-term potentiation and depression. *Neuron* **21**: 1353–1361.

Chomienne, C., Ballerini, P., Balitrand, N. *et al.* (1990) All-*trans* retinoic acid in acute promyelocytic leukemias. II. *In vitro* studies: structure–function relationship. *Blood* **76**: 1710–1717.

Christianson, A.M., King, D.L., Hatzivassiliou, E. *et al.* (1992) DNA binding and hetero-merization of the *Drosophila* transcription factor chorion factor 1/ultraspiracle. *Proc. Natl. Acad. Sci. USA* **89**: 11 503–11 507.

Chung, A.C., Durica, D.S., Clifton, S.W., Roe, B.A. and Hopkins, P.M. (1998) Cloning of crustacean ecdysteroid receptor and retinoid-X receptor gene homologs and elevation of retinoid-X receptor mRNA by retinoic acid. *Mol. Cell. Endocrinol.* **139**: 209–227.

Chytil, F. and Ong, D.E. (1979) Cellular retinol- and retinoic acid-binding proteins in vitamin A action. *Fed. Proc.* **38**: 2510–2514.

Clark, B.R. and Keating, A. (1995) Biology of bone marrow stroma. *Ann. N. Y. Acad. Sci.* **770**: 70–78.

Cooney, A.J. and Tsai, M.-J. (1994) Nuclear receptor–DNA interaction. In: M.-J. Tsai and B.W. O'Malley (eds) *Mechanism of Steroid Hormone Regulation of Gene Transcription.* R.G. Landes, Austin, 25–58.

Cooney, A.J., Tsai, S.Y., O'Malley, B.W. and Tsai, M.-J. (1992) Chicken ovalbumin upstream promoter. *Mol. Cell. Biol.* **12**: 4153–4163.

Darwiche, N., Celli, G. and De Luca, L.M. (1994) Specificity of retinoid receptor gene expression in mouse cervical epithelia. *Endocrinology* **134**: 2018–2025.

Darwiche, N., Celli, G., Tennenbaum, T., Glick, A.B., Yuspa, S.H. and De Luca, L.M. (1995) Mouse skin tumor progression results in differential expression of retinoic acid and retinoid X receptors. *Cancer Res.* **55**: 2774–2782.

Darwiche, N., Scita, G., Jones, C. *et al.* (1996) Loss of retinoic acid receptors in mouse skin and skin tumors is associated with activation of the *ras*(Ha) oncogene and high risk for premalignant progression. *Cancer Res.* **56**: 4942–4949.

de The, H., Marchio, A., Tollais, P. and Dejean, A. (1989) Differential expression and ligand regulation of retinoic acid receptor α and β genes. *EMBO J.* **8**: 429–433.

de The, H., Vivanco-Ruiz, M.M., Tiollais, P., Stunnenberg, H. and Dejean, A. (1990a) Identification of a retinoic acid responsive element in the retinoic acid receptor β gene. *Nature* **343**: 177–180.

de The, J., Chomienne, C., Lanotte, M., Degos, L. and Dejean, A. (1990b) The t(15;17) translocation of acute promyelocytic leukemia fuses the retinoic acid receptor α gene to a novel transcribed locus. *Nature* **347**: 558–561.

de The, H., Lavau, C., Marchio, A., Chomienne, C., Degos, L. and Dejean, A. (1991) The PML–RARα fusion mRNA generated by the t(15;17) translocation in acute promyelocytic leukemia encodes a functionally altered RAR. *Cell* **66**: 675–684.

De Vos, S., Dawson, M.I., Holden, S. *et al.* (1997) Effects of retinoid X receptor-selective ligands on proliferation of prostate cancer cells. *Prostate* **32**: 115–121.

Dejean, A., Bougueleret, L., Grzeschik, K.H. and Tiollais, P. (1986) Hepatitis B virus DNA integration in a sequence homologus to v-*erb*-A and steroid hormone receptor gene in a hepatocellular carcinoma. *Nature* **322**: 70–72.

Ding, X.F., Anderson, C.M., Ma, H. *et al.* (1998) Nuclear receptor-binding sites of coactivators glucocorticoid receptor interacting protein 1 (GRIP1) and steroid receptor coactivator 1 (SRC-1): multiple motifs with different binding specificities. *Mol. Endocrinol.* **12**: 302–313.

Dolle, P., Ruberte, E., Kastner, P. *et al.* (1989) Differential expression of genes encoding α, β and γ retinoic acid receptors and CRABP in the developing limbs of the mouse. *Nature* **342**: 702–705.

Dolle, P., Ruberte, E., Leroy, P., Morriss-Kay, G. and Chambon, P. (1990) Retinoic acid receptors and cellular retinoid binding proteins. I. A systematic study of their differential pattern of transcription during mouse organogenesis. *Development* **110**: 1133–1151.

Dong, S., Zhu, J., Reid, A. *et al.* (1996) Amino-terminal protein–protein interaction motif (POZ-domain) is responsible for activities of the promyelocytic leukemia zinc finger-retinoic acid receptor-α fusion protein. *Proc. Natl. Acad. Sci. USA* **93**: 3624–3629.

Durand, B., Saunders, M., Leroy, P., Leid, M. and Chambon, P. (1992) All-*trans* and 9-*cis* retinoic acid induction of CRABPII transcription is mediated by RAR–RXR hetero-dimers bound to DR1 and DR2 repeated motifs. *Cell* **71**: 73–85.

Durand, B., Saunders, M., Gaudon, T., Roy, B., Losson, R. and Chambon, P. (1994) Activation function 2 (AF-2) of retinoic acid receptor and 9-*cis* retinoic acid receptor: the presence of a conserved autonomous constitutive activating domain and influence of the nature of the response element on AF-2 activity. *EMBO J.* **13**: 5370–5382.

Durston, A.J., Timmermans, J.P., Hage, W.J. *et al.* (1989) Retinoic acid causes an anteroposterior transformation in the developing central nervous system. *Nature* **340**: 140–144.

Durston, A.J., van der Wees, J., Pijnappel, W.W.M. and Godsave, S.F. (1998) Retinoids and related signals in early development of the vertebrate central nervous system. *Curr. Top. Dev. Biol.* **40**: 111–175.

Edwards, D.P. (1999) Coregulatory proteins in nuclear hormone receptor action. *Vitam. Horm.* **55**: 165–218.

Eichele, G. (1997) Retinoids: from hindbrain patterning to Parkinson disease. *Trends Genet.* **13**: 343–345.

Elder, J.T., Fisher, G.J., Zhang, Q.Y. *et al.* (1991) Retinoic acid receptor gene expression in human skin. *J. Invest. Dermatol.* **96**: 423–433.

Elder, J.T., Astrom, A., Pettersson, U. *et al.* (1992) Differential regulation of retinoic acid receptors and binding proteins in human skin. *J. Invest. Dermatol.* **98**: 673–679.

Elliott, S., Taylor, K., White, S. *et al.* (1992) Proof of differentiative mode of action of all-*trans* retinoic acid in acute promyelocytic leukemia using X-linked clonal analysis. *Blood* **79**: 1916–1919.

Escriva, H., Safi, R., Hanni, C. *et al.* (1997) Ligand binding was acquired during evolution of nuclear receptors. *Proc. Natl. Acad. Sci. USA* **94**: 6803–6808.

Esgleyes-Ribot, T., Chandraratna, R.A.S., Lew-Kaya, D.A., Sefton, J. and Duvic, M. (1994) Response of psoriasis to a new topical retinoid, AGN 190168. *J. Am. Acad. Dermatol.* **30**: 581–590.

Evans, R.M. (1988) The steroid and thyroid hormone receptor superfamily. *Science* **240**: 889–895.

Fanjul, A., Dawson, M.I., Hobbs, P.D. *et al.* (1994) A new class of retinoids with selective inhibition of AP-1 inhibits proliferation. *Nature* **372**: 107–111.

Fanjul, A.N., Bouterfa, H., Dawson, M. and Pfahl, M. (1996) Potential role for retinoic acid receptor γ in the inhibition of breast cancer cells. *Cancer Res.* **56**: 1571–1577.

Fiorella, P.D. and Napoli, J.L. (1991) Expression of cellular retinoic acid binding protein (CRABP) in *Escherichia coli*: characterization and evidence that holo-CRABP is a substrate in retinoic acid metabolism. *J. Biol. Chem.* **266**: 16 572–16 579.

Fiorella, P.D. and Napoli, J.L. (1994) Microsomal retinoic acid metabolism: effects of cellular retinoic acid binding protein (type I) and C18-hydroxylation as an initial step. *J. Biol. Chem.* **269**: 10 538–10 544.

Fiorella, P.D., Giguere, V. and Napoli, J.L. (1993) Expression of cellular retinoic acid binding proteins (type II) in *Escherichia coli*: characterization and comparison to cellular retinoic acid binding protein (type I). *J. Biol. Chem.* **268**: 21 545–21 552.

Fisher, G.J. and Voorhees, J.J. (1996) Molecular mechanisms of retinoid actions in skin. *FASEB J.* **10**: 1002–1013.

Fisher, G.J., Talwar, H.S., Xiao, J.H. *et al.* (1994) Immunological identification and functional quantitation of retinoic acid and retinoid X receptor protein in human skin. *J. Biol. Chem* **269**: 20 629–20 635.

Fisher, G.J., Reddy, A.P., Datta, S.C. *et al.* (1995) All-*trans* retinoic acid induces cellular retinol-binding protein in human skin *in vivo*. *Invest. Dermatol.* **105**: 80–86.

Fisher, G.J., Wang, Z.Q., Datta, S.C., Varani, J., Kang, S. and Voorhees, J.J. (1997) Pathophysiology of premature skin aging induced by ultraviolet light. *N. Engl. J. Med.* **337**: 1419–1428.

Fisher, G.J., Talwar, H.S., Lin, J. *et al.* (1998) Retinoic acid inhibits induction of c-Jun protein by ultraviolet radiation that occurs subsequent to activation of mitogen-activated protein kinase pathways in human skin *in vivo*. *J. Clin. Invest.* **101**: 1432–1440.

Fogh, K., Voorhees, J.J. and Astrom, A. (1993) Expression, purification, and binding properties of human cellular retinoic acid-binding protein type I and type II. *Arch. Biochem. Biophys.* **300**: 751–755.

Folkers, G.E., van der Leede, B.-M. and van der Saag, P.T. (1993) The retinoic acid receptor-β2 contains two separate cell-specific transactivation domains at the N-terminus and in the ligand binding domain. *Mol. Endocrinol.* **7**: 616–627.

Forman, B.M. (1998) Orphan nuclear receptors and their ligands. In: L.P. Freedom (ed.) *Molecular Biology of Steroid and Nuclear Hormone Receptors*, pp. 281–305. Birkhauser, Boston.

Forman, B.M., Yange, C.-R. Au, M., Casanova, J., Ghysdael, J. and Samuels, H.H. (1989) A domain containing leucine-zipper-like motifs mediate novel *in vivo* interaction between the thyroid hormone and retinoic acid receptors. *Mol. Endocrinol.* **3**: 1610–1626.

Forman, B.M., Umesono, K., Chen, J. and Evans, R.M (1995) Unique response pathways are established by allosteric interactions among nuclear hormone receptors. *Cell* **81**: 541–550.

Gaboli, M., Gandini, D., Delva, L., Wang, Z.-G. and Pandolfi, P.P. (1998) Acute promyelocytic leukemia as a model for cross-talk between interferon and retinoic acid pathways: from molecular biology to clinical applications. *Leuk. Lymph.* **30**: 11–22.

Gebert, J.F., Moghal, N., Frangioni, J.V., Sugarbaker, D.J. and Neel, B.G. (1991) High frequency of retinoic acid receptor *β* abnormalities in human lung cancer. *Oncogene* **6**: 1859–1868.

Georgiades, P., Wood, J. and Brickell, P.M. (1998) Retinoid X receptor-*γ* gene expression is developmentally regulated in the embryonic rodent peripheral nervous system. *Anat. Embryol.* **197**: 477–484.

Geradts, J., Chen, J.-Y., Russell, E.K., Yankaskas, J.R., Nievers, L. and Minna, J.D. (1993) Human lung cancer cell lines exhibit resistance to retinoic acid treatment. *Cell Growth Differ.* **4**: 799–809.

Giannini, F., Maestro, R., Vukosavljevic, T., Pomponi, F. and Boiocchi, M. (1997a) All-*trans*, 13-*cis* and 9-*cis* retinoic acids induce a fully reversible growth inhibition in HNSCC cell lines: implications for *in vivo* retinoic acid use. *Int. J. Cancer* **70**: 194–200.

Giannini, G., Dawson, M.I., Zhang, X. and Thiele, C.J. (1997b) Activation of three distinct RXR/RAR heterodimers induces growth arrest and differentiation of neuroblastoma cells. *J. Biol. Chem* **272**: 26 693–26 701.

Giguere, V. (1994) Retinoic acid receptors and cellular retinoid binding proteins: complex interplay in retinoid signaling. *Endocr. Rev.* **15**: 61–79.

Giguere, V., Ong, E.S., Segui, P. and Evans, R.M. (1987) Identification of a receptor for the morphogen retinoic acid. *Nature* **330**: 624–629.

Giguere, V., Lyn, S., Yip, P., Siu, C.H. and Amin, S. (1990) Molecular cloning of a cDNA encoding a second cellular retinoic acid binding protein. *Proc. Natl. Acad. Sci. USA* **87**: 6233–6237.

Glass, C. (1996) Some new twists in the regulation of gene expression by thyroid hormone and retinoic acid receptors. *J. Endocrinol.* **150**: 349–357.

Glass, C.K., Devary, O.V. and Rosenfeld, M.G. (1990) Multiple cell type-specific proteins differentially regulate target sequence recognition by the α retinoic acid receptor. *Cell* **63**: 729–738.

Gottardis, M.M., Lamph, W.W., Shalinsky, D.R., Wellstein, A. and Heyman, R.A. (1996) The efficacy of 9-*cis* retinoic acid in experimental models of cancer. *Breast Cancer Res. Treat.* **38**: 85–96.

Graupner, G., Malle, G., Maignan, J., Lang, G., Prunieras, M. and Pfahl, M. (1991) 6′-Substituted naphthalene-2-carboxylic acid analogs, a new class of retinoic acid receptor subtype-specific ligands. *Biochem. Biophys. Res. Commun.* **179**: 1554–1561.

Green, S. and Chambon, P. (1988) Nuclear receptors enhance our understanding of transcription regulation. *Trends Genet.* **4**: 309–314.

Grignani, F., De Matteis, S., Nervi, C. *et al.* (1998) Fusion proteins of the retinoic acid receptor-α recruit histone deacetylase in promyelocytic leukaemia. *Nature* **391**: 815–818.

Grimwade, D. and Solomon, E. (1997) Characterization of the PML/RAR α rearrangement associated with t(15;17) acute promyelocytic leukemia. *Curr. Top. Microbiol. Immunol.* **220**: 81–112.

Grimwade, D., Howe, K., Langabeer, S. *et al.* (1996) Establishing the presence of the t(15:17) in suspected acute promyelocytic leukemia: cytogenetic, molecular and PML immuno-fluorescence assessment of patients entered into the MRC ATRA. *Br. J. Haematol.* **94**: 557–573.

Grimwade, D., Langabeer, S., Howe, K. and Solomon, E. (1998) RT-PCR in diagnosis and disease monitoring of acute promyelocytic leukemia (APL). *Methods Mol. Biol.* **89**: 333–358.

Grondona, J.M., Kastner, P., Gansmuller, A., Decimo, D., Chambon, P. and Mark, M. (1996) Retinal dysplasia and degeneration in RARβ2/RARγ2 compound mutant mice. *Development* **122**: 2173–2188.

Guo, X., Xu, Q., Harmon, M. *et al.* (1998) Isolation of two functional retinoid X receptor subtypes from the ixodid tick, *Ambylyomma americanum. Mol. Cell Endocrinol.* **139**: 45–60.

Hassig, C.A., Fleischer, T.C., Billin, A.N., Schreiber, S.L. and Ayer, D.E. (1997) Histone deacetylase activity is required for full transcriptional repression by mSin3A. *Cell* **89**: 341–347.

He, L.Z., Guidez, F., Triboli, C. *et al.* (1998) Distinct interactions of PML–RARα and PLZF–RARα with co-repressors determine differential responses to RA in APL. *Nat. Genet.* **18**: 126–135.

Heery, D.M., Zacharewski, T., Pierrat, B., Gronemeyer, H., Chambon, P. and Losson, R. (1993) Efficient transactivation by retinoic acid receptors in yeast requires retinoid X receptors. *Proc. Natl. Acad. Sci. USA* **90**: 4281–4285.

Heery, D.M., Pierrat, B., Gronemeyer, H., Chambon, P. and Losson, R. (1994) Homo- and heterodimers of the retinoid X receptor (RXR) activated in yeast. *Nucleic Acids Res.* **22**: 726–731.

Heery, D.M., Kalkoven, E., Hoare, S. and Parker, M.G. (1997) A signature motif in coactivators mediates binding to nuclear receptors. *Nature* **387**: 733–736.

Hewison, M., Dabrowski, M., Vadher, S. *et al.* (1996) Antisense inhibition of vitamin D receptor expression induces apoptosis in monoblastoid U937 cells. *J. Immunol.* **156**: 4391–4400.

Heyman, R.A., Mangelsdorf, D.J., Dysk, J.A. *et al.* (1992) 9-*cis* retinoic acid is a high affinity ligand for the retinoid X receptor. *Cell* **68**: 397–406.

Hollenberg, A.N., Monden, T., Madura, J.P., Lee, K. and Wondisford, F.E. (1996) Function of nuclear co-repressor protein on thyroid hormone response elements is regulated by the receptor A/B domain. *J. Biol. Chem.* **271**: 28 516–28 520.

Holmbeck, S.M.A., Dyson, H.J. and Wright, P.E. (1998a) DNA-induced conformational changes are the basis for cooperative dimerization by the DNA binding domain of the retinoid X receptor. *J. Mol. Biol.* **284**: 533–539.

Holmbeck, S.M.A., Foster, M.P., Casimiro, D.R., Sem, D.S., Dyson, H.J. and Wright, P.E. (1998b) High-resolution solution structure of the retinoid X receptor DNA-binding domain. *J. Mol. Biol.* **281**: 271–284.

Hong, W.K. and Itri, L.M. (1994) Retinoids and human cancer. In: M.B. Sporn, A.B. Roberts and D.S. Goodman (eds) *The Retinoids*, pp. 597–658. Raven Press, New York.

Hong, H., Kohli, K., Garabedian, M.J. and Stallcup, M.R. (1997a) GRIP1, a transcriptional coactivator for the AF-2 transactivation domain of steroid, thyroid, retinoid, and vitamin D receptors. *Mol. Cell. Biol.* **17**: 2735–2744.

Hong, S.H., David, G., Wong, C.W., Dejean, A. and Privalsky, M.L. (1997b) SMRT corepressor interact with PLZF and with the PML–retinoic acid receptor α (RARα) and PLZF-RARα oncoproteins associated with acute promyelocytic leukemia. *Proc. Natl. Acad. Sci. USA* **94**: 9028–9033.

Hong, S.H., Wong, C.W. and Privalsky, M.L. (1998) Signaling by tyrosine kinases negatively regulates the interaction between transcription factors and SMRT (silencing mediator of retinoic acid and thyroid hormone receptor) co-repressor. *Mol. Endocrinol.* **12**: 1161–1171.

Hoopes, C.W., Taketo, M., Ozato, K. *et al.* (1992) Mapping of the RXR loci encoding nuclear receptors RXRα, RXRβ and RXRγ. *Genomics* **14**: 611–617.

Horlein, A.J., Naar, A.M., Heinzel, T. *et al.* (1995) Ligand-independent repression by the thyroid hormone receptor mediated by a nuclear receptor corepressor. *Nature* **377**: 397–404.

Houle, B., Leduc, F. and Bradley, W.E. (1991) Implication of RAR-β in epidermoid (squamous) lung cancer. *Genes Chromosom. Cancer* **3**: 358–366.

Hsu, C.A., Rishi, A.K., Su-Li, X. *et al.* (1997) Retinoid induced apoptosis in leukemia cells through a retinoic acid nuclear-receptor-independent pathway. *Blood* **89**: 4470–4479.

Huang, C., Ma, W.Y., Dawson, M.I., Rincon, M., Flavell, R.A. and Dong, Z. (1997) Blocking activator protein-1 activity, but not activating retinoid response elements, is required for the antitumor promotion effect of retinoic acid. *Proc. Natl. Acad. Sci. USA* **94**: 5826–5830.

Huang, M.-E., Ye, Y.-C., Chen, S.-R. *et al.* (1988) Use of all-*trans* retinoic acid in the treatment of acute promyelocytic leukemia. *Blood* **72**: 567–572.

Ishikawa, A., Umesono, K., Mangelsdorf, D.J. *et al.* (1990) A functional retinoic acid receptor encoded by a gene on human chromosome 12. *Mol. Endocrinol.* **4**: 837–844.

James, S.Y., Williams, M.A., Newland, A.C. and Colston, K.W. (1999) Leukemia cell differentiation: cellular and molecular interactions of retinoids and vitamin D. *Gen. Pharmacol.* **32**: 143–154.

Jeyakumar, M., Tanen, M.R. and Bagchi, M.K. (1997) Analysis of the functional role of steroid receptor coactivator-1 in ligand-induced transactivation by thyroid hormone receptor. *Mol. Endocrinol.* **11**: 755–767.

Jindra, M., Huang, J.-Y., Hiruma, K. and Riddiford, L.M. (1997) Identification and mRNA developmental profiles of two ultraspiracle isoforms in the epidermis and wings of *Manduca sexta. Insect Mol. Biol.* **6**: 41–53.

Jing, Y., Zhang, J., Bleiweiss, I.J., Waxman, S., Zelent, A. and Mira-Y-Lopez, R. (1996) Defective expression of cellular retinoid binding protein type I and retinoic acid receptors α2, β2, and γ2 in human breast cancer cells. *FASEB J.* **10**: 1064–1070.

Johnson, K.J. and Scadding, S.R. (1991) Effects of vitamin A and other retinoids on the differentiation and morphogenesis of the integument and limbs of vertebrates. *Can. J. Zool.* **69**: 263–273.

Jones, B.B., Ohno, C.K., Allenby, G. *et al.* (1995) New retinoid X receptor subtypes in zebra fish (*Danio rerio*) differentially modulate transcription and do not bind 9-*cis* retinoic acid. *Mol. Cell. Biol.* **15**: 5226–5234.

Jones, G. and Sharp, P.A. (1997) Ultraspiracle: an invertebrate nuclear receptor for juvenile hormones. *Proc. Natl. Acad. Sci. USA* **94**: 13 499–13 503.

Kakizuka, A., Miller, W.H., Jr., Umesono, K. *et al.* (1991) Chromosomal translocation t(15;17) in human acute promyelocytic leukemia fuses RARα with a novel putative transcription factor, PML. *Cell* **66**: 663–674.

Kalemkerian, G.P. and Ramnath, N. (1996) Retinoids and apoptosis in cancer therapy. *Apoptosis* **1**: 11–24.

Kamei, Y., Xu, L., Heinzel, T. *et al.* (1996) A CBP integrator complex mediates transcriptional activation and AP-1 inhibition by nuclear receptors. *Cell* **85**: 403–414.

Kanai, M., Raz, A. and Goodman, D.S. (1968) Retinol-binding protein: the transport protein for vitamin A in human plasma. *J. Clin. Invest.* **47**: 2025–2044.

Karin, M., Liu, Z.G. and Zandi, E. (1997) AP-1 function and regulation. *Curr. Opin. Cell. Biol.* **9**: 240–246.

Kastner, P., Kurst, A., Mendelsohn, C. *et al.* (1990) Murine isoforms of the mouse retinoic acid receptor γ with specific patterns of expression. *Proc. Natl. Acad. Sci. USA* **87**: 2700–2704.

Kastner, P. Leid, M. and Chambon, P. (1994) Role of nuclear retinoic acid receptors in the regulation of gene expression. In: R. Blomhoff (ed.) *Vitamin A in Health and Disease*, pp. 189–238. Marcel Dekker, New York.

Kastner, P., Mark, M. and Chambon, P. (1995) Nonsteroid nuclear receptors: what are genetic studies telling us about their role in real life? *Cell* **83**: 859–869.

Kastner, P., Mark, M., Leid, M. *et al.* (1996) Abnormal spermatogenesis in RXRβ mutant mice. *Genes Dev.* **10**: 80–92.

Kastner, P., Mark, M., Ghyselinck, N. *et al.* (1997a) Genetic evidence that the retinoid signal is transduced by heterodimeric RXR/RAR functional units during mouse development. *Development* **124**: 313–326.

Kastner, P., Messaddeq, N., Mark, M. *et al.* (1997b) Vitamin A deficiency and mutations of RXRα, RXRβ and RARα lead to early differentiation of embryonic ventricular cardiomyocytes. *Development* **124**: 4749–4758.

Kawasaki, H., Eckner, R., Yao, T.P. *et al.* (1998) Distinct roles of the co-activators p300 and CBP in retinoic-acid-induced F9-cell differentiation. *Nature* **393**: 284–289.

Kersten, S., Dawson, M.I., Lewis, B.A. and Noy, N. (1996) Individual subunits of heterodimers comprised of retinoic acid and retinoid X receptors interact with their ligands independently. *Biochemistry* **35**: 3816–3824.

Kitareewan, S., Burka, L.T., Tomer, K.B. *et al.* (1996) Phytol metabolites are circulating dietary factors that activate the nuclear RXR. *Mol. Biol. Cell.* **7**: 1153–1166.

Kizaki, M., Dawson, M.I., Heyman, R. *et al.* (1996) Effects of novel retinoid X receptor-selective ligands on myeloid leukemia differentiation and proliferation *in vitro*. *Blood* **87**: 1977–1984.

Klaholz, B.P., Renaud, J.P., Mitschler, A. *et al.* (1998) Conformational adaptation of agonists at the human nuclear receptor RARγ. *Nat. Struct. Biol.* **5**: 199–202.

Kliewer, S.A., Umesono, K., Mangelsdorf, D.J. and Evans, R.M. (1992a) Retinoid X receptor interacts with nuclear receptors in retinoic acid, thyroid hormone and vitamin D$_3$ signalling. *Nature* **355**: 446–449.

Kliewer, S.A., Umesono, K., Heyman, R.A., Mangelsdorf, D.J., Dyck, J.A. and Evans, R.M. (1992b) Retinoid X receptor–COUP-TF interactions modulate retinoic acid signaling. *Proc. Natl. Acad. Sci. USA* **89**: 1448–1452.

Kliewer, S.A., Umesono, K., Noonan, D.J., Heyman, R.A. and Evans, R.M. (1992c) Convergence of 9-*cis* retinoic acid and peroxisome proliferator signalling pathways through heterodimer formation of their receptors. *Nature* **358**: 771–774.

Knegtel, R.M.A., Katahia, M., Schilthuis, J.G. *et al.* (1993) The solution structure of the human retinoic acid receptor β DNA-binding domain. *J. Biomol. NMR* **3**: 1–17.

Kok, K., Osinga, J., Carritt B. *et al.* (1987) Deletion of a DNA sequence at the chromosomal region 3p21 in all major types of lung cancer. *Nature* **330**: 578–581.

Koller, E., Krieger, O., Kasparu, H. and Lutz, D. (1991) Restoration of all-*trans* retinoic acid sensitivity by interferon in acute promyelocytic leukemia. *Lancet* **338**: 1154–1155.

Kopan, R., Traska, G. and Fuchs, E. (1987) Retinoids as important regulators of terminal differentiation: examining keratin expression in individual epidermis cells at various stages of keratinization. *J. Cell Biol.* **105**: 427–440.

Korzus, E., Torchia, J., Rose, D.W. *et al.* (1998) Transcription factor specific requirements for coactivators and their acetyltransferase functions. *Science* **279**: 703–707.

Kostrouch, Z., Kostrouchova, M., Love, W., Jannini, E., Piatigorsky, J. and Rall, J.E. (1998)

Retinoic acid X receptor in the diploblast, *Tripedalia cystophora. Proc. Natl. Acad. Sci. USA* **95**: 13 442–13 447.

Krezel, W., Dupe, V., Mark, M., Dierich, A., Kastner, P. and Chambon, P. (1996) RXR null mice are apparently normal and compound RXRα$^{+/-}$/RXRβ$^{-/-}$/RXRγ mutant mice are viable. *Proc. Natl. Acad. Sci. USA* **93**: 9010–9014.

Krezel, W., Ghyselnick, N., Samad, T.A. *et al.* (1998) Impaired locomotion and dopamine signaling in retinoid receptor mutant mice. *Science* **279**: 863–867.

Krust, A., Kastner, P., Petkovich, M., Zelent, A. and Chambon, P. (1989) A third human retinoic acid receptor, hRAR-γ. *Proc. Natl. Acad. Sci. USA* **86**: 5310–5314.

Kumar, R., Shoemaker, A.R. and Verma, A.K. (1994) Retinoic acid nuclear receptors and tumor promotion: decreased expression of retinoic acid nuclear receptors by the tumor promoter 12-*O*-tetradecanoylphorbol-13-acetate. *Carcinogenesis* **15**: 701–705.

Kurokawa, R., Yu, V.C., Naar, A. *et al.* (1993) Differential orientations of the DNA-binding domain and carboxy-terminal dimerization interface regulate receptor heterodimers. *Genes Dev.* **7**: 1423–1435.

Kurokawa, R., DiRenzo, J., Boehm, M. *et al.* (1994) Regulation of retinoid signalling by receptor polarity and allosteric control of ligand binding. *Nature* **371**: 528–531.

Kurokawa, R., Soderstrom, M., Horlein, A. *et al.* (1995) Polarity-specific activities of retinoic acid receptors determined by a corepressor. *Nature* **375**: 451–454.

Lala, D.S., Mukherjee, R., Schulman, I.G. *et al.* (1996) Activation of specific RXR heterodimers by an antagonist of RXR homodimers. *Nature* **383**: 450–453.

Laudet, V. (1997) Evolution of the nuclear receptor superfamily: early diversification from an ancestral orphan receptor. *J. Mol. Endocrinol.* **19**: 207–226.

Lazzarino, M., Corso, A., Regazzi, M.B., Iacona, I. and Bernasconi, C. (1995) Modulation of all-*trans* retinoic acid pharmacokinetics in acute promyelocytic leukemia by prolonged interferon-α therapy. *Br. J. Haematol.* **90**: 928–930.

LeDouarin, B., VomBaur, E., Zechel, C. *et al.* (1996) Ligand-dependent interaction of nuclear receptors with potential transcriptional intermediary factors (mediators). *Phil. Trans. R. Soc. Lond. B.* **351**: 569–578.

Lee, M.S., Kliewer, S.A., Provencal, J., Wright, P.E. and Evans, R.M. (1993a) Structure of the retinoid X receptor α DNA binding domain: a helix required for homodimeric DNA binding. *Science* **260**: 1117–1121.

Lee, J.S., Newman, R.A., Lippman, S.M. *et al.* (1993b) Phase I evaluation of all-*trans*-retinoic acid in adults with solid tumors. *J. Clin. Oncol.* **11**: 959–966.

Lee, M.-O., Hobbs, P.D., Zhang, X.-K., Dawson, M.I. and Pfhal, M. (1994) A new retinoid antagonist inhibits the HIV-1 promoter. *Proc. Natl. Acad. Sci. USA* **91**: 5632–5636.

Lehmann, J.M., Dawson, M.I., Hobbs, P.D., Husmann, M. and Pfahl, M. (1991) Indentification of retinoids with nuclear receptor subtype-selective activities. *Cancer Res.* **51**: 4804–4808.

Lehmann, J.M., Jong, L., Fanjul, A. *et al.* (1992a) Retinoids selective for retinoid X receptor response pathways. *Science* **258**: 1944–1946.

Lehmann, J.M., Zhang, X.K. and Pfahl, M. (1992b) RARγ2 expression is regulated through a retinoic acid response element embedded in Sp1 sites. *Mol. Cell. Biol.* **12**: 2976–2985.

Lehmann, J.M., Zhang, X.-K., Graupner, G. *et al.* (1993) Formation of retinoid X-receptor homodimers leads to repression of T3 response: hormonal crosstalk by ligand-induced squelching. *Mol. Cell. Biol.* **13**: 7698–7707.

Leid, M., Kastner, P. and Chambon, P. (1992) Multiplicity generates diversity in the retinoic acid signalling pathways. *Trends Biochem. Sci.* **17**: 427–433.

Leid, M., Kastner, P., Lyons, R. *et al.* (1995) Purification, cloning, and RXR identity of the

HeLa cell factor with which RAR or TR heterodimerizes to bind target sequences efficiently. *Cell* **68**: 377–395.

Lemotte, P.K., Keidel, S. and Apfel, C.M. (1996) Phytanic acid is a retinoid X receptor ligand. *Eur. J. Biochem.* **236**: 328–333.

Leng, X., Blanco, J., Tsai, S.Y., Ozato, K., O'Malley, B.W. and Tsai, M.-J. (1995) Mouse retinoid X receptor contains a separable ligand binding and transactivation domain in its E region. *Mol. Cell. Biol.* **15**: 255–263.

Lenny, N., Westendorf, J.J. and Hiebert, A.S.W. (1997) Transcriptional regulation during myelopoiesis. *Mol. Biol. Rep.* **24**: 157–168.

Leo, M.A., Lasker, J.M., Raucy, J.K., Kim, C.I., Black, M. and Lieber, C.S. (1989) Metabolism of retinol and retinoic acid by human liver cytochrome P450IIC8. *Arch. Biochem. Biophys.* **269**: 305–312.

Leroy, P., Nakshatri, H. and Chambon, P. (1991) Mouse retinoic acid receptor α2 isoform transcribed from a promoter that contains a retinoic acid response element. *Proc. Natl. Acad. Sci. USA* **88**: 10 138–10 142.

Levin, A.A., Sturgenbecker, L.J., Kazmer, S. *et al.* (1992) 9-*cis* retinoic acid stereoisomer binds and activates the nuclear receptor RXRα. *Nature* **355**: 359–361.

Li, H. and Chen, J.D. (1998) The receptor-associated activator 3 activates transcription through CREB-binding protein recruitment and autoregulation. *J. Biol. Chem.* **273**: 5948–5954.

Li, X.-Y., Xiao, J.-H., Feng, X. and Voorhees, J.J. (1997a) Retinoid X receptor-specific ligands synergistically upregulate 1,25-dihydroxyvitamin D_3-dependent transcription in epidermal keratinocytes *in vitro* and *in vivo*. *J. Invest. Dermatol.* **108**: 506–512.

Li, H., Gomes, P.J. and Chen, J.D. (1997b) RAC3, a steroid/nuclear receptor-associated coactivator that is related to SRC-1 and TIF2. *Proc. Natl. Acad. Sci. USA* **94**: 8479–8484.

Li, Y., Dawson, M.I., Agadir, A. *et al.* (1998) Regulation of RAR β expression by RAR- and RXR-selective retinoids in human lung cancer cell lines: effect on growth inhibition and apoptosis induction. *Int. J. Cancer* **75**: 88–95.

Licht, J.D., Chomienne, C., Goy, A. *et al.* (1995) Clinical and molecular characterization of a rare syndrome of acute promyelocytic leukemia associated with translocation (11;17) *Blood* **85**: 1083–1094.

Lin, R.J., Nagy, L., Inoue, S., Shao, W., Miller, W.H., Jr. and Evans, R.M. (1998) Role of the histone deacetylase complex in acute promyelocytic leukaemia. *Nature* **391**: 811–814.

Lin, R.J., Egan, D.A. and Evans, R.M. (1999) Molecular genetics of acute promyelocytic leukemia. *Trends Genet.* **15**: 179–184.

Lippman, S.M., Kavanagh, J.J., Paredes-Espinoza, M. *et al.* (1992a) 13-*cis*-retinoic acid plus interferon α-2a: highly active systemic therapy for squamous cell carcinoma of the cervix. *J. Natl. Cancer Inst.* **84**: 241–245.

Lippman, S.M., Parkinson, D.R., Itri, L.M. *et al.* (1992b) 13-*cis*-retinoic acid and interferon α-2a: effective combination therapy for advanced squamous cell carcinoma of the skin. *J. Natl. Cancer Inst.* **84**: 235–240.

Lippman, S.M., Glisson, B.S., Kavanagh, J.J. *et al.* (1993) Retinoic acid and interferon combination studies in human cancer. *Eur. J. Cancer* **29A** (Suppl. 5): S9–13.

Lippman, S.M., Benner, S.E. and Hong, W.K. (1994a) Cancer chemoprevention. *J. Clin. Oncol.* **12**: 851–873.

Lippman, S.M., Benner, S.E. and Hong, W.K. (1994b) Retinoid chemoprevention studies in upper aerodigestive tract and lung carcinogenesis. *Cancer Res.* **54** (Suppl.): 2025s–2028s.

Lippman, S.M., Lotan, R. and Schleuniger, U. (1997) Retinoid-interferon therapy of solid tumors. *Int. J. Cancer* **70**: 481–483.

Liu, Q. and Linney, E. (1993) The mouse retinoid X receptor γ gene: evidence for functional isoforms. *Mol. Endocrinol.* **7**: 651–658.

Liu, Y., Lee, M.O., Wang, H.G. *et al.* (1996) Retinoic acid receptor β mediates the growth-inhibitory effect of retinoic acid by promoting apoptosis in human breast cancer cells. *Mol. Cell. Biol.* **16**: 1138–1149.

Lohnes, D., Dierich, A., Ghyselinck, N. *et al.* (1992) Retinoid receptors and binding proteins. *J. Cell Sci. Suppl.* **16**: 69–76.

Lohnes, D., Kastner, P., Dierich, A., Mark, M., LeMeur, M. and Chambon, P. (1993) Function of retinoic acid receptor γ in the mouse. *Cell* **73**: 643–658.

Lohnes, D., Mark, M., Mendelsohn, C. *et al.* (1994) Function of the retinoic acid receptors (RARs) during development I: Craniofacial and skeletal abnormalities in RAR double mutants. *Development* **120**: 2723–2748.

Lohnes, D., Mark, M., Mendelsohn, C. *et al.* (1995) Developmental roles of the retinoic acid receptors. *J. Steroid Biochem. Mol. Biol.* **53**: 475–486.

Lotan, R. (1997) Roles of retinoids and their nuclear receptors in the development and prevention of upper aerodigestive tract cancers. *Environ. Health Perspect.* **105**: S985–988.

Lotan, R., Xu, X.C., Lippman, S.M. *et al.* (1995) Suppression of retinoic acid receptor β in oral premalignant lesions and its upregulation by isotretinoin. *N. Engl. J. Med.* **332**: 1405–1410.

Lufkin, T., Lohnes, D., Mark, M. *et al.* (1993) High postnatal lethality and testis degeneration in retinoic acid receptor α mutant mice. *Proc. Natl. Acad. Sci. USA* **90**: 7225–7229.

Lundblad., J.K., Kwok, R.P., Laurance, M.E., Hartner, M.L. and Goodman, R.H. (1995) Adenoviral E1A-associated protein p300 as a functional homologue of the transcription coactivator CBP. *Nature* **374**: 85–88.

MacDonald, P.N., Dowd, D.R., Nakajima, S. *et al.* (1993) Retinoid X receptors stimulate and 9-*cis* retinoic acid inhibits 1,25-dihydroxyvitamin D_3-activated expression of the rat osteocalcin gene. *Mol. Cell. Biol.* **13**: 5907–5917.

Maden, M. (1982) Vitamin A and pattern formation in the regenerating limb. *Nature* **295**: 672–675.

Maden, M., Sonneveld, E., van der Saag, P.T. and Gale, E. (1998) The distribution of endogenous retinoic acid in the chick embryo: implications for developmental mechanisms. *Development* **125**: 4133–4144.

Mader, S., Leroy, P., Chen, J.Y. and Chambon, P. (1993) Multiple parameters control the selectivity of nuclear receptors for their response elements, selectivity and promiscuity in response element recognition by retinoic acid receptors and retinoid X receptors. *J. Biol. Chem.* **286**: 591–600.

Makishima, M., Shudo, K. and Honma, Y. (1999) Greater synergism of retinoic acid receptor (RAR) agonists with vitamin D_3 than that of reinoid X receptor (RXR) agonists with regard to growth inhibition and differentiation induction in monoblastic leukemia cells. *Biochem. Pharmacol.* **57**: 521–539.

Mangelsdorf, D.J. and Evans, R.M. (1995) The RXR heterodimers and orphan receptors. *Cell* **83**: 841–850.

Mangelsdorf, D.J. Ong, E.S., Dyck, J.A. and Evans, R.M. (1990) Nuclear receptor that identifies a novel retinoic acid-response pathway. *Nature* **345**: 224–229.

Mangelsdorf, D.J., Umesono, K., Kliewer, S.A., Borgmeyer, U., Ong, E.S. and Evans, R.M. (1991) A direct repeat in the cellular retinol-binding protein type II gene confers differential regulation by RXR and RAR. *Cell* **66**: 555–561.

Mangelsdorf, D.J., Borgmeyer, U., Heyman, R.A. *et al.* (1992) Characterization of three RXR genes that mediate the action of 9-*cis*-retinoic acid. *Genes Dev.* **6**: 329–344.

Mangelsdorf, D.J., Umesono, K. and Evans, R.M. (1994) The retinoid receptors. In: M.B. Sporn, A.B. Roberts and D.S. Goodman (eds) *The Retinoids: Biology, Chemistry, and Medicine*, 2nd edn., pp. 319–349. Raven Press, New York.

Mannervik, M., Nibu, Y., Zhang, H. and Levine, M. (1999) Transcriptional coregulators in development. *Science* **284**: 606–609.

Mascrez, B., Mark, M., Dierich, A., Ghyselinck, N.B., Kastner, P. and Chambon, P. (1998) The RXRα ligand-dependent activation function 2 (AF-2) is important for mouse development. *Development* **125**: 4691–4707.

Mattei, M.-G., Riviere, M., Krust, A. *et al.* (1991) Chromosomal assignment of retinoic acid receptor (RAR) genes in the human, mouse, and rat genomes. *Genomics* **10**: 1061–1069.

McCaffery, P. and Drager, U.C. (1994) High levels of a retinoic acid-generating dehydrogenase in the mesotelecephalic dopamine system. *Proc. Natl. Acad. Sci. USA* **91**: 7772–7776.

McCormick, D.L., Rao, K.V., Dooley, L. *et al.* (1998) Influence of *N*-methyl-*N*-nitrosourea, testosterone, and *N*-(4-hydroxyphenyl)-all-*trans*-retinamide on prostate cancer induction in Wistar–Unilever rats. *Cancer Res.* **58**: 3282–3288.

Mendelsohn, C., Lohnes, D., Decimo, D. *et al.* (1994) Function of the retinoic acid receptors (RARs) during development II: multiple abnormalities at various stages of organogenesis in RAR double mutants. *Development* **120**: 2749–2771.

Minucci, S., Leid, M., Toyama, R. *et al.* (1997) Retinoid X receptor (RXR) within the RXR–retinoic acid receptor heterodimer binds its ligand and enhances retinoid-dependent gene expression. *Mol. Cell. Biol.* **17**: 644–655.

Montano, M.M., Muller, V., Trobaugh, A. and Katzenellenbogen, B.S. (1995) The carboxy-terminal F domain of the human estrogen receptor: role in the transcriptional activity of the receptor and effectiveness of antiestrogens as estrogen antagonists. *Mol. Endocrinol.* **9**: 814–825.

Moon, R.C., Mehta, R.G. and Rao, K.V.N. (1994) Retinoids and cancer in experimental animals. In: M.B. Sporn, A.B. Roberts and D.S. Goodman (eds) *The Retinoids: Biology, Chemistry, and Medicine*, pp. 573–595. Raven Press, New York.

Moore, D.M., Kalvakolanu, D.V., Lippman, S.M. *et al.* (1994) Retinoic acid and interferon in human cancer: mechanistic and clinical studies. *Semin. Hematol.* **31**: 31–37.

Morriss-Kay, G.M. and Ward, S.J. (1999) Retinoids and mammalian development. *Int. Rev. Cytol.* **188**: 73–131.

Mukherjee, R., Davies, P.J., Crombie, D.L. *et al.* (1997) Sensitization of diabetic and obese mice to insulin by retinoid X receptor agonists. *Nature* **386**: 407–410.

Muto, A., Kizaki, M., Yamato, K. *et al.* (1999) 1,25-Dihydroxyvitamin D_3 induces differentiation of a retinoic acid-resistant acute promyelocytic leukemia cell line (UF-1) associated with expression of p21(WAF1/CIP1) and p27(KIP1) *Blood* **93**: 2225–2233.

Naar, A.M., Boutin, J.M., Lipkin, S.M. *et al.* (1991) The orientation and spacing of core DNA-binding motifs dictate selective transcriptional responses to three nuclear receptors. *Cell* **65**: 1267–1279.

Nagpal, S., Zelent, A. and Chambon, P. (1992a) RAR-β4, a retinoic acid receptor isoform, is generated from RAR-β2 by alternative splicing and usage of a CUG initiator codon. *Proc. Natl. Acad. Sci. USA* **89**: 2718–2722.

Nagpal, S., Saunders, M., Kastner, P., Durand, B., Nakshatri, H. and Chambon, P. (1992b) Promoter context- and response element-dependent specificity of the transcriptional activation and modulating functions of retinoic acid receptors. *Cell* **70**: 1007–1019.

Nagpal, S., Frinat, S., Nakshatri, H. and Chambon, P. (1993) RARs and RXRs: evidence for two autonomous transactivation functions (AF-1 and AF-2) and heterodimerization *in vivo*. *EMBO J.* **12**: 2349–2360.

Nagpal, S., Athanikar, J. and Chandraratna, R.A.S. (1995) Separation of transactivation and AP-1 antagonism functions of retinoic acid receptor α. *J. Biol. Chem* **13**: 923–927.

Nagy, L., Thomazy, V.A., Shipley, G.L. *et al.* (1995) Activation of retinoid X receptors induces apoptosis in HL-60 cell lines. *Mol. Cell. Biol.* **15**: 3540–3551.

Nagy, L., Thomazay, V.A., Heyman, R.A. and Davies, P.J.A. (1998) Retinoid-induced apoptosis in normal and neoplastic tissues. *Cell Death Differ.* **5**: 11–19.

Napoli, J.L., Posch, K.C., Fiorella, P.D., Boerman, M.H.E.M., Salerno, G.J. and Burns, R.D. (1993) Roles of cellular retinol-binding protein and cellular retinoid acid-binding protein in the metabolic channeling of retinoids. In: M.A. Livrea and L. Packer (eds) *Retinoids: Progress in Research and Clinical Application*, pp. 29–48. Marcel Dekker, New York.

Nason-Burchenal, K., Gandini, D., Botto, M. *et al.* (1996) Interferon augments PML and PML-RARα expression in normal myeloid and acute promyelocytic cells and cooperates with all-*trans* retinoic acid to induce maturation of a retinoid resistant promyelocytic cell line. *Blood* **889**: 3926–3936.

Niederreither, K., McCaffery, P., Drager, U.C., Chambon, P. and Dolle, P. (1997) Restricted expression and retinoic acid-induced downregulation of the retinaldehyde dehydrogenase type 2 (RALDH-2) gene during mouse development. *Mech. Dev.* **62**: 67–78.

Niederreither, K., Subbarayan, V., Dolle, P. and Chambon, P. (1999) Embryonic retinoic acid synthesis is essential for early mouse post-implantation development. *Nat. Genet.* **21**: 444–448.

Niles, R.M. (1998) Control of retinoid nuclear receptors function. *Subcell. Biochem.* **30**: 3–28.

Noji, S., Yamsai, T., Koyama, E. *et al.* (1989) Expression of retinoic acid receptor genes in keratinizing front skin. *FEBS Lett.* **259**: 86–90.

Nolte, R.T., Wisely, G.B., Westin, S. *et al.* (1998) Ligand binding and coactivator assembly of the peroxisome proliferator-activated receptor γ. *Nature* **395**: 137–143.

Norum, K.R. (1994) Retinoids and acute myeloid leukemia. In: R. Blomhoff (ed.) *Vitamin A in Health and Disease*, pp. 485–501. Marcel Dekker, New York.

Nunez, S.B., Medin, J.A., Braissant, O. *et al.* (1997) Retinoid X receptor and peroxisome proliferator-activated receptor activate an estrogen responsive gene independent of the estrogen receptor. *Mol. Cell. Endocrinol.* **127**: 27–40.

Olsson, I., Bergh, G., Ehinger, M. and Gullberg, U. (1996) Cell differentiation in acute myeloid leukemia. *Eur. J. Haematol.* **57**: 1–16.

Onate, S.A., Tsai, S.Y., Tsai, M.-J. and O'Malley, B.W. (1995) Sequence and characterization of a coactivator for the steroid hormone receptor superfamily. *Science* **270**: 1354–1357.

Ong, D.E. and Chytil, F. (1975) Specificity of cellular retinol-binding protein for compounds with vitamin A activity. *Nature* **255**: 74–75.

Ong, D.E. and Chytil, F. (1978) Cellular retinoic acid-binding protein in rat testis: Purification and characterization. *J. Biol. Chem* **253**: 4551–4554.

Ong, D.E., Newcomer, M.E. and Chytil, F. (1994) Cellular retinoid-binding proteins. In: M.B. Sporn, A.B. Roberts and D.S. Goodman (eds) *The Retinoids: Biology, Chemistry and Medicine*, 2nd edn, pp. 283–317. Raven Press, New York.

Orfanos, C.E., Zouboulis, C.C., Almond-Roesler, B. and Geilen, C.C. (1997) Current use and future potential role of retinoids in dermatology. *Drugs* **53**: 258–388.

Oridate, N., Zou, C.P., Mitchell, M.F., Hong, W.K. and Lotan, R. (1994) Activation by retinoic acid of native retinoic acid receptor β-2 promoter is suppressed in human oral squamous cell carcinoma SqCC/Y1 cell. *Mol. Cell Differ.* **2**: 413–431.

Oridate, N., Higuchi, M., Suzuki, S., Shroot, B., Hong, W.K. and Lotan, R. (1997) Rapid induction of apoptosis in human C33A cervical carcinoma cells by the synthetic retinoid 6-[3-(1-adamantyl)hydroxyphenyl]-2-naphtalene carboxylic acid (CD437) *Int. J. Cancer* **70**: 484–487.

Oro, A.E., McKeown, M. and Evans, R.M. (1990) Relationship between the product of the *Drosophila ultraspiracle* locus and the vertebrate retinoid X receptor. *Nature* **347**: 298–301.

Pasquali, D., Bellastella, A., Valente, A. *et al.* (1997) Retinoic acid receptor α, β, and γ, and cellular retinol binding protein-I expression in breast fibrocystic disease and cancer. *Eur. J. Endocrinol.* **137**: 410–414.

Pastorino, U., Infante, M., Maioli, M. *et al.* (1993) Adjuvant treatment of stage I lung cancer with high-dose vitamin A. *J. Clin. Oncol.* **11**: 1216–1222.

Perlmann, T. and Jansson L. (1995) A novel pathway for vitamin A signaling mediated by RXR heterodimerization with NGFI-B and NURR1. *Genes Dev.* **9**: 769–782.

Perlmann, T., Rangarajan, P.N., Umesono, K. and Evans, R.M. (1993) Determinants for selective RAR and TR recognition of direct repeat HREs. *Genes Dev.* **7**: 1411–1422.

Peruzzi, D., DeBlasio, T., Warell, R.P.J. and Pandolfi, P.P. (1996) Highly sensitive RT-PCR assay for detection of minimal residual disease in acute promyeloid leukemia. *Blood* **88**: 366a.

Petkovitch, M., Brand, N.J., Krust, A. and Chambon, P. (1987) A human retinoic acid receptor which belongs to the family of nuclear receptors. *Nature* **330**: 440–450.

Pfahl, M., Apfel, R., Bendik, I. *et al.* (1994) Nuclear retinoid receptors and their mechanism of action. *Vitam. Horm.* **49**: 327–382.

Pijnappel, W.W.M., Hendriks, H.F.J., Folkers, G.E. *et al.* (1993) The retinoid ligand 4-oxo-retinoic acid is a highly active modulator of positional specification. *Nature* **366**: 340–344.

Posch, K.C., Boerman, M.H.E.M., Burns, R.D. and Napoli, J.L. (1991) Holocellular retinol binding protein as a substrate for microsomal retinal synthesis. *Biochemistry* **30**: 6224–6230.

Posch, K.C., Burns, R.D. and Napoli, J.L. (1992) Biosynthesis of all-*trans* retinoic acid from retinal: recognition of retinal bound to cellular retinoal binding protein (type I) as substrate by a purified cytosolic dehydrogenase. *J. Biol. Chem.* **267**: 19 676–19 682.

Puzianowska-Kuznicka, M., Damjanovski, S. and Shi, Y.B. (1997) Both thyroid hormone and 9-*cis* retinoic acid receptors are required to efficiently mediate the effects of thyroid hormone on embryonic development and specific gene regulation in *Xenopus laevis*. *Mol. Cell. Biol.* **17**: 4738–4749.

Rastinejad, F., Perlmann, T., Evans, R.M. and Sigler, P.B. (1995) Structural determinants of nuclear receptor assembly on DNA direct repeats. *Nature* **375**: 203–211.

Redner, R.L., Rush, E.A., Faas, S., Rudert, W.A. and Corey, S.J. (1996) The t(5;17) variant of acute promyelocytic leukemia expresses a nucleophosmin–retinoic acid receptor fusion. *Blood* **3**: 882–886.

Renaud, J.-P., Rochel, N., Ruff, M. *et al.* (1995) Crystal structure of the RAR-γ ligand-binding domain bound to all-*trans* retinoic acid. *Nature* **378**: 681–689.

Resche-Rigon, M. and Gronemeyer, H. (1998) Therapeutic potential of selective modulators of nuclear receptor action. *Curr. Opin. Chem. Biol.* **2**: 501–507.

Richard, S. and Zingg, H.H. (1991) Identification of a retinoic acid response element in the human oxytocin promoter. *Biol. Chem.* **266**: 21428–21433.

Rizvi, N.A., Marshall, J.L., Ness, E. *et al.* (1998) Phase I study of 9-*cis* retinoic acid (ALRT1057 capsules) in adults with advanced cancer. *Clin. Cancer Res.* **4**: 1437–1442.

Rochel, N., Renaud, J.-P., Ruff, M. *et al.* (1997) Purification of the human RAR-α ligand-binding domain and crystallization of its complex with all-*trans* retinoic acid. *Biochem. Biophys. Res. Commun.* **230**: 293–296.

Rogers, M.B. (1997) Life-and-death decisions influenced by retinoids. *Curr. Top. Dev. Biol.* **35**: 1–46.

Roman, S.D., Clarke, C.L., Hall, R.E., Alexander, I.E. and Sutherland, R.L. (1992) Expression and regulation of retinoid acid receptors in human breast cancer cells. *Cancer Res.* **52**: 2236–2242.

Roman, S.D., Ormandy, C.J., Manning, D.L. *et al.* (1993) Estradiol induction of retinoic acid receptors in human breast cancer cells. *Cancer Res.* **15**: 5940–5945.

Rosati, R., Ramnath, N., Adil, M.R. *et al.* (1998) Activity of 9-*cis*-retinoic acid and receptor-selective retinoids in small cell lung cancer cell lines. *Anticancer Res.* **18**: 4071–4075.

Rosen, E.D., Beningof, E.G. and Koenig, R.J. (1993) Dimerization interfaces of thyroid hormone, retinoic acid, vitamin D, and retinoid-X receptors. *J. Biol. Chem* **268**: 11 534–11 541.

Rowe, A., Eagerm, N.S.C. and Brickell, P.M. (1991) A member of the RXR nuclear receptor family is expressed in neural crest derived cells of the developing chick peripheral nervous system. *Development* **111**: 771–778.

Rowley, J.D., Golomb, H.M. and Dougherty, C. (1997) 15/17 translocation, a consistent chromosomal change in acute promyelocytic leukemia. *Lancet* **i**: 549–550.

Ruberte, E., Dolle, P., Chambon, P. and Morriss-Kay, G. (1990) Retinoic acid receptors and cellular retinoid binding proteins. II. Their differential patterns of transcription during early morphogenesis in mouse embryos. *Development* **111**: 45–60.

Ruberte, E., Dolle, P., Krust, A., Zelent, A., Morriss-Kay, G. and Chambon, P. (1991) Specific spatial and temporal distribution of retinoic acid receptor γ transcripts during mouse embryogenesis. *Development* **108**: 213–222.

Ruberte, E., Friedrich, V., Chambon, P. and Morris-Kay, G. (1993) Retinoic acid receptors and cellular retinoid binding protein III: Their differential transcript distribution during mouse nervous system development. *Development* **118**: 267–282.

Rubin, M., Fenig, E., Rosenauer, A. *et al.* (1994) 9-*cis* retinoic acid inhibits growth of breast cancer cells and down-regulates estrogen receptor RNA and protein. *Cancer Res.* **54**: 6549–6556.

Ruthardt, M., Testa, U., Nervi, C. *et al.* (1997) Opposite effects of the acute promyelocytic leukemia PML–retinoic acid receptor alpha (RARα) and PLZF-RARα fusion proteins on retinoic acid signalling. *Mol. Cell. Biol.* **17**: 4859–4869.

Saga, Y., Kobayashi, M., Ohta, H. *et al.* (1999) Impaired extrapyramidal function caused by the targeted disruption of retinoid X receptor RXRγ1 isoform. *Genes Cells* **4**: 219–228.

Saiardi, A., Bozzi, Y., Baik, J.-H. and Borrelli, E. (1997) Antiproliferative role of dopamine: loss of D2-receptor causes hormonal dysfunction and pituitary hyperplasia. *Neuron* **19**: 115–126.

Sakashita, A., Kizaki, M., Pakkala, S. *et al.* (1993) 9-*cis*-retinoic acid: effects on normal and leukemic hematopoiesis *in vitro*. *Blood* **81**: 1009–1016.

Samad, T.A., Krezel, W., Chambon, P. and Borrelli, E. (1997) Regulation of dopaminergic pathways by retinoids: activation of the D_2 receptor promoter by members of the retinoic acid receptor–retinoid X receptor family. *Proc. Natl. Acad. Sci. USA* **94**: 14 349–14 354.

Sande, S. and Privalsky, M.L. (1996) Identification of TRACs (T_3 receptor-associating cofactors), a family of cofactors that associate with, and modulate the activity of nuclear hormone receptors. *Mol. Endocrinol.* **10**: 813–825.

Sani, B.P. and Hill, D.L. (1974) Retinoic acid: a binding protein in chick embryo metatarsal skin. *Biochem. Biophys. Res. Commun.* **61**: 1276–1281.

Schadendorf, D., Kern, M.A., Artuc, M. *et al.* (1996) Treatment of melanoma cells with the synthetic retinoid CD437 induces apoptosis via activation of AP-1 *in vitro*, and causes growth inhibition in xenografts *in vivo*. *J. Cell Biol.* **135**: 1889–1898.

Schulman, I.G., Shao, G. and Heyman, R.A. (1998) Transactivation by retinoid X receptor–peroxisome proliferator-activated receptor γ (PPARγ) heterodimers: intermolecular synergy requires only the PPARγ hormone-dependent activation function. *Mol. Cell. Biol.* **18**: 3483–3494.

Schwabe, J.W.R. and Rhodes, D. (1991) Beyond zinc fingers: steroid hormone receptors have a novel structural motif for DNA recognition. *Trends Biochem. Sci.* **16**: 291–296.

Seale, J.R.C., Varma, S., Swirsky, D., Pandolfi, P.P., Goldman, J.M. and Cross, N.C.P. (1996) Quantification of PML–RARα transcripts in acute promyelocytic leukemia: explanation for the lack of sensitivity of RT-PCR for the detection of minimal residual disease and induction of detection of the leukemia specific mRNA by α interferon. *Br. J. Hematol.* **95**: 95–101.

Shalinsky, D.R., Bischoff, E.D., Lamph, W.W. *et al.* (1997) A novel retinoic acid receptor-selective retinoid, ALRT1550, has potent antitumor activity against human oral squamous carcinoma xenografts in nude mice. *Cancer Res.* **57**: 162–168.

Shao, M.Z., Dawson, M.I., Su Li, X. *et al.* (1995) p53 independent G0/G1 arrest and apoptosis induced by a novel retinoid in human breast cancer cells. *Oncogene* **11**: 493–504.

Shao, W., Fanelli, M., Ferrara, F.F. *et al.* (1998) Arsenic trioxide as an inducer of apoptosis and loss of PML/RARα protein in acute promyelocytic leukemia cells. *J. Natl. Cancer Inst.* **90**: 124–133.

Slack, J.L. and Gallagher, R.E. (1999) The molecular biology of acute promyelocytic leukemia. *Cancer Treat. Res.* **99**: 75–124.

Smith, W.C., Nakshatri, H., Leroy, P., Rees, J. and Chambon, P. (1991) A retinoic acid response element is present in the mouse cellular retinol binding protein I (mCRBPI) promoter. *EMBO J.* **10**: 2223–2230.

Solomin, L., Johansson, C.B., Zetterstrom, R.H. *et al.* (1998) Retinoid-X receptor signalling in the developing spinal cord. *Nature* **395**: 398–402.

Soprano, D.R., Pickett, C.B., Smith, J.E. and Goodman, D.S. (1981) Biosynthesis of plasma retinol-binding protein in liver as a larger molecular weight precursor. *J. Biol. Chem* **256**: 8256–8258.

Sporn, M. and Newton, D.L. (1979) Chemoprevention of cancer with retinoids. *Fed. Proc.* **38**: 2528–2534.

Sporn, M.B., Roberts, A.B. and DeWitt, S.G. (1994) *The Retinoids: Biology, Chemistry and Medicine*, 2nd edn. Raven Press, New York.

Stadler, M., Chelbi-Alix, M.K., Kohen, M.H. *et al.* (1995) Transcriptional induction of the PML growth suppressor gene by interferons is mediated through an ISRE and a GAS element. *Oncogene* **11**: 2565–2573.

Stunnenberg, H.G., Garcia-Jimenez, C. and Betz, J.L. (1999) Leukemia: the sophisticated subversion of hematopoiesis by nuclear receptor oncoproteins. *Biochim. Biophys. Acta* **1423**: F15–33.

Sucov, H.M., Murakami, K.K. and Evans, R.M. (1990) Characterization of an autoregulated response element in the mouse retinoic acid receptor type β gene. *Proc. Natl. Acad. Sci. USA* **87**: 5392–5396.

Sucov, H.M., Dyson, E., Gumeringer, C.L., Prive, J., Chien, K.R. and Evans, R.M. (1994) RXRα mutant mice establish a genetic basis for vitamin A signaling in heart morphogenesis. *Genes Dev.* **8**: 1007–1018.

Sutherland, J.D., Kozlova, T., Tzertzinis, G. and Kafatos, F.C. (1995) Drosophila hormone receptor 38: a second partner for Drosophila USP suggests an unexpected role for nuclear receptors of the nerve growth factor-induced protein B type. *Proc. Natl. Acad. Sci. USA* **92**: 7966–7970.

Szondy. Z., Reichert, U. and Fesus, L. (1998) Retinoic acids regulate apoptosis of T lymphocytes through an interplay between RAR and RXR receptors. *Cell Death Differ.* **5**: 4–10.

Takeshita, A., Yen, P.M., Misiti, S., Cardona, G.R., Liu, Y. and Chin, W.W. (1996) Molecular cloning and properties of a full-length putative thyroid hormone receptor coactivator. *Endocrinology* **137**: 3594–3597.

Tallman, M.S. (1998) Therapy of acute promyelocytic leukemia: all-*trans* retinoic acid and beyond. *Leukemia* **12**: S37–S40.

Tallman, M.S., Andersen, J.W., Schiffer, C.A., *et al.* (1997) All-*trans*-retinoic acid in acute promyelocytic leukemia. *N. Engl. J. Med.* **337**: 1021–1028.

Tate, B.F., Allenby, G., Perez, J.R., Levin, A.A. and Grippo, J.F. (1996) A systematic analysis of the AF-2 domain of human retinoic acid receptor α reveals amino acids critical for transcriptional activation and conformational integrity. *FASEB J.* **10**: 1524–1531.

Thacher, S.M., Nagpal, S., Klein, E.S. *et al.* (1999) Cell type and gene-specific activity of the retinoid inverse agonist AGN 193109: divergent effects from agonist at retinoic acid receptor γ in human keratinocytes. *Cell Growth Differ.* **10**: 255–262.

Thaller, C. and Eichele, G. (1990) Isolation of 3,4-didehydroretinoic acid, a novel morphogenetic signal in the chick wing bud. *Nature* **345**: 815–819.

Thomas, H.E., Stunnenberg, H.G. and Stewart. A.F. (1993) Heterodimerization of the *Drosophila* ecdysone receptor with retinoid X receptor and ultraspiracle. *Nature* **362**: 471–475.

Tini, M., Otulakowski, G., Breitman, M.L., Tsui, L.C. and Giguere, V. (1993) An everted repeat mediates retinoic acid induction of the γF-crystallin gene: evidence of a direct role for retinoids in lens development. *Genes Dev.* **7**: 295–307.

Torchia, J., Rose, D.W., Inostrosa, J. *et al.* (1997) The transcriptional coactivator p/CIP binds CBP and mediates nuclear-receptor function. *Nature* **387**: 677–684.

Torchia, J., Glass, C.K. and Rosenfeld, M.G. (1998) Coactivators and corepressors in the integration of transcriptional responses. *Curr. Opin. Cell Biol.* **10**: 373–383.

Torma, H., Asselineau, D., Andersson, E. *et al.* (1994) Biologic activities of retinoic acid and 3,4-didehydroretinoic acid in human keratinocytes are similar and correlate with receptor affinities and transactivation properties. *Invest. Dermatol.* **102**: 49–54.

Trans, C.M. and Sucov, H.M. (1998) The *RXRα* gene functions in a non-cell-autonomous manner during mouse cardiac morphogenesis. *Development* **125**: 1951–1956.

Tsai, M.-J. and O'Malley, B.W. (1994) Molecular mechanisms of action of steroid/thyroid receptor superfamily members. *Annu. Rev. Biochem.* **63**: 451–486.

Tzukerman, M.T., Esty, A., Santiso-Mere, D. *et al.* (1994) Human estrogen receptor transactivational capacity is determined by both cellular and promoter context and mediated by two functionally distinct intramolecular regions. *Mol. Endocrinol.* **8**: 21–30.

Umesono, K., Murakami, K.K., Thompson, C.C. and Evans, R.M. (1991) Direct receptors as selective response elements for the thyroid hormone, retinoic acid, and vitamin D_3 receptors. *Cell* **65**: 1255–1266.

Vahlquist, A. (1994) Role of retinoids in normal and diseased skins. In: R. Blomhoff (ed.) *Vitamin A in Health and Disease*, pp. 365–424. Marcel Dekker, New York.

Valdenaire, O., Vernier, P., Maus, M., Dumas Milne Edwards, J.B. and Mallet, J. (1994) Transcription of the rat dopamine-D_2-receptor from two promoters. *Eur. J. Biochem.* **220**: 511–584.

Verschoore, M., Langner, A., Wolska, H., Jablonska, S., Czernielewski, J. and Schaefer, H.

(1991) Efficacy and safety of topical CD 271 alcoholic gels. A new treatment candidate for acne vulgaris. *Br. J. Dermatol.* **124**: 368–371.

Vivat, V., Zechel, C., Wurtz, J.M. *et al.* (1997) A mutation mimicking ligand-induced conformational change yields a constitutive RXR that senses allosteric effects in heterodimers. *EMBO J.* **16**: 5697–5709.

Voegel, J.J., Heine, M.J.S., Zechel, C., Chambon, P. and Gronemeyer, H. (1996) TIF2, a 160 kDa transcriptional mediator for the ligand-dependent activation function AF-2 of nuclear receptors. *EMBO J.* **15**: 3667–3675.

Wang, X., Penzes, P. and Napoli, J.L. (1996) Cloning of a cDNA encoding an aldehyde dehydrogenase and its expression in *Escherichia coli*. Recognition of retinal as substrate. *J. Biol. Chem.* **271**: 16 288–16 293.

Wang, Z., Boudjelal, M., Kang, S., Voorhees, J.J. and Fisher, G.J. (1999) Ultraviolet irradiation of human skin causes functional vitamin A deficiency, preventable by all-*trans* retinoic acid pre-treatment. *Nat. Med.* **5**: 418–422.

Warrell, R.P., Jr., Frankel, S.R., Miller, W.H., Jr. *et al.* (1991) Differentiation therapy of acute promyelocytic leukemia with tretinoin (all-*trans*-retinoic acid). *N. Engl. J. Med.* **324**: 1385–1393.

Warrell, R.P., Jr., de The, H., Wang, Z.Y. and Degos, L. (1993) Acute promyelocytic leukemia. *N. Engl. J. Med.* **329**: 177–189.

Warrell, R.P., Jr., Maslak, P., Eardley, A., Heller, G., Miller, W.H., Jr. and Frankel, S.R. (1994) Treatment of acute promyelocytic leukemia with all-*trans* retinoic acid: an update of the New York experience. *Leukemia* **8**: 929–933.

Warrell, R.P., Jr., He, L.Z., Richon, V., Calleja, E. and Pandolfi, P.P. (1998) Therapeutic targeting of transcription in acute promyelocytic leukemia by use of an inhibitor of histone deacetylase. *J. Natl. Cancer Inst.* **90**: 1621–1625.

Wells, R.A., Hummel, J.L., De Koven, A. *et al.* (1996) A new variant translocation in acute promyelocytic leukemia: molecular characterization and clinical correlation. *Leukemia* **10**: 735–740.

Wells, R.A., Catzavelos, C. and Kamel-Reid, S. (1997) Fusion of retinoic acid receptor α to NuMA, the nuclear mitotic apparatus protein, by a variant translocation in acute promyelocytic leukaemia. *Nat. Genet.* **17**: 109–113.

Wendling, O., Chambon, P. and Mark, M. (1999) Retinoid X receptors are essential for early mouse development and placentogenesis. *Proc. Natl. Acad. Sci. USA* **96**: 547–551.

Westin, S., Kurokawa, R., Nolte, R.T. *et al.* (1998) Interactions controlling the assembly of nuclear-receptor heterodimers and co-activators. *Nature* **395**: 199–202.

Widschwendter, M., Daxenbichler, G., Culig, Z. *et al.* (1997) Activity of retinoic acid receptor-γ selectively binding retinoids alone and in combination with interferon-γ in breast cancer cell lines. *Int. J. Cancer* **71**: 497–504.

Williams, G.R., Harney, J.W., Moore, D.D., Larsen, P.R. and Brent, G.A. (1992) Differential capacity of wild type promoter elements for binding and *trans*-activation by retinoic acid and thyroid hormone receptors. *Mol. Endocrinol.* **6**: 1527–1537.

Willy, P.J. and Mangelsdorf, D.J. (1997) Unique requirements for retinoid-dependent transcriptional activation by the orphan receptor LXR. *Genes Dev.* **11**: 289–298.

Willy, P.J. and Mangelsdorf, D.J. (1998) Nuclear orphan receptors: the search for novel ligands and signal pathways. *Horm. Signal.* **1**: 307–358.

Willy, P.J., Umesono, K., Ong, E.S., Evans, R.M., Heyman, R.A. and Mangelsdorf, D.J. (1995) LXR, a nuclear receptor that defines a distinct retinoid response pathway. *Genes Dev.* **9**: 1033–1045.

Wilson, T.E., Paulsen, R.E., Padgett, K.A. and Millbrandt, J. (1992) Participation of non-zinc finger residues in DNA binding by two nuclear orphan receptors. *Science* **256**: 107–110.

Wolbach, S.B. (1956) Effects of vitamin A deficiency and hypervitaminosis in animals. In: W.H. Sebrell and R.S. Harris (eds) *The Vitamins*, Vol. 1, pp. 106–137. Academic Press, New York.

Wolbach, S.B. and Howe, P.R. (1925) Tissue changes following deprivation of fat-soluble A vitamin. *J. Exp. Med.* **62**: 753–777.

Wolf, G. (1998) Vitamin A functions in the regulation of the dopaminergic system in the brain and pituitary gland. *Nutr. Rev.* **56**: 354–355.

Wu, Q., Dawson, M. I., Zheng, Y. *et al.* (1997a) Inhibition of *trans*-retinoic acid-resistant human breast cancer cell growth by retinoid X receptor-selective retinoids. *Mol. Cell. Biol.* **17**: 6598–6608.

Wu, Q., Li, Y., Liu, R. *et al.* (1997b) Modulation of retinoic acid sensitivity in lung cancer cells through dynamic balance of orphan receptors nur77 and COUP-TF and their heterodimerization. *EMBO J.* **16**: 1656–1669.

Wurtz, J.-M., Bourguet, W., Renaud, J.-P. *et al.* (1996) A canonical structure for the ligand-binding domain of nuclear receptors. *Nat. Struct. Biol.* **3**: 87–94.

Xiao, J.H., Durand, B., Chambon, P. and Voorhees, J.J. (1995) Endogenous retinoic acid receptor–retinoid X receptor heterodimers are the major functional forms regulating retinoid-responsive elements in adult human keratinocytes. *J. Biol. Chem* **270**: 3001–3011.

Xu, L., Glass, C.K. and Rosenfeld, M.G. (1999) Coactivator and corepressor complexes in nuclear receptor function. *Curr. Opin. Genet. Dev.* **9**: 140–147.

Xu, X.C., Ro, J.Y., Lee, J.S., Shin, D.M., Hong, W.K. and Lotan, R. (1994) Differential expression of nuclear retinoic acid receptors in normal, premalignant, malignant head and neck tissues. *Cancer Res.* **54**: 3580–3587.

Yang, X.J., Ogryzko, W., Nishikawa, J., Howard, B.H. and Nakatani, Y. (1996) A p300/CBP-associated factor that competes with the adenoviral oncoprotein E1A. *Nature* **382**: 319–324.

Yao, T.P., Segraves, W.A., Oro, A.E., McKeown, M. and Evans, R.M. (1992) *Drosophila* ultraspiracle modulates ecdysone receptor function via heterodimer formation. *Cell* **71**: 63–72.

Yao, T.P., Forman, B.M., Jiang, J. *et al.* (1993) Functional ecdysone receptor is the product of EcR and ultraspiracle genes. *Nature* **366**: 476–479.

Yao, T.P., Oh, S.P., Fuchs, M. *et al.* (1998) Gene dosage-dependent embryonic development and proliferation defects in mice lacking the transcriptional integrator p300. *Cell* **93**: 361–372.

Ylikomi, T., Bocquel, M.T., Berry, M., Gronemeyer, H. and Chambon, P. (1992) Cooperation of proto-signals for nuclear accumulation of estrogen and progesterone receptor. *EMBO J.* **11**: 3681–3694.

Zamir, I., Zhang, J. and Lazar, M.A. (1997) Stoichiometric and steric principles governing repression by nuclear hormone receptors. *Genes Dev.* **11**: 835–846.

Zechel, C., Shen, X.Q., Chambon, P. and Gronemeyer, H. (1994a) Dimerization interfaces formed between the DNA binding domains determine the cooperative binding of RXR/RAR and RXR/TR heterodimers to DR5 and DR5 elements. *EMBO J.* **13**: 1414–1424.

Zechel, C., Shen, X.-O. Chen, J.-Y., Chen, Z.P., Chambon, P. and Gronemeyer, H. (1994b) The dimerization interfaces formed between the DNA binding domains of RXR, RAR, and TR determine the binding specificity and polarity of the full-length receptors to direct repeats. *EMBO J.* **13**: 1425–1433.

Zelent, A., Krust, A., Petkovich, M., Kastner, P. and Chambon, P. (1989) Cloning of murine α and β retinoic acid receptors and a novel receptor γ predominantly express in skin. *Nature* **339**: 714–717.

Zelent, A., Mendelsohn, C., Kastner, P. *et al.* (1991) Differentially expressed isoforms of the mouse retinoic acid receptor β are generated by usage of two promoters and alternative splicing. *EMBO J.* **10**: 71–81.

Zetterstrom, R.H., Solomin, L., Jansson, L., Hoffer, B.J., Olson, L. and Perlmann, T. (1997) Dopamine neuron agenesis in Nurr1-deficient mice. *Science* **276**: 248–250.

Zetterstrom, R.H., Lindqvist, E., de Urquiza, A.M. *et al.* (1999) Role of retinoids in the CNS: differential expression of retinoid binding proteins and receptors and evidence for presence of retinoic acid. *Eur. J. Neurosci.* **11**: 407–416.

Zhang, Y.K., Hoffmann, B., Tran, P.B.V., Graupner, G. and Pfahl, M. (1992) Retinoid X receptor is an auxiliary protein for thyroid hormone and retinoic acid receptors. *Nature* **355**: 441–445.

Zheng, A., Savolainen, E.R. and Koistinen, P. (1996) All-*trans* retinoic acid combined with interferon-α effectively inhibits granulocyte–macrophage colony formation in chronic myeloid leukemia. *Leuk. Res.* **20**: 243–248.

Zheng, Y., Kramer, P.M., Olson, G. *et al.* (1997) Prevention by retinoids of azoxymethane-induced tumors and aberrant crypt foci and their modulation of cell proliferation in the colon of rats. *Carcinogenesis* **18**: 2119–2125.

Zhu, Y., Qi, C., Calandra, C., Rao, M.S. and Reddy, J.K. (1996) Cloning and identification of mouse steroid receptor coactivator (mSRC-1), as a coactivator of peroxisome-proliferator-activated receptor γ. *Gen. Exp.* **6**: 185–195.

Chapter 8

Mineralocorticoid and Glucocorticoid Receptors

Tomoshige Kino, Alessandra Vottero and George P. Chrousos

INTRODUCTION

Two adrenal corticosteroids, the mineralocorticoid aldosterone and the glucocorticoid cortisol, exert profound influences on many physiologic functions by virtue of their diverse roles in growth, development and maintenance of homeostasis (Munck *et al.*, 1984; Clark *et al.*, 1992). Their actions are mediated by intracellular receptor proteins, the mineralocorticoid (MR) and glucocorticoid (GR) receptors, which function as hormone-activated transcription factors regulating the expression of, respectively, the mineralocorticoid and glucocorticoid target genes (Beato and Sanchez-Pacheco, 1996).

The MR mediates the sodium-retaining effects of aldosterone in the kidney, salivary glands, sweat glands and colon (Jorgensen, 1986). In addition, the MR located in the central nervous system (CNS), also called glucocorticoid receptor type I, appears to have a role in regulation of the stress response and feedback control of the hypothalamic–pituitary–adrenal (HPA) axis (de Kloet *et al.*, 1990). Recently, inactivating mutations in the MR were shown to cause pseudohypoaldosteronism type 1 (PHA-1), i.e. mineralo-corticoid resistance (Geller *et al.*, 1998). This disease, however, is overwhelmingly due to loss-of-function mutations in the subunits of the amiloride-sensitive sodium channel (ASSC), which represent a post-MR step in the signaling cascade of aldosterone in its target tissues (Chang *et al.*, 1996; Strautnieks *et al.*, 1996).

The GR is expressed ubiquitously in almost all human tissues and organs. The presence of glucocorticoids is crucial for the integrity of CNS function and for maintenance of cardiovascular, metabolic and immune homeostasis. Increased gluco-corticoid secretion during stress alters CNS function, assists with adjustments in energy expenditures, and modulates the inflammatory/immune response (Chrousos, 1995). As glucocorticoids possess a broad array of life-sustaining functions, only partial or incomplete glucocorticoid resistance has been reported so far, suggesting that complete inability of glucocorticoids to exert their effects on their target tissues is incompatible with human life. Over 10 kindreds and individual patients suffering from glucocorticoid resistance have been described to date and the molecular mechanisms of their resistance have been analyzed in some of them (Chrousos, 1995; Arai and Chrousos, 1995; de Castro and Chrousos, 1997).

Nuclear Receptors and Genetic Disease
ISBN 0-12-146160-2

STRUCTURE AND ACTIONS OF THE MINERALOCORTICOID AND GLUCOCORTICOID RECEPTORS

The MR and the GR are members of the steroid/sterol/thyroid/retinoid/orphan receptor superfamily of nuclear transactivating factors, with over 150 members currently cloned and characterized across species (Mangelsdorf *et al.*, 1995). Together with the progesterone and androgen receptors, MR and GR form the steroid receptor subfamily (Fig. 8.1). Steroid receptors display a modular structure comprised of five to six regions (A–F), with the N-terminal A/B region harboring an autonomous activation function (AF-1), and the C and E regions corresponding to the DNA- and ligand-binding domains. The MR and the GR consist of 984 and 777 amino acids respectively; they have almost identical DNA-binding domains (94%) and very similar ligand-binding domains (57%), but divergent N-terminal A/B regions (< 15%) (Arai and Chrousos, 1995). The MR and GR in their unliganded state are located primarily in the cytoplasm, as part of heterooligomeric complexes containing heat shock proteins 90, 70 and 50, and possibly other proteins (Beato and Sanchez-Pacheco, 1996).

After binding to their respective ligand, the MR and GR undergo conformational changes, dissociate from the heat shock proteins, homodimerize, and translocate into the nucleus where they interact with hormone-responsive elements and/or other transcription factors in the promoter regions of target genes. Both the MR and GR bind to and modulate transcription driven by the glucocorticoid-responsive element (GRE)-containing murine mammary tumor virus (MMTV) promoter. No specific mineralocorticoid-responsive elements (MREs) have been characterized in the regulatory regions of genes physiologically regulated by aldosterone as yet (Rupprecht *et al.*, 1993). Active GREs, on the other hand, are present in the promoter regions of many glucocorticoid-responsive genes. The GR as a dimer or monomer also modulates the transcription rates of non-GRE-containing genes regulated by other transcription factors, such as activator protein-1 (AP-1) (Schule *et al.*, 1990), nuclear factor (NF)-κB (Caldenhoven *et al.*, 1995) and signal transducer and activator of transcription (STAT)5 (Stocklin *et al.*, 1996), through protein–protein interactions with these factors.

The human GR complementary DNA (cDNA) was isolated by expression cloning in 1985 (Hollenberg *et al.*, 1985). The cDNA for the human MR was subsequently isolated by low-stringency hybridization, using the human GR cDNA as a probe in 1987 (Arriza *et al.*, 1987). The genes of the MR and GR consist of nine exons each; their loci are on chromosome 4 and 5 respectively (Fig. 8.1). For the GR, there are two 3′ splicing variants, GRα and β, from alternative use of a different terminal exon 9α or 9β. GRα is the classic GR, which binds to glucocorticoids and transactivates or transrepresses glucocorticoid-responsive promoters. On the other hand, GRβ does not bind glucocorticoids and functions as a weak dominant negative inhibitor of GRα on GRE-containing glucocorticoid-responsive promoters (Bamberger *et al.*, 1995). For the MR, alternative promoters of the MR gene have been reported regulating production of the same final receptor protein; the functional significance of this is not clear (Zennaro *et al.*, 1994).

The MR has a high affinity for both aldosterone and cortisol, and the circulating levels of cortisol are over 100 times higher than those of aldosterone. The MR of the distal convoluted tubule, and possibly other mineralocorticoid target tissues, is protected from the actions of cortisol by expression of 11β-hydroxysteroid dehydrogenase type 2, which converts cortisol into the inactive cortisone (Funder, 1997).

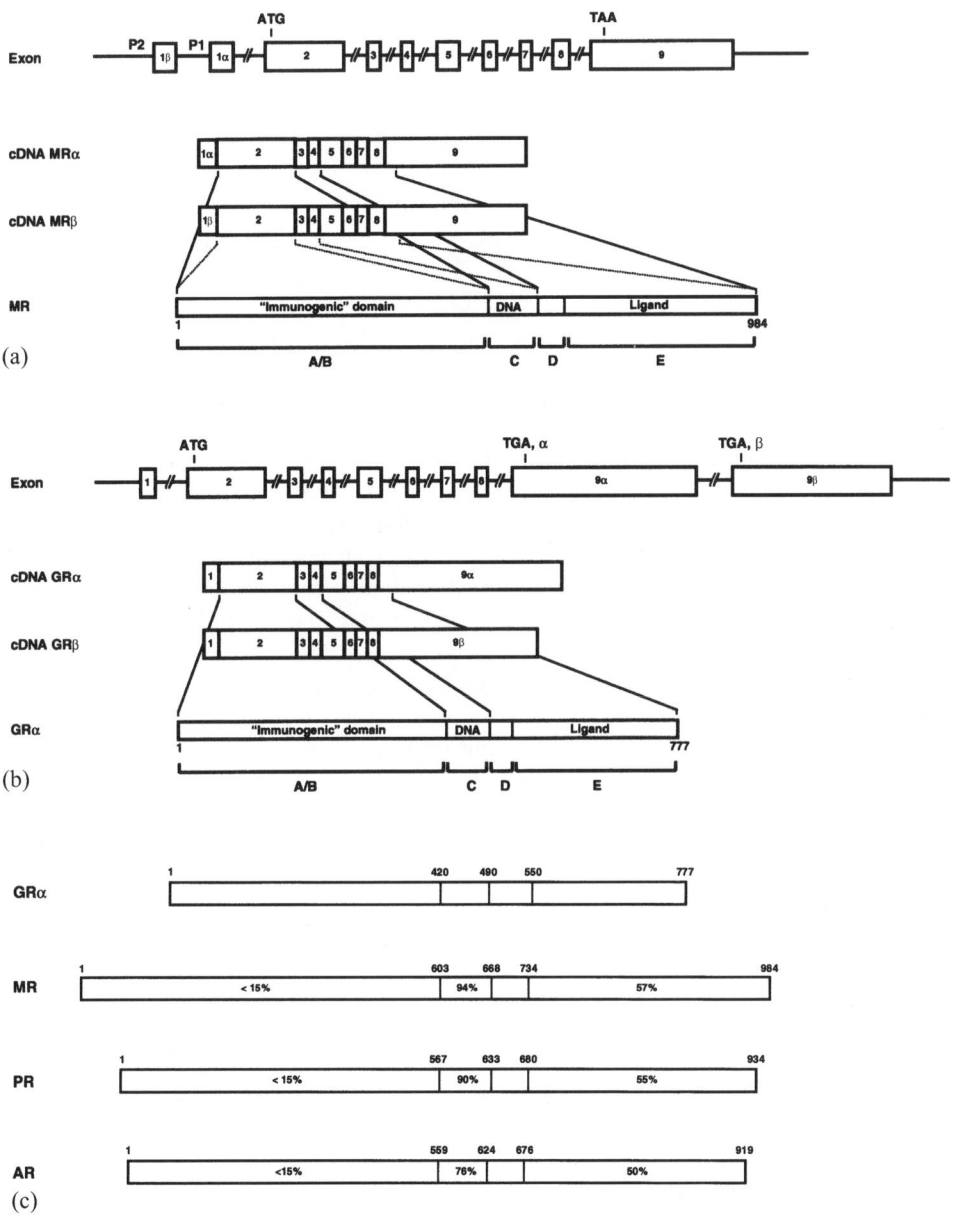

Fig. 8.1 Genomic and complementary DNA (cDNA), and protein structures of (a) the human mineralocorticoid receptor (MR), (b) the glucocorticoid receptor (GR) and (c) their homologies to other steroid receptors. (a and b) Both MR and GR genes consist of 10 exons. The MR has two exons 1 (exon 1α and 1β), each with an alternative promoter; however, the final translated MR protein is identical. Both exon 1α and 1β are untranslated, exon 2 codes for the immunogenic domain (A/B), exons 3 and 4 for the DNA-binding domain (C), and exons 5–9 for the hinge region (D) and the ligand-binding domain (E). The GR gene contains two terminal exons 9 (exon 9α and 9β), alternatively spliced to produce the classic GRα and the nonligand-binding GRβ. (c) Homologies between steroid receptors expressed as percentage identity with the primary sequence of the GR. PR, progesterone receptor; AR, androgen receptor.

PATHOPHYSIOLOGY AND CLINICAL PRESENTATION

Mineralocorticoid receptor mutations

The mechanism by which aldosterone stimulates sodium transport in its target tissues may involve the synthesis of a protein associated with the function of the ASSC. The latter is located in the apical membrane of epithelial cells of the renal distal convoluted tubule, and in the plasma membranes of cells in other tissues involved with salt conservation (Canessa *et al.*, 1993, 1994). The phenotype of patients with loss-of-function mutations of the MR mimics that of patients with defects in the subunits of the ASSC (Geller *et al.*, 1998), who, however, represent the bulk of patients with PHA-1 (Chang *et al.*, 1996; Strautnieks *et al.*, 1996).

Cheek and Perry first reported PHA-1 in an infant with severe salt-wasting syndrome in 1958; PHA-1 was subsequently reported in more than 70 patients (Speiser and Stoner, 1986). This syndrome usually presents in infancy with urinary salt wasting and failure to thrive. Plasma renin activity and aldosterone concentrations are markedly raised. Approximately one-fifth of these cases are familial (Arai and Chrousos, 1995). All patients have renal tubular unresponsiveness to aldosterone, whereas some have multiple mineralocorticoid target tissue involvement, including the sweat and salivary glands and the colonic epithelium (Oberfield *et al.*, 1977; Savage *et al.*, 1982; Armanini *et al.*, 1985; Caufriez *et al.*, 1986).

In kindreds with PHA-1, both an autosomal dominant and recessive form of genetic transmission have been observed. The autosomal recessive form was associated with severe disease, with manifestations persisting into adulthood. The authors and others have failed to find pathologic mutations in the MR gene in sporadic and familial cases with autosomal recessive PHA-1, and concluded that, most likely, this condition was due to a defect in a post-MR step of aldosterone action (Zennaro *et al.*, 1994; Arai *et al.*, 1995). Indeed, in 1996, PHA-1 was found to be caused by loss-of-function mutations in genes encoding subunits of the ASSC (Chang *et al.*, 1996; Strautnieks *et al.*, 1996). However, more recently, Geller *et al.* (1998) identified heterozygotic MR gene loss-of-function mutations in one sporadic case and four autosomal dominant cases of PHA-1 (Fig. 8.2, Table 8.1). These included two frameshift mutations, each deleting a single base pair in exon 2; the resultant frameshifts resulted in a gene product lacking the entire DNA- and hormone-binding domains, as well as a dimerization motif. Two families had an identical mutation, introducing a premature termination codon in exon 2 at position 537. One case showed a single base-pair deletion in the intron-5 splice donor site. We recently suggested that double heterozygocity between MR and ASSC subunit gene mutations might also result in PHA-1 (Arai *et al.*, 1999).

Glucocorticoid receptor mutations

A complex negative feedback system exists in the human CNS that regulates glucocorticoid homeostasis. Glucocorticoids exert negative feedback effects on both hypothalamic corticotropic-releasing hormone (CRH) and arginine vasopressin (AVP) secretion and inhibit pituitary adrenocorticotropic hormone (ACTH) secretion itself. In addition, glucocorticoids influence the activity of suprahypothalamic centers that control the activity of CRH and AVP neurons (Chrousos, 1995). This complex regulatory system is activated in patients with loss-of-function GR mutations, resulting in compensatory

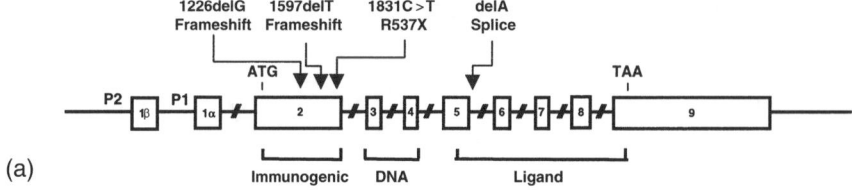

(a)

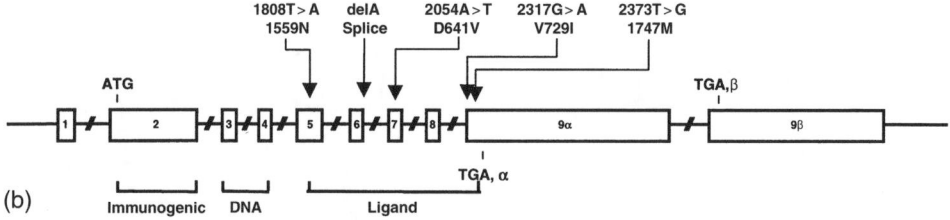

(b)

Fig. 8.2 Location of known mutations of (a) the mineralocorticoid receptor and (b) the glucocorticoid receptor in the genomic structure.

Table 8.1 Mutations of the MR causing mineralocorticoid resistance

	Mutation position			Genotype and	Clinical
	cDNA	Amino acid	Localization	transmission	phenotype
Case 1	1226delG	409 Frameshift	A/B	Heterozygote, sporadic	PHA-1
Kindred 2	1597delT	533 Frameshift	A/B	Heterozygote, autosomal dominant	PHA-1
Kindred 3	1831C > T	R537X	A/B	Heterozygote, autosomal dominant	PHA-1
Kindred 4	DelA at intron 5 splice donor site		E	Heterozygote, autosomal dominant	PHA-1

From Geller *et al.* (1998). PHA-1, pseudohypoaldosteronism type 1.

increases in ACTH and cortisol secretion (Fig. 8.3). Although adequate compensation is apparently achieved by increased cortisol concentrations in the great majority of the patients described, excess ACTH secretion also results in increased production of adrenal steroids with salt-retaining activity and enhanced secretion of adrenal androgens. The former, together with cortisol, is responsible for mineralocorticoid excess, whereas the latter cause varying manifestations of hyperandrogenism.

The syndrome of familial glucocorticoid resistance, first reported in 1976, is a disorder characterized by hypercorticosolism without cushingoid features (Vingerhoeds *et al.*, 1976; Chrousos *et al.*, 1982). Since then, more than 10 kindreds and sporadic cases have been reported (Iida *et al.*, 1985; Lamberts *et al.*, 1986, 1992; Nawata *et al.*, 1987; Vecsei *et al.*, 1989; Hurley *et al.*, 1991; Karl *et al.*, 1993, 1996; Malchoff *et al.*, 1993). Abnormal-

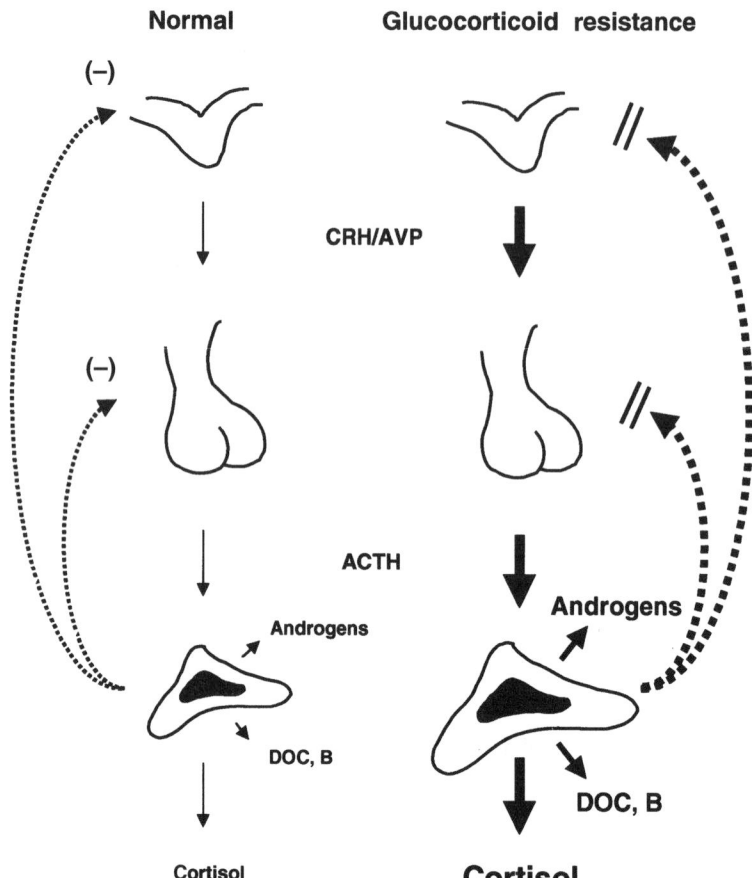

Fig. 8.3 Pathophysiologic mechanism of glucocorticoid resistance induced by loss-of-function and/or dominant negative glucocorticoid receptor mutations. The elaborate negative feedback mechanism responsible for maintenance of glucocorticoid homeostasis compensates for tissue insensitivity to glucocorticoids by resetting the hypothalamic–pituitary–adrenal axis at a higher level. Thus, corticotropic-releasing hormone (CRH)/arginine vasopressin (AVP), adrenocorticotropic hormone (ACTH) and cortisol secretion are increased. The compensatory increase in ACTH production augments the secretion of cortisol and glucocorticoid precursors with mineralocorticoid activity (DOC, deoxycorticosterone; B, corticosterone), as well as the secretion of several adrenal androgens including Δ4-androstenedione, which has considerable androgen activity.

ities of GR number, affinity for glucocorticoids, stability, and translocation into the nucleus have been described. The molecular defects of four kindreds and one sporadic case have been elucidated so far (Fig. 8.2, Table 8.2).

The propositus of the original kindred was found by Hurley *et al.* (1991) to be a homozygote for a single nonconservative point mutation, replacing aspartic acid with valine at amino acid 641 in the hormone-binding domain of the GR; this mutation reduced the binding affinity for dexamethasone by threefold and caused loss of transactivation activity on the MMTV promoter. The proposita of the second family

Table 8.2 Mutations of the GR causing glucocorticoid resistance

Reference	Mutation position		Localization	Biochemical phenotype	Genotype and transmission
	cDNA	Amino acid			
Hurley et al. (1991)	2054A > T	D461V	E	Affinity↓ Transactivation↓	Homozygote, autosomal recessive
Karl et al. (1993)	Δ4 at the 3' boundary of exon and intron 6		E	Number↓ Inactivation of the affected allele	Heterozygote, autosomal dominant
Malchoff et al. (1993)	2317G > A	V7291	E	Affinity↓ Transactivation↓	Homozygote, autosomal recessive
Karl et al. (1996)	1808T > A	1559N	D	Number↓ Transactivation↓ Dominant negative activity on wild-type	Heterozygote, sporadic
Vottero et al. (1999)	2373T > G	1747M	E	Affinity↓ Transactivation↓ Dominant negative activity on wild-type	Heterozygote, autosomal dominant

had four-base deletion at the 3' boundary of exon 6, removing a donor splice site. This resulted in complete ablation of one of the GR alleles in affected family members (Karl *et al.*, 1993). The propositus of the third kindred had a single homozygotic point mutation at amino acid 729 (valine to isoleucine; V729I) in the hormone binding domain, which reduced both the affinity and transactivation activity of the GR (Malchoff *et al.*, 1993). There was also an interesting sporadic case of a man with a *de novo*, germline, heterozygotic GR mutation at amino acid 559 (isoleucine to asparagine; I559N) in the hormone-binding domain, close to the DNA-binding domain at the so-called hinge region of the GR. This mutant GR bound no ligand but exerted dominant negative activity upon the wild-type receptor and was associated with later development of Cushing disease due to an ACTH-secreting pituitary adenoma (Karl *et al.*, 1996). Recently, the present authors studied a fifth case/kindred with glucocorticoid resistance, and a heterozygotic GR mutation in the ligand-binding domain (amino acid 747, replacing isoleucine with methionine; I747M) was determined; the mutant receptor had reduced affinity for dexamethasone and decreased transactivational activity; interestingly, it also had dominant negative activity upon the wild-type receptor.

DIAGNOSTIC EVALUATION AND THERAPY

Mineralocorticoid resistance

Patients with MR mutations present with dehydration, low to normal levels of serum sodium, hyperkalemia, acidosis, hyperreninemia, and 'paradoxically' increased plasma and urinary aldosterone levels, especially in infancy. A mild presentation and improvement with age are usually seen in patients with MR mutations, compared with patients carrying mutations in the ASSC subunits, who also present with a similar early phenotype but have a more severe presentation and course (Geller *et al.*, 1998). The improvement observed in PHA-1 of either etiology with age is consistent with reduced dependence of the patients on aldosterone action as they grow older (Rosler, 1984).

Patients with PHA-1 require supplemental sodium chloride, which usually corrects the hyponatremia and hyperkalemia, improves symptoms, and enhances growth. After infancy, the disorder typically abates sufficiently to permit reduction or discontinuation of sodium chroride supplements, but the condition may recur during periods of dietary salt restriction (Funder *et al.*, 1990). Treatment with high doses of the synthetic mineralocorticoid fluorocortisone and/or the 11β-hydroxysteroid dehydrogenase type 2 inhibitor carbenoxolone was tried with normalization of the serum electrolyte concentration in a patient with PHA-1 (Arai and Chrousos, 1995). These therapeutic methods could be tried also in patients with PHA-1 carrying mutations of the MR that only partially affect its ability to function.

Glucocorticoid resistance

Owing to the compensatory increase of ACTH secretion to achieve raised circulating cortisol concentrations, patients with loss-of-function GR mutations present with manifestations resulting from increased production of adrenal steroids with salt-retaining and/or androgenic activity. The former cause symptoms and signs of mineralocorticoid excess, such as hypertension and/or hypokalemic alkalosis. The latter lead to symptoms and signs of androgen excess in women, such as acne, hirsutism, male pattern baldness, menstrual irregularities and infertility. Precocious puberty has been seen in a child due to early and excessive prepubertal adrenal androgen secretion. In the adult male, oligospermia and infertility have been observed, possibly as a result of disturbances in follicle-stimulating hormone regulation caused by excessive levels of adrenal androgens.

The spectrum of clinical manifestations in patients with GR mutations is quite broad, as a large number of subjects are asymptomatic and show only biochemical changes. The hallmark of the diagnostic evaluation is hypercortisolism without Cushing stigmata. The patients retain the circadian rhythm and responsiveness of cortisol to stress and are resistant to single or multiple doses of dexamethasone.

Patients with generalized glucocorticoid resistance are treated with high doses of mineralocorticoid-sparing synthetic glucocorticoids. The goal is to suppress the increased levels of ACTH, which cause overproduction of mineralocorticoids and androgens (Fig. 8.3). As all cases described thus far have had partial inactivation of GR activity, the administration of synthetic potent glucocorticoids with minimal intrinsic mineralocorticoid activity (e.g. dexamethasone) is a rational approach to achieve sufficient activation of the mutated GR and to suppress the compensatory increase in ACTH activity. The patients should be treated with high, individualized doses of oral dexamethasone, which would be pharmacologic for the normal population (1–3 mg daily). Dexamethasone does indeed suppress ACTH and therefore endogenous cortisol, deoxycorticosterone, corticosterone and adrenal androgen secretion, so correcting the mineralocorticoid and androgen excess states of these patients.

REFERENCES

Arai, K. and Chrousos, G.P. (1995) Syndromes of glucocorticoid and mineralocorticoid resistance. *Steroids* **60**: 173–179.

Arai, K., Tsigos, C., Suzuki, Y. *et al.* (1995) No apparent mineralocorticoid receptor defect in a series of sporadic cases of pseudohypoaldosteronism. *J. Clin. Endocrinol. Metab.* **80**: 814–817.

Arai, K., Zachman, K., Shibasaki, T. and Chrousos, G.P. (1999) Polymorphisms of amiloride-sensitive sodium channel subunits in five sporadic cases of pseudohypoaldosteronism: do they have pathologic potential? *J. Clin. Endocrinol. Metab.* **84**: 2434–2437.

Armanini, D., Kuhnle, U., Strasser, T. *et al.* (1985) Aldosterone-receptor deficiency in pseudohypoaldosteronism. *N. Engl. J. Med.* **313**: 1178–1181.

Arriza, J.L., Weinberger, C., Cerelli, G. *et al.* (1987) Cloning of human mineralocorticoid receptor complementary DNA: structural and functional kinship with the glucocorticoid receptor. *Science* **237**: 268–275.

Bamberger, C.M., Bamberger, A.M., de Castro, M. and Chrousos, G.P. (1995) Glucocorticoid receptor β, a potential endogenous inhibitor of glucocorticoid action in humans. *J. Clin. Invest.* **95**: 2435–2441.

Beato, M. and Sanchez-Pacheco, A. (1996) Interaction of steroid hormone receptors with the transcription initiation complex. *Endocr. Rev.* **17**: 587–609.

Caldenhoven, E., Liden, J., Wissink, S. *et al.* (1995) Negative cross-talk between RelA and the glucocorticoid receptor: a possible mechanism for the antiinflammatory action of glucocorticoids. *Mol. Endocrinol.* **9**: 401–412.

Canessa, C.M., Horisberger, J.D. and Rossier, B.C. (1993) Epithelial sodium channel related to proteins involved in neurodegeneration. *Nature* **361**: 467–470.

Canessa, C.M., Schild, L., Buell, G. *et al.* (1994) Amiloride-sensitive epithelial Na$^+$ channel is made of three homologous subunits. *Nature* **367**: 463–467.

Caufriez, A., Golstein, J., Tadjerouni, A. *et al.* (1986) Modulation of immunoreactive somatomedin-C levels by sex steroids. *Acta Endocrinol. (Copenh.)* **112**: 284–289.

Chang, S.S., Grunder, S., Hanukoglu, A. *et al.* (1996) Mutations in subunits of the epithelial sodium channel cause salt wasting with hyperkalaemic acidosis, pseudohypoaldosteronism type 1. *Nat. Genet.* **12**: 248–253.

Chrousos, G.P. (1995) The hypothalamic–pituitary–adrenal axis and immune-mediated inflammation. *N. Engl. J. Med.* **332**: 1351–1362.

Chrousos, G.P., Vingerhoeds, A., Brandon, D. *et al.* (1982) Primary cortisol resistance in man. A glucocorticoid receptor-mediated disease. *J. Clin. Invest.* **69**: 1261–1269.

Clark, J.K., Schrader, W.T. and O'Malley, B.W. (1992) Mechanism of steroid hormones. In: J.D. Wilson and D.W. Foster (eds) *Williams Textbook of Endocrinology*, pp. 35–90. WB Saunders, Philadelphia.

de Castro, M. and Chrousos, G.P. (1997) Glucocorticoid resistance. *Curr. Ther. Endocrinol. Metab.* **6**: 188–189.

de Kloet, E.R., Reul, J.M. and Sutanto, W. (1990) Corticosteroids and the brain. *J. Steroid Biochem. Mol. Biol.* **37**: 387–394.

Funder, J.W. (1997) Glucocorticoid and mineralocorticoid receptors: biology and clinical relevance. *Annu. Rev. Med.* **48**: 231–240.

Funder, J.W., Pearce, P.T., Myles, K. and Roy, L.P. (1990) Apparent mineralocorticoid excess, pseudohypoaldosteronism, and urinary electrolyte excretion: toward a redefinition of mineralocorticoid action. *FASEB J.* **4**: 3234–3238.

Geller, D.S., Rodriguez-Soriano, J., Vallo Boado, A. *et al.* (1998) Mutations in the mineralocorticoid receptor gene cause autosomal dominant pseudohypoaldosteronism type I. *Nat. Genet.* **19**: 279–281.

Hollenberg, S.M., Weinberger, C., Ong, E.S. *et al.* (1985) Primary structure and expression of a functional human glucocorticoid receptor cDNA. *Nature* **318**: 635–641.

Hurley, D.M., Accili, D., Stratakis, C.A. *et al.* (1991) Point mutation causing a single amino acid substitution in the hormone binding domain of the glucocorticoid receptor in familial glucocorticoid resistance. *J. Clin. Invest.* **87**: 680–686.

Iida, S., Gomi, M., Moriwaki, K. *et al.* (1985) Primary cortisol resistance accompanied by a reduction in glucocorticoid receptors in two members of the same family. *J. Clin. Endocrinol. Metab.* **60**: 967–971.

Jorgensen, P.L. (1986) Structure, function and regulation of Na,K-ATPase in the kidney. *Kidney Int.* **29**: 10–20.

Karl, M., Lamberts, S.W., Detera-Wadleigh, S.D. *et al.* (1993) Familial glucocorticoid resistance caused by a splice site deletion in the human glucocorticoid receptor gene. *J. Clin. Endocrinol. Metab.* **76**: 683–689.

Karl, M., Lamberts, S.W., Koper, J.W. *et al.* (1996) Cushing's disease preceded by generalized glucocorticoid resistance: clinical consequences of a novel, dominant-negative glucocorticoid receptor mutation. *Proc. Assoc. Am. Phys.* **108**: 296–307.

Lamberts, S.W., Poldermans, D., Zweens, M. and de Jong, F.H. (1986) Familial cortisol resistance: differential diagnostic and therapeutic aspects. *J. Clin. Endocrinol. Metab.* **63**: 1328–1333.

Lamberts, S.W., Koper, J.W., Biemond, P., den Holder, F.H. and de Jong, F.H. (1992) Cortisol receptor resistance: the variability of its clinical presentation and response to treatment. *J. Clin. Endocrinol. Metab.* **74**: 313–321.

Malchoff, D.M., Brufsky, A., Reardon, G. *et al.* (1993) A mutation of the glucocorticoid receptor in primary cortisol resistance. *J. Clin. Invest.* **91**: 1918–1925.

Mangelsdorf, D.J., Thummel, C., Beato, M. *et al.* (1995) The nuclear receptor superfamily: the second decade. *Cell* **83**: 835–839.

Munck, A., Guyre, P.M. and Holbrook, N.J. (1984) Physiological functions of glucocorticoids in stress and their relation to pharmacological actions. *Endocr. Rev.* **5**: 25–44.

Nawata, H., Sekiya, K., Higuchi, K., Kato, K. and Ibayashi, H. (1987) Decreased deoxyribonucleic acid binding of glucocorticoid–receptor complex in cultured skin fibroblasts from a patient with the glucocorticoid resistance syndrome. *J. Clin. Endocrinol. Metab.* **65**: 219–226.

Oberfield, S.E., Levine, L.S., Carey, R.M., Bejar, R. and New, M.I. (1977) Pseudohypoaldosteronism: multiple target organ unresponsiveness to mineralocorticoid hormones. *J. Clin. Endocrinol. Metab.* **48**: 228–234.

Rosler, A. (1984) The natural history of salt-wasting disorders of adrenal and renal origin. *J. Clin. Endocrinol. Metab.* **59**: 689–700.

Rupprecht, R., Arriza, J.L., Spengler, D. *et al.* (1993) Transactivation and synergistic properties of the mineralocorticoid receptor: relationship to the glucocorticoid receptor. *Mol. Endocrinol.* **7**: 597–603.

Savage, M.O., Jefferson, I.G., Dillon, M.J., Milla, P.J., Honour, J.W. and Grant, D.B. (1982) Pseudohypoaldosteronism: severe salt wasting in infancy caused by generalized mineralocorticoid unresponsiveness. *J. Pediatr.* **101**: 239–242.

Schule, R., Rangarajan, P., Kliewer, S. *et al.* (1990) Functional antagonism between oncoprotein c-Jun and the glucocorticoid receptor. *Cell* **62**: 1217–1226.

Speiser, P.W., Stoner, E. and New, M.I. (1986) Pseudohypoaldosteronism: a review and report of two new cases. *Adv. Exp. Med. Biol.* **196**: 173–195.

Stocklin, E., Wissler, M., Gouilleux, F. and Groner, B. (1996) Functional interactions between Stat5 and the glucocorticoid receptor. *Nature* **383**: 726–728.

Strautnieks, S.S., Thompson, R.J., Gardiner, R.M. and Chung, E. (1996) A novel splice-site mutation in the γ subunit of the epithelial sodium channel gene in three pseudohypoaldosteronism type 1 families. *Nat. Genet.* **13**: 248–250.

Vecsei, P., Frank, K., Haack, D. *et al.* (1989) Primary glucocorticoid receptor defect with likely familial involvement. *Cancer Res.* **49**: 2220s–2221s.

Vingerhoeds, A.C.M., Thijssen, J.H.H. and Schwarts, F. (1976) Spontaneous hypercortisolism without Cushing's syndrome. *J. Clin. Endocrinol. Metab.* **43**: 1128–1133.

Vottero, A., Combe, H., Lecomte, P., Longvil, C.A. and Chrousos, G.P. (1999) 81st annual meeting of the Endocrine Society. 12–15 June, San Diego, CA.

Zennaro, M.C., Borensztein, P., Jeunemaitre, X., Armanini, D. and Soubrier, F. (1994) No alteration in the primary structure of the mineralocorticoid receptor in a family with pseudohypoaldosteronism. *J. Clin. Endocrinol. Metab.* **79**: 32–380.

Chapter 9

Hepatocyte Nuclear Factor 4α

Frances M. Sladek and Shawn D. Seidel

INTRODUCTION AND HISTORICAL PERSPECTIVE

Hepatocyte nuclear factor 4 (HNF4*; NR2A1 in the unified nomenclature system (Nuclear Receptor Nomenclature Committee, 1999)) is an orphan member of the nuclear receptor superfamily since a ligand for it has not been definitively identified. It is also one of several transcription factors essential for liver-specific gene expression. It is a fairly abundant, constitutively acting positive factor that plays a role in regulating the transcription of many of the genes essential to the functioning of the liver. In addition to liver, HNF4 is also present at high levels in the kidney, small intestine and colon, and at lower levels in the pancreas and stomach. While much has been learned about HNF4 in the 10 years since it was first mentioned in the literature, much more remains to be elucidated about this intriguing factor. As an extensive review covering the first 5 years of HNF4 research already exists (Sladek, 1994), this review emphasizes what has been learned about HNF4 during the past 5 years, particularly in relation to human disease.

Purification and cloning of rat HNF4α1

HNF4 was originally identified as an activity in crude rat liver nuclear extracts that bound DNA elements required for the transcription of two liver-specific genes, transthyretin (TTR) and apolipoprotein CIII (apoCIII) (Costa *et al.*, 1989; Sladek *et al.*, 1990). Only a 5000-fold purification was required to obtain sufficient HNF4 protein for amino acid sequencing and subsequent cloning of the complementary DNA (cDNA). Calculations based on the amount of HNF4 protein obtained from a given number of rat livers suggest that there may be as many as 80 000 molecules of HNF4 per adult rat hepatocyte. It is perhaps not too surprising then that HNF4 is one of the few nuclear receptors isolated by protein purification. The original HNF4 clone was subsequently renamed HNF4α1.

The sequence of HNF4α1 immediately identified it as a member of the nuclear

---

*'HNF4' is preferred to 'HNF-4', the original designation, as the identification of additional HNF4 genes and isoforms has made the name more complex.

Nuclear Receptors and Genetic Disease
ISBN 0-12-146160-2

receptor superfamily as it contained the conserved zinc finger/DNA-binding domain (DBD) as well as the large hydrophobic ligand-binding domain (LBD) (Fig. 9.1). The mammalian receptors most similar to HNF4 in amino acid sequence are the retinoid X receptor α (RXRα) with close to 60% amino acid sequence identity in the DBD and over 35% identity in the LBD. Evolutionarily, the HNF4 subfamily is also most similar to the RXR family (Nuclear Receptor Nomenclature Committee, 1999). Orphan receptors chick ovalbumin upstream promoter-transcription factor (COUP-TF) I and II (also known as apolipoprotein regulatory protein-1 (ARP-1)) are also very similar in amino acid sequence to HNF4α1 (Mietus-Snyder *et al.*, 1992).

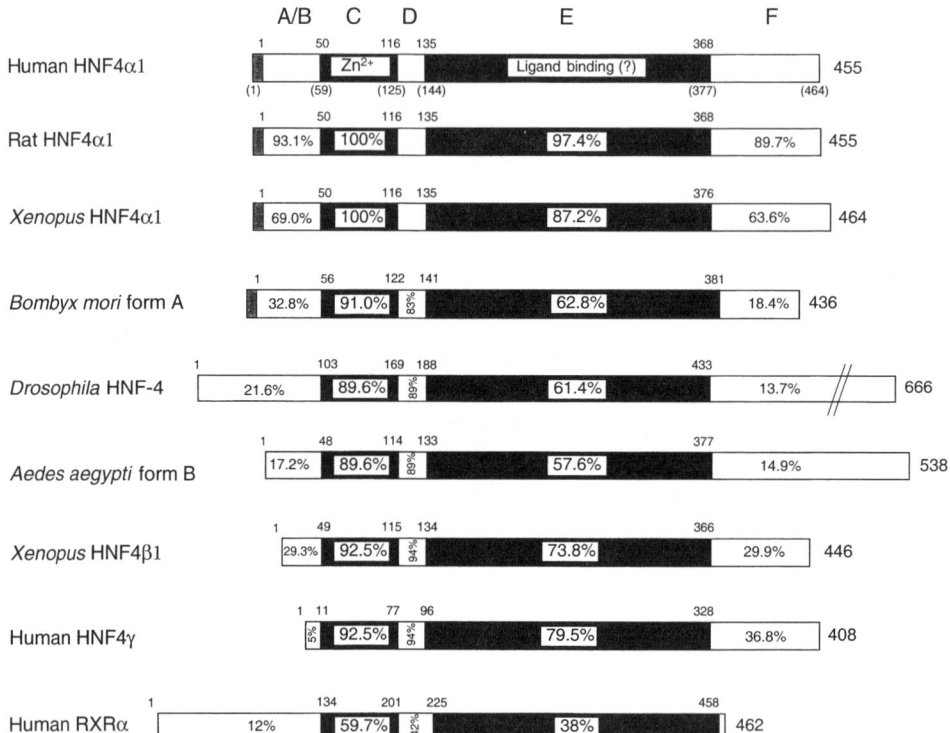

Fig. 9.1 Conservation of hepatocyte nuclear factor (HNF)4α amino acid sequence. The domain structure and amino acid sequence similarity in percentage identity of HNF4α genes from a variety of species are shown. Only the isoforms most similar to human HNF4α1 are shown. The original amino acid numbering corresponding to the start and end of the DNA-binding domain (Zn^{2+}, domain C) and the ligand-binding domain (Ligand Binding?, domain E) is given on top. The boundaries correspond to the greatest sequence conservation across species. Numbering with the nine amino acid N-terminal extension (shaded box) is given in parentheses for human (see text for details). The N-terminal extension was included when calculating the percentage identity for all genes. Not shown is mouse HNF4α1, which differs from rat HNF4α1 by four amino acids. The hinge region (domain D) of rat and *Xenopus* are 100% identical to that of human. GenBank accession numbers: human HNF4α1 (X87870); rat HNF4α1 (X57133); *Xenopus* HNF4α1 (Z37526); *Bombyx mori* form A (U63843); *Drosophila* HNF-4 (U70874); *Aedes aegypti* form B (AF059027); *Xenopus* HNF4β1 (Z49827); human HNF4γ (Z49826); human RXRα (X52773).

Conservation of amino acid sequence and tissue distribution of HNF4α

HNF4α1 is one of the most evolutionarily conserved nuclear receptors. In addition to rat, HNF4α cDNAs have also been cloned from human, mouse, frog and several insects: *Bombyx mori* (silkmoth), *Drosophila melanogaster* (fruitfly) and *Aedes aegypti* (mosquito) (see Fig. 9.1). The vertebrate HNF4α1 genes all have 100% amino acid sequence identity in the DBD and 87% or more identity in the LBD. Insects (*Drosophila* and *A. aegypti*) are over 89% identical to human HNF4α in the DBD and over 58% identical in the LBD. This high degree of conservation suggests that if there is a ligand for HNF4α1 it too will be highly conserved.

Like the amino acid sequence, the tissue distribution of HNF4α is also conserved between vertebrates and invertebrates: HNF4α is primarily found in the vertebrate liver, kidney and intestine, and in the invertebrate fat body, malpighian tubules and gut, which are the functionally equivalent tissues (Sladek *et al.*, 1990; Zhong *et al.*, 1993; Swevers and Iatrou, 1998). Two other HNF4 genes have been cloned in addition to HNF4α. HNF4β was cloned from *Xenopus* and HNF4γ from human (Fig. 9.1).

Human HNF4γ, consisting of 12 exons and spanning 35 kilobases (kb) on chromosome 8, encodes a 408 amino acid protein with 70% overall identity to the human HNF4α1 protein (Fig. 9.1). The most notable features of HNF4γ are a truncated A/B domain in the N-terminus and an F domain similar but not identical to HNF4α1. The DBD and LBD are remarkably similar to those of HNF4α. (The first HNF4γ cDNA identified contained an inordinately long A/B domain in the N-terminus (Drewes *et al.*, 1996), which was later determined to be a cloning artefact (Plengvidhya *et al.*, 1999).) HNF4γ RNA appears to be 10-fold less abundant than HNF4α1 RNA by Northern blot analysis (Drewes *et al.*, 1996). The data on tissue distribution are sketchy but there are reports of HNF4γ RNA in the liver, kidney, pancreas and possibly intestine, brain, lung and testis. There is also mention of kidney-specific expression of exon 2A on HNF4γ (Drewes *et al.*, 1996; Plengvidhya *et al.*, 1999). The only published functional data on HNF4γ was performed with the N-terminal cloning artefact although even that construct activated transcription well, which is what would be expected from what is known about HNF4α1 function (Drewes *et al.*, 1996; Sladek *et al.*, 1999). Polymorphisms have been identified in the HNF4γ gene but have not yet been linked to diabetes or any other disease (Plengvidhya *et al.*, 1999).

Xenopus HNF4β has a much wider tissue distribution than *Xenopus* HNF4α, with the protein being expressed most abundantly in ovary and testes and to a lesser extent in liver, kidney, intestine, lung and stomach. Like HNF4α, *Xenopus* HNF4β is a maternal protein and is present throughout embryogenesis, although the RNA levels vary somewhat and are lower than those seen in the adult liver. *Xenopus* HNF4α is expressed during early oogenesis and is absent in egg, whereas HNF4β is first detected in the latest stages of oogenesis and is present in the egg and early cleavage stages. HNF4β also activates transcription from an HNF4α response element, although not as well as HNF4α (Holewa *et al.*, 1997). As little more is currently known about the HNF4β and HNF4γ genes, the remainder of this review will concern only HNF4α.

Structure of the HNF4α gene and multiple splicing variants

The original HNF4 clone was renamed HNF4α1 when other isoforms derived from alternative splicing events in the HNF4α gene were identified. Several naturally occurring splicing events have been identified thus far in the HNF4α mammalian RNA

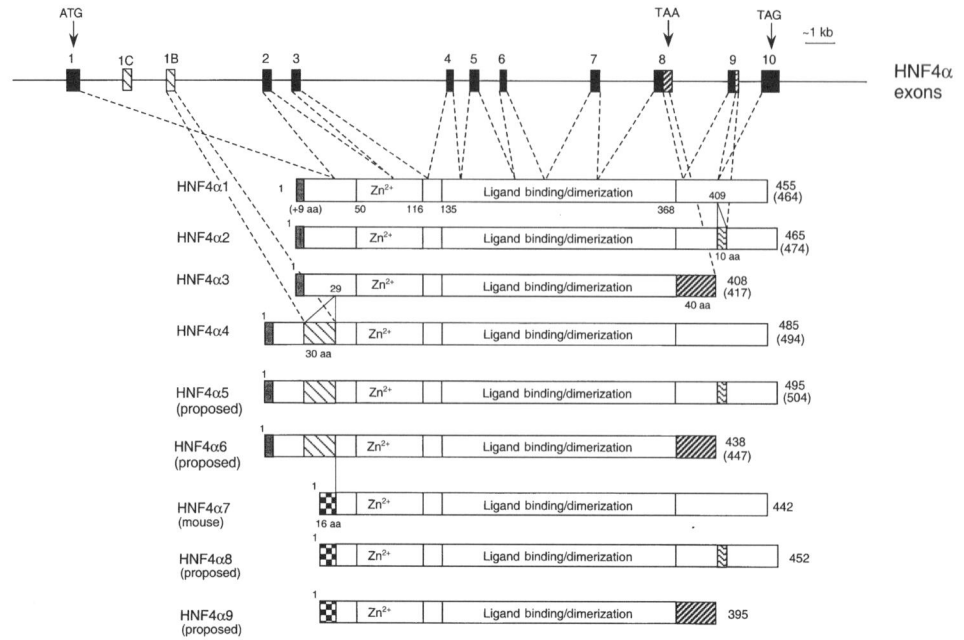

Fig. 9.2 Structure of the human hepatocyte nuclear factor (HNF)4α gene and isoforms generated by alternative splicing events. Shown are the 12 exons of the human HNF4α gene spanning more than 25 kb; exons 1 and 10 also contain 5′ and 3′ untranslated regions, respectively (not shown). Numbering of amino acids is as in Fig. 9.1; the number of residues in the inserts are indicated. Alternative names for the isoforms found in the literature are: HNF4α1 (HNF4, HNF-4, HNF4A), HNF4α2 (HNF4B, HNF4CL4) and HNF4α3 (HNF4C). The HNF4α4 complementary DNA (cDNA) is shown as containing only the sequence from exon 1B, as originally reported by Drewes *et al.* (1996). Furuta *et al.* (1997) indicated that the sequence from exon 1C is joined with that from exon 1B in HNF4α4. HNF4α5, HNF4α6, HNF4α8 and HNF4α9 transcripts have not yet been identified *in vivo*, hence the designation 'proposed'. To date, HNF4α7 has been found only in mouse as a cDNA. The mouse HNF4α gene has essentially an identical structure. Adapted from Furuta *et al.* (1997) and Taraviras *et al.* (1994).

which have the potential to yield a total of at least nine distinct isoforms (HNF4α1 through HNF4α9) (see Fig. 9.2). The cloning of HNF4α1 messenger RNA (mRNA) from species other than rat also suggested that there might be an alternative start methionine nine amino acids upstream of the start codon identified in the original cloning (Drewes *et al.*, 1996). Recent, unpublished, work from the authors' laboratory indicates that those nine amino acids are indeed translated *in vitro* and *in vivo*. However, it is not known which methionine is the preferred start codon *in vivo* and most of the literature on HNF4α, including that on naturally occurring mutations in the human HNF4α gene, uses the original numbering system. Therefore, that system will be used for the purposes of this review unless indicated otherwise.

The HNF4α gene (gene symbol, *TCF14*) maps to human chromosome 20 q13.1-13.2 between *PLCG1* and *D20S17* (Argyrokastritis *et al.*, 1997). The human gene consists of at least 12 exons and spans 30 kb (Furuta *et al.*, 1997). It is unique among nuclear receptors in that the second zinc finger is encoded by two exons, whereas in other receptors there is one finger per exon (Taraviras *et al.*, 1994). This unique genomic structure is conserved with the *Drosophila* gene (Zhong *et al.*, 1993), suggesting that the gene existed before the divergence of arthropod and vertebrate lineages.

The sequence of the human HNF4α gene indicates that there are at least nine different splicing variants or isoforms, several of which were previously identified by cloning. In the A/B domain, two additional exons were noted in the human gene structure compared with the mouse gene (Taraviras *et al.*, 1994). One exon (1B) encodes a 30 amino acid insertion identified in the human HNF4α4 cDNA (Drewes *et al.*, 1996). The sequence of the other exon (1C) is predicted to encode a further 21 amino acids and is reportedly found in transcripts from the liver but not other tissues (Furuta *et al.*, 1997); it has not been reported in any published cDNA sequence although polymorphisms in it in certain human populations have been observed (Lehto *et al.*, 1999b). A mouse cDNA, termed HNF4α7, was also found to contain a novel N-terminal sequence followed by amino acids 30–50 of HNF4α1 (Nakhei *et al.*, 1998); that sequence has not yet been located in the human genomic sequence. In the F domain, two isoforms were found: one contains a 10 amino acid insertion at codon 409 (HNF4α2) (Chartier *et al.*, 1994; Hata *et al.*, 1992, 1995) and another contains a completely distinct F domain (HNF4α3) (Kritis *et al.*, 1996). Using alternative combinations of these splicing events, there are at least another four possible isoforms (HNF4α5, HNF4α6, HNF4α8 and HNF4α9), the existence of which remains to be determined. At the protein level, Western blot analysis with an antiserum recognizing either the very N- or C-terminus of HNF4α1 also suggests that there are multiple HNF4α isoforms in liver and kidney, although all the various isoforms could not be definitively identified (L. Nepomuceno and F.M. Sladek, unpublished results).

The existence of alternative splicing variants is not unique to HNF4α since nearly every nuclear receptor examined thus far undergoes splicing events, particularly in the 5′ region of the gene (Gronemeyer and Laudet, 1995). However, the splicing events in the 3′ region which alter the F domain are thus far unique to HNF4α. The phenomenon of alternative splicing is also conserved among species as multiple HNF4 isoforms in the A/B and F domains have also been detected in the silkmoth and mosquito, including one in mosquito that deletes the DBD, resulting in a dominant negative form of the receptor (Kapitskaya *et al.*, 1998; Swevers and Iatrou, 1998). Three HNF4β splice variants have also been identified in *Xenopus* (Holewa *et al.*, 1997).

Tissue distribution of HNF4α isoforms

The tissue distribution of the vertebrate isoforms that have been analyzed thus far is very similar with liver, kidney, small intestine and colon being the predominant tissues for most of the isoforms tested (HNF4α1, HNF4α2, HNF4α3, HNF4α4 and HNF4α7) (Sladek *et al.*, 1990; Hata *et al.*, 1992; Drewes *et al.*, 1996; Kritis *et al.*, 1996; Nakhei *et al.*, 1998). There are also detectable amounts in pancreas and stomach, and a barely detectable amount of some HNF4α RNA in testis and possibly skeletal muscle (Drewes *et al.*, 1996; Kritis *et al.*, 1996). Whereas much of the work on tissue distribution consists of Northern blot analysis which does not distinguish between the different isoforms, some reverse transcriptase–polymerase chain reaction (RT-PCR) and RNAase protection work has been performed that more specifically identifies the different isoforms. For example, it is known that HNF4α2 RNA is more abundant than HNF4α1 RNA in rat, mouse and human liver and kidney and the human hepatocarcinoma cell line, HepG2 (Hata *et al.*, 1992, 1995; Chartier *et al.*, 1994). In contrast, HNF4α3 is much less abundant in all tissues analyzed (Kritis *et al.*, 1996). Certain isoforms appear to be differentially expressed in different tissues. For example, HNF4α1, HNF4α2 and HNF4α3 mRNA were detected by RT-PCR in human pancreas and liver but the insert in the A/B domain of HNF4α4, HNF4α5 and HNF4α6 was detected only in liver (Furuta *et al.*, 1997). HNF4α7, reportedly absent in kidney but present at high levels in stomach, was also found in a dedifferentiated cell line (F9 cells) and at trace amounts in several other tissues typically lacking HNF4α (ovary, heart, bladder, brain), prompting the authors to speculate that HNF4α7 might be present in a stem cell population (Nakhei *et al.*, 1998). In *Bombyx mori*, differential tissue distribution of isoforms is also noted, with one isoform expressed at high levels in the testes (Swevers and Iatrou, 1998). Finally, EHS-matrix overlay of primary rat hepatocytes may regulate the expression of HNF4α1 and HNF4α2 proteins (Runge *et al.*, 1999).

The significance of the different HNF4α isoforms in human disease is not yet known. All the isoforms tested bind consensus HNF4 response elements, which is to be expected since they are identical in the DNA binding and dimerization domains. It is also anticipated that they will all be able to heterodimerize with one another. Several of the isoforms (HNF4α1, HNF4α2 and HNF4α3) activate transcription well in a constitutive fashion (i.e. in the absence of an exogenously added ligand) whereas others do not (HNF4α4 and HNF4α7) (Sladek *et al.*, 1990, 1999; Drewes *et al.*, 1996; Kritis *et al.*, 1996; Nakhei *et al.*, 1998). Also, more subtle differences have been noted, such as HNF4α2 activating transcription somewhat better than HNF4α1 in certain cell lines (Sladek *et al.*, 1999) but not in others (Suaud *et al.*, 1999), and HNF4α7 transactivation being a function of the cell type used in the transfection assays (Nakhei *et al.*, 1998). Therefore, it is conceivable that different isoforms will play different roles in the liver and other organs.

PHYSIOLOGY

Role of HNF4α in development

Numerous studies indicate that HNF4α plays a critical role in nutrition and metabolism from very early in embryonic development through the adult. They also indicate that this role is conserved throughout evolution. Since, by and large, the probes used in the developmental studies recognize the central portion of HNF4α that is common to all the isoforms, it is not known which HNF4α isoforms are the most important, or indeed whether there are significant differences between the isoforms. Even less is known about the role of HNF4β and HNF4γ in development. These caveats should be kept in mind in the following discussion.

The earliest that HNF4α is detected in any organism is in the oocyte of the silkmoth (Swevers and Iatrou, 1998) and the ovary of the mosquito where the levels of HNF4α mRNA are increased significantly after a blood meal (Kapitskaya *et al.*, 1998). *Drosophila* and *Xenopus* HNF4α are present in the fertilized egg as a maternal RNA and/or protein (Zhong *et al.*, 1993; Holewa *et al.*, 1996). In contrast, in the mouse, HNF4α RNA is not detected by *in situ* hybridizations until embryonic day 4.5, when it appears in the primitive endoderm. It is now considered to be one of the earliest known markers of that tissue (Coucouvanis and Martin, 1999). The primitive endoderm goes on to form the visceral endoderm, which is an extraembryotic tissue that develops into the yolk sac. The yolk sac expresses many of the same genes as does the liver, including

HNF4α and hepatocyte nuclear factor 1α (HNF1α), another transcription factor important in liver differentiation. The yolk sac is also important for the maternofetal exchange of nutrients before the functioning of the placenta and is the site of embryonic hematopoeisis before liver. From embryonic day 5.5 through 8.5, HNF4α mRNA is expressed very strongly in columnar visceral endodermal cells of the yolk sac (Duncan *et al.*, 1994).

In contrast to the early appearance in the extraembryonic tissue, mouse HNF4α RNA is not expressed in the embryonic tissue until embryonic day 8.5, where it is first present in the liver bud and the hindgut. This appearance coincides with the proliferation of liver cells and the expression of liver-specific genes such as albumin and α-fetoprotein (AFP) (Duncan *et al.*, 1994). Interestingly, neither of these genes is directly activated by HNF4α although they are by the pou-homeo transcription factor HNF1α, which is itself a target for HNF4α (Kuo *et al.*, 1992). At embryonic day 9.5, expression of HNF4α is greatly enhanced in the liver and is apparent in the midgut, including the primordium of the pancreas, and the gallbladder. At embryonic day 10.5, in addition to liver and pancreas, HNF4α RNA is also seen in the stomach and mesonephric tubules which go on to form the cortex of the kidney, where HNF4α is seen in the adult (Mietus-Snyder *et al.*, 1992). This pattern is maintained through embryonic day 13.5, when expression of HNF4α RNA is also seen from the midgut through the hindgut (the intestine) stopping only at the rectal–anus border (Duncan *et al.*, 1994). In another study employing Northern blot analysis, rat HNF4α RNA expression was shown to peak in the liver at embryonic day 18. At embryonic day 20 there was a large drop in HNF4α expression and by birth (day 0) only very low levels of HNF4α RNA were detected. Days 2 and 4 after birth exhibited even lower levels of HNF4α RNA while the adult liver expressed a level close to that seen at birth (Nagy *et al.*, 1994). Much less is known about levels of HNF4α protein during development: in one study, HNF4α DNA-binding activity was detectable by gel shift analysis at embryonic day 8.5 in the yolk sac and at day 14.5 in the liver, while only very small amounts were detectable in the kidney at embryonic day 17.5. Earlier times were not examined (Cereghini *et al.*, 1992).

HNF4α-deficient mice show that HNF4α plays a crucial if indirect role in gastrulation. In HNF4α$^{-/-}$ mice, an abnormal phenotype was first evident at embryonic day 6.5, when gastrulation starts. Gastrulation was retarded and failed to progress beyond the primitive streak stage. The embryos overall showed a dramatic decrease in growth and little mesoderm was observed. Since gastrulation takes place before HNF4α is expressed in embryonic tissue (embryonic day 6.5 versus 8.5 respectively), it was thought that the defect in gastrulation was due to lack of a properly functioning visceral endoderm where HNF4α is highly expressed (Chen *et al.*, 1994c). This model was supported by a system using embryoid bodies which showed that a fully differentiated visceral endoderm is mandatory for murine gastrulation and that HNF4α was mandatory for proper functioning of the visceral endoderm. However, HNF4α is not required for early differentiation of the visceral endoderm as visceral endoderm markers such as transcription factors GATA4 and HNF1β (vHNF1) were not affected by the presence or absence of HNF4α. HNF4α, however, was required for the expression of secreted proteins such as AFP, transferrin, apolipoprotein (Apo) AI, ApoB and transthyretin (TTR). Presumably, expression in the visceral endoderm of secreted proteins is required to provide nutrients to the developing embryo (Duncan *et al.*, 1997). Growth of a murine embryonal carcinoma cell line in a serum-free medium that contains only transferrin and high-density lipoprotein (HDL) and low-density lipoprotein (LDL) suggests that at least some of the required nutrients may be iron and lipids (Heath and Deller, 1983).

Other critical proteins secreted from the visceral endoderm, such as TTR and retinol-binding protein, may be required to transport thyroid hormone and vitamin A, respectively, from the mother to the developing embryo (Duncan *et al.*, 1994).

What relevance do these developmental data have for human disease? First and foremost, they show that HNF4α is an essential gene and that an organism must have at least one good copy of the gene to survive to birth. This finding is supported by data from human populations in which only wild-type homozygotes or mutant heterozygotes have been detected; i.e. no individual containing mutations in both copies of the HNF4α gene have been identified thus far, and the mouse knockout data suggest that they will not be found. In contrast, certain other receptors, such as androgen and thryoid hormone receptor, exhibit homozygous mutations in certain human diseases (see Chapters 2 and 4). Other receptors, such as retinoic acid receptor (RAR), require that more than one gene must be deleted in mice in order to observe a significant phenotype (reviewed in Kastner *et al.*, 1995).

HNF4α as a downstream effector of the TGFβ/SMAD pathway

HNF4α was first observed to be part of a transcriptional hierarchy when it was found to regulate the expression of another liver-enriched transcription factor, HNF1α (Tian and Schibler, 1991; Kuo *et al.*, 1992). Since then, that hierarchy has grown tremendously in both a linear fashion upstream and horizontally in a branched fashion containing positive and negative feedback loops. As with the case of the role of HNF4α in development, however, as more and more connections are found between HNF4α and other systems, we realize that the system is much more complex than previously imagined.

Figure 9.3 outlines the transforming growth factor β (TGFβ) signaling pathway and the role that HNF4α may play in it (for reviews on the TGFβ pathway, see Massague, 1998; Christian and Nakayama, 1999; Janknecht and Hunter, 1999). Whereas it has been known for quite some time that TGFβ plays an important role in cell proliferation and differentiation, our understanding of the mechanism by which this occurs has literally exploded in the last few years. Most relevant to this review is the fact that HNF4α appears to be a very important piece of the puzzle and is, in fact, commonly used as a critical downstream marker in certain systems. Whereas there are many studies that contribute to various aspects of this hierarchy, only that information most pertinent to HNF4α is reviewed here.

Members of the TGFβ superfamily such as TGFβ, activins, nodal and bone morphogenic proteins (BMPs), are extracellular, secreted proteins that elicit a broad spectrum of cellular responses from cell proliferation and differentiation to specification of important developmental processes. They form complexes with transmembrane serine/threonine kinase receptors (type I and type II) which in turn phosphorylate, and thereby activate, *C. elegans* Sma and *Drosophila* Mothers Against Dpp (SMAD) proteins (SMAD 1, 2, 3 and 5). Recruitment of SMADs to the receptor is aided by proteins termed Smad AnchoR for Activation (SARA) (Tsukazaki *et al.*, 1998).

The activated SMAD proteins dissociate from the receptors to form complexes with SMAD4. The SMAD complexes then translocate to the nucleus and bind to target promoters in association with DNA-binding cofactors. One such cofactor, considered to be the prototypic SMAD partner is forkhead activin signal transducer-1 (FAST-1), a forkhead domain protein that binds to activin response elements (AREs) (Labbé *et al.*, 1998). The C-terminal portion of the non-SMAD4 partner then recruits coactivators, such as p300/CBP creb binding protein (creb = cAMP response element binding

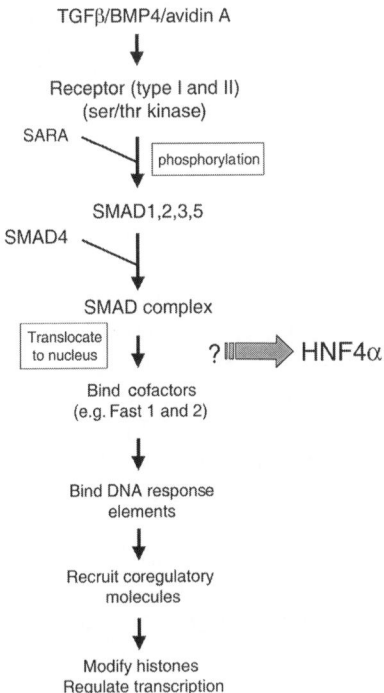

Fig. 9.3 Hepatocyte nuclear factor (HNF)4α as a downstream target of the transforming growth factor (TGFβ) *C. elegans* Sma and *Drosophila* mothers against Dpp (SMAD) signaling pathway. Genetic evidence indicates that TGFβ family members, their receptors and SMADs are all upstream of HNF4α, but it is not known exactly how, or at which point (?), the pathway activates HNF4α gene expression (see text for details).

protein), or corepressors, such as 5'TG3'-interacting factor (TGIF). Coactivators often contain or are associated with histone acetylase activity, which leads to an open chromatin conformation and an increase in transcription, while the corepressors associate with histone deacetylase activity, causing a closed conformation and an inhibition of transcription. The net result is regulation of a variety of genes responsive to TGFβ family members (Janknecht *et al.*, 1998; Wotton *et al.*, 1999).

Considering the prevalence of HNF4α in the visceral endoderm and its role in signaling gastrulation, a connection to the TGFβ family is not all that surprising in retrospect. Inactivation of nodal or BMP-4, which is known to induce differentiation of the visceral endoderm, results in severe early defects and impaired mesoderm formation, one of the hallmarks of gastrulation (reviewed in Derynck, 1998). In mice deficient in HNF4α the appearance of mesodermal cells is also delayed and severely impaired (Chen *et al.*, 1994c). More direct evidence comes from studies showing that addition of BMP protein to cultures of S2 embryoid bodies induces expression of HNF4α and other visceral endoderm markers (Coucouvanis and Martin, 1999). TGFβ, in the presence of epidermal growth factor (EGF), has also been shown to modulate terminal maturation of fetal hepatocytes and to cause an increase in HNF4α and HNF1α DNA-binding activity (Sanchez *et al.*, 1998). Finally, activin A, a potent mesoderm differentiation signal in *Xenopus*, has been shown to increase HNF1α RNA levels in the frog embryo via an HNF4α response element in the HNF1α gene promoter (Weber *et al.*, 1996; Pogge v. Strandmann *et al.*, 1997). Opposing animal to vegetal gradients of HNF4α protein and activin A have also been observed in the frog embryo (Holewa *et al.*, 1996).

Other studies concerning more downstream points in the pathway further support the notion that HNF4α is a target of TGF*β* signaling. For example, mice deficient in activin receptor type I, subset A (ActRIA), a type I TGF*β* receptor expressed primarily in the extraembryonic visceral endoderm before gastrulation and in both embryonic and extraembryonic tissues after gastrulation (as is HNF4α), have reduced levels of HNF4α RNA. They also exhibit a phenotype similar to, although somewhat more severe than, that of the HNF4α-deficient mice, supporting the notion that ActRIA is upstream of HNF4α in a signaling cascade (Gu *et al.*, 1999). Finally, in another study, expression of a dominant negative BMP receptor in mice inhibited cavitation of the early embryo as well as HNF4α expression in the visceral endoderm (Coucouvanis and Martin, 1999).

The immediate downstream target of TGF*β* receptors, SMAD2, is also known to be essential for embryonic mesoderm formation (Derynck, 1998), and a deficiency in its partner SMAD4 results in a highly abnormal visceral endoderm, a defect in gastrulation, growth retardation due to a decrease in cell proliferation, and a decreased expression of HNF4α. The embryos die at embryonic day 7.5, a few days earlier than the HNF4α-deficient embryos (Sirard *et al.*, 1998). Finally, SMAD 1, 2 and 4 are present in the maternal pool of RNA in *Xenopus* and *Drosophila* embryos, as is HNF4α (Massague *et al.*, 1997).

These studies indicate that TGF*β*-like molecules and SMADs are clearly upstream of HNF4α in a signaling cascade. What is not known, however, is the mechanism by which SMADs activate the HNF4α gene. Do they bind the HNF4α promoter, alone or in conjunction with other proteins? Whereas SMADs have been reported to contain DNA-binding activity, they also often act by forming complexes with other factors that bind DNA. The prototypic factor is FAST-1 which is a forkhead/winged helix protein. The founding member of the forkhead family is hepatocyte nuclear factor 3 (HNF3) (Lai *et al.*, 1991). Whereas HNF3 is clearly upstream of HNF4α in a transcription cascade (see Fig. 9.4), and whereas a consensus site for HNF3 has been reported in the HNF4α

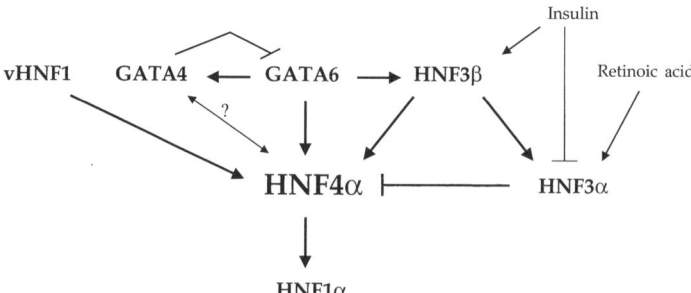

Fig. 9.4 Hepatocyte nuclear factor (HNF)4α as part of a transcriptional network. Genetic and molecular biologic experiments indicate that HNF4α is part of a complex transcriptional network involving factors from at least three transcription factor families: GATA, HNF3/forkhead and pou-homeodomain HNF1α. Insulin, retinoic acid and other compounds not shown also affect the network. Arrows, positive regulation; blunted lines, negative regulation, ?, conflicting data (see text for details).

promoter (Furuta *et al.*, 1997), it is not yet known whether HNF3 responds to TGFβ signaling.

HNF4α in a complex transcriptional network with GATA, HNF3/forkhead and HNF1α

In addition to, or perhaps in conjunction with, the TGFβ/SMAD pathway, HNF4α gene expression is regulated by another cascade of transcription factors: GATAs and HNF3/forkhead proteins. GATA1–6 are members of a family of zinc finger proteins distinct from the nuclear receptors that are also important in early development and in differentiation of adult tissue, particularly in the hematopoietic system (GATA1–3) and in cardiac and gut tissue (GATA4–6) (Jordan and Van Zant, 1998; Charron and Nemer, 1999). HNF3 is a winged helix protein that was purified from rat liver at the same time as HNF4α (Lai *et al.*, 1991) and that was simultaneously identified as a forkhead gene important in *Drosophila* development (Weigel and Jäckle, 1990). It will be useful to keep in mind during this discussion the intriguing suggestion that both GATA and HNF3/forkhead transcription factors play a key role in gut endoderm development by imparting developmental competence to pluripotent precursor cells. They do this apparently by acting as genetic potentiators, binding target sites on potentially active genes and thereby increasing the responsiveness of the gene to signals later on in development (Zaret, 1998, 1999). (For reviews covering additional aspects of gene expression during early liver development, see Zaret, 1994, 1996.)

Very recently several different groups have published findings on the role of GATA4 and GATA6 in development. The results are all in agreement that HNF4α expression is influenced by GATA4 and GATA6 expression and that there is a very complex network of regulation between all these transcription factors. However, whereas certain connections are consistent between the different systems used, namely knockout mice and embryoid bodies, other connections are less clear. For example, GATA6 appears to regulate HNF4α, HNF3β and GATA4 gene expression positively, while GATA4 appears to downregulate GATA6. However, what is more ambiguous is whether there is a direct connection between GATA4 and HNF4α (Soudais *et al.*, 1995; Duncan *et al.*, 1997; Kuo *et al.*, 1997; Molkentin *et al.*, 1997; Morrisey *et al.*, 1998).

Analysis of embryoid bodies deficient in HNF3β indicates that HNF3β positively regulates both HNF4α and HNF3α. The data also suggest that, since the expression of several HNF4α target genes is decreased under forced expression of HNF3α, HNF3α may actually inhibit HNF4α expression (Duncan *et al.*, 1998). Both HNF3α and HNF3β are expressed in the embryo earlier than HNF4α and HNF1α, supporting the notion that HNF3 is upstream of HNF4α in a transcription hierarchy (reviewed in Cereghini, 1996; Zaret, 1996). There is also a conserved putative HNF3-binding site in the promoter of HNF4α (Furuta *et al.*, 1997). In addition, retinoic acid, presumably via the RAR, has been shown to activate HNF3α gene expression (Roach *et al.*, 1994; Jacob *et al.*, 1999). This could explain the findings that retinoic acid downregulates HNF4α and HNF1α gene expression in hepatocellular carcinoma cells in culture (Magee *et al.*, 1998). Finally, in embryonic stem cells, insulin has been shown to upregulate HNF3β gene expression and to downregulate HNF3α expression (Duncan *et al.*, 1998), both of which would serve to turn on HNF4α gene expression. Whereas it remains to be determined whether HNF4α gene expression is dependent on insulin in adult tissues, it is an intriguing possibility that should be kept in mind, especially in discussion of HNF4α and diabetes.

Many studies indicate that HNF1α is upregulated at least in part by HNF4α (Tian and

Schibler, 1991; Kuo *et al.*, 1992; Bulla and Fournier, 1994; Duncan *et al.*, 1997; Spath and Weiss, 1997), a scenario that is evolutionarily conserved (Holewa *et al.*, 1996). There are also a few studies of liver cell lines that indicate that HNF1α positively regulates the expression of the HNF4α gene (Bulla and Fournier, 1994; Spath and Weiss, 1997; Bailly *et al.*, 1998), and at least one study in which HNF1α synergizes with HNF4α to activate its own promoter (Miura and Tanaka, 1993). Whereas there is a HNF1α consensus site in the HNF4α promoter that plays an important role in liver-specific expression in transient transfection assays, it was not found to be sufficient for liver-specific expression in transgenic mice (Zhong *et al.*, 1994). Furthermore, HNF1α-deficient mice display normal levels of HNF4α mRNA and a very different phenotype from the HNF4α-deficient mice (Poutoglio *et al.*, 1996).

Another possible role for the HNF1α site in the HNF4α promoter is to bind vHNF1 (also referred to as HNF1β). vHNF1 is a homeodomain-containing protein structurally related to HNF1α which heterodimerizes and shares DNA response elements with HNF1α. Two independent groups have recently shown that mice deleted in the vHNF1 gene lack HNF4α mRNA. The tissue distribution of vHNF1 (present in the liver, kidney, pancreas and lung in the adult, and in the visceral endoderm in the early embryo) and the phenotype of the vHNF1 knockout mice are both consistent with vHNF1 being upstream of HNF4α in a transcription cascade (Barbacci *et al.*, 1999; Coffinier *et al.*, 1999a,b). Curiously, however, while vHNF1 appears to control the level of HNF4α1 RNA, expression of the HNF4α7 isoform, which has a distinct N-terminus, is unaffected in the vHNF1$^{-/-}$ mice, suggesting that HNF4α7 may be under the control of a different promoter than the other HNF4α isoforms (Barbacci *et al.*, 1999).

Other aspects of HNF4α gene regulation

Whereas the HNF4α gene promoter contains HNF1α and HNF3 binding sites, it lacks a TATA box, an apparently common feature of nuclear receptor genes (Taraviras *et al.*, 1994). Consensus binding sites for myogenic protein D (MyoD), Wilm's tumor 1 (WT1), glucocorticoid receptor (GR), nuclear factor 1 (NF-1), activation protein 1 (AP-1) and HNF-6 have also been identified in the mouse and human HNF4α promoter, although those factors have not yet been shown to play a functional role in HNF4α gene expression (Taraviras *et al.*, 1994; Furuta *et al.*, 1997). Furthermore, any analysis of the HNF4α promoter must be tempered by the finding that distal enhancer regions more than 5.5 kb upstream of the HNF4α transcription start site are required for HNF4α-like expression of a β-galactosidase reporter gene in transgenic mice (Zhong *et al.*, 1994).

HNF4α, along with HNF1α, is considered to be one of the primary determinants of the liver phenotype, necessary and perhaps even sufficient for expression of the liver phenotype (Herbst *et al.*, 1991; Tonjes *et al.*, 1992; Griffo *et al.*, 1993; Bulla and Fournier, 1994; Nagy *et al.*, 1994; Spath and Weiss, 1997, 1998). A recent study suggests that this ability to regulate the expression of the liver phenotype may be related to the ability of HNF4α and HNF1α to alter the chromatin structure of liver-specific genes (Rollini and Fournier, 1999). It is not surprising, therefore, that HNF4α mRNA and protein levels have been monitored in a variety of systems under a myriad of conditions, both natural and artificial. In primary hepatocytes, some sort of extracellular matrix (e.g. collagen or EHS-gel) is apparently the most critical factor for maintaining HNF4α RNA levels (Oda *et al.*, 1995; Gomez-Lechon *et al.*, 1998; Runge *et al.*, 1998). Other naturally occurring compounds that have been found to increase HNF4α RNA or protein expression include hepatocyte growth factor (HGF), EGF (Runge *et al.*, 1998), dexamethasone (Oda *et al.*, 1995), interleukin (IL) 6, TGF*β* and activin (Mizuguchi *et al.*, 1998). However, the effects of these compounds can be cell-type and concentration specific. For example, under different experimental conditions TGF*β* has been shown to decrease HNF4α expression (Sanchez *et al.*, 1999). Exogenous compounds that increase HNF4α expression include dimethylsulfoxide (DMSO) and sodium butyrate, which are known to induce differentiation in general (Mizuguchi *et al.*, 1998; Zvibel *et al.*, 1998). Compounds that decrease HNF4α RNA levels include hypolipidemic peroxisome proliferators (Hertz *et al.*, 1995) and dioxin after partial hepatectomy (Bauman *et al.*, 1995).

In conclusion, it is not known at this time whether there is any link between the TGF*β*/ SMAD and the GATA and HNF3 pathways, or between these pathways and HNF4α-related human disease, although there is plenty of room for speculation. For example, it is conceivable that a mutation or polymorphism in any one of the genes upstream of HNF4α could alter HNF4α protein levels, thereby bringing on a disease state. Likewise, whereas physiologic relevance remains to be established for any of the modulators of HNF4α expression mentioned above, these too could potentially cause or treat a disease state dependent on HNF4α protein levels.

HNF4α target genes

Since HNF4α is a transcription factor, it is reasonable to assume that any pathology resulting from alterations in HNF4α will be due not to the loss of HNF4α function *per se* but rather to the loss of some activity encoded by a gene that HNF4α regulates. In other words, a good method of identifying potential HNF4α-related diseases is to survey HNF4α target genes. In Table 9.1, we have compiled a list of all known HNF4α1 target genes, grouping them by their function (e.g. lipid and retinol transport versus lipid and steroid metabolism). As to be expected, this list contains several more sites than a similar list that was compiled in 1994, although interestingly the breadth of target genes is not greatly changed (Sladek, 1994). Whereas relatively little work has been done with HNF4*β* or HNF4*γ*, as those genes display a very high level of amino acid sequence similarity with HNF4α in the DNA-binding domain (see Fig. 9.1), it can be assumed that at least a sizeable fraction of these target genes will also be regulated by the other HNF4 genes. Furthermore, because no differences in DNA-binding specificity have yet been detected between the different HNF4α isoforms, all of these HNF4α1 target genes must also be considered as potential targets for the other HNF4α isoforms.

The majority of the target genes are expressed primarily, if not exclusively, in the liver, although there are some exceptions. For example, erythropoietin is expressed primarily in the kidney and to a lesser degree in the liver and spleen (Fandrey and Bunn, 1993). Cellular retinol-binding protein II (CRBPII) and intestinal fatty acid-binding protein (I-FABP) are expressed exclusively in the intestine in the adult (Blomhoff *et al.*, 1991; Gordon *et al.*, 1992). However, this bias toward liver-specific target genes most likely reflects the fact that, in general, many fewer kidney- or intestine-specific genes have been identified than liver-specific ones, and will almost certainly change as more genes are identified.

The grouping of the genes clearly indicates that, as in the visceral endoderm, HNF4α plays an important role in the adult organism in the transport of nutrients, for example cholesterol and lipids by the apolipoproteins; iron by transferrin; and thyroid hormone T_4 by transthyretin (TTR). HNF4α also regulates genes that play a major role in several different aspects of metabolism, such as glucose by phosphoenolpyruvate carboxykinase

Table 9.1 HNF4α target genes and binding sites

Gene	Location	Binding site	Full gene name	Reference
Nutrient transport				
Lipid and retinol transport				
hApoAI	−117/−132	5′-tag AGTTCA a GGATCA-3′	Apolipoprotein AI	Chan et al. (1993)
hApoAI	−195/−213	5′-gca GGGGTC a AGGGTT cag-3′	Apolipoprotein AI	Hardon et al. (1988)
hApoAII	−718/−736	5′-acc AGGGTA a AGGTTG aag-3′	Apolipoprotein AII	Ladias et al. (1992)
hApoAIV	−143/−126	5′-agg GTCACA a AAGTCC aa-3′	Apolipoprotein AIV	Ktistaki et al. (1994)
hApoB	−63/−81	5′-aaa GGTCCA a AGGGCG cct-3′	Apolipoprotein B	Metzger et al. (1993)
hApoCII	−81/−96	5′-AGGCCA a AGTCCT ggc-3′	Apolipoprotein CII	Kardassis et al. (1998)
hApoCIII	−67/−85	5′-cgc TGGGCA a AGGTCA cct-3′	Apolipoprotein CIII	Sladek et al. (1990)
hApoCIII	−734/−716	5′-gtg GGTCCA g AGGGCA aaa-3′	Apolipoprotein CIII	Kardassis et al. (1997)
hApoCIII	−2880/−2929	5′-gtc AGTCCA g AGGTCA gag-3′	Apolipoprotein CIII	Vergnes et al. (1997)
rCRBPII	−59/−77	5′-aca GAGTCA a AGGTCA taa-3′	Cellular retinol-binding protein II	Nakshatri and Chambon (1994)
rFabpi	−85/−69	5′-a AGTTCA a AGTTCA aga-3′	Intestinal fatty acid-binding protein	Rottman and Gordon (1993)
Other serum proteins				
hTransferrin	−71/−53	5′-ggg AGGTCA a AGATTG cgc-3′	Transferrin	Schaeffer et al. (1993)
mTTR	−154/−136	5′-cta GGCAAG g TTCATA ttt-3′	Transthyretin	Sladek et al. (1990)
hα-1-AT	−106/−124	5′-aca GGGGCT a AGTCCA ctg-3′	α$_1$-Antitrypsin	Hardon et al. (1988)
hSHBG	−20/−38	5′-gga GGGTTA a AGGTTG ccc-3′	Sex hormone-binding globulin	Janne and Hammond (1998)
hSHBG	−85/−67	5′-cag GGGTCA a GGGTCA gtg-3′	Sex hormone-binding globulin	Janne and Hammond (1998)
Nutrient metabolism				
Lipid and steroid metabolism				
hMCAD	−326/−308	5′-ctc CGGGTA a AGGTGA agg-3′	Medium chain acyl-coenzyme A dehydrogenase	Carter et al. (1993)
rACO	−555/−573	5′-acc AGGACA a AGGTCA cgt-3′	Acyl-CoA oxidase	Tugwood et al. (1992)
rmitHMG-CoA	−92/−104	5′-GGGCCA a AGGTCT-3′	3-Hydroxy-3-methylglutaryl-CoA	Rodriguez et al. (1998)
rHD	−2931/−2950	5′-taa AGTTCA at AGGTCA aag-3′	Acyl-CoA hydratase-dehydrogenase	Winrow et al. (1994)
mCYP2A4	−61/−49	5′-AGACCA a AGTCCG-3′	Cytochrome 2A4	Yokomori et al. (1997)
rabCYP2C1	−120/−102	5′-agt GGACCA a AGTCCA tcc-3′	Cytochrome 2C1	Chen et al. (1994b)

Gene	Position	Sequence	Protein	Reference
rabCYP2C2	−115/−97	5′-agt GGTCCA a AGTCCA ctc-3′	Cytochrome 2C2	Chen et al. (1994a)
rabCYP2C3	−99/−81	5′-ata AGACCA a AGTGCA atg-3′	Cytochrome 2C3	Chen et al. (1994b)
hCYP2C9	−135/−152	5′-ctt TCCTGA a ACTGGG tg-3′	Cytochrome 2C9	Ibeanu and Goldstein (1995)
hCYP2D6	−55/−43	5′-AGGGCA a AGGCCA-3′	Cytochrome 2D6	Cairns et al. (1996)
rCYP3A1	−105/−88	5′-ga GTACCA a AGTCCA cgt-3′	Cytochrome 3A1	Ogino et al. (1999)
rCYP7	−149/−131	5′-cta TGGACT t AGTTCA agg-3′	Cytochrome 7	Crestani et al. (1998)
hCYP7	−147/−129	5′-gaa AGGGCA a TGACGT ccc-3′	Cytochrome 7	Cooper et al. (1997)
Glucose metabolism				
rPEPCK	−436/−454	5′-cca CGGCCA a AGGTCA tga-3′	Phospho-enol-pyruvate carboxykinase	Hall et al. (1992)
rL-PK	−127/−145	5′-tgg GGGGCA g AGTTCA gga-3′	Liver-type pyruvate kinase	Diaz Guerra et al. (1993)
rAldolaseB	2148/2165	5′-aa GGAGTA a AGTTCA tta-3′	Aldolase B	Gregori et al. (1998)
Amino acid metabolism				
rTAT	−3583/−3601	5′-tac AGATCA a AGAGCA gca-3′	Tyrosime aminotransferase	Nitsch et al. (1993)
rOTC (enh)	98/80	5′-aaa GGTTTA a AGTTCA tct-3′	Ornithine transcarbamylase	Nishiyori et al. (1994)
rOTC (enh)	183/164	5′-tag AGTTCA g AGGTTA agc-3′	Ornithine transcarbamylase	Nishiyori et al. (1994)
rOTC	−15/−33	5′-agg GGATCA a AGGTCC tac-3′	Ornithine transcarbamylase	Kimura et al. (1993)
rOTC	−115/−97	5′-tta GGCTTA a AGTTCA agt-3′	Ornithine transcarbamylase	Kimura et al. (1993)
hmitALDH2	−324/−306	5′-ttg GGGTCA a AGGCAC aca-3′	Aldehyde dehydrogenase 2	Stewart et al. (1998)
Blood maintenance				
mFactor VII	−32/−47	5′-ggg AGGGCA a AGGTCA-3′	Factor VII	Stauffer et al. (1998)
hFactor VII	−1/−19	5′-tga CGGGCA a AGTTCT ctg-3′	Factor VII	Hung and High (1993)
hFactor VIII	−279/−297	5′-gta GGGGCA t AAGTCT gct-3′	Factor VIII	Figueiredo and Brownlee (1995)
hFactor IX	15/33	5′-ctg CTAGCA a AGGTTA tgc-3′	Factor IX	Naka and Brownlee (1996)
hFactor IX	6/−10	5′-AGTGGT a AGGTCG att-3′	Factor IX	Naka and Brownlee (1996)
hFactor IX	−12/−30	5′-gtt GTACCA a AGTACA agc-3′	Factor IX	Reijnen et al. (1992)
hFactor X	−45/−61	5′-ct GGAGCA a AGTCCA cg-3′	Factor X	Miao et al. (1992)
hEPO	162/144	5′-gta GGGTCG ag AGGTCA ga-3′	Erythropoietin	Galson et al. (1995)
hEPO	3499/3481	5′-ggt AGGTCG a GAGGTC aga-3′	Erythropoietin	Blanchard et al. (1992)

continued overleaf

Table 9.1 *continued*

hATIII	−108/−126	5′-caa AGTGTA g AGCCCA gtg-3′	Antithrombin III	Fernandez-Rachubinski et al. (1996)
hATIII	−70/−88	5′-ctg AGGTCA a AGGCTG atg-3′	Antithrombin III	Tremp et al. (1995)
Liver differentiation				
hHNF1α	−66/−48*	5′-tga AGTCCA a AGTTCA gtc-3′	Hepatocyte nuclear factor 1α	Gragnoli et al. (1997)
XnLFB1/HNF1	−270/−252	5′-tgg GGTCCA a AGTTCA gta-3′	Hepatocyte nuclear factor 1α	Zapp et al. (1993)
Immune function				
Immune system				
mBF	−290/−308	5′-gat GGAGCA a AGTCCA tcc-3′	Factor B	Garnier et al. (1996)
hMSP	19/35	5′-tg GGGTCA c AGTGCA gc-3′	Macrophage-stimulating protein	Ueda et al. (1998)
hMSP	−85/−69	5′-cc AGGTCT c AGGTCA gg-3′	Macrophage-stimulating protein	Ueda et al. (1998)
hAMBP	−2755/−2773	5′-atg CTGCCA a GGGCCA ctt-3′	α_1-Microglobulin and bikumin	Rouet et al. (1995)
hAMBP	−2786/−2773	5′-c AGTCAA a AGTTCA-3′	α_1-Microglobulin and bikumin	Rouet et al. (1998)
hAMBP	−2923/−2936	5′-g GTCTAA g AGTCCA-3′	α_1-Microglobulin and bikumin	Rouet et al. (1998)
Viral genes				
hHBV enh I	1154/1136	5′-aac GGGGTA a AGGTTC agt-3′	Hepatitis B virus enhancer I	Garcia et al. (1993)
hHBV enh II	1675/1660	5′-g AGTCCA a GAGTCC tc-3′	Hepatitis B virus enhancer II	Guo et al. (1993)
hHBVnucleocapsid	1754/1772	5′-att AGGTTA a AGGTCT ttg-3′	Hepatitis B virus nucleocapsid	Raney et al. (1997)
wWHVEnII	1820/1803	5′-agg AGTCCA a AGGTCC tt-3′	Woodchuck hepatitis enhancer II	Ueda et al. (1996)
HIV LTR	−356/−343	5′-a GGGCCA a GGGTCA-3′	Human immunodeficiency virus long terminal repeat	Ladias (1994)
Growth factors				
hHGFL	−124/−111	5′-AGGTCA g GGTCCA g-3′	Hepatocyte growth factor-like protein	Waltz et al. (1996)
rPRLR	7/25	5′-aag GGGGCA a AGTCAA gcg-3′	Prolactin receptor	Moldrup et al. (1996)
Synthetic site				
Direct repeat 1 (DRiG)		5′-AGGTCA g AGGTCA-3′		Nakshatri and Chambon (1994)
Direct repeat 2 (DR2AA)		5′-cgc AGGTCA aa AGGTCA cct-3′		Jiang and Sladek (1997)

Final consensus

```
GGGTCA A AGGTCA
A tCt    g g TCtg
   aG        g c
              t
123456    123456
 1st        2nd
```

Half-site

HNF4α target genes and binding sites are grouped according to the function of the gene products. Included are only those sites that have been shown to bind HNF4α in any one of several ways: supershift with HNF4α1 antibody in EMSA, EMSA or footprinting using *in vitro* translated NHF4α1 or extracts from HNF4α1-transfected cells, or transient cotransfection of the site in question with HNF4α1 cDNA. Only one species for each binding site is shown: human or the species most closely related to human. Not included are sites that are based only on similarities to the consensus and/or competition in EMSA. Capital letters in final consensus sequence represent nucleotides that appear in the position 25% or more of the time, lower-case letters represent nucleotides that appear between 10% and 25% of the time. Species designations are: h, human; hmit, human mitochondria; r, rat; rmit, rat mitochondria; rab, rabbit; m, mouse; w, woodchuck; Xn, *Xenopus laevis*.

* Based on subsequent information from *Xenopus* (Zapp *et al.*, 1993) and the position of the MODY-3 mutation at −58 (Gragnoli *et al.*, 1997), this binding site is shifted over by 7 bp from the site previously listed in the mouse HNF1α promoter (Sladek, 1994). See text for details.

(PEPCK), L-pyruvate kinase (L-PK) and aldolase B; amino acids by tyrosine amino-transferase (TAT) and ornithine transcarbamylase (OTC); and lipids and steroids by cytochrome P450 family members (*cyp* genes). The third largest grouping of target genes, that containing blood coagulation and anticoagulation factors, appears to be unrelated to the transport and metabolism groupings at first glance. However, if one considers that these genes play an important role in maintaining the blood supply (i.e. the delivery system of the nutrients), then the predominance of these target genes is more under-standable. Furthermore, the visceral endoderm, where HNF4α is highly expressed in early development, is also the major site of hematopoiesis before the liver takes over.

HNF4α regulates most of the major vitamin K-dependent blood coagulation genes either directly (Factor VII–XII) or indirectly via HNF1 (e.g. fibrinogen (Factor I) and prothrombin (Factor II)) (Zakin *et al.*, 1994). HNF4α regulates at least one anti-coagulation gene, antithrombin III, suggesting that via its target genes HNF4α is responsible both for repairing 'leaks' in the delivery system as well as for removing 'clogs' in it. One could also apply the delivery system analogy to erythropoietin, a protein hormone that controls delivery of oxygen to organs by stimulating red blood cell production. In response to hypoxic conditions, HNF4α induces the expression of erythropoietin by acting in concert with hypoxia inducible factor 1 (HIF-1), a hetero-dimer containing two basic helix-loop-helix proteins in the per-arnt-sim (PAS) family, one of which is the constitutively expressed Ah (romatic hydrocarbon) receptor nuclear transporter (ARNT), the other which is hypoxia inducible factor 1α (HIF-1α) (Guille-min and Krasnow, 1997). Recombinant human erythropoietin is used widely to prevent anemia in patients undergoing cancer therapy (Smyth *et al.*, 1996) and is often abused by athletes, sometimes fatally, to artifically enhance their performance (Gareau *et al.*, 1996).

The only category of target genes that is not clearly related to the metabolism and transport of nutrients is that of the proteins involved in the immune system such as factor B (Bf), involved in complement activation, and macrophage-stimulating factor (MSP). The reason for these genes being HNF4α1 targets is even more obtuse when one considers that HNF4α is highly conserved in organisms such as *Drosophila* which have immune systems very different from that found in mammals. In contrast, many of the metabolism target genes are highly conserved across evolutionary boundaries. The category of viral genes, primarily hepatitis B viral genes, is understandable in that the virus has evolved to make use of one of the most abundant liver-enriched transcription factors available in order to transcribe its genes: HNF4α.

It is of interest to note that, whereas the overall themes of nutrient transport and metabolism are consistent with each other, many of the individual HNF4α target genes carry out apparently opposing functions in the body, such as the blood coagulation and anticoagulation factors; PEPCK which controls gluconeogenesis and L-PK which is central to glycolysis; ApoB which is the protein component of LDL (the so-called 'bad' cholesterol) and ApoAI which is the protein component of HDL (the so-called 'good' cholesterol). How or even why HNF4α might be controlling opposing processes in the cell remains an important area of investigation. In a previous review of HNF4α (Sladek, 1994) it was proposed that phosphorylation may play a role. Since then extensive serine/threonine (Jiang *et al.*, 1997b) and possible tyrosine phosphorylation (Ktistaki *et al.*, 1995) of HNF4α1 has been reported, although only one site has been mapped definitively (Viollet *et al.*, 1997) (discussed further below). It remains to be determined whether differential phosphorylation of HNF4α can result in differential activation of target genes, although that is certainly a likely possibility. Another possible explanation for HNF4α activating target genes with apparently opposing effects is that different HNF4α isoforms will have different effects on the different genes. Finally, since it is now known that corepressor, coactivator and mediator complexes all influence transcription by nuclear receptors (reviewed in Freedman, 1999; McKenna et al., 1999; Xu et al., 1999), and that there are various forms of each of these transcription regulators, it is very possible that the coregulators will have differential effects on HNF4α on different target genes.

It is also important to note that HNF4α regulates many of the target genes in concert with other liver-enriched factors, such as basic region/leucine zipper (bZIP) protein CAAT enhancer binding protein (C/EBP) (*ApoB*) (Metzger et al., 1993), forkhead protein HNF3β (*ApoAI*) (Harnish et al., 1996) and pou homeodomain protein HNF1α (α_1-antitrypsin, HNF1α) (Ktistaki and Talianidis, 1997b; Hu and Perlmutter, 1999). HNF4α also synergizes with ubiquitous factors to regulate transcription, such as orphan receptor COUP-TF (*cyp7a*, *HNF1α*) (Ktistaki and Talianidis, 1997a; Crestani et al., 1998) and basic helix-loop-helix leucine zipper proteins TFE3 (transcription factor muE3) and USF2a (upstream stimulatory factor-2a) (L-PK, ApoAII) (Moriizumi et al., 1998; Ribeiro et al., 1999). Therefore, whereas systems examining isolated elements are required to establish HNF4α response elements initially, one must keep in mind that the final regulation of any gene is the result of a combination of transcription factors and coregulators.

Another observation is that HNF4α can be instrumental in executing a transcriptional response to extracellular signals such as insulin (TAT) (Nitsch et al., 1993), glucocorticoids (PEPCK) (Hall et al., 1995), estrogen (ApoAI) (Farsetti et al., 1998; Harnish et al., 1998) and possibly polyunsaturated fatty acids (L-PK) (Liimatta et al., 1994). Interestingly, however, these signals apparently do not act directly upon HNF4α but rather upon other activators that bind nearby DNA response elements (for a partial review see Sladek, 1994). These signals add an additional layer of complexity to the regulation of the HNF4α target genes, a complexity that in turn offers many opportunities for specifically altering the expression of one gene but not that of another.

HNF4α-binding site, a consensus sequence and human diseases

There are several important observations about the sequence of the individual response elements. One of the most striking features is that, while there is a consensus sequence, there is only one incident of HNF4α-binding sites from two different target genes being identical, and there is no obvious relationship between the two genes (human Factor X and murine factor B). This suggests that different HNF4α-binding sites perform different functions in different genes. This notion is supported by the observation that there are at least a few examples of the sequence of a given response element at a particular position in a target gene being conserved across species, whereas two response elements in the same target gene in the same species, albeit in different positions, are not conserved. For example, the HNF4α response element in the human ApoAI gene at -120 (AGTTCA a GGATCA) is identical to the response element at -120 in the rat ApoAI gene (AGTTCA a GGATCA) but different from the response element at -198 (GGGGTC a AGGGTT) in the human gene. This suggests not only that the sequence of a particular response element has a specific function but also that the position of the response element in the promoter may have a particular function.

The addition of more naturally occurring target genes for HNF4α has allowed the consensus HNF4α-binding site to be refined somewhat (Table 9.1). The consensus, based on 64 nonspecies-redundant native elements and two synthetic elements, agrees well with that found by Fraser and colleagues (1998) based on 23 synthetic elements using a PCR strategy. The most important features of the HNF4α consensus-binding site are the following: (1) HNF4α prefers sites composed of two direct repeats of a half-site separated by one nucleotide (DR1), although it also binds sites separated by two nucleotides (DR2) (Jiang and Sladek, 1997); (2) both half-sites fit the AGGTCA sequence common to many nuclear receptors, although there are some differences between the two half-sites; (3) a defining feature of a HNF4α-binding site is the three As in the middle of the site; (4) the preference for triple As in the middle of the site results in less variability of the sixth position of the first half-site and the first position in the second half-site; and (5) the second position of both half-sites is almost always a G (> 86%), whereas the fourth position in both half-sites is highly variant.

HNF4α-binding sites can often, but not always, be identified by comparison to the consensus sequence. For example, Kuo *et al.* (1992) originally identified a 30-nucleotide region in the mouse HNF1α promoter that was important for HNF4α binding and transactivation. From that region the putative HNF4α site was identified and noted in the 1994 compilation of HNF4α target sites as − 104 to − 92 (AGGCTG a AGTCCA) (Sladek, 1994). However, an inherited mutation causing a form of noninsulin-dependent diabetes (*MODY3*) in the human HNF1α promoter at − 58 A > C now suggests that the sequence noted in 1994 may not be the relevant HNF4α-binding site (Gragnoli *et al.*, 1997). If the site identified in 1994 were a HNF4α-binding site, then the − 58 mutation would be in the last position of the second half-site (AGGCTG a AGTCC**A**, the mouse and human numbering differ by 34 nucleotides). However, according to the consensus sequence, this position is quite variable and not expected to influence HNF4α binding greatly. In contrast, the relevant HNF4α-binding site may in fact be seven nucleotides downstream. This new site matches the consensus sequence better, and if mutated at − 58 is predicted to cause a much more drastic change in HNF4α binding since it affects the central three As (AGTCC**A** a AGTTCA) (see Table 9.2 below). Whereas more experiments are clearly needed to determine which is the relevant HNF4α-binding site in the HNF1α promoter, this analysis none the less raises an important issue: naturally occurring mutations in HNF4α-binding sites that lead to significant phenotypes *in vivo* can provide valuable information about HNF4α DNA-binding properties. These are intriguing examples that reverse the usual flow of information from the *in vitro* to the *in vivo* situation that are likely to be seen more often in the future as more and more mutations are mapped in human diseases.

STRUCTURE AND MECHANISM OF ACTION OF HNF4α

HNF4α is an unique member of the nuclear receptor superfamily for several reasons. It displays unique DNA-binding properties, such as strong homodimerization, and unique transactivation properties, such as activation of transcription in a constitutive fashion (i.e. in the absence of an exogenous ligand).

DNA-binding properties of HNF4α

Considering that HNF4α was originally purified based on its ability to bind DNA, it is perhaps not too surprising that it displays a relatively high affinity for DNA in electrophoretic mobility shift assays (EMSAs), (K_d 5.2 nmol L^{-1}) (Jiang *et al.*, 1997b). This high-affinity binding is due, at least in part, to the fact that HNF4α exists in solution as a stable homodimer as a result of strong dimerization motifs in the LBD (Jiang *et al.*, 1995, 1997a). To date, HNF4α has been found to bind DNA only as a homodimer; HNF4α does not form heterodimers with other receptors, at least in the classic sense, although it is possible that a heterodimerization partner may yet be found.

This exclusive homodimerization of HNF4α on DNA and in solution is reminiscent of the steroid receptors. However, unlike the steroid receptors, HNF4α is found exclusively in the nucleus, is not complexed with heat shock protein, and binds response elements consisting of direct repeats, not indirect repeats as do the steroid receptors. For these reasons it is proposed that HNF4α forms a separate subclass of receptors (Jiang *et al.*, 1995) (see Fig. 9.5). In contrast, the closest mammalian receptor to HNF4α, retinoid X receptor α (RXRα), promiscuously forms heterodimers with a variety of other receptors and binds many of the same response elements as HNF4α. A construct containing just the DBD of HNF4α forms a heterodimeric complex on DNA with a RXRα DBD, indicating that it is regions of the LBD that prevent heterodimerization (Jiang and Sladek, 1997).

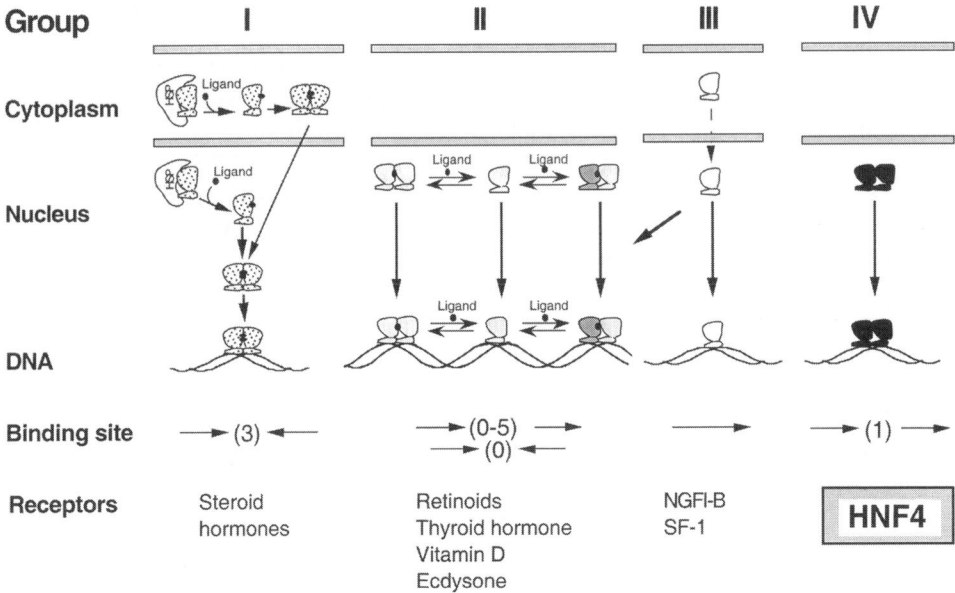

Fig. 9.5   Hepatocyte nuclear factor (HNF)4α defines a novel subclass of nuclear receptors. Receptors are grouped according to dimerization potential, DNA-binding specificity and subcellular localization. Binding sites are depicted by orientation of and nucleotide spacing between the half-sites. Representative receptor types are given for each subclass. Group I receptors are released from heat shock proteins (HSP) upon binding ligand, and bind indirect repeats as homodimers. Group II receptors bind direct repeats preferentially as heterodimers but can also bind as homodimers and monomers; the dimerization state can be influenced by the presence of ligands. Group III receptors bind primarily as monomers, although some have been shown to dimerize with group II receptors, primarily retinoid X receptor. Group IV receptors, typified by HNF4α, HNF4β, HNF4γ, bind direct repeats only as homodimers. NGFI-B, nerve growth factor induced gene; SF-1, steroidogenic factor 1. Adapted from Jiang *et al.*, 1995.

The authors have recently identified a charge incompatibility between HNF4α and RXRα in the ninth and tenth helix in the C-terminal portion of the LBD that can partially explain the lack of heterodimerization between HNF4α and other known receptors. The X-ray crystallographic structure of the RXRα LBD indicates that there is a salt bridge between a negatively charged glutamic acid residue at 390 (E390$^-$) in one monomer subunit and a positively charged lysine residue at 417 (K417$^+$) in the other monomer subunit. Whereas all receptors known to heterodimerize with RXRα have the same charges in these positions, all of the HNF4 genes have the opposite charges (e.g. K300$^+$ and E327$^-$ in rat HNF4α1). Furthermore, whereas mutation of the HNF4α residues to the RXRα residues still did not allow dimerization with RXR, it did prevent dimerization with the native HNF4α1. Therefore, it appears that the charge incompatibility in the ninth and tenth helix, in addition to other residues in the N-terminal portion of the LBD, prevents HNF4α from heterodimerizing with other receptors (F.M. Sladek *et al.*, personal communication). Finally, since the HNF4α charge compatibility is maintained in all HNF4 genes identified thus far, and since evolutionarily distinct HNF4α proteins (rat and *Drosophila*) heterodimerize with one another (Zhong *et al.*, 1993), it is anticipated that HNF4α will be able to heterodimerize with HNF4β and HNF4γ proteins.

Even though HNF4α does not appear to heterodimerize with other receptors, it does share its response elements with several other receptors, most notably RXR and its heterodimerization partners RAR and peroxisome proliferator-activated receptor (PPAR) and orphan receptors COUP-TFI and COUP-TFII (ARP-1). However, thus far there is no evidence of co-occupancy of the response element by HNF4α and the other receptors, and no significant protein–protein interactions, in either the absence or presence of DNA, have been noted (Jiang *et al.*, 1995; Lee *et al.*, 1998). The one notable exception is COUP-TFI/II, rather ubiquitous orphan receptors that bind DNA both as homodimers and as heterodimers with RXR (Tsai and Tsai, 1997). They exhibit a DNA-binding specificity very similar to that of HNF4α, but on most promoters do not activate transcription. Therefore, COUP-TFs often appear to compete for control of a promoter with HNF4α by competing for binding to their shared response element (ApoCIII, ApoB, ApoAI, ApoAII, ApoAIV) (Ladias *et al.*, 1992; Mietus-Snyder *et al.*, 1992; Ochoa *et al.*, 1993; Sladek, 1993; Hargrove *et al.*, 1999). However, on certain promoters, such as HNF1α, α$_1$-antitrypsin and TTR, instead of COUP-TFs inhibiting activation by HNF4α they actually augment transactivation in a synergistic manner. This synergy appears to depend on interactions between HNF4α and COUP-TFs in the N-terminal portion of the LBD and, at least on certain promoters, appears to be independent of COUP-TF binding DNA. It is proposed that COUP-TFs act as auxiliary cofactors, orienting the HNF4α activation domain in a more efficient configuration to achieve enhanced transcriptional activity (Ktistaki and Talianidis, 1997a).

Finally, HNF4α has been found to interact with the receptor lacking a DBD, short heterodimer partner (SHP) (Seol *et al.*, 1996), although not as a heterodimer *per se*. The interactions appear to be via activation function (AF)2 of HNF4α and cause SHP to compete with coactivators for binding HNF4α (Lee *et al.*, 2000).

Transactivation properties of HNF4α

When HNF4α was originally cloned, it was readily shown to activate transcription in liver cell lines, such as HepG2, as well as nonliver cell lines, such as Jurkat, HeLa and COS (Sladek *et al.*, 1990; Kuo *et al.*, 1992; Mietus-Snyder *et al.*, 1992; Metzger *et al.*,

1993). Furthermore, transactivation was observed in the presence of stripped serum as well as in the absence of serum, suggesting that, if a ligand for HNF4α is required, it is present as an endogenous factor in the cell (Jiang *et al.*, 1995 and unpublished data). More recently, several groups have shown that HNF4α interacts either *in vitro* or in yeast cells with p160 (glucocorticoid receptor interacting protein 1 (GRIP-1), steroid receptor coactivator 1 (SRC-1)) and p300/CBP coactivators in the absence of exogenously added ligand, arguing for at least some ligand-independent transactivation function of HNF4α (Yoshida *et al.*, 1997; Green *et al.*, 1998; Wang *et al.*, 1998; Dell and Hadzopoulou-Cladaras, 1999; Sladek *et al.*, 1999).

Whereas to date there are no reports in the published literature of exogenously added compounds altering the transactivation function of HNF4α in a significant fashion (e.g., more than two-fold; see discussion of a potential ligand below), there is one condition that reproducibly augments transactivation by HNF4α. Removal of the unusually large F domain of HNF4α stimulates transcription five- to 10-fold in transient transfection assays and enhances transactivation by HNF4α in yeast cells as well as interactions with p160 coactivators GRIP-1 and SRC-1 (Hadzopoulou-Cladaras *et al.*, 1997; Sladek *et al.*, 1999; Suaud *et al.*, 1999). As with other receptors, the conserved AF-2 region at the C-terminal end of the LBD is required for transactivation (Hadzopoulou-Cladaras *et al.*, 1997; Dell and Hadzopoulou-Cladaras, 1999). This suggests that the F domain inhibits access to the AF-2, and possibly other regions, by coactivators that are known to interact with the AF-2 (Sladek *et al.*, 1999; Suaud *et al.*, 1999) (see Fig. 9.6).

A repressor region has been identified in the F domain that contains a high degree of amino acid sequence similarity with a similar region in certain of the steroid receptors (progesterone, glucocorticoid, mineralcorticoid and androgen) (Iyemere *et al.*, 1998; Sladek *et al.*, 1999). Whereas the crystal structure of the progesterone receptor LBD indicates that the repressor region makes direct contact with the LBD (Williams and Sigler, 1998), it remains to be determined whether a similar mechanism is involved with the HNF4α repressor region. Finally, given the very strong transactivation ability of

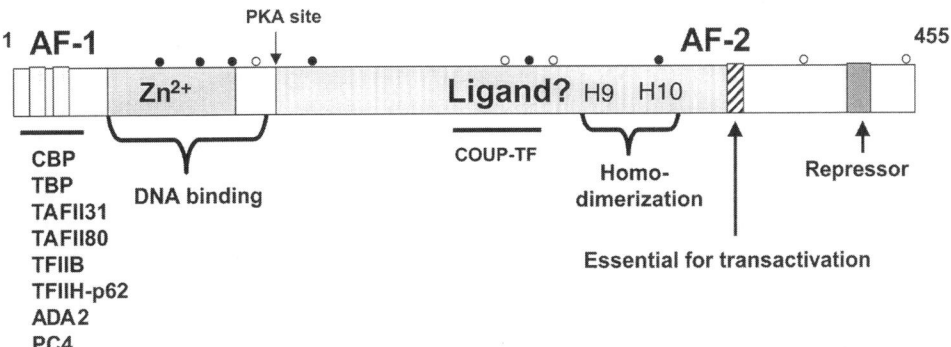

Fig. 9.6 Functional domains of hepatocyte nuclear factor (HNF)4α1. Indicated are activation function 1 (AF-1) and 2 (AF-2), helices 9 and 10 (H9, H10) in the ligand-binding domain, the location of serine 134 (protein kinase A (PKA) site) and residues mutated in patients with maturity-onset diabetes of the young (MODY) 1 (●, definitively linked to diabetes; ○, possibly linked to diabetes; see Table 9.3 for details on mutants). Brackets and arrows indicate functional regions; lines mark regions that bind indicated proteins. See text for details and abbreviations.

HNF4α constructs lacking the F domain (as high as that of other receptors in the presence of ligand), one possible role of a potential ligand could be to introduce some sort of conformational change in HNF4α that displaces the F domain. In this regard, it is of interest to note that the 10 amino acid insert in the F domain of HNF4α2 can lead to enhanced transactivation and response to coactivator, which may be due to an altered receptor conformation (Sladek *et al.*, 1999). The F domains of HNF4α3 and HNF4γ have also been shown to inhibit transactivation by HNF4, although to a lesser extent than that of HNF4α1. These different F domains also appear to have a slightly different effect of HNF4α transactivation depending on the HNF4α-binding site used (Suaud *et al.*, 1999). All this suggests that, once a ligand is found for HNF4, it will be important to analyze the effect that it has on the function of the F domain.

In addition to the requirement for the AF-2 for transactivation, other regions in HNF4α also make contact with coactivators and mediators, at least *in vitro*. For example, in the absence of the F domain, a truncated HNF4α construct lacking the AF-2 binds GRIP-1 and CBP *in vitro*. This interaction is independent of at least two of the three LXXLL motifs in GRIP-1 that are known to contact the AF-2 (Dell and Hadzopoulou-Cladaras, 1999; Sladek *et al.*, 1999). Whereas LXXLL/AF-2-independent interactions with coactivators are being uncovered in other receptors (Hong *et al.*, 1999; Kraichely *et al.*, 1999), one can anticipate that HNF4α may display some unique interactions.

Whereas the F domain of HNF4α represses transactivation, the A/B domain is required for full transcriptional activity. When this region, also referred to as the AF-1, is deleted, transactivation is reproducibly decreased in a number of systems by approximately 50% (Hadzopoulou-Cladaras *et al.*, 1997; Green *et al.*, 1998; Sladek *et al.*, 1999). The AF-1 region has been shown to interact with a number of coactivators, such as CBP, alteration/deficiency in activation (ADA)2 and positive cofactor (PC)4, as well as members of the basal transcription machinery, such as transcription factor (TF)IIB, TATA binding protein (TBP), TBP associated factor (TAF)II31, TAFII80, and TFIIH-p62 (see Fig. 9.6). Extensive analysis of the AF-1 by site-directed mutagenesis shows that the aromatic and bulky hydrophobic residues are essential for AF-1 function. The AF-1 was also shown to consist of two regions, both of which share structural motifs with other transcription activators, such as tumor suppressor p53 and nuclear factor (NF)-κB-p65, suggesting that HNF4α might also share similar mechanisms of action (Green *et al.*, 1998). Finally, it will be of interest to determine whether the ability of HNF4α to interact with so many components of the basal transcription machinery is in any way related to the fact that HNF4α activates the transcription of several TATA-less promoters, and to the fact that the HNF4α-binding site in those promoters is very close to the transcriptional start site (e.g. Factor VII, Factor IX and sex hormone-binding globulin) (see Table 9.1). In other words, could HNF4α in some situations actually substitute for the function of the TATA-binding protein (TBP)?

Protein phosphorylation of HNF4α

Nearly every nuclear receptor examined to date has been shown to be phosphorylated (Orti *et al.*, 1992; Kuiper and Brinkmann, 1994). In this regard, HNF4α is no exception. However, HNF4α is perhaps somewhat unusual in that it appears to be very heavily phosphorylated, at least when isolated from transfected COS cells. Two-dimensional phosphopeptide mapping identified 13 potential serine and threonine phosphorylation sites (Jiang *et al.*, 1997b). There is also one report of potential tyrosine phosphorylation of HNF4α (Ktistaki *et al.*, 1995). Only one potential phosphorylation site has been characterized thus far. Serine 134, located in the A box, which plays a role in DNA binding, is conserved in all HNF4 genes, including *Xenopus* HNF4β, human HNF4γ and *Drosophila* HNF4α, and is in the same position as a phosphorylated serine present in several species of NGF-IB. One report suggested that this serine may be phosphorylated by protein kinase A (PKA), resulting in a decrease in DNA binding and transactivation function. It also suggested that there may be a physiologic role for this phosphorylation event in that, under conditions of starvation that result in an induction of cyclic adenosine monophosphate (cAMP), and hence of PKA activity, there is significantly less HNF4α DNA-binding activity found in the liver of treated animals (Viollet *et al.*, 1997). However, a follow-up study in which S134 was mutated could not confirm that S134 is responsive to PKA activity (Gourdon *et al.*, 1999). In another study, dietary protein restriction caused a 40% decrease in HNF4α DNA-binding activity without a change in total HNF4α protein, suggesting that a post-translational modification, such as phosphorylation, may be involved (Marten *et al.*, 1996). Clearly, more research on phosphorylation of HNF4α is required, particularly considering that there appears to be as yet unidentified factors working in conjunction with HNF4α to cause diseases such as diabetes (see below). The possibility must be considered that some of these factors may modulate HNF4α function via altering its phosphorylation state.

GENETICS

There are currently three ways in which HNF4α is known to cause human disease; two involve inherited mutations in human populations and the third involves viral infections. The first type of genetic linkage between HNF4α and disease consists of mutations in HNF4α response elements in the promoter regions of target genes; the second involves mutations in the HNF4α gene itself. It is also conceivable that mutations in the promoter/enhancer region of HNF4α might also lead to disease, although no such mutations have been found as of yet.

Inherited mutations in HNF4α-binding sites

The naturally occurring mutations in HNF4α-binding sites that have been identified thus far are shown in Table 9.2. Although the number of mutations is low, there are a few observations that are worth mentioning. The first is that the mutations are not randomly distributed throughout the site. Four of the eight mutations are located in the central three As which is not too surprising considering that those As are quite conserved in the consensus sequence (see Table 9.1). In contrast, the other four mutations are all in the fifth position of the second half-site and are to nucleotides (T or G) that are found in other native HNF4α-binding sites. Whereas more mutations are needed before any conclusions can be drawn, the data thus far suggest that there is a high degree of specificity in the sequence of the HNF4α-binding site in any given promoter. They also suggest that certain specific variations from that sequence can result in severe physiologic effects.

   The first human disease found related directly to HNF4α was hemophilia B Leyden, an X chromosome-linked recessive bleeding disorder that is characterized by mutations between −40 and +20 of the factor IX promoter (Xq26.3-27.1). Factor IX is a vitamin K-dependent glycoprotein that plays a key role in the blood coagulation cascade

Table 9.2. Naturally occurring mutations in HNF4α-binding sites that lead to disease states

Disease or gene	Mutation*	Orientation†	Native site	Mutated site‡	HNF4α binding§	Reference
Hemophilia						
hFactor VII	−11 T > G	Rev comp	CGGGCA a AGTTCT	CGGGCA a **C**GTTCT	Abolish	Arbini et al. (1997)
hFactor IX	−6 G > A	Rev comp	AGTGGT a AGGTCG	AGTGGT a AGGT**T**G	Weak disrupt	Naka and Brownlee (1996)
	−6 G > C	Rev comp	AGTGGT a AGGTCG	AGTGGT a AGGT**G**G	Abolish	Naka and Brownlee (1996)
	−20 T > A	Rev comp	GTACCA a AGTACA	GTACC**T** a AGTACA	Disrupt	Reijen et al. (1992)
	−21 T > G	Rev comp	GTACCA a AGTACA	GTACCA **c** AGTACA	Disrupt	Reijen et al. (1993)
	−26 G > C	Rev comp	GTACCA a AGTACA	GTACCA a AGTA**G**A	Abolish	Reijen et al. (1992)
	−26 G > A	Rev comp	GTACCA a AGTACA	GTACCA a AGTA**T**A	Disrupt	Morgan et al. (1997)
Diabetes						
HNF1α (MODY-3)	−58 A > C	Native	AGTCCA a AGTTCA	AGTCC**C** a AGTTCA	Disrupt	Gragnoli et al. (1997); Lausen et al. (2000)

Consensus binding site¶

```
GGGTCA A AGGTCA
A tCt  g g TCtg
  aG       g c
             t
123456 x 123456
 1st      2nd

    Half-site
```

* Numbering is with respect to start site of transcription. † The site orientation that best fits the consensus is shown (Rev comp, reverse complement). ‡ Mutation is shown in bold and is underlined. § Effect of mutation on binding HNF4α protein. ¶ The HNF4α consensus binding site from Table 9.1 is shown at the bottom for comparison purposes.

(Bowman, 1993). The first patients identified with hemophilia B Leyden had one of two point mutations in the HNF4α-binding site (-20 T → A or -26 G → C), a mutation in a nearby C/EBPα site or a mutation in a site binding an unknown factor (Crossley et al., 1992; Reijnen et al., 1992). Additional mutations at -6 (G → A, G → C), -21 (T → G) and -26 (G → A) associated with this type of hemophilia were subsequently found (Reijnen et al., 1993; Naka and Brownlee, 1996; Morgan et al., 1997). Mutations at positions -20 and -21 in the HNF4α-binding site display the clinical phenotype of improving after puberty whereas mutations at the -26 position do not (the latter are sometimes referred to as the Brandenburg phenotype (Heit et al., 1999)). The apparent reason for the difference in the effects of puberty is that there is a response element to which the androgen receptor (AR) binds and transactivates immediately upstream of the HNF4α site. Mutations at -26 abolish AR binding as well as HNF4α binding (Morgan et al., 1997). The implication is that AR, in response to an increase of testosterone during puberty, can compensate for a lack of binding by HNF4α due to mutations at -20 or -21 but not due to mutations at -26 because the latter also abolishes AR binding.

Another type of inherited bleeding disorder is caused by a mutation in an HNF4α-binding site in the promoter of another blood coagulation factor, Factor VII (13q34-qter). Factor VII deficiency, an autosomal recessive disorder, occurs in the general population with an estimated incidence of one in 500 000 and is most commonly associated with mutations in the coding region of the Factor VII gene. However, there is one case of an American girl of French Canadian ancestry who showed signs of excess bleeding at 2 months and who was put on chronic factor VII treatment by 8 months. She had less then 1% of the normal levels of Factor VII in her serum. By age 6 years, a homozygous mutation in the HNF4α-binding site was identified in the promoter of her factor VII gene (-61T > G).† Both the patient's parents were heterozygous for the same mutation and had 30–44% of the normal levels of factor VII in their plasma pool but no clinical symptoms (Arbini et al., 1997). This example, as well as those with Factor IX, dramatically shows that a single point mutation in an HNF4α-binding site can have very severe effects for the patient. It is of interest to note that both the Factor VII and Factor IX promoters are TATA-less promoters and the HNF4α-binding site is located very close to the transcription start site. It would be of interest to determine whether this unique position in the HNF4α-binding site plays a role in the severity of the effects caused by the mutations.

As indicated in Table 9.1, there are several other genes involved in coagulation that are also regulated by HNF4α, although those sites have not yet been specifically linked to a disease. For example, inherited deficiencies in antithrombin III, a serine protease inhibitor (serpin) that inhibits active forms of thrombin, Factor IX and Factor X, lead to thrombosis and affect one person in 5000, accounting for 1–2% of all patients with venous thromboembolic disease. Many different mutations in various parts of the antithrombin III gene, including the regulatory region, have been identified, although to date none has been mapped to the HNF4α-binding site (reviewed in Bowman, 1993).

In addition to genes involved in blood coagulation, HNF4α also plays a critical role in the expression of several other genes that are linked to human disease (Fig. 9.7). For nearly every example given below, single point mutations introduced artificially into the HNF4α-binding site have been shown to have drastic effects on the regulation of these genes in tissue culture systems. Therefore, whereas naturally occurring mutations in the

---

† Numbering in the text is given with respect to the start site of translation; numbering is also sometimes given in the literature with respect to the transcription start site, which in this case is -11.

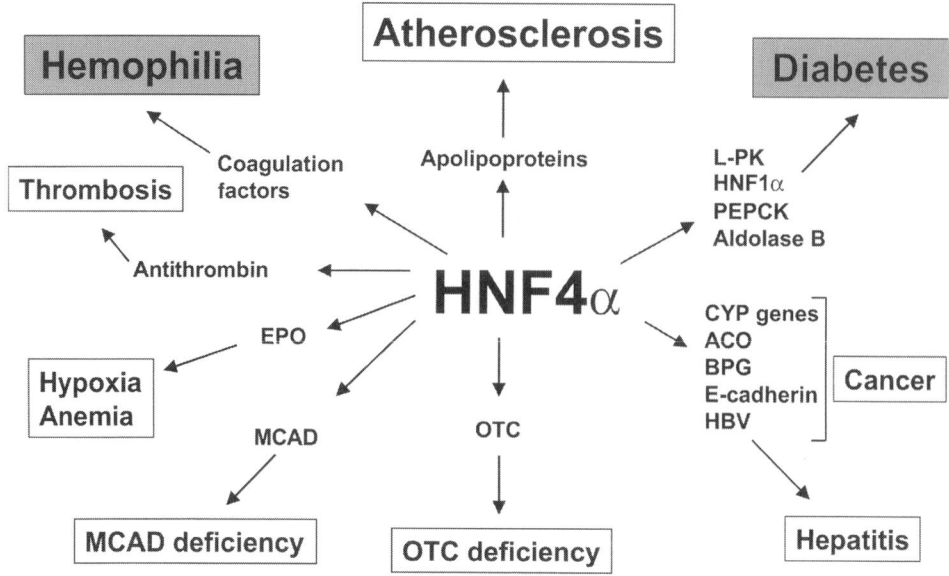

Fig. 9.7 Hepatocyte nuclear factor (HNF)4α and human disease. The relationship between HNF4α, selected target genes and human diseases is shown. Shaded box, proven connection to diseases indicated; open box, possible connection. See text for details and abbrevations.

HNF4α-binding site of these genes have yet to be identified, they might well be in the future in a scenario analogous to the blood coagulation factors. For example, HNF4α binds the liver factor (LF) A1 site in the α_1-antitrypsin gene and positively regulates the expression of the α_1-antitrypsin gene. α_1-Antitrypsin, one of the main protease inhibitors in human serum and evolutionarily related to the serpin antithrombin III, is thought to help control the inflammatory response by inhibiting excess elastase and collagenase released from leukocytes. Single point mutations in α_1-antitrypsin can result in early-onset emphysema and severe lung and liver disease. Whereas many different mutations have been mapped in the coding and noncoding region of the α_1-antitrypsin gene, there are no published mutations in the HNF4α-binding site (reviewed in Bowman, 1993; Sladek, 1994). Another example is medium chain acyl-coenzyme A dehydrogenase (MCAD; EC 1.3.99.3), a HNF4α target gene important in fatty acid β-oxidation. Deficiency in MCAD is a common inherited metabolic defect and an important cause of childhood Reye-like syndrome, hypoglycemia and sudden death. Finally, deficiency in ornithine transcarbamylase (OTC; EC 2.1.3.3), a HNF4α target gene in the urea cycle that detoxifies ammonia, manifests itself shortly after birth or later on in life and is accompanied by symptoms of lethargy, hypothermia, encephalopathy and behavioral abnormalities (reviewed in Sladek, 1994). Again, naturally occurring mutations in the HNF4α-binding site in either MCAD or OTC have not been identified but could be in the future.

Finally, HNF4α is known to play a very critical and yet complex role in apolipoprotein gene regulation. Apolipoproteins are synthesized primarily in the liver and to a lesser extent in the intestine and then secreted into the bloodstream where they associate with lipids to form large particles of varying densities called plasma lipoproteins. These particles are responsible for delivering lipids, including cholesterol, to cells in other tissues in the body and returning excess lipids to the liver for clearance. Since altered levels of the lipoprotein particles and of the apolipoproteins themselves are associated with an increased risk of atherosclerosis and coronary heart disease, an intense effort has been put into dissecting the promoter elements responsible for the regulation of the apolipoproteins. HNF4α is known to be important in regulating the expression of several of the apolipoprotein genes associated with clinical disorders of genetic origin (apoAI, AII, AIV, B, CII and CIII; see Table 9.1) (Breslow, 1988; Sladek, 1994), although inherited mutations in the HNF4α-binding sites of these genes have not yet been identified.

Inherited mutations in the HNF4α gene

The second type of genetic linkage between HNF4α and human disease involves mutations in the HNF4α gene itself. Thus far only one disease has been mapped to the coding region of the HNF4α gene – maturity onset diabetes of the young 1 (MODY-1), a form of noninsulin-dependent diabetes mellitus (NIDDM, also referred to as type II diabetes). Patients with MODY, who account for 1–5% of all diabetics, inherit the disease in an autosomal dominant pattern and usually develop symptoms before 25 years of age. They exhibit normal insulin sensitivity and liver and kidney function but exhibit a defect in glucose-stimulated insulin secretion from the pancreatic β cells. Whereas MODY-1 was the first type of MODY described (in 1958), it represents only a small fraction of diabetes cases (< 0.0001% of all NIDDM and 2–4% of all MODY) (Malecki *et al.*, 1999).

Four other types of MODY in addition to MODY-1 (MODY 2–5) have been identified, one of which is also related to HNF4α, MODY-3 (reviewed in Froguel, 1998; Velho and Froguel, 1998). MODY-3 consists of mutations in the pou-homeo-domain transcription factor HNF1α which is a downstream target of HNF4α in a transcriptional hierarchy in the liver (see Fig. 9.4 and discussion above on the mutation in the HNF4α-binding site in the HNF1α promoter). HNF4α-binding sites have not yet been found in the promoter regions of the other MODY genes: MODY-2, glucokinase, a glycolytic enzyme that plays a key role in glucose-sensing by pancreatic β cells and integration of hepatic glucose metabolism (Matschinsky, 1990); MODY-4, insulin promoter factor 1 or PDX-1, a homeodomain protein that regulates pancreatic development and the expression of various β-cell genes, including insulin (Stoffers *et al.*, 1997); or MODY-5, HNF1β (also referred to as vHNF1), a heterodimeric partner of HNF1α (Horikawa *et al.*, 1997), which appears to be in a separate transcriptional hierarchy from HNF4α (Power and Cereghini, 1996).

Thus far, no mutations have been identified in the promoter/enhancer region of HNF4α gene that are linked to MODY-1, or any other disease. However, there is evidence that another diabetes-predisposing gene is located near HNF4α on chromosome 20. The mutations could be in HNF3β, which maps near HNF4α and which positively regulates HNF4α gene expression (see Fig. 9.4) (Ghosh *et al.*, 1999), or possibly in an extended region of the HNF4α promoter (Moller *et al.*, 1999). Studies of the HNF4α promoter show that a much larger portion is required for appropriate liver-specific gene expression (approximately 6.5 kb) (Zhong *et al.*, 1994) than is typically screened when looking for mutations. Therefore, mutations in the HNF4α promoter/enhancer may yet be identified. Finally, several polymorphisms have been found in the HNF4α gene that are not associated with MODY-1. They include nucleotide changes in the promoter, introns and coding regions of HNF4α, including T130I (Furuta *et al.*, 1997; Moller *et al.*, 1997; Hani *et al.*, 1998).

The prevalence of different forms of MODY appears to differ among ethnic and racial groups. HNF1α (MODY-3) mutations are the most common mutations identified to date among patients in the UK (73%), Germany (25–35%) and Japan (8%). In France, glucokinase mutations (MODY-2; 50%) are more prevalent than HNF1α mutations (25%) (discussed in Lindner *et al.*, 1997). Mutations in HNF4α are much less prevalent in all populations tested and thus far no significant variations between different racial or ethnic groups have been identified, although that could be due to the low number of MODY-1 mutations in general.

PATHOPHYSIOLOGY

Before discussing the pathophysiology of the MODY-1 mutation in diabetes, we will discuss the other major diseases linked to HNF4α: hepatitis and liver cancer. Hemophilia, atherosclerosis and other diseases potentially linked to HNF4α via its target sites have been discussed above.

HNF4α, hepatitis B virus and liver cancer

HNF4α is potentially linked to cancer, particularly liver cancer, in several ways, the most notable of which is via hepatitis and hepatitis B viral (HBV) infections. HBV is an extremely common infectious agent that frequently results in chronic hepatitis, cirrhosis of the liver and hepatocellular carcinoma. Approximately 350 million people worldwide are infected with the virus and up to one million persons die each year from HBV-related chronic liver disease (Mahoney, 1999). As the risk of developing liver cancer has been found to be up to 200 times greater in those infected with HBV than in those who are not infected, HBV and chronic hepatitis is considered one of the major causative agents of liver cancer in the world today (Levine, 1992). Massive immunization programs against HBV have been initiated recently in more than 80 countries and will hopefully reduce the number of HBV infections (Mahoney, 1999). Nonetheless, the mechanism of action of HBV remains a very active field of research and is providing new insights into transcription regulation.

The mechanism by which HBV causes liver cancer is not known but is thought to be due at least in part to the mitogenic effect in the liver as a result of chronic HBV infection. HNF4α binds and transactivates one of four HBV promoters and two enhancer regions (see Table 9.1). HNF4α sites in enhancer II and the core/nucleocapsid promoter give rise to the core (capsid) protein (HBcAg), surface antigen (HBsAg) and the viral DNA polymerase, and is therefore linked to liver-specific viral replication. The HNF4α site in enhancer I is instrumental in the activation of the pX gene, which may be the most relevant in terms of liver cancer. pX (also referred to as the X protein or HBx) is capable of transforming normal rodent cells, causes liver cancer in certain strains of transgenic mice, and has recently been found to interact with members of the basal transcription machinery as well as the tumor suppressor gene p53 (Feitelson and Duan, 1997).

It is of interest to note that, in contrast to the other diseases discussed above, in the case of HBV it is the activity of the wild-type HNF4α protein, and not that of a mutated form of HNF4α or its binding site, that is linked to human disease. Furthermore, what is learned from the role of HNF4α, and other liver-enriched transcription factors, in HBV may eventually be applicable to other viruses associated with hepatitis, namely hepatitis C, D, F and G viruses (HCV, HDV, HFV, HGV). For example, it was noted in 1999, that an estimated 170 million people worldwide were infected with HCV and that the number of annuals deaths in the USA from HCV-caused liver damage and cancer may soon overtake deaths caused by acquired immune deficiency syndrome (Cohen, 1999). Therefore, whereas HCV is a different type of virus from HBV, it will be important to determine whether it employs some of the same tactics as HBV to achieve liver-specific expression, and hence hepatitis and liver cancer, namely, recruitment of liver-enriched transcription factors such as HNF4α.

Other potential links between HNF4α and cancer

HNF4α is potentially related to cancer, particularly liver cancer, in at least three other ways. The first is via cytochrome P450 genes which are regulated by HNF4α (cyp2A4, 2C1, 2C2, 2C3, 2C9, 2D6, 3A1 and 7; see Table 9.1). This family of genes plays a critical role not just in lipid and steroid metabolism but also in biotransformation. The *cyp* genes encode phase I enzymes that metabolize xenobiotics and result in the activation or inactivation of potential carcinogens (Klaassen *et al.*, 1996). Therefore, via regulation of the *cyp* genes, HNF4α could potentially influence the production of carcinogenic compounds in the liver, which may then act *in situ* or elsewhere in the body.

The second connection is via acylcoenzyme A oxidase (ACO; EC 1.3.3.6), another HNF4α target gene. ACO performs the rate-limiting step in β oxidation and is the most widely used marker of peroxisome proliferation that is linked to liver hypertrophy and ultimately hepatocellular carcinoma. HNF4α shares its response element on ACO with PPARα, another member of the nuclear receptor superfamily (reviewed in Sladek, 1994; see also Chapter 10).

The final potential link between HNF4α and cancer is via the TGFβ/SMAD pathway, which, as discussed above, regulates HNF4α gene expression during early development. Mutations in TGFβ receptors and SMADs are associated with tumor progression and invasiveness in a number of tissues, including colon and pancreas. For example, mutations in TGFβ type II receptors are found in patients with hereditary nonpolyposis colonic cancer (HNPCC) due to microsatellite instability. Homozygous mutations in SMAD4 (also known as DPC4, *d*eleted in *p*ancreatic *c*arcinoma) are found in up to 50% of human pancreatic carcinomas and 30% of human colorectal cancers as well as other cancers. SMAD2 is also frequently mutated in colonic cancer (reviewed in Massague *et al.*, 1997; Hata *et al.*, 1998; White, 1998).

The role of HNF4α in these TGFβ–SMAD-induced cancers, if any, is not known. However, if the TGFβ/SMAD pathway plays a similar role in controlling HNF4α gene expression in the adult as it does in the embryo, then the possibility must be considered that inactivating mutations in TGFβ/SMAD pathway members may also adversely affect HNF4α gene expression in neoplastic or pre-neoplastic tissue. The question then arises as to how altered HNF4α gene expression could lead to tumor progression. One possible answer is E-cadherin. E-cadherin, mutated in inherited gastric cancer, mediates an important cell–cell adhesion function that is lost during the development of most epithelial cancers, including those of HNF4α-expressing tissues: colon, prostate, stomach, liver and kidney (Guilford *et al.*, 1998; Christofori and Semb, 1999). Finally, it is of interest to note that overexpression of HNF4α in dedifferentiated rat hepatoma cells activates the expression of E-cadherin (Spath and Weiss, 1998). Could it be that one of the relevant downstream targets of the TGFβ/SMAD pathway is E-cadherin and that HNF4α is somehow involved?

Finally, a discussion of HNF4α and cancer would not be complete without discussing what is known about HNF4α RNA and protein levels in tumors, and may be particularly relevant if the scenario described above with the TGFβ/SMAD pathway turns out to be relevant. However, attempts thus far to correlate HNF4α levels with tumorigenicity are few and not very conclusive, especially in terms of human cancer. HNF4α protein levels have been shown to be abnormally high in one pancreatic endocrine tumor, but the tumor was from a transgenic mouse, raising the issue of physiologic relevance (Miquerol et al., 1994). HNF4α DNA-binding activity was found to vary greatly between different rat hepatocellular carcinomas (Stumpf et al., 1995) and only a very modest reduction in HNF4α RNA (20–30%) was seen in mouse liver tumors (Kalkuhl et al., 1996), arguing against a link between this tumor type and HNF4α. Human kidney tumors showed a more consistent decrease in HNF4α protein or DNA-binding activities although only a few samples were examined (Sel et al., 1996). Finally, none of these studies examined the possibility of different forms of HNF4α (either splicing variants or posttranslation modifications) being involved. Therefore, at this point, it can only be said that members of the TGFβ pathway both upstream and downstream of HNF4α are directly implicated in tumor progression in tissues where HNF4α is found, and that more studies looking directly at HNF4α are needed to know whether it itself is involved in any of those tumor types.

HNF4α and MODY-1: potential gene targets

Patients with mutations in the HNF4α gene (MODY-1) tend to have severe diabetes, requiring treatment with either insulin or oral hypoglycemic drugs. And, whereas they often have complications such as proliferative retinopathy and macrovascular and microvascular disease (coronary heart disease), they do not exhibit insulin resistance or show signs of renal dysfunction (nephropathy) or liver malfunction (Furuta et al., 1997; Lindner et al., 1997; Iwasaki et al., 1998). Therefore, whereas the liver and kidney both play an important role in glucose homeostasis and both express much more HNF4α than the adult pancreas, the preponderance of evidence thus far suggests that the pancreas, and β cells in particular, are the primary target of mutations in MODY-1. In contrast, an argument has been made that abnormalities in glucose metabolism in the liver, in addition to that in the pancreas, may play an important role in the pathogenesis of patients with MODY-2 (Velho and Froguel, 1998).

One possibility is that the pancreas as a whole organ is affected. For example, in nondiabetic subjects with MODY-1 and MODY-3, diminished insulin secretion precedes the development of hyperglycemia. Since HNF4α is expressed at high levels in the pancreas during development, it is possible that the defect in individuals with MODY-1 is caused by a reduction in the mass of pancreatic islets, both β cells which secrete insulin and α cells which secrete glucagon (Herman et al., 1997).

The second possibility is that the expression of one or more HNF4α target genes in the pancreas is affected as a result of MODY-1 mutations. To date, one of the most likely candidates is the HNF1α gene. Whereas very little work has been done on the role of HNF4α and HNF1α in pancreatic function, several pieces of evidence suggest that a similar HNF4α → HNF1α hierarchy may exist in the pancreas as it does in the liver (see Fig. 9.4 and discussion above). For example, only 11 mutations have been mapped thus far to the HNF4α gene in MODY-1, while over 70 mutations have been mapped to the

HNF1α gene in MODY-3 (Yoshiuchi *et al.*, 1999). Furthermore, as mentioned above, one of the MODY-3 mutations maps to the HNF4α-binding site in the HNF1α promoter (Gragnoli *et al.*, 1997). Finally, and perhaps most telling, is the observation that patients with MODY-1 and MODY-3 exhibit a similar clinical profile. They have a similar age of onset (postpuberty), they both progress to severe diabetes with about 30% of the patients eventually requiring insulin, and they both present frequent diabetes-associated complications, such as retinopathy and microvascular complications later in life. In contrast, patients with other forms of MODY, such as MODY-2 (glucokinase) and MODY-4 (insulin promoter factor 1 or PDX-1), exhibit a different age of onset, present a more mild form of diabetes and rarely develop complications (Velho and Froguel, 1998). There are, however, some differences in the symptoms of patients with MODY-1 and MODY-3, most notably in the kidney where those with MODY-3 and MODY-5 but not patients with MODY-1 have reduced kidney function (Iwasaki *et al.*, 1998).

Since the primary pathology of patients with MODY-1 is a decrease in insulin secretion, another obvious candidate as an HNF4α target would be the insulin gene itself. However, the insulin gene does not appear to be an HNF4α target. No HNF4α sites have been reported in the insulin gene promoter in the literature, although a COUP-TF site was reported some time ago (Hwung *et al.*, 1988). However, despite the fact that HNF4α and COUP-TF have very similar DNA-binding specificities (Sladek, 1993), that site does not bind HNF4α *in vitro* (F.M. Sladek and Y. Dallas-Yang, unpublished results). In contrast, a weak HNF1α site has been reported in the insulin promoter (Emens *et al.*, 1992), so it is possible that HNF4α could affect insulin gene expression via its control of HNF1α expression. Other potential targets for HNF4α in the pancreas that might be related to diabetes are L-pyruvate kinase (L-PK; EC 2.7.1.40) and glucose transporter 2 (glut2). L-PK plays a central role in the glycolytic pathway and is a known HNF4α target gene in the liver and pancreas (Miquerol *et al.*, 1994). Glut2 encodes a glucose transporter present in liver and pancreas (Kim *et al.*, 1998) and, although no HNF4α-binding site has been identified in the glut2 promoter, there is genetic evidence indicating that HNF4α controls glut2 RNA levels either directly or indirectly (Stoffel and Duncan, 1997).

Finally, identification of the appropriate HNF4α target gene could be important in examining the functional properties of HNF4α mutations. As discussed below, conflicting biochemical phenotypes of a given mutation in HNF4α are sometimes reported by different groups. The differences could be due to the use of different HNF4α response elements which may bind HNF4α mutants differentially. Since mutation of the HNF4α-binding site in the HNF1α promoter is known to cause MODY-3, the HNF1α promoter may at this point be the best target gene to use in analyzing the functional properties of HNF4α mutations.

Effect of mutations in the HNF4α gene in MODY-1

Thus far 11 mutations in the HNF4α coding region have been reported in the literature to be linked to MODY-1 (see Table 9.3 and Fig. 9.6). Three mutations are in the DBD (domain C), one is in the hinge region (domain E), five are in the LBD (domain E) and two are in the F domain (see Fig. 9.6). Curiously, no mutations have been reported in the A/B domain in the N-terminal portion of the protein, although polymorphisms have been noted (Lehto *et al.*, 1999b). However, only six of the mutations are indisputably the cause of diabetes. Four of those mutations (F75fs, K99fs, R154X, Q268X) result in prematurely truncated proteins that are anticipated to obliterate the normal HNF4α

Table 9.3 Mutations in HNF4α resulting in maturity-onset diabetes of the young 1 (MODY-1)

Mutation-aa	Mutation-nt	Location	Biochemical phenotype	References§	Additional comments¶
F75fs	Exon 2 224delT	C domain Zn finger	Protein truncated at aa117 Does not bind DNA or transactivate†	Moller et al. (1999)	Age of onset 11–40 years
K99fs	Exon 3 295–296delAA	C domain Zn finger	Protein truncated at aa122 Does not bind DNA or transactivate†	Lehto et al. (1999a)	Age of onset 18 years Lower triglycerides due to lower ApoCIII and ApoB, higher ApoAI/HDL
G115S	Exon 3 343G > A	C domain	ND‡	Malecki et al. (1999)	Age of onset 14–50 years Mutation is in the completely conserved GM motif
R127W	Exon 4 379C > T	D domain T box	Activates transcription Binds DNA	Furuta et al. (1997) Navas et al. (1999)	Age of onset 16–90 years Variable penetration of allele Linkage to NIDDM not certain
R154X	Exon 4 460C > T	E domain	Truncated protein May be able to bind DNA†	Lindner et al. (1997)	Age on onset 15–28 years Increase in serum lipoprotein(a) levels
V255M	Exon 7 763G > A	E domain	Activates transcription	Moller et al. (1997) Navas et al. (1999)	Linkage to NIDDM questioned Mutation is in potential ligand-binding pocket
Q268X	Exon 7 802C > T	E domain	Does not bind DNA or activate transcription Dimerizes with wild-type Abnormal subcellular localization	Yamagata et al. (1996) Stoffel and Duncan (1997) Sladek et al. (1998)	Variable age of onset Late-onset NIDDM also present in family First MODY-1 mutation mapped

E276Q	Exon 7 826G > C	E domain	Binds DNA Dimerizes with wild-type? Activates transcription?	Bulman et al. (1997) Navas et al. (1999) Suaud et al. (1999)	Age of onset 21–41 years Controversy over linkage to NIDDM
V328–329ins	Exon 8 985insGTN*	E domain	ND	Lehto et al. (1999b)	Age of onset 18–21 years
V393I	Exon 9 1177G > A	F domain	Decreased transactivation	Hani et al. (1998)	Age of onset 40–55 years Other factors may be involved in causing NIDDM
I454V	Exon 10 1360A > G	F domain	ND	Malecki et al. (1999)	Age of onset 13–64 years Linkage to NIDDM not certain

aa, amino acid; nt, nucleotide based on human HNF4α2 cDNA sequence using the original start methionine as +1 (accession number X87871); fs, frameshift.
* The nucleotide sequence of the insert was not reported; the possible codons for the inserted valine are shown. † Phenotype based on results from in vitro studies of related constructs (Jiang and Sladek, 1997). ‡ ND, in vitro functional properties not determined. § The first reference is to the original identification of the mutation; other references are to functional characterization. ¶ Age of onset refers to age of diagnosis and could be earlier as some patients already have complications when they are first diagnosed with diabetes; abnormal functions and questions concerning linkage to NIDDM are noted.

function. The fifth mutation changes the glycine (G) to a serine (S) (G115S) in the invariant GM motif found at the C-terminal end of the zinc finger in every nuclear receptor; this mutation is anticipated to have a severe effect on HNF4α function although it has not yet been examined *in vitro*. The sixth mutation (V328–329ins) inserts a valine between two leucines in helix 10, a region known to be important in homodimerization (see discussion above). The other five mutations of questionable linkage to MODY-1 (R127W, V255M, E276Q, V393I, I454V) are all missense mutations that have less severe (and even questionable) effects on HNF4α function *in vitro*. We will discuss in detail just one of the truncated proteins and several of the missense mutations to highlight some of the issues involved in examining HNF4α mutations.

The first mutation identified in MODY-1 was found in a rather large pedigree (R-W) of Americans of German descent (Yamagata *et al.*, 1996). It introduced a stop codon at residue 268 (Q268X) that results in a truncated protein containing the DBD and part of the LBD.[‡] Since Q268X lacks the AF-2 transactivation region, it was originally thought that it might act in a dominant negative fashion. However, we and another group independently showed that Q268X neither binds DNA *in vitro* nor affects transactivation of normal HNF4α1 in transient transfection assays (Stoffel and Duncan, 1997; Sladek *et al.*, 1998). We also showed that Q268X binds wild-type HNF4α1 in solution and that Q268X is differentially localized in the nucleus with respect to the normal protein (possibly in the nuclear membrane), which could explain why it does not display a dominant negative effect either *in vitro* or *in vivo* in patients. Furthermore, transient transfection studies using a range of nonmutated HNF4α1 and Q268X DNA concentrations to mimic the *in vivo* situation in heterozygotes indicated that the primary effect of this (and possibly other) MODY-1 mutation is one of dosage (Sladek *et al.*, 1998). Therefore, we propose that the Q268X mutation does not cause detectable liver or kidney malfunctioning because there is excess HNF4α in those tissues. In contrast, in the pancreas where there are limiting amounts of HNF4α RNA, and presumably protein, reduction of one-half the functional protein has a severe effect. The same argument can be made for the role of HNF4α during early development where it is expressed at high levels in the visceral endoderm: heterozygote individuals with MODY-1 mutations develop normally because there is excess HNF4α present. Finally, it is not too surprising that all the MODY-1 mutations definitively identified and characterized thus far result in nonfunctional proteins that do not interfere with the wild-type protein; a naturally occurring mutation in HNF4α that acted in a dominant negative fashion would be expected to cause embryonic lethality as seen in the HNF4α null mice, or at the very least severe liver and kidney damage (Chen *et al.*, 1994c).

E276Q, the first missense mutation identified in MODY-1, is one of the most intriguing MODY-1 mutations isolated thus far. Two groups have analyzed it for functional properties and found very different results. Whereas both groups found that, when synthesized *in vitro*, E276Q binds DNA, one group found that it could heterodimerize with wild-type HNF4α2 (Suaud *et al.*, 1999), while the other group found that it could not (Navas *et al.*, 1999). In transient transfection assays, Suaud *et al.* (1999) found that E276Q activated transcription in a manner similar to the normal protein except that it could not synergize with orphan receptor COUP-TFII on the HNF1α promoter. This supports the previous finding that COUP-TF interacts with HNF4α in this region

---

[‡] All of the MODY-1 mutations have been identified by DNA sequencing only. HNF4α protein analysis from tissues of affected individuals has not been reported for any of the mutations.

(Ktistaki and Talianidis, 1997a). In contrast, Navas *et al.* (1999) found that E276Q did not activate transcription at all and in fact was not very stable *in vivo*, producing a 40-kDa fragment consisting of the C-terminal portion of HNF4α; Suaud *et al.* (1999) did not observe such a proteolytic fragment in their transfected cells. Both groups report a modest dominant negative activity on an ApoCIII reporter construct. The difference between the two groups' results is not known but could be due at least in part to different reporter constructs used. A less likely possibility is that Suaud *et al.* introduced the E276Q mutation into the human HNF4α2 sequence while Navas *et al.* introduced it into the rat HNF4α1 sequence. It would be very surprising if the different species were responsible for the effects seen since they exhibit such a high degree of sequence similarity, and as yet no difference has been noted between rat and human HNF4α function. On the other hand, since functional differences between HNF4α1 and HNF4α2 have been reported (Sladek *et al.*, 1999), a direct comparison of the E276Q mutation in the two isoforms may prove enlightening.

Navas *et al.* (1999) also raise the important issue of distinguishing between a mutation linked to MODY-1 and a cosegregating rare polymorphism in a patient pool too small for statistical analysis. For example, two of the missense mutations identified in HNF4α in patients with MODY-1 (R127W and V255M) were observed in families in which not all diabetic family members carried the mutation, suggesting that there may be an additional cause of diabetes in those families (Furuta *et al.*, 1997; Moller *et al.*, 1997). Furthermore, no obvious defects in DNA binding or transactivation could be noted in either R127W or V255M, although only a limited number of experimental conditions was examined (Navas *et al.*, 1999).

HNF4α and late-onset diabetes

NIDDM is increasing in the USA at a distressingly high rate. Whereas most of the diabetes is among older populations in which a tendency toward obesity and a more sedentary lifestyle is typical, there are some data showing that diabetes is also on the rise among certain younger populations (Burke *et al.*, 1999; Rosenbloom *et al.*, 1999). Therefore, it is becoming increasingly important to identify any and all factors that might be associated with an increased risk of diabetes. Although mutations in HNF4α in MODY-1 account for only a very small fraction of all diabetics, there may be a connection between HNF4α and other types of diabetes. For example, incomplete penetrance of a mutation in HNF4α and a mild effect on the activity of the HNF4α protein both seem to be associated with a later age of onset of diabetes. For example, in family members with the V393I mutation, the average age of onset is 45 years (compared with 25 years or younger for typical MODY), and the V393I mutation results in only a 25–50% reduction in transactivation ability in transient transfection studies. This would amount to a 12.5–25% reduction in a heterozygote individual compared with up to a 50% reduction seen for other mutations such as Q268X. Therefore, it has been suggested that other factors that accumulate with time are required for diabetes to occur in patients with the V393I mutation (Hani *et al.*, 1998). Furthermore, there are at least two cases of individuals carrying MODY-1 mutations who do not have diabetes: Q268X (Yamagata *et al.*, 1996) and E276Q (Bulman *et al.*, 1997); both of these individuals are remarked upon as being lean and physically active.

These cases suggest that, at least in certain patients with MODY-1, there are factors that work in conjunction with mutations in HNF4α that trigger the disease. Interestingly, some of those factors (e.g. obesity and lack of physical activity) are the same as those associated with late-onset polygenic NIDDM. This raises the very important issue of whether as yet unidentified polymorphisms in HNF4α and/or ligand-like modulators of HNF4α might also be involved in at least some cases of late-onset NIDDM. There is already an example of a polymorphism in HNF1α being associated with reduced insulin response to an oral glucose challenge in healthy young subjects, suggesting that polymorphisms in that gene may be linked to late onset NIDDM in certain populations (Velho and Froguel, 1998). Furthermore, another nuclear receptor, PPARγ, which responds to circulating levels of fatty acids presumably derived from the diet, has also been linked to type II diabetes via its responsiveness to the antidiabetic compounds thiazolidinediones (reviewed in Kliewer *et al.*, 1999). Therefore, the precedence exists for both polymorphisms and nuclear receptor ligands to be involved in late-onset diabetes. It must now be determined whether those principles can also be applied to HNF4α.

DIAGNOSIS

Currently the only method to determine whether HNF4α is linked to a human disease is to sequence the HNF4α gene or the promoter of the gene in question and look for mutations. Single-stranded conformational polymorphism (SSCP) can and have been used to screen for potential mutations, but eventually sequencing must be done (Ghosh *et al.*, 1999). For diseases in which the HNF4α gene is altered, it is conceivable that an antibody-based assay could be used, although to date there is no such test available. Such a test would require a biopsy of one of the patient's organs, most likely the liver due to its regenerative powers. The most useful type of antibody would be one that detects the N-terminus of HNF4α1 since that would be able to detect truncated as well as full-length products; about half of the HNF4α mutations identified thus far in MODY-1 are predicted to produce truncated products. Detection of the missense mutations would require specific antibodies to those mutations. Ideally, an antiserum that detected all of the various HNF4α isoforms would also be used.

In terms of detecting mutations in HNF4α-binding sites, it must be known or suspected that a given gene is related to a given disease. For a gene whose promoter is well characterized, a simple literature search combined with programs found on the Web to identify potential transcription factor binding sites should indicate whether the site is a known or potential HNF4α-binding site. Table 9.1 will be useful for previously identified sites and the consensus site can serve as a guide for identifying new sites. However, it is important to note that there can be quite a bit of variability in HNF4α-binding sites, so that sometimes the only way to be certain that a site binds HNF4α is to test it directly. In that case, a gel shift assay using *in vitro* translated HNF4α with the appropriate competitor oligonucleotides must be done. If liver nuclear extracts are to be used, antisera specific to HNF4α must be used because there are several nuclear receptors found in the liver that bind DNA with a similar specificity as HNF4α, namely, RXR, RAR, PPARα and the COUP-TFs. Finally, transient transfections can be done with a reporter construct containing the promoter in question linked to a reporter gene in the presence and absence of a cotransfected HNF4α expression vector. Needless to say, these latter methods are for the initial identification of HNF4α-binding sites and not for screening mutations in known sites.

The therapeutic implications of testing for HNF4α mutations are potentially very considerable, although they have not been utilized to date. For example, one could identify the members of an affected family who carry a given HNF4α mutation and who are therefore at risk of developing the disease. These individuals could then be monitored at regular intervals so that therapeutic measures could be taken early in the course of the disease, thereby reducing long-term complications, such as microvascular and macro-vascular complications in diabetes (Furuta *et al.*, 1997).

PHARMACOLOGY

A ligand for HNF4α?

Since, to date, no ligand has been definitively identified for HNF4α, there is no pharmacology *per se*. However, it has been proposed that fatty acylcoenzyme A thioesters are ligands for HNF4α (Hertz *et al.*, 1998). These compounds are very attractive potential ligands in that they play a central role in both fatty acid and glucose metabolism (Gurr and Harwood, 1991), two pathways regulated by HNF4α (see Table 9.1). However, the effect of these compounds on HNF4α-mediated transcription *in vivo* is very small (approximately two-fold) and the effects on HNF4α *in vitro* could be due to a nonspecific detergent-like action of the compounds. Indeed, other compounds such as fatty acid-free bovine serum albumin and the nonionic detergent Tween 20 have a much greater effect on HNF4α DNA binding *in vitro* than do fatty acylcoenzymeA thioesters (F.M. Sladek *et al.*, personal communication). Furthermore, fatty acylcoenzymeA thioesters are known to bind nonspecifically to many proteins, including the thyroid hormone receptor (Li *et al.*, 1993), and are much larger than any known ligand and/or ligand-binding pocket in a nuclear receptor (Bogan *et al.*, 1998). Finally, recent computer modeling of the HNF4α LBD based on the crystal structure of other receptor LBDs indicates that, whereas a cavity may indeed be present in the HNF4α LBD, it could not accommodate compounds as large as fatty acylcoenzymeA thioesters (A.A. Bogan *et al.*, personal communication).

As it is likely that a ligand will eventually be found for HNF4α, some thoughts on the effects that a ligand might have on HNF4α may be useful. HNF4α binds DNA *in vitro* and activates transcription *in vitro* and *in vivo* in the absence of exogenously added ligand. Therefore, in one sense, it could be said that HNF4α does not 'need' a ligand, although it could always be argued that an endogenous ligand is present in the cells or extracts. The extreme conservation of HNF4α in the LBD across species, and the well conserved AF-2 region at the end of the LBD, suggest that HNF4α might indeed respond to a ligand. Furthermore, perhaps the most convincing argument for a ligand for HNF4α is the fact that, if you remove the large F domain, HNF4α activates transcription at least 10-fold better (see discussion above). This is the same order of magnitude seen for other receptors upon addition of ligand. The enhanced activation of transcription upon truncation of the F domain indicates that the AF-2/LBD is capable of very strong transcriptional activation and that the F domain modulates that activation. Two questions then arise. (1) Why would the HNF4α LBD evolve to be capable of strongly activating transcription if that potential was not used at least under certain circumstances? (2) Could it be that the purpose of the proposed ligand is not to position the AF-2 in the appropriate place to bind coactivator, as it apparently does in other receptors, but rather to displace the F domain and expose the AF-2 region?

Whether or not there is an endogenous ligand for HNF4α, it can be expected that HNF4α will respond to some sort of lipophilic compound as do other nuclear receptors. The challenge for the future, therefore, will be to develop compounds that specifically affect one desired aspect of HNF4α function without adversely affecting its myriad of other critical functions. The recent progress being made with selective estrogen receptor modifiers (SERMs) and the ER suggest that this may indeed be possible (see Chapter 3).

CONCLUSION

Like the other nuclear receptors, HNF4α is a relatively large molecule with a variety of important properties – DNA binding, protein dimerization and phosphorylation, ligand binding and coregulator interaction – that all contribute to specificity. It is a system just waiting to be tweaked in any one of many different ways in order to yield the desired effect. Indeed, the complexity of physiologic pathways and the role of HNF4α in them suggest that this is already happening *in vivo*.

Acknowledgments

The authors thank Michael D. Ruse, Jr. for critical reading of the manuscript and Lorilei Nepomuceno for library assistance. This work was funded by a Public Health Service grant DK53892-01 from the National Institute of Diabetes and Digestive and Kidney Disease to F.M.S.

REFERENCES

Arbini, A.A., Pollak, E.S., Bayleran, J.K., High, K.A. and Bauer, K.A. (1997) Severe factor VII deficiency due to a mutation disrupting a hepatocyte nuclear factor 4 binding site in the factor VII promoter. *Blood* **89**: 176–182.

Argyrokastritis, A., Kamakari, S., Kapsetaki, M., Kritis, A., Talianidis, I. and Moschonas, N.K. (1997) Human hepatocyte nuclear factor-4 (hHNF-4) gene maps to 20q12-q13.1 between PLCG1 and D20S17. *Hum. Genet.* **99**: 233–236.

Bailly, A., Spath, G., Bender, V. and Weiss, M.C. (1998) Phenotypic effects of the forced expression of HNF4 and HNF1α are conditioned by properties of the recipient cell. *J. Cell Sci.* **111**: 2411–2421.

Barbacci, E., Reber, M., Ott, M.-O., Breillat, C., Huetz, F. and Cereghini, S. (1999) Variant hepatocyte nuclear factor 1 is required for visceral endoderm specification. *Development* **126**: 4795–4805.

Bauman, J.W., Goldsworthy, T.L., Dunn, C.S. and Fox, T.R. (1995) Inhibitory effects of 2,3,7,8-tetrachlorodibenzo-*p*-dioxin on rat hepatocyte proliferation induced by 2/3 partial hepatectomy. *Cell Prolif.* **28**: 437–451.

Blanchard, K.L., Acquaviva, A.M., Galson, D.L. and Bunn, H.F. (1992) Hypoxia induction of the human erythropoietin gene: cooperation between the promoter and enhancer, each of which contains steroid receptor response elements. *Mol. Cell. Biol.* **12**: 5373–5385.

Blomhoff, R., Green, M., Green, J., Berg, T. and Norum, K. (1991) Vitamin A metabolism: new perspectives on absorption, transport and storage. *Physiol. Rev.* **71**: 951–990.

Bogan, A.A., Cohen, F.E. and Scanlan, T.S. (1998) Natural ligands of nuclear receptors have conserved volumes. *Nat. Struct. Biol.* **5**: 679–681.

Bowman, B.H. (1993) *Hepatic Plasma Proteins: Mechanisms of Function and Regulation.* Academic Press, San Diego.

Breslow, J. (1988) Apolipoprotein genetic variation and human disease. *Physiol. Rev.* **68**: 85–132.

Bulla, G.A. and Fournier, R.E. (1994) Genetic analysis of a transcriptional activation pathway by using hepatoma cell variants. *Mol. Cell. Biol.* **14**: 7086–7094.

Bulman, M.P., Dronsfield, M.J., Frayling, T. *et al.* (1997) A missense mutation in the hepatocyte nuclear factor 4α gene in a UK pedigree with maturity-onset diabetes of the young. *Diabetologia* 40: 859–862.

Burke, J.P., Williams, K., Gaskill, S.P., Hazuda, H.P., Haffner, S.M. and Stern, M.P. (1999) Rapid rise in the incidence of type 2 diabetes from 1987 to 1996: results from the San Antonio Heart Study. *Arch. Intern. Med.* **159**: 1450–1456.

Cairns, W., Smith, C., McLaren, A.W. and Wolf, C.R. (1996) Characterization of the human cytochrome P4502D6 promoter. A potential role for antagonistic interactions between members of the nuclear receptor family. *J. Biol. Chem.* **271**: 25 269–25 276.

Carter, M.E., Gulick, T., Raisher, B.D. *et al.* (1993) Hepatocyte nuclear factor-4 activates medium chain acyl-CoA dehydrogenase gene transcription by interacting with a complex regulatory element. *J. Biol. Chem.* **268**: 13 805–13 810.

Cereghini, S. (1996) Liver-enriched transcription factors and hepatocyte differentiation. *FASEB J.* **10**: 267–282.

Cereghini, S., Ott, M.O., Power, S. and Maury, M. (1992) Expression patterns of vHNF1 and HNF1 homeoproteins in early postimplantation embryos suggest distinct and sequential developmental roles. *Development* 116: 783–797.

Chan, J., Nakabayashi, H. and Wong, N.C. (1993) HNF-4 increases activity of the rat Apo A1 gene. *Nucleic Acids Res.* **21**: 1205–1211.

Charron, F. and Nemer, M. (1999) GATA transcription factors and cardiac development. *Semin. Cell Dev. Biol.* **10**: 85–91.

Chartier, F.L., Bossu, J.-P., Laudet, V., Fruchart, J.-C. and Laine, B. (1994) Cloning and sequencing of cDNAs encoding the human hepatocyte nuclear factor 4 indicate the presence of two isoforms in human liver. *Gene* **147**: 269–272.

Chen, D., Lepar, G. and Kemper, B. (1994a) A transcriptional regulatory element common to a large family of hepatic cytochrome P450 genes is a functional binding site of the orphan receptor HNF-4. *J. Biol. Chem.* **269**: 5420–5427.

Chen, D., Park, Y. and Kemper, B. (1994b) Differential protein binding and transcriptional activities of HNF-4 elements in three closely related CYP2C genes. *DNA Cell Biol.* **13**: 771–779.

Chen, W.S., Manova, K., Weinstein, D.C. *et al.* (1994c) Disruption of the HNF-4 gene, expressed in visceral endoderm, leads to cell death in embryonic ectoderm and impaired gastrulation of mouse embryos. *Genes Dev.* **8**: 2466–2477.

Christian, J.L. and Nakayama, T. (1999) Can't get no SMADisfaction: SMAD proteins as positive and negative regulators of TGF-β family signals. *Bioessays* 21: 382–390.

Christofori, G. and Semb, H. (1999) The role of the cell-adhesion molecule E-cadherin as a tumour-suppressor gene. *Trends Biochem. Sci.* **24**: 73–76.

Coffinier, C., Barra, J., Babinet, C. and Yaniv, M. (1999a) Expression of the vHNF1/HNF1β homeoprotein gene during mouse organogenesis. *Mech. Dev.* **87**: 1–3.

Coffinier, C., Thepot, D., Babinet, C., Yaniv, M. and Barra, J. (1999b) Essential role for the homeoprotein vHNF1/HNF1β in visceral endoderm differentiation. *Development* 126: 4785–4794.

Cohen, J. (1999) The scientific challenge of hepatitis C. *Science* **285**: 26–30.

Cooper, A.D., Chen, J., Botelho, Y.M., Cao, Y., Taniguchi, T. and Levy, W.B. (1997)

Characterization of hepatic-specific regulatory elements in the promoter region of the human cholesterol 7α-hydroxylase gene. *J. Biol. Chem.* **272**: 3444–3452.

Costa, R.H., Grayson, D.R. and Darnell, J.E., Jr. (1989) Multiple hepatocyte-enriched nuclear factors function in the regulation of transthyretin and α₁-antitrypsin genes. *Mol. Cell. Biol.* **9**: 1415–1425.

Coucouvanis, E. and Martin, G.R. (1999) BMP signaling plays a role in visceral endoderm differentiation and cavitation in the early mouse embryo. *Development* **126**: 535–546.

Crestani, M., Sadeghpour, A., Stroup, D., Galli, G. and Chiang, J.Y. (1998) Transcriptional activation of the cholesterol 7α-hydroxylase gene (CYP7A) by nuclear hormone receptors. *J. Lipid Res.* **39**: 2192–2200.

Crossley, M., Ludwig, M., Stowell, K.M., Vos, P.D., Olek, K. and Brownlee, G.G. (1992) Recovery from hemophilia B Leyden: an androgen-responsive element in the factor IX promoter. *Science* **257**: 377–379.

Dell, H. and Hadzopoulou-Cladaras, M. (1999) CREB-binding protein is a transcriptional coactivator for hepatocyte nuclear factor-4 and enhances apolipoprotein gene expression. *J. Biol. Chem.* **274**: 9013–9021.

Derynck, R. (1998) SMAD proteins and mammalian anatomy. *Nature* **393**: 737–739.

Diaz Guerra, M., Bergot, M.-O., Martinez, A., Cuif, M.-H., Kahn, A. and Raymondjean, M. (1993) Functional characterization of the L-type pyruvate kinase gene glucose response element. *Mol. Cell. Biol.* **13**: 7725–7735.

Drewes, T., Senkel, S., Holewa, B. and Ryffel, G.U. (1996) Human hepatocyte nuclear factor 4 isoforms are encoded by distinct and differentially expressed genes. *Mol. Cell. Biol.* **16**: 925–931.

Duncan, S., Manova, K., Chen, W. *et al.* (1994) Expression of transcription factor HNF-4 in the extraembryonic endoderm, gut and nephrogenic tissue of the developing mouse embryo: HNF-4 is a marker for primary endoderm in the implanting blastocyst. *Proc. Natl. Acad. Sci. USA* **91**: 7598–7602.

Duncan, S.A., Nagy, A. and Chan, W. (1997) Murine gastrulation requires HNF-4 regulated gene expression in the visceral endoderm: tetraploid rescue of Hnf-4$^{(-/-)}$ embryos. *Development* **124**: 279–287.

Duncan, S.A., Navas, M.A., Dufort, D., Rossant, J. and Stoffel, M. (1998) Regulation of a transcription factor network required for differentiation and metabolism. *Science* **281**: 692–695.

Emens, L.A., Landers, D.W. and Moss, L.G. (1992) Hepatocyte nuclear factor 1α is expressed in a hamster insulinoma line and transactivates the rat insulin I gene. *Proc. Natl. Acad. Sci. USA* **89**: 7300–7304.

Fandrey, J. and Bunn, H.F. (1993) *In vivo* and *in vitro* regulation of erythropoietin mRNA: measurement by competitive polymerase chain reaction. *Blood* **81**: 617–623.

Farsetti, A., Moretti, F., Narducci, M. *et al.* (1998) Orphan receptor hepatocyte nuclear factor-4 antagonizes estrogen receptor α-mediated induction of human coagulation factor XII gene. *Endocrinology* **139**: 4581–4589.

Feitelson, M.A. and Duan, L.X. (1997) Hepatitis B virus X antigen in the pathogenesis of chronic infections and the development of hepatocellular carcinoma. *Am. J. Pathol.* **150**: 1141–1157.

Fernandez-Rachubinski, F.A., Weiner, J.H. and Blajchman, M.A. (1996) Regions flanking exon 1 regulate constitutive expression of the human antithrombin gene. *J. Biol. Chem.* **271**: 29 502–29 512.

Figueiredo, M.S. and Brownlee, G.G. (1995) *cis*-Acting elements and transcription factors involved in the promoter activity of the human factor VIII gene. *J. Biol. Chem.* **270**: 11 828–11 838.

Fraser, J.D., Martinez, V., Straney, R. and Briggs, M.R. (1998) DNA binding and transcription activation specificity of hepatocyte nuclear factor 4. *Nucleic Acids Res.* **26**: 2702–2707.

Freedman, L.P. (1999) Increasing the complexity of coactivation in nuclear receptor signaling. *Cell* **97**: 5–8.

Froguel, P. (1998) Nuclear factors and type 2 diabetes. *Schweiz. Med. Wochenschr.* **128**: 1936–1939.

Furuta, H., Iwasaki, N., Oda, N. *et al.* (1997) Organization and partial sequence of the hepatocyte nuclear factor-4α/MODY1 gene and identification of a missense mutation, R127W, in a Japanese family with MODY. *Diabetes* **46**: 1652–1657.

Galson, D.L., Tsuchiya, T., Tendler, D.S. *et al.* (1995) The orphan receptor hepatic nuclear factor 4 functions as a transcriptional activator for tissue-specific and hypoxia-specific erythropoietin gene expression and is antagonized by EAR3/COUP-TF1. *Mol. Cell. Biol.* **15**: 2135–2144.

Garcia, A.D., Ostapchuk, P. and Hearing, P. (1993) Functional interaction of nuclear factors EF-C, HNF-4 and RXRα with hepatitis B virus enhancer I. *J. Virol.* **67**: 3940–3950.

Gareau, R., Audran, M., Baynes, R.D. *et al.* (1996) Erythropoietin abuse in athletes. *Nature* **380**: 113.

Garnier, G., Circolo, A. and Colten, H.R. (1996) Constitutive expression of murine complement factor B gene is regulated by the interaction of its upstream promoter with hepatocyte nuclear factor 4. *J. Biol. Chem.* **271**: 30 205–30 211.

Ghosh, S., Watanabe, R.M., Hauser, E.R. *et al.* (1999) Type 2 diabetes: evidence for linkage on chromosome 20 in 716 Finnish affected sib pairs. *Proc. Natl. Acad. Sci. USA* **96**: 2198–2203.

Gomez-Lechon, M.J., Jover, R., Donato, T. *et al.* (1998) Long-term expression of differentiated functions in hepatocytes cultured in three-dimensional collagen matrix. *J. Cell. Physiol.* **177**: 553–562.

Gordon, J., Schmidt, G. and Roth, K. (1992) Studies of intestinal stem cells using normal, chimeric and transgenic mice. *FASEB J.* **6**: 3039–3050.

Gourdon, L., Luo, D.Q., Raymondjean, M., Vasseur-Cognet, M. and Kahn, A. (1999) Negative cyclin AMP response elements in the promoter of the L-type pyruvate kinase gene. *FEBS Lett.* **459**: 9–14.

Gragnoli, C., Lindner, T., Cockburn, B.N. *et al.* (1997) Maturity-onset diabetes of the young due to a mutation in the hepatocyte nuclear factor-4α binding site in the promoter of the hepatocyte nuclear factor-1α gene. *Diabetes* **46**: 1648–1651.

Green, V.J., Kokkotou, E. and Ladias, J.A.A. (1998) Critical structural elements and multitarget protein interactions of the transcriptional activator AF-1 of hepatocyte nuclear factor 4. *J. Biol. Chem.* **273**: 29 950–29 957.

Gregori, C., Porteu, A., Lopez, S., Kahn, A. and Pichard, A.L. (1998) Characterization of the aldolase B intronic enhancer. *J. Biol. Chem.* **273**: 25 237–25 243.

Griffo, G., Hamon-Benais, C., Angrand, P. *et al.* (1993) HNF4 and HNF1 as well as a panel of hepatic functions are extinguished and reexpressed in parallel in chromosomally reduced rat hepatoma–human fibroblast hybrids. *J. Cell Biol.* **121**: 887–898.

Gronemeyer, H. and Laudet, V. (1995) Transcription factors 3: nuclear receptors. *Protein Profile* **2**: 1173–1308.

Gu, Z., Reynolds, E.M., Song, J. *et al.* (1999) The type I serine/threonine kinase receptor

ActRIA (ALK2) is required for gastrulation of the mouse embryo. *Development* **126**: 2551–2561.

Guilford, P., Hopkins, J., Harraway, J. *et al.* (1998) E-cadherin germline mutations in familial gastric cancer. *Nature* **392**: 402–405.

Guillemin, K. and Krasnow, M.A. (1997) The hypoxic response: huffing and HIFing. *Cell* **89**: 9–12.

Guo, W., Chen, M., Yen, T.S.B. and Ou, J.-H. (1993) Hepatocyte-specific expression of the hepatitis B virus core promoter depends on both positive and negative regulation. *Mol. Cell. Biol.* **13**: 443–448.

Gurr, M.I. and Harwood, J.L. (1991) *Lipid Biochemistry: An Introduction*, 4th edn. Chapman and Hall, London.

Hadzopoulou-Cladaras, M., Kistanova, E., Evagelopoulou, C., Zeng, S., Cladaras, C. and Ladias, J.A. (1997) Functional domains of the nuclear receptor hepatocyte nuclear factor 4. *J. Biol. Chem.* **272**: 539–550.

Hall, R.K., Scott, D.K., Noisin, E.L., Lucas, P.C. and Granner, D.K. (1992) Activation of the phosphoenolpyruvate carboxykinase gene retinoic acid response element is dependent on a retinoic acid receptor/coregulator complex. *Mol. Cell. Biol.* **12**: 5527–5535.

Hall, R.K., Sladek, F.M. and Granner, D.K. (1995) The orphan receptors COUP-TF and HNF-4 serve as accessory factors required for induction of phosphoenolpyruvate carboxykinase gene transcription by glucocorticoids. *Proc. Natl. Acad. Sci. USA* **92**: 412–416.

Hani, E.H., Suaud, L., Boutin, P. *et al.* (1998) A missense mutation in hepatocyte nuclear factor-4α, resulting in a reduced transactivation activity, in human late-onset non-insulin-dependent diabetes mellitus. *J. Clin. Invest.* **101**: 521–526.

Hardon, E.M., Frain, M., Paonessa, G. and Cortese, R. (1988) Two distinct factors interact with the promoter regions of several liver-specific genes. *EMBO J.* **7**: 1711–1719.

Hargrove, G.M., Junco, A. and Wong, N.C. (1999) Hormonal regulation of apolipoprotein AI. *J. Mol. Endocrinol.* **22**: 103–111.

Harnish, D.C., Malik, S., Kilbourne, E., Costa, R. and Karathanasis, S.K. (1996) Control of apolipoprotein AI gene expression through synergistic interactions between hepatocyte nuclear factors 3 and 4. *J. Biol. Chem.* **271**: 13 621–13 628.

Harnish, D.C., Evans, M.J., Scicchitano, M.S., Bhat, R.A. and Karathanasis, S.K. (1998) Estrogen regulation of the apolipoprotein AI gene promoter through transcription cofactor sharing. *J. Biol. Chem.* **273**: 9270–9278.

Hata, S., Tsukamoto, T. and Osumi, T. (1992) A novel isoform of rat hepatocyte nuclear factor 4 (HNF-4). *Biochim. Biophys. Acta* **1131**: 211–213.

Hata, S., Inoue, T., Kosuga, K., Nakashima, T., Tsukamoto, T. and Osumi, T. (1995) Identification of two splice isoforms of mRNA for mouse hepatocyte nuclear factor 4 (HNF-4). *Biochim. Biophys. Acta* **1260**: 55–61.

Hata, A., Shi, Y. and Massagué, J. (1998) TGF-β signaling and cancer: structural and functional consequences of mutations in SMADs. *Mol. Med. Today* **4**: 257–262.

Heath, J.K. and Deller, M.J. (1983) Serum-free culture of PC13 murine embryonal carcinoma cells. *J. Cell. Physiol.* **115**: 225–230.

Heit, J.A., Ketterling, R.P., Zapata, R.E., Ordonez, S.M., Kasper, C.K. and Sommer, S.S. (1999) Haemophilia B Brandenberg-type promoter mutation. *Haemophilia* **5**: 73–75.

Herbst, R.S., Nielsch, U., Sladek, F., Lai, E., Babiss, L.E. and Darnell, J.E. (1991) Decreased transcription factor concentration explains the limited hepatocyte-like transcription in hepatoma cells. *New Biol.* **3**: 289–296.

Herman, W.H., Fajans, S.S., Smith, M.J., Polonsky, K.S., Bell, G.I. and Halter, J.B. (1997)

Diminished insulin and glucagon secretory responses to arginine in nondiabetic subjects with a mutation in the hepatocyte nuclear factor-4α/MODY1 gene. *Diabetes* **46**: 1749–1754.

Hertz, R., Bishara, S.J. and Bar, T.J. (1995) Mode of action of peroxisome proliferators as hypolipidemic drugs. Suppression of apolipoprotein C-III. *J. Biol. Chem.* **270**: 13 470–13 475.

Hertz, R., Magenheim, J., Berman, I. and Bar, T.J. (1998) Fatty acyl-CoA thioesters are ligands of hepatic nuclear factor-4α. *Nature* **392**: 512–516.

Holewa, B., Strandmann, E.P., Zapp, D., Lorenz, P. and Ryffel, G.U. (1996) Transcriptional hierarchy in *Xenopus* embryogenesis: HNF4 a maternal factor involved in the developmental activation of the gene encoding the tissue specific transcription factor HNF1α (LFB1). *Mech. Dev.* **54**: 45–57.

Holewa, B., Zapp, D., Drewes, T., Senkel, S. and Ryffel, G.U. (1997) HNF4β, a new gene of the HNF4 family with distinct activation and expression profiles in oogenesis and embryogenesis of *Xenopus laevis*. *Mol. Cell. Biol.* **17**: 687–694.

Hong, H., Darimont, B.D., Ma, H., Yang, L., Yamamoto, K.R. and Stallcup, M.R. (1999) An additional region of coactivator GRIP1 required for interaction with the hormone-binding domains of a subset of nuclear receptors. *J. Biol. Chem.* **274**: 3496–3502.

Horikawa, Y., Iwasaki, N., Hara, M. *et al.* (1997) Mutation in hepatocyte nuclear factor-1β gene (TCF2) associated with MODY. *Nat. Genet.* **17**: 384–385.

Hu, C. and Perlmutter, D. (1999) Regulation of α$_1$-antitrypsin gene expression in human intestinal epithelial cell line caco-2 by HNF-1α and HNF-4. *Am. J. Physiol.* **276**: G1181–1194.

Hung, H.-L. and High, K.A. (1993) Hepatocyte nuclear factor 4 binds to the promoters of factors VII, IX and X. *Thromb. Haemost.* **69**: 1237.

Hwung, Y., Crowe, D., Wang, L., Tsai, S. and Tsai, M. (1988) The COUP transcription factor binds to an upstream promoter element of the rat insulin II gene. *Mol. Cell. Biol.* **8**: 2070–2077.

Ibeanu, G.C. and Goldstein, J.A. (1995) Transcriptional regulation of human CYP2C genes: functional comparison of CYP2C9 and CYP2C18 promoter regions. *Biochemistry* **34**: 8028–8036.

Iwasaki, N., Ogata, M., Tomonaga, O. *et al.* (1998) Liver and kidney function in Japanese patients with maturity-onset diabetes of the young. *Diabetes Care* **21**: 2144–2148.

Iyemere, V.P., Davies, N.H. and Brownlee, G.G. (1998) The activation function 2 domain of hepatic nuclear factor 4 is regulated by a short C-terminal proline-rich repressor domain. *Nucleic Acids Res.* **26**: 2098–2104.

Jacob, A., Budhiraja, S. and Reichel, R.R. (1999) The HNF-3α transcription factor is a primary target for retinoic acid action. *Exp. Cell Res.* **250**: 1–9.

Janknecht, R. and Hunter, T. (1999) Nuclear fusion of signaling pathways. *Science* **284**: 443–444.

Janknecht, R., Wells, N.J. and Hunter, T. (1998) TGF-β-stimulated cooperation of SMAD proteins with the coactivators CBP/p300. *Genes Dev.* **12**: 2114–2119.

Janne, M. and Hammond, G.L. (1998) Hepatocyte nuclear factor-4 controls transcription from a TATA-less human sex hormone-binding globulin gene promoter. *J. Biol. Chem.* **273**: 34105–34114.

Jiang, G. and Sladek, F.M. (1997) The DNA binding domain of hepatocyte nuclear factor 4 mediates cooperative, specific binding to DNA and heterodimerization with the retinoid X receptor α. *J. Biol. Chem.* **272**: 1218–1225.

Jiang, G., Nepomuceno, L., Hopkins, K. and Sladek, F.M. (1995) Exclusive homodimerization of orphan receptor hepatocyte nuclear factor 4 defines a new subclass of nuclear receptors. *Mol. Cell. Biol.* **15**: 5131–5143.

Jiang, G., Lee, U. and Sladek, F.M. (1997a) Proposed mechanism for the stabilization of nuclear receptor DNA binding via protein dimerization. *Mol. Cell. Biol.* **17**: 6546–6554.

Jiang, G., Nepomuceno, L., Yang, Q. and Sladek, F.M. (1997b) Serine/threonine phosphorylation of orphan receptor hepatocyte nuclear factor 4. *Arch. Biochem. Biophys.* **340**: 1–9.

Jordan, C.T. and Van Zant, G. (1998) Recent progress in identifying genes regulating hematopoietic stem cell function and fate. *Curr. Opin. Cell Biol.* **10**: 716–720.

Kalkuhl, A., Kaestner, K., Buchmann, A. and Schwarz, M. (1996) Expression of hepatocyte-enriched nuclear transcription factors in mouse liver tumours. *Carcinogenesis* **17**: 609–612.

Kapitskaya, M.Z., Dittmer, N.T., Deitsch, K.W. *et al.* (1998) Three isoforms of a hepatocyte nuclear factor-4 transcription factor with tissue- and stage-specific expression in the adult mosquito. *J. Biol. Chem.* **273**: 29801–29810.

Kardassis, D., Tzameli, I., Hadzopoulou, C. M., Talianidis, I. and Zannis, V. (1997) Distal apolipoprotein C-III regulatory elements F to J act as a general modular enhancer for proximal promoters that contain hormone response elements. Synergism between hepatic nuclear factor-4 molecules bound to the proximal promoter and distal enhancer sites. *Arterioscler. Thromb. Vasc. Biol.* **17**: 222–232.

Kardassis, D., Sacharidou, E. and Zannis, V.I. (1998) Transactivation of the human apolipoprotein CII promoter by orphan and ligand-dependent nuclear receptors. The regulatory element CIIC is a thyroid hormone response element. *J. Biol. Chem.* **273**: 17810–17816.

Kastner, P., Mark, M. and Chambon, P. (1995) Nonsteroid nuclear receptors: what are genetic studies telling us about their role in real life? *Cell* **83**: 856–869.

Kim, J.W., Kim, Y.K. and Ahn, Y.H. (1998) A mechanism of differential expression of GLUT2 in hepatocyte and pancreatic β-cell line. *Exp. Mol. Med.* **30**: 15–20.

Kimura, A., Nishiyori, A., Murakami, T. *et al.* (1993) COUP-TF represses transcription from the promoter of the gene for ornithine transcarbamylase in a manner antagonistic to HNF-4. *J. Biol. Chem.* **268**: 11125–11133.

Klaassen, C.D., Amdur, M.O. and Doull, J. (1996) *Casarett and Doull's Toxicology: The Basic Science of Poisons*, 5th edn. McGraw Hill, New York.

Kliewer, S.A., Lehmann, J.M. and Willson, T.M. (1999) Orphan nuclear receptors: shifting endocrinology into reverse. *Science* **284**: 757–760.

Kraichely, D.M., Collins, J.J., III, DeLisle, R.K. and MacDonald, P.N. (1999) The autonomous transactivation domain in helix H3 of the vitamin D receptor is required for transactivation and coactivator interaction. *J. Biol. Chem.* **274**: 14352–14358.

Kritis, A.A., Argyrokastritis, A., Moschonas, N.K. *et al.* (1996) Isolation and characterization of a third isoform of human hepatocyte nuclear factor 4. *Gene* **173**: 275–280.

Ktistaki, E. and Talianidis, I. (1997a) Chicken ovalbumin upstream promoter transcription factors act as auxiliary cofactors for hepatocyte nuclear factor 4 and enhance hepatic gene expression. *Mol. Cell. Biol.* **17**: 2790–2797.

Ktistaki, E. and Talianidis, I. (1997b) Modulation of hepatic gene expression by hepatocyte nuclear factor 1. *Science* **277**: 109–112.

Ktistaki, E., Lacorte, J., Katrakili, N., Zannis, V. and Talianidis, I. (1994) Transcriptional regulation of the apolipoprotein A-IV gene involves synergism between a proximal orphan receptor response element and a distant enhancer located in the upstream promoter region of the apolipoprotein C-III gene. *Nucleic Acids Res.* **22**: 4689–4696.

Ktistaki, E., Ktistakis, N.T., Papadogeorgaki, E. and Talianidis, I. (1995) Recruitment of hepatocyte nuclear factor 4 into specific intranuclear compartments depends on tyrosine phosphorylation that affects its DNA-binding and transactivation potential. *Proc. Natl. Acad. Sci. USA* **92**: 9876–9880.

Kuiper, G.G. and Brinkmann, A.O. (1994) Steroid hormone receptor phosphorylation: is there a physiological role? *Mol. Cell. Endocrinol.* **100**: 103–107.

Kuo, C.J., Conley, P.B., Chen, L., Sladek, F.M., Darnell, J.E., Jr. and Crabtree, G.R. (1992) A transcriptional hierarchy involved in mammalian cell-type specification. *Nature (Lond.)* **355**: 457–461.

Kuo, C.T., Morrisey, E.E., Anandappa, R. *et al.* (1997) GATA4 transcription factor is required for ventral morphogenesis and heart tube formation. *Genes Dev.* **11**: 1048–1060.

Labbé, E., Silvestri, C., Hoodless, P.A., Wrana, J.L. and Attisano, L. (1998) SMAD2 and SMAD3 positively and negatively regulate TGFβ-dependent transcription through the forkhead DNA-binding protein FAST2. *Mol. Cell* **2**: 109–120.

Ladias, J. (1994) Convergence of multiple nuclear receptor signaling pathways onto the long terminal repeat of human immunodeficiency virus-1. *J. Biol. Chem.* **269**: 5944–5951.

Ladias, J.A.A., Hadzopoulou-Cladaras, M., Kardassis, D. *et al.* (1992) Transcriptional regulation of human apolipoprotein genes apoB, apoCIII and apoAII by members of the steroid hormone receptor superfamily HNF-4, ARP-1, EAR-2 and EAR-3. *J. Biol. Chem.* **267**: 15 849–15 860.

Lai, E., Prezioso, V.R., Tao, W.F., Chen, W.S. and Darnell, J.E., Jr. (1991) Hepatocyte nuclear factor 3α belongs to a gene family in mammals that is homologous to the *Drosophila* homeotic gene forkhead. *Genes Dev.* **5**: 416–427.

Lee, Y.K., Dell, H., Dowhan, D.H. *et al.* (2000) The orphan nuclear receptor SHP inhibits hepatocyte nuclear factor 4 and retinoid X receptor transactivation: two mechanisms for repression. *Molecular and Cellular Biology* **20**: 187–195.

Lee, S.K., Na, S.Y., Kim, H.J., Soh, J., Choi, H.S. and Lee, J.W. (1998) Identification of critical residues for heterodimerization within the ligand-binding domain of retinoid X receptor. *Mol. Endocrinol.* **12**: 325–332.

Lehto, M., Bitzen, P., Isomaa, B. *et al.* (1999a) Mutation in HNF-4α gene affects insulin secretion and triglyceride metabolism. *Diabetes* **48**: 423–425.

Lehto, M., Wipemo, C., Ivarsson, S. *et al.* (1999b) High frequency of mutations in MODY and mitochondrial genes in Scandinavian patients with familial early-onset diabetes. *Diabetologia* **42**: 1131–1137.

Levine, A.J. (1992) *Viruses.* Scientific American Library, New York.

Li, Q., Yamamoto, N., Morisawa, S. and Inoue, A. (1993) Fatty acyl-CoA binding activity of the nuclear thyroid hormone receptor. *J. Cell. Biochem.* **51**: 458–464.

Liimatta, M., Towle, H.C., Clarke, S. and Jump, D.B. (1994) Dietary polyunsaturated fatty acids interfere with the insulin/glucose activation of L-type pyruvate kinase gene transcription. *Mol. Endocrinol.* **8**: 1147–1153.

Lindner, T., Gragnoli, C., Furuta, H. *et al.* (1997) Hepatic function in a family with a nonsense mutation (R154X) in the hepatocyte nuclear factor-4α/MODY1 gene. *J. Clin. Invest.* **100**: 1400–1405.

Magee, T.R., Cai, Y., El-Houseini, M.E., Locker, J. and Wan, Y.J. (1998) Retinoic acid mediates down-regulation of the α-fetoprotein gene through decreased expression of hepatocyte nuclear factors. *J. Biol. Chem.* **273**: 30 024–30 032.

Mahoney, F.J. (1999) Update on diagnosis, management and prevention of hepatitis B virus infection. *Clin. Microbiol. Rev.* **12**: 351–366.

Malecki, M.T., Yang, Y., Antonellis, A., Curtis, S., Warram, J.H. and Krolewski, A.S. (1999) Identification of new mutations in the hepatocyte nuclear factor 4α gene among families with early onset type 2 diabetes mellitus. *Diabetes Med.* **16**: 193–200.

Marten, N.W., Sladek, F.M. and Straus, D.S. (1996) Effect of dietary protein restriction on liver transcription factors. *Biochem. J.* **317**: 361–370.

Massague, J. (1998) TGF-*β* signal transduction. *Ann. Rev. Biochem.* **67**: 753–791.

Massague, J., Hata, A. and Liu, F. (1997) TGF-*β* signalling through the SMAD pathway. *Trends Cell Biol.* **7**: 187–192.

Matschinsky, F.M. (1990) Glucokinase as glucose sensor and metabolic signal generator in pancreatic *β*-cells and hepatocytes. *Diabetes* **39**: 647–652.

McKenna, N.J., Lanz, R.B. and O'Malley, B.W. (1999) Nuclear receptor coregulators: cellular and molecular biology. *Endocr. Rev.* **20**: 321–344.

Metzger, S., Halaas, J.L., Breslow, J.L. and Sladek, F.M. (1993) Orphan receptor HNF-4 and bZip protein C/EBPα bind to overlapping regions of the apolipoprotein B gene promoter and synergistically activate transcription. *J. Biol. Chem.* **268**: 16831–16838.

Miao, C., Leytus, S., Chung, D. and Davie, E. (1992) Liver-specific expression of the gene coding for human factor X, a blood coagulation factor. *J. Biol. Chem.* **267**: 7395–7401.

Mietus-Snyder, M., Sladek, F.M., Ginsburg, G.S. *et al.* (1992) Antagonism between apolipoprotein regulatory protein 1, Ear3/Coup-TF, and hepatocyte nuclear factor 4 modulates apolipoprotein CIII gene expression in liver and intestinal cells. *Mol. Cell. Biol.* **12**: 1708–1718.

Miquerol, L., Lopez, S., Cartier, N., Tulliez, M., Raymondjean, M. and Kahn, A. (1994) Expression of the L-type pyruvate kinase gene and of the hepatocyte nuclear factor 4 transcription factor in exocrine and endocrine pancreas. *J. Biol. Chem.* **269**: 8944–8951.

Miura, N. and Tanaka, K. (1993) Analysis of the rat hepatocyte nuclear factor (HNF) 1 gene promoter: synergistic activation by HNF4 and HNF1 proteins. *Nucleic Acids Res.* **21**: 3731–3736.

Mizuguchi, T., Mitaka, T., Hirata, K., Oda, H. and Mochizuki, Y. (1998) Alteration of expression of liver-enriched transcription factors in the transition between growth and differentiation of primary cultured rat hepatocytes. *J. Cell. Physiol.* **174**: 273–284.

Moldrup, A., Ormandy, C., Nagano, M. *et al.* (1996) Differential promoter usage in prolactin receptor gene expression: hepatocyte nuclear factor 4 binds to and activates the promoter preferentially active in the liver. *Mol. Endocrinol.* **10**: 661–671.

Molkentin, J.D., Lin, Q., Duncan, S.A. and Olson, E.N. (1997) Requirement of the transcription factor GATA4 for heart tube formation and ventral morphogenesis. *Genes Dev.* **11**: 1061–1072.

Moller, A.M., Urhammer, S.A., Dalgaard, L.T. *et al.* (1997) Studies of the genetic variability of the coding region of the hepatocyte nuclear factor-4α in Caucasians with maturity onset NIDDM. *Diabetologia* **40**: 980–983.

Moller, A.M., Dalgaard, L.T., Ambye, L. *et al.* (1999) A novel Phe75fsdelT mutation in the hepatocyte nuclear factor-4α gene in a Danish pedigree with maturity-onset diabetes of the young. *J. Clin. Endocrinol. Metab.* **84**: 367–369.

Morgan, G.E., Rowley, G., Green, P.M., Chisholm, M., Giannelli, F. and Brownlee, G.G. (1997) Further evidence for the importance of an androgen response element in the factor IX promoter. *Br. J. Haematol.* **98**: 79–85.

Moriizumi, S., Gourdon, L., Lefrancois-Martinez, A.M., Kahn, A. and Raymondjean, M. (1998) Effect of different basic helix-loop-helix leucine zipper factors on the glucose response unit of the L-type pyruvate kinase gene. *Gene Expr.* **7**: 103–113.

Morrisey, E.E., Tang, Z., Sigrist, K. *et al.* (1998) GATA6 regulates HNF4 and is required for differentiation of visceral endoderm in the mouse embryo. *Genes Dev.* **12**: 3579–3590.

Nagy, P., Bisgaard, H. and Thorgeirsson, S. (1994) Expression of hepatic transcription factors during liver development and oval cell differentiation. *J. Cell Biol.* **126**: 223–233.

Naka, H. and Brownlee, G.G. (1996) Transcriptional regulation of the human factor IX promoter by the orphan receptor superfamily factors, HNF4, ARP1 and COUP/Ear3. *Br. J. Haematol.* **92**: 231–240.

Nakhei, H., Lingott, A., Lemm, I. and Ryffel, G.U. (1998) An alternative splice variant of the tissue specific transcription factor HNF4α predominates in undifferentiated murine cell types. *Nucleic Acids Res.* **26**: 497–504.

Nakshatri, H. and Chambon, P. (1994) The directly repeated RG(G/T)TCA motifs of the rat and mouse cellular retinol-binding protein II genes are promiscuous binding sites for RAR, RXR, HNF-4 and ARP-1 homo- and heterodimers. *J. Biol. Chem.* **269**: 890–902.

Navas, M.A., Munoz-Elias, E.J., Kim, J., Shih, D. and Stoffel, M. (1999) Functional characterization of the MODY1 gene mutations HNF4(R127W), HNF4(V255M) and HNF4(E276Q). *Diabetes* **48**: 1459–1465.

Nishiyori, A., Tashiro, H., Kimura, A., Akagi, K., Yamamura, K.M.M. and Takiguchi, M. (1994) Determination of tissue specificity of the enhancer by combinatorial operation of tissue-enriched transcription factors: both HNF-4 and C/EBPβ are required for liver-specific activity of the ornithine transcarbamylase enhancer. *J. Biol. Chem.* **269**: 1323–1331.

Nitsch, D., Boshart, M. and Schutz, G. (1993) Activation of the tyrosine aminotransferase gene is dependent on synergy between liver-specific and hormone-responsive elements. *Proc. Natl. Acad. Sci. USA* **90**: 5479–5483.

Nuclear Receptor Nomenclature Committee (1999) A unified nomenclature system for the nuclear receptor superfamily. *Cell* **97**: 161–163.

Ochoa, A., Bovard-Houppermans, S. and Zakin, M.M. (1993) Human apolipoprotein A-IV gene expression is modulated by members of the nuclear hormone receptor superfamily. *Biochim. Biophys. Acta* **1210**: 41–47.

Oda, H., Nozawa, K., Hitomi, Y. and Kakinuma, A. (1995) Laminin-rich extracellular matrix maintains high level of hepatocyte nuclear factor 4 in rat hepatocyte culture. *Biochem. Biophys. Res. Commun.* **212**: 800–805.

Ogino, M., Nagata, K., Miyata, M. and Yamazoe, Y. (1999) Hepatocyte nuclear factor 4-mediated activation of rat *CYP3A1* gene and its modes of modulation by apolipoprotein AI regulatory protein I and v-ErbA-related protein 3. *Arch. Biochem. Biophys.* **362**: 32–37.

Orti, E., Bodwell, J. and Munck, A. (1992) Phosphorylation of steroid hormone receptors. *Endocr. Rev.* **13**: 105–128.

Plengvidhya, N., Antonellis, A., Wogan, L.T. *et al.* (1999) Hepatocyte nuclear factor-4γ: cDNA sequence, gene organization, and mutation screening in early-onset autosomal-dominant type 2 diabetes. *Diabetes* **48**: 2099–2102.

Pogge v. Strandmann, E., Nastos, A., Holewa, B., Senkel, S., Weber, H. and Ryffel, G.U. (1997) Patterning expression of a tissue-specific transcription factor in embryogenesis: HNF1α gene activation during *Xenopus* development. *Mech. Dev.* **64**: 7–17.

Pontoglio, M., Barra, J., Hadchouel, M. *et al.* (1996) Hepatocyte nuclear factor 1 inactivation results in hepatic dysfunction, phenylketonuria and renal Fanconi syndrome. *Cell* **84**: 575–585.

Power, S.C. and Cereghini, S. (1996) Positive regulation of the vHNF1 promoter by the orphan receptors COUP-TF1/Ear3 and COUP-TFII/Arp1. *Mol. Cell. Biol.* **16**: 778–791.

Raney, A.K., Johnson, J.L., Palmer, C.N. and McLachlan, A. (1997) Members of the nuclear receptor superfamily regulate transcription from the hepatitis B virus nucleocapsid promoter. *J. Virol.* **71**: 1058–1071.

Reijnen, M.J., Sladek, F.M., Bertina, R.M. and Reitsma, P.H. (1992) Disruption of a binding site for hepatocyte nuclear factor 4 results in hemophilia B. Leyden. *Proc. Natl. Acad. Sci. USA* **89**: 6300–6303.

Reijnen, M.J., Peerlinck, K., Maasdam, D., Bertina, R.M. and Reitsma, P.H. (1993) Hemophilia B Leyden: substitution of thymine for guanine at position -21 results in a disruption of a hepatocyte nuclear factor 4 binding site in the factor IX promoter. *Blood* **82**: 151–158.

Ribeiro, A., Pastier, D., Kardassis, D., Chambaz, J. and Cardot, P. (1999) Cooperative binding of upstream stimulatory factor and hepatic nuclear factor 4 drives the transcription of the human apolipoprotein A-II gene. *J. Biol. Chem.* **274**: 1216–1225.

Roach, S., Schmid, W. and Pera, M. (1994) Hepatocytic transcription factor expression in human embryonal carcinoma and yolk sac carcinoma cell lines: expression of HNF-3α in models of early endodermal cell differentiation. *Exp. Cell Res.* **1994**: 189–198.

Rodriguez, J.C., Ortiz, J.A., Hegardt, F.G. and Haro, D. (1998) The hepatocyte nuclear factor 4 (HNF-4) represses the mitochondrial HMG-CoA synthase gene. *Biochem. Biophys. Res. Commun.* **242**: 692–696.

Rollini, P. and Fournier, R.E.K. (1999) The HNF-4/HNF-1α transactivation cascade regulates gene activity and chromatin structure of the human serine protease inhibitor gene cluster at 14q32.1. *Proc. Natl. Acad. Sci. USA* **96**: 10 308–10 313.

Rosenbloom, A.L., Joe, J.R., Young, R.S. and Winter, W.E. (1999) Emerging epidemic of type 2 diabetes in youth. *Diabetes Care* **22**: 345–354.

Rottman, J.N. and Gordon, J.I. (1993) Comparison of the patterns of expression of rat intestinal fatty acid binding protein/human growth hormone fusion genes in cultured intestinal epithelial cell lines and in the gut epithelium of transgenic mice. *J. Biol. Chem.* **268**: 11 994–12 002.

Rouet, P., Raguenez, G., Tronche, F., Mfou'ou, V. and Salier, J.P. (1995) Hierarchy and positive/negative interplays of the hepatocyte nuclear factors HNF-1, -3 and -4 in the liver-specific enhancer for the human α$_1$-microglobulin/bikunin precursor. *Nucleic Acids Res.* **23**: 395–404.

Rouet, P., Raguenez, G., Ruminy, P. and Salier, J.P. (1998) An array of binding sites for hepatocyte nuclear factor 4 of high and low affinities modulates the liver-specific enhancer for the human α$_1$-microglobulin/bikunin precursor. *Biochem. J.* **334**: 577–584.

Runge, D., Runge, D.M., Drenning, S.D., Bowen, W.C., Jr., Grandis, J.R. and Michalopoulos, G.K. (1998) Growth and differentiation of rat hepatocytes: changes in transcription factors HNF-3, HNF-4, STAT-3, and STAT-5. *Biochem. Biophys. Res. Commun.* **250**: 762–768.

Runge, D., Runge, D.M., Daskalakis, N., Lubecki, K.A., Bowen, W.C. and Michalopoulos, G.K. (1999) Matrix-mediated changes in the expression of HNF-4α isoforms and in DNA-binding activity of ARP-1 in primary cultures of rat hepatocytes. *Biochem. Biophys. Res. Commun.* **259**: 651–655.

Sanchez, A., Pagan, R., Alvarez, A.M. *et al.* (1998) Transforming growth factor-β (TGF-β) and EGF promote cord-like structures that indicate terminal differentiation of fetal hepatocytes in primary culture. *Exp. Cell Res.* **242**: 27–37.

Sanchez, A., Alvarez, A.M., Lopez Pedrosa, J.M., Roncero, C., Benito, M. and Fabregat, I. (1999) Apoptotic response to TGF-β in fetal hepatocytes depends upon their state of differentiation. *Exp. Cell Res.* **252**: 281–291.

Schaeffer, E., Guillou, F., Part, D. and Zakin, M.M. (1993) A different combination of transcription factors modulates the expression of the human transferrin promoter in liver and Sertoli cells. *J. Biol. Chem.* **268**: 23 399–23 408.

Sel, S., Ebert, T., Ryffel, G.U. and Drewes, T. (1996) Human renal cell carcinogenesis is accompanied by a coordinate loss of the tissue specific transcription factors HNF4α and HNF1α. *Cancer Lett.* **101**: 205–210.

Seol, W., Choi, H.-S. and Moore, D. (1996) An orphan nuclear hormone receptor that lacks a DNA binding domain and heterodimerizes with other receptors. *Science* **272**: 1336–1339.

Sirard, C., de la Pompa, J., Elia, A. *et al.* (1998) The tumor suppressor gene SMAD4/DPC4 is required for gastrulation and later for anterior development of the mouse embryo. *Genes Dev.* **12**: 107–119.

Sladek, F.M. (1993) Orphan receptor HNF-4 and liver-specific gene expression. *Receptor* **3**: 223–232.

Sladek, F.M. (1994) Hepatocyte nuclear factor 4 (HNF-4). In: F. Tronche and M. Yaniv (eds) *Liver Gene Expression*, pp. 207–230. R.G. Landes, Austin, Texas.

Sladek, F.M., Zhong, W., Lai, E. and Darnell, J.E. (1990) Liver-enriched transcription factor HNF-4 is a novel member of the steroid hormone receptor superfamily. *Genes Dev.* **4**: 2353–2365.

Sladek, F.M., Dallas-Yang, Q. and Nepomuceno, L. (1998) MODY1 mutation Q268X in hepatocyte nuclear factor 4α allows for dimerization in solution but causes abnormal subcellular localization. *Diabetes* **47**: 985–990.

Sladek, F.M., Ruse, M.D., Nepomuceno, L., Huang, S.-M. and Stallcup, M.R. (1999) Modulation of transcriptional activation and co-activator interaction by a splicing variation in the F domain of nuclear receptor hepatocyte nuclear factor 4α1. *Mol. Cell. Biol.* **19**: 6509–6522.

Smyth, J.F., Boogaerts, M.A. and Ehmer, B.R.-M. (1996) *rh Erythropoietin in Cancer Supportive Treatment.* Marcel Dekker, New York.

Soudais, C., Bielinska, M., Heikinheimo, M. *et al.* (1995) Targeted mutagenesis of the transcription factor GATA-4 gene in mouse embryonic stem cells disrupts visceral endoderm differentiation *in vitro. Development* **121**: 3877–3888.

Spath, G.F. and Weiss, M.C. (1997) Hepatocyte nuclear factor 4 expression overcomes repression of the hepatic phenotype in dedifferentiated hepatoma cells. *Mol. Cell. Biol.* **17**: 1913–1922.

Spath, G.F. and Weiss, M.C. (1998) Hepatocyte nuclear factor 4 provokes expression of epithelial marker genes, acting as a morphogen in dedifferentiated hepatoma cells. *J. Cell. Biol.* **140**: 935–946.

Stauffer, D.R., Chukwumezie, B.N., Wilberding, J.A., Rosen, E.D. and Castellino, F.J. (1998) Characterization of transcriptional regulatory elements in the promoter region of the murine blood coagulation factor VII gene. *J. Biol. Chem.* **273**: 2277–2287.

Stewart, M.J., Dipple, K.M., Estonius, M., Nakshatri, H., Everett, L.M. and Crabb, D.W. (1998) Binding and activation of the human aldehyde dehydrogenase 2 promoter by hepatocyte nuclear factor 4. *Biochim. Biophys. Acta* **1399**: 181–186.

Stoffel, M. and Duncan, S.A. (1997) The maturity-onset diabetes of the young (MODY1) transcription factor HNF4α regulates expression of genes required for glucose transport and metabolism. *Proc. Natl. Acad. Sci. USA* **94**: 13 209–13 214.

Stoffers, D.A., Ferrer, J., Clarke, W.L. and Habener, J.F. (1997) Early-onset type-II diabetes mellitus (MODY4) linked to IPF1. *Nat. Genet.* **17**: 138–139.

Stumpf, H., Senkel, S., Rabes, H.M. and Ryffel, G.U. (1995) The DNA binding activity of the liver transcription factors LFB1 (HNF1) and HNF4 varies coordinately in rat hepatocellular carcinoma. *Carcinogenesis* **16**: 143–145.

Suaud, L., Formstecher, P. and Laine, B. (1999) The activity of the activation function 2 of the human hepatocyte nuclear factor 4 (HNF-4α) is differently modulated by F domains from various origins. *Biochem. J.* **340**: 161–169.

Swevers, L. and Iatrou, K. (1998) The orphan receptor BmHNF-4 of the silkmoth *Bombyx mori*: ovarian and zygotic expression of two mRNA isoforms encoding polypeptides with different activating domains. *Mech. Dev.* **72**: 3–13.

Taraviras, S., Monaghan, A.P., Schutz, G. and Kelsey, G. (1994) Characterization of the mouse HNF-4 gene and its expression during mouse embryogenesis. *Mech. Dev.* **48**: 67–79.

Tian, J.M. and Schibler, U. (1991) Tissue-specific expression of the gene encoding hepatocyte nuclear factor 1 may involve hepatocyte nuclear factor 4. *Genes Dev.* **5**: 2225–2234.

Tonjes, R., Xanthopoulos, K., Darnell, J. and Paul, D. (1992) Transcriptional control in hepatocytes of normal and c*12CoS* albino deletion mice. *EMBO J.* **11**: 127–133.

Tremp, G.L., Duchange, N., Branellec, D. *et al.* (1995) A 700-bp fragment of the human antithrombin III promoter is sufficient to confer high, tissue-specific expression on human apolipoprotein A-II in transgenic mice. *Gene* **156**: 199–205.

Tsai, S.Y. and Tsai, M.J. (1997) Chick ovalbumin upstream promoter-transcription factors (COUP-TFs): coming of age. *Endocr. Rev.* **18**: 229–240.

Tsukazaki, T., Chiang, T.A., Davison, A.F., Attisano, L. and Wrana, J.L. (1998) SARA, a FYVE domain protein that recruits SMAD2 to the TGFβ receptor. *Cell* **95**: 779–791.

Tugwood, J.D., Issemann, I. anderson, R.G., Bundell, K.R., McPheat, W.L. and Green, S. (1992) The mouse peroxisome proliferator activated receptor recognizes a response element in the 5′ flanking sequence of the rat acyl CoA oxidase gene. *EMBO J.* **11**: 433–439.

Ueda, A., Takeshita, F., Yamashiro, S. and Yoshimura, T. (1998) Positive regulation of the human macrophage stimulating protein gene transcription. Identification of a new hepatocyte nuclear factor-4 (HNF-4) binding element and evidence that indicates direct association between NF-Y and HNF-4. *J. Biol. Chem.* **273**: 19 339–19 347.

Ueda, K., Wei, Y. and Ganem, D. (1996) Cellular factors controlling the activity of woodchuck hepatitis virus enhancer II. *J. Virol.* **70**: 4714–4723.

Velho, G. and Froguel, P. (1998) Genetic, metabolic and clinical characteristics of maturity onset diabetes of the young. *Eur. J. Endocrinol.* **138**: 233–239.

Vergnes, L., Taniguchi, T., Omori, K., Zakin, M.M. and Ochoa, A. (1997) The apolipoprotein A-I/C-III/A-IV gene cluster: ApoC-III and ApoA-IV expression is regulated by two common enhancers. *Biochim. Biophys. Acta* **1348**: 299–310.

Viollet, B., Kahn, A. and Raymondjean, M. (1997) Protein kinase A-dependent phosphorylation modulates DNA-binding activity of hepatocyte nuclear factor 4. *Mol. Cell. Biol.* **17**: 4208–4219.

Waltz, S.E., Gould, F.K., Air, E.L., McDowell, S.A. and Degen, S.J. (1996) Hepatocyte nuclear factor-4 is responsible for the liver-specific expression of the gene coding for hepatocyte growth factor-like protein. *J. Biol. Chem.* **271**: 9024–9032.

Wang, J.C., Stafford, J.M. and Granner, D.K. (1998) SRC-1 and GRIP1 coactivate transcription with hepatocyte nuclear factor 4. *J. Biol. Chem.* **273**: 30 847–30 850.

Weber, H., Holewa, B., Jones, E.A. and Ryffel, G.U. (1996) Mesoderm and endoderm differentiation in animal cap explants: identification of the HNF4-binding site as an activin A responsive element in the *Xenopus* HNF1α promoter. *Development* **122**: 1975–1984.

Weigel, D. and Jäckle, H. (1990) The forkhead domain: a novel DNA binding motif of eukaryotic transcription factors? *Cell* **63**: 455–456.

White, R.L. (1998) Tumor suppressing pathways. *Cell* **92**: 591–592.

Williams, S.P. and Sigler, P.B. (1998) Atomic structure of progesterone complexed with its receptor. *Nature* **393**: 392–396.

Winrow, C.J., Marcus, S.L., Miyata, K.S., Zhang, B., Capone, J.P. and Rachubinski, R.A. (1994) Transactivation of the peroxisome proliferator-activated receptor is differentially modulated by hepatocyte nuclear factor-4. *Gene Expr.* **4**: 53–62.

Wotton, D., Lo, R.S., Lee, S. and Massagué, J. (1999) A SMAD transcriptional corepressor. *Cell* **97**: 29–39.

Xu, L., Glass, C.K. and Rosenfeld, M.G. (1999) Coactivator and corepressor complexes in nuclear receptor function. *Curr. Opin. Genet. Dev.* **9**: 140–147.

Yamagata, K., Furuta, H., Oda, N. *et al.* (1996) Mutations in the hepatocyte nuclear factor-4α gene in maturity-onset diabetes of the young (MODY1) *Nature* **384**: 458–460.

Yokomori, N., Nishio, K., Aida, K. and Negishi, M. (1997) Transcriptional regulation by HNF-4 of the steroid 15α-hydroxylase P450 (Cyp2a-4) gene in mouse liver. *J. Steroid Biochem. Mol. Biol.* **62**: 307–314.

Yoshida, E., Aratani, S., Itou, H. *et al.* (1997) Functional association between CBP and HNF4 in trans-activation. *Biochem. Biophys. Res. Commun.* **241**: 664–669.

Yoshiuchi, I., Yamagata, K., Yang, Q. *et al.* (1999) Three new mutations in the hepatocyte nuclear factor-1α gene in Japanese subjects with diabetes mellitus: clinical features and functional characterization. *Diabetologia* **42**: 621–626.

Zakin, M.M., Bovard-Houppermans, S. and Ochoa, A. (1994) Liver-specific expression of plasma protein genes. In: F. Tronche and M. Yaniv (eds) *Liver Gene Expression*, pp. 53–61. R.G. Landes, Austin, Texas.

Zapp, D., Bartkowski, S., Holewa, B., Zoidl, C., Klein-Hitpass, L. and Ryffel, G.U. (1993) Elements and factors involved in tissue-specific and embryonic expression of the liver transcription factor LFB1 in *Xenopus laevis*. *Mol. Cell. Biol.* **13**: 6416–6426.

Zaret, K.S. (1994) Genetic control of hepatocyte differentiation. In: I.M. Arias, J.L. Boyer, N. Fausto, W.B. Jakoby, D.A. Schachter and D.A. Shafritz (eds) *The Liver: Biology and Pathobiology*, pp. 53–68. Raven Press, New York.

Zaret, K.S. (1996) Molecular genetics of early liver development. *Annu. Rev. Physiol.* **58**: 231–251.

Zaret, K. (1998) Early liver differentiation: genetic potentiation and multilevel growth control. *Curr. Opin. Genet. Dev.* **8**: 526–531.

Zaret, K. (1999) Developmental competence of the gut endoderm: genetic potentiation by GATA and HNF3/forkhead proteins. *Dev. Biol.* **209**: 1–10.

Zhong, W., Sladek, F.M. and Darnell, J.E. (1993) The expression pattern of a *Drosophila* homologue to the mouse transcription factor HNF-4 suggests a determinative role in gut formation. *EMBO J.* **12**: 537–544.

Zhong, W., Mirkovitch, J. and Darnell, J.J. (1994) Tissue specific regulation of mouse HNF-4 expression. *Mol. Cell. Biol.* **14**: 7276–7284.

Zvibel, I., Fiorino, A.S., Brill, S. and Reid, L.M. (1998) Phenotypic characterization of rat hepatoma cell lines and lineage-specific regulation of gene expression by differentiation agents. *Differentiation* **63**: 215–223.

Chapter 10

Peroxisome Proliferator-Activated Receptors

Alex Elbrecht, Alan Adams and David E. Moller

INTRODUCTION AND HISTORICAL PERSPECTIVES

There are three members in this subfamily of nuclear receptors: peroxisome proliferator-activated receptor (PPAR)α, PPARδ (also referred to as NUC1 or PPARβ) and PPARγ. The designation of all three as PPARs is somewhat unfortunate as only PPARα, the first member of this subfamily to be cloned and characterized, plays a key role in peroxisome proliferation. However, DNA sequence analysis shows that the three are closely related, sharing greater than 60% sequence identity, and thus they form a distinct branch on the nuclear receptor family tree. Furthermore, all three can bind and be activated by similar classes of ligands, such as fatty acids. Although peroxisome proliferation does not reflect the roles played by all three receptors, it is at the heart of their discovery and characterization.

Structure and function of peroxisomes

Peroxisomes are best known for the role they play in controlling cell damage caused by oxygen free radicals, and indeed they are named with regard to their ability to produce hydrogen peroxide. Importantly, peroxisomes are also involved in the metabolism of drugs and xenobiotics (Masters, 1998), energy homeostasis through lipid metabolism (Wanders and Tager, 1998), cell differentiation and carcinogenesis (Masters and Crane, 1998), and possibly aging (Perichon *et al.*, 1998).

Peroxisomes were first identified microscopically in mouse kidney cells as rounded organelles ranging in size from 0.2 to 1 μm in diameter and were referred to as microbodies (De Duve and Baudhuin, 1966). This particular view of their architecture may have resulted from the examination of individual tissue sections and, in fact, serial sections would suggest the existence of a peroxisomal network (Yamamoto and Fahimi, 1987), although it might also suggest that peroxisomes are formed by budding from existing peroxisomes. All cells, with the exception of erythrocytes, are thought to contain peroxisomes, and treatment of rodents with a diverse group of compounds including hypolipidemic drugs, herbicides or pthalate plasticizers leads to induction of peroxisomes in liver parenchymal cells (Reddy and Lalwai, 1983; Reddy, 1990). This process,

Nuclear Receptors and Genetic Disease
ISBN 0-12-146160-2

referred to as peroxisome proliferation, which in rodents is closely associated with liver growth and tumors, does not seem to apply to humans, and thus is the subject of considerable research interest. In addition to liver, peroxisomes are particularly abundant in tissues active in lipid metabolism including sebaceous glands, brown fat and myelin-producing oligodendrocytes (Adamo *et al.*, 1986).

Peroxisomes are bounded by a single lipid membrane and those in the liver may have a crystalline or granular core. Unlike mitochondria, peroxisomal proteins are encoded by nuclear genes and are imported into the organelle. Peroxisome biogenesis is initiated by the synthesis of new membrane proteins and is followed by production of matrix components (Luers *et al.*, 1990). The numerous biochemical functions expressed in peroxisomes include peroxisomal oxidation and respiration, glyoxylate metabolism, cholesterol and dolichol metabolism, fatty acid β-oxidation, and ether-lipid synthesis (reviewed by Bosch *et al.*, 1992).

The key observation leading to our understanding of the role of peroxisomes in lipid metabolism was made by Brown *et al.* (1982), who noted raised levels of very-long-chain fatty acids in the plasma and skin fibroblasts of patients with Zellweger syndrome. These individuals exhibited multiple craniofacial abnormalities, generalized hypotonia, cortical renal cysts and hepatomegaly, leading to the original term of cerebrohepatorenal syndrome for this state. The absence of peroxisomes in patients with cerebrohepatorenal syndrome had been noted previously by Versmold *et al.* (1977). The establishment of this link between peroxisomes and fatty acid metabolism opened the doors for focused research in this area. Although peroxisomes contain enzymes capable of catalyzing many different reactions, most of these are associated with lipid metabolism (Wanders and Tager, 1998) and are discussed later in this chapter.

Receptors

To date, three members of the PPAR subfamily of nuclear receptors have been identified: PPARα, PPARγ and PPARδ. Human genome sequencing is expected to be completed in the year 2003 (Collins *et al.*, 1998). Until scientists have completed analysis of the sequence, it is tempting to speculate that there might be additional members in the PPAR family, which might account for the divergent effects seen with some ligands (discussed in the section on PPARγ below).

In general, the PPARs can bind and be activated by fatty acids (reviewed by Wolf, 1998) and, as altered fatty acid levels are associated with the development of diabetes, obesity, hypertension and atherosclerosis, these receptors could function to monitor and regulate metabolic pathways affected in several important disease states. As such, the receptors are good candidates for targets of pharmaceutical intervention.

Characterization and cloning of PPARα

Peroxisomal proliferation is a process that can be initiated in livers of rodents by treatment with small molecules termed peroxisome proliferators, as well as by exposure to high-fat diets, starvation and hypothermia. Although the peroxisome proliferators share little structural similarity, with the exception of an acidic moiety (see Pharmacology section), Green (1995) suggested that they might exert their effects via a receptor-mediated process. The receptor was appropriately named PPAR (now known as PPARα), and was shown to be a member of the family of nuclear receptors (Issemann and Green, 1990). DNA sequences for PPARα have now been obtained for several different species including: human (Sher *et al.*, 1993), amphibian (Dreyer *et al.*, 1992), rabbit (Guan *et al.*, 1997), mouse (Issemann and Green, 1990) and rat (Gottlicher *et al.*, 1992). In mice, the gene was localized to chromosome 15 (Jones *et al.*, 1995). Although it is expressed in other tissues, the highest level of expression is in the liver (Jones *et al.*, 1995; Braissant *et al.*, 1996).

Characterization of PPARα using binding assays has proven to be difficult, and to date there is relatively little published information (Devchand *et al.*, 1996), however, the interaction of diverse peroxisome proliferators with PPARα has been demonstrated indirectly by showing that the putative ligands are capable of inducing conformational changes in the receptor (Dowell *et al.*, 1997), which apparently also results in an increased affinity of the receptor for DNA response elements (Forman *et al.*, 1997) and protein coactivators (Krey *et al.*, 1997). Although these methods do not provide us with a better understanding of ligand binding, they are consistent in demonstrating the spectrum of PPARα ligands, which include natural fatty acids, eicosanoids and hypolipidemic agents.

Characterization and cloning of PPARγ

In 1992, Dreyer *et al.* cloned three, closely related, members of the nuclear receptor family from a *Xenopus* complementary DNA (cDNA) library. Since all three receptors were capable of activating the acylcoenzyme A oxidase gene, involved in peroxisomal fatty acid β-oxidation, the receptors were termed PPARα, β and γ. PPARγ was subsequently cloned from mouse (Zhu *et al.*, 1993), hamster (Aperlo *et al.*, 1995) and human (Greene *et al.*, 1995) cells. There are two PPARγ isoforms, γ1 and γ2, in mouse (Zhu *et al.*, 1995) and human (Elbrecht *et al.*, 1996), which differ only in that γ2 has an additional 30 N-terminal amino acids. For the murine gene, the two isoforms were shown to be products of alternative promoter usage from a single gene located on chromosome 6 E3-F1 (Zhu *et al.*, 1995).

A prostaglandin J_2 metabolite has been identified as a putative (low affinity) natural, ligand for PPARγ (Kliewer *et al.*, 1995); however, the physiologic role of this interaction is unclear. A more notable discovery was that the antidiabetic, synthetic thiazolidine-diones were high-affinity ligands for PPARγ (Lehmann *et al.*, 1995) and that activation of this receptor improves hyperglycemia and hypertriglyceridemia *in vivo* (Willson *et al.*, 1996; Berger *et al.*, 1999). The identification of specific ligands for PPARγ has led to further characterization of the multiple pathways affected by this receptor.

Characterization and cloning of PPARδ

The third member of this subfamily, PPARδ, is also referred to in the literature as PPARβ (Dreyer *et al.*, 1992) or NUC1 (Schmidt *et al.*, 1992); however, DNA sequence analysis shows that they all form a single cluster distinct from PPARα and PPARγ. The mouse PPARδ gene is located on chromosome 17 (Jones *et al.*, 1995) and it is expressed in many different cell types and tissues (Schmidt *et al.*, 1992; Braissant *et al.*, 1996). Although the consequences of physiologic activation of this receptor are unknown, it probably does not play a direct role in peroxisome proliferation.

PHYSIOLOGY

PPARα-mediated regulation of lipid metabolism

An excellent review of lipid metabolism in peroxisomes has been published recently by Wanders and Tager (1998); a summary of the various activities is presented below.

Peroxisomal enzymes are involved in fatty acid β-oxidation, fatty acid α-oxidation, synthesis of isoprenoids including cholesterol, ether phospholipid (plasmalogen) synthesis and the biosynthesis of polyunsaturated fatty acids. As PPARα has been shown to regulate the transcription of many of the genes encoding these enzymes, activators of PPARα have the potential to modulate fatty acid and cholesterol metabolism.

Peroxisomal and mitochondrial β-oxidation enzymes exhibit similar activities; however, there are distinct genes for each set of activities. An acyl-CoA oxidase catalyzes the first step in peroxisomal β-oxidation and is rate limiting for the process. It, as well as the genes for several other peroxisomal β-oxidation enzymes, contains PPAR response elements (PPREs) in their upstream regulatory sequences, and can be activated by PPARα ligands (Green and Wahli, 1994). Peroxisomal fatty acid oxidation is an incomplete process, shortening carboxylate chains by two-carbon units to a point where they become substrates for mitochondrial β-oxidation. Interestingly, transcription of the rate-limiting enzyme for mitochondrial β-oxidation, carnitine palmitoyl transferase, is also regulated by PPARα in cardiac myocytes (Brandt et al., 1998).

Dicarboxylic acids are also targeted to peroxisomes for oxidation. These fatty acids are produced in the endoplasmic reticulum by ω-oxidation involving a cytochrome P450-dependent hydroxylase. PPARα-mediated regulation of cytochrome P450 4A1 has been demonstrated (Demoz et al., 1994) and may reflect a mechanism to accommodate conditions of fatty acid overload. Indeed, long-chain dicarboxylic acids, formed via the P450 4A1 pathway, could regulate peroxisomal fatty acid β-oxidation (Kaikaus et al., 1993), possibly through activation of PPARα.

Peroxisomes are necessary for normal cholesterol synthesis (Krisans, 1996) and contain the enzymes responsible for conversion of acetyl-CoA to farnesyl pyrophosphate in cholesterol biosynthesis (Biardi and Krisans, 1996). Although treatment with fibrates results in a modest decrease in low-density lipoprotein (LDL) cholesterol and an increase in high-density lipoprotein (HDL) cholesterol, this effect is believed to occur through regulation of apolipoprotein synthesis (reviewed in Staels et al., 1998). The regulation of peroxisomal sterol-metabolizing enzymes by PPARα has not yet been examined, and the physiologic role remains unclear.

Fibrates, including fenofibrate and gemfibrozil, are weak (5–20 μmol L^{-1}) but specific PPARα agonists (see Pharmacology section). Circulating levels of fibrates in humans are compatible with this mode of action. In addition, the major components of fish oils (such as eicosapentanoic acid) which exert beneficial effects on lipids that are similar to the fibrates, also have weak (10–20 μmol L^{-1}) PPARα agonist activity. Beneficial effects of fibrates on triglycerides are ablated in PPARα null mice. The mechanism by which fibrates reduce triglycerides and increase HDL is believed to involve suppression of apolipoprotein (apo) CIII and induction of apoAI gene expression respectively (Staels et al., 1998). In addition, upregulation of hepatic lipoprotein lipase gene expression may have a role in augmenting clearance of triglyceride-rich particles.

Role of PPARγ in adipogenesis

The seminal observation connecting PPARγ to its role in adipocyte differentiation was made by workers in Spiegelman's laboratory when they showed that ARF6, an adipocyte transcription factor complex, consisted of the heterodimeric pair PPARγ–retinoid X receptor (RXR) α (Tontonoz et al., 1994a). Using retrovirally mediated expression of PPARγ in fibroblasts, the group then demonstrated that PPARγ2 could play a role in the differentiation of the adipocyte cell lineage (Tontonoz et al., 1994b).

Indeed, the highest level of expression for PPARγ is in differentiated fat cells (Jones *et al.*, 1995; Braissant *et al.*, 1996). The distribution of the two isoforms and their regulation was investigated by Vidal-Puig *et al.* (1996), who showed that expression was probably not altered in rodent models of obesity, but was affected by diet.

Potential role of PPARγ in glucose metabolism

The identification of thiazolidinedione (TZD) antidiabetic compounds as ligands for PPARγ (Lehmann *et al.*, 1995) has led to a global research effort in this area. Several lines of evidence implicate PPARγ activation as the predominant mechanism of TZD insulin sensitizer action: (1) *in vivo* efficacy in rodents generally correlates with *in vitro* PPARγ activity; (2) non-TZD PPARγ agonists of several structural types also exert antihyperglycemic effects in rodent models of noninsulin-dependent diabetes mellitus (NIDDM); (3) structurally distinct compounds that function as selective RXR ligands activate PPARγ–RXR heterodimers and cause *in vivo* insulin sensitization in rodent NIDDM models (Mukherjee *et al.*, 1997); and (4) transcriptional activation of at least one gene (lipoprotein lipase) by PPARγ, has been linked to one of the classic *in vivo* effects of TZD – triglyceride lowering (Schoonjans *et al.*, 1996a). Although several other PPARγ-regulated genes have been identified, there is no comprehensive model available to account for the insulin-sensitizing effects. Genes regulated by PPARγ and the resultant effects are shown in Table 10.1.

Other potential roles for PPARγ

Recently published studies suggest that PPARγ may also act as a negative regulator of macrophage activation (Ricote *et al.*, 1998) and inhibit the production of monocyte inflammatory cytokines (Jiang *et al.*, 1998). Interestingly, TZDs have also been reported to induce terminal differentiation of malignant breast epithelial cells (Mueller *et al.*, 1998), suggesting that activators of the receptor could be used in the treatment of breast cancer. Recently, it was shown that ligands for PPARγ are potent inhibitors of angiogenesis (Xin *et al.*, 1999); however, the dose–response characteristics do not agree with those for receptor binding and activation (Lehmann *et al.*, 1995; Elbrecht *et al.*, 1996).

As the potency of selected compounds as PPARγ ligands does not always correlate with their activity in mediating the diverse physiologic effects noted above, it is likely that some effects are mediated by other mechanisms (such as cell surface prostanoid receptors in the case of the prostaglandin J_2 metabolite).

STRUCTURE AND MECHANISM OF ACTION

Transcriptional regulation

In general, nuclear receptors exert their effects by regulating the transcription of target genes (reviewed in Beato *et al.*, 1995; Mangelsdorf and Evans, 1995; Mangelsdorf *et al.*, 1995). It is interesting to note that, although the models for steroid hormone action have become more complex, the mechanism of action proposed more than 20 years ago for the progesterone receptor (PR) by O'Malley and Schrader (1976) still applies. The more recent identification of coactivators and corepressors has provided additional

Table 10.1 Potential mechanisms of insulin sensitization

Defined genes	Effects in cells	Effects in tissues	Net *in vivo* effects	References
↑*Cbl*-associated protein (WAT)	↑Glycogen (WAT)	↑Glucose uptake, glycogen synthesis (WAT, muscle)	↓FFA	Ribon *et al.* (1998); Combettes-Souverain and Issad (1998)
↑UCP-1(BAT)	↓TNFα activity (WAT)	↓Glucose-6-phosphatase (liver)	↓Triglyceride	Kelly *et al.* (1998); Souza *et al.* (1998); Foster *et al.* (1997)
↑LPL (WAT)	Adipose cell differentiation	↑Glucokinase (liver)	↓Hepatic glucose output	Schoonjans *et al.* (1996a); Spiegelman (1998); Kahn (1997)
↑PEPCK (WAT)	↑PI-3-kinase (WAT)	↓Gluconeogenesis (liver)	↑Glycolysis	Glorian *et al.* (1998); Combettes-Souverain and Issad (1998); Chen (1998)
↑aP2 (WAT)	↓IRS Ser kinase	↓Glycogenolysis (liver)	↑Glucose disposal	Zhang *et al.* (1996); Le Marchand-Brustel (1999); Qiao *et al.* (1999); Bollen *et al.* (1998)
↑FATP (WAT)				Frohnert *et al.* (1999)
↑CD36 (WAT)				Aitman *et al.* (1999)
↑Acyl-CoA synthase (WAT)				Martin *et al.* (1997)
↓Stearoyl-CoA desaturase (WAT)				Jones *et al.* (1998)

WAT, white adipose tissue; BAT, brown adipose tissue; UCP, uncoupling protein; LPL, lipoprotein lipase; PEPCK, phosphoenol pyruvate carboxykinase; IRS, insulin receptor substrate; FFA, free fatty acids; TNF, tumour necrosis factor; PI-3-kinase, phosphatidylinositol 3-kinase.

components which may account for the diversity of responses observed for a single ligand. Although these cofactors are described in more detail below, the model suggests that tissue-specific gene expression is not due solely to the presence of receptor and ligand, but rather is a reflection of the components in the complex made up of receptor, polymerase and cofactors.

Nuclear receptors exhibit a modular structure consisting of six regions named A–F (Laudet *et al.*, 1992). The most highly conserved regions are the DNA-binding domain (DBD) or C domain, and the ligand-binding domain (LBD) or E domain. The DBD consists of two zinc fingers which recognize specific DNA sequences (hormone response elements or HREs) in the promoters of regulated genes (reviewed by Mangelsdorf *et al.*, 1995). Two activation domains, termed activation function (AF) 1 and 2, have been characterized and are located at opposite poles of the receptor. AF-1 is ligand independent and is located near the N-terminus in the A/B domain, while the AF-2 function, located in the E domain, is necessary and sufficient for ligand-dependent activation.

In addition to activation by ligand, nuclear receptors can be regulated by phosphorylation. In the case of PPARγ2, and consistent with the observation that mitogens inhibit adipocyte differentiation, phosphorylation of Ser112 is mediated by the mitogen-activated/extracellular receptor-regulated kinase (MEK)/extracellular signal-regulated kinase (ERK) signaling pathway, resulting in inhibition of adipogenesis (Hu *et al.*, 1996). This is in contrast to PPARα, where insulin stimulation causes serine phosphorylation and activation of the receptor (Juge-Aubry *et al.*, 1999). The serine residues are located in the AF-1 region and might be involved in interactions with corepressors.

RXR heterodimers

The RXR was first identified in 1990 by Mangelsdorf *et al.* It was subsequently shown that, unlike steroid hormone receptors that function as homodimers, the thyroid hormone receptor (TR) and the retinoic acid receptor (RAR) form heterodimers with RXR (reviewed by Mangelsdorf and Evans, 1995). All three PPARs fall in the type II subgroup of nuclear receptors and pair with RXR, which also exists in three isotypes: RXRα, β and γ (Mangelsdorf *et al.*, 1992). Evidence for an *in vivo* interaction of PPAR and RXR came from coexpression studies in yeast, which are devoid of nuclear receptors (Miyata *et al.*, 1994). The amino acids involved in dimerization have been identified for PPARα and are located in the LBD. Surprisingly, mutation of several of these residues permitted selective activation of PPARα-regulated genes (Gorla-Bajszczak *et al.*, 1999), suggesting the possibility of other heterodimeric partners for the PPARs. As RXR can be activated by its own ligand, 9-*cis* retinoic acid (Allenby *et al.*, 1993), it can function as either a silent partner or an active partner. The latter is the case with PPARγ, where Mukherjee *et al.* (1997) have used RXR-selective agonists, referred to as rexinoids, to replicate the antidiabetic activity of PPARγ ligands. Thus, rexinoids can function as insulin sensitizers decreasing hyperglycemia, hypertriglyceridemia and hyperinsulinemia in mouse models of NIDDM.

Hormone response elements

The ability of nuclear receptors to bind DNA is central to their role as ligand-activated transcription factors. The specific DNA-binding sites, or HREs, are usually located in the 5'-flanking regions of target genes. HREs consist of hexanucleotide elements separated by spacers of one to five nucleotides. The consensus half-site for the type II RXR-binding receptors is -AGGTCA-, and the specific PPAR response element (PPRE) is a direct repeat of this sequence with a single nucleotide spacer (direct repeat (DR1)).

However, DR1 elements are somewhat promiscuous and can bind RAR–RXR hetero-dimers and RXR homodimers as well (Nakshatri and Chambon, 1994). Additional specificity for binding of PPARs may be provided by sequences that flank the HRE (Juge-Aubry *et al.*, 1997). Furthermore, at least in some cases, the polarity of the PPAR–RXR heterodimer is reversed on the DR1 element, when compared with RXR–TR and RXR–RAR for DR4 and DR5 elements, respectively (IJpeuberg *et al.*, 1997).

Coactivators and corepressors

With the identification of coactivators and corepressors, and their ability to regulate chromosomal structure through histone modification (reviewed by Torchia *et al.*, 1998), it is now possible to envisage complexes capable of producing tissue-specific effects, and even gene-specific effects. Many of these transcriptional cofactors are not expressed in a tissue-specific manner, suggesting that subtle regulatory effects can be achieved by discrete alterations in their relative abundance. This was suggested by Xu *et al.* (1998) with the steroid receptor coactivator (SRC) 1 knockout mouse which exhibited a blunted response to treatment with steroid hormones. However, TIF-2, also referred to as glucocorticoid receptor-interacting protein (GRIP) 1 and SRC-2, were upregulated in these animals, which could account for the partial responses observed.

SRC-1 null mice were used to show that this specific coactivator is not necessary for PPARα-regulated gene expression in liver (Qi *et al.*, 1999). The situation for PPARγ and SRC-1 is less clear; a ligand-dependent interaction has been demonstrated *in vitro* (Zhu *et al.*, 1996; Zhou *et al.*, 1998) but *in vivo* effects in SRC-1 knockout mice have not been determined. PPARγ has been shown to interact with other coactivators as well. PPARγ-binding protein (PBP) was cloned using a yeast two-hybrid system with Gal4–PPARγ as the bait, but, as is the case with other coactivators, it is somewhat promiscuous and also binds to PPARα, RARα, RXR and TRβ1 (Zhu *et al.*, 1997). Similarly, PGC-1, which was cloned from a brown fat cDNA library (Puigserver *et al.*, 1998), can increase PPARγ-induced UCP-1 gene expression (Kelly *et al.*, 1998) and thus may play a role in adaptive thermogenesis. It is clear that PPARγ is capable of interacting with a number of different coactivators, but it should be kept in mind that the affinities for each are likely to be different. Using an *in vitro* coactivator association assay, Zhou *et al.* (1998) were able to show that PPARγ exhibits a clear preference for CBP over SRC-1.

The coactivators interact with nuclear receptors through a conserved LXXLL (where X denotes any amino acid) motif (Heery, *et al.*, 1997; Torchia *et al.*, 1997). Interestingly, all PPARs sequenced to date contain a single LXXLL motif which overlaps the AF-2 domain (A. Elbrecht, unpublished results). The AF-2 domain plays a key role in receptor activation and its position shifts in response to ligand binding, suggesting that the PPAR LXXLL motif may be involved in recruitment of cofactors.

Several potential mechanisms have been described for repression of gene transcription by PPARs. Receptor activity may be attenuated by interaction with a heterodimeric partner other than RXRα, or by interaction with corepressors. An example of negative regulation involved the effect of the corepressor SMRT to abolish activation of the lipoprotein lipase promoter by the combined actions of chick ovalbumin upstream promoter-transcription factor (COUP-TF) II (ARP-1) and PPARγ (Robinson *et al.*, 1999).

Structure and conformation

The term 'ligand-activated transcription factor' comes from the ability of steroids and other small molecules to bind and activate their cognate nuclear receptors. McDonnell *et al.* (1995), using the estrogen receptor (ER), were able to describe distinct receptor conformations when activated by different ligands. Changes in receptor conformation can be determined by resistance to digestion by proteases (Allan *et al.*, 1992); in general, ligand binding results in a more compact, protease-resistant core. This technique has been applied to characterize ligands for all three PPARs (Elbrecht *et al.*, 1996; Dowell *et al.*, 1997; Berger *et al.*, 1999), although it may not be sufficiently sensitive to distinguish full agonists from partial agonists (Elbrecht *et al.*, 1999). Of course, this approach provides a very low level of resolution for structural characterization of the receptor. Crystal structures have recently been obtained for the LBDs of PPARγ (Uppenberg *et al.*, 1998) and PPARδ (Xu *et al.*, 1999). These structures reveal that the PPAR LBDs have a typical structure consisting of 13 α helices and a small four-stranded β sheet. Interestingly, the ligand-binding pocket of apo-PPARγ is relatively large, which may account for the diversity of small molecules capable of entering and interacting with the receptor (Nolte *et al.*, 1998).

GENETICS

Genetic variation in PPARγ

Given its probable roles in regulation of body adiposity, glucose homeostasis and lipid metabolism, the potential for genetic variation involving PPARγ has received considerable attention. The human and murine PPARγ genes have been cloned and characterized (Zhu *et al.*, 1995; Beamer *et al.*, 1997). Of nine exons, the coding region spans seven exons, one of which encodes the additional N-terminal amino acids corresponding to the PPARγ splice variant (Zhu *et al.*, 1995); in addition there are two separate promoters that mediate regulated expression of both PPARγ isoforms. Using DNA samples derived from a number of human subjects, several investigators have reported single nucleotide sequence polymorphisms within the coding exons of the PPARγ gene (Yen *et al.*, 1997; Deeb *et al.*, 1998; Meirhaeghe *et al.*, 1998; Ristow *et al.*, 1998). To date, no significant genetic variants in PPARγ have been reported in nonhuman species.

Two groups have reported the presence of a silent polymorphism (1431C > T) in the sixth exon common to PPARγ1 and PPARγ2 (Yen *et al.*, 1997; Meirhaeghe *et al.*, 1998). In one study (Meirhaeghe *et al.*, 1998), multiple metabolic parameters were examined in 820 French subjects genotyped for variation at the exon 6 1431C > T polymorphism. Although no association between this polymorphism and body mass index (BMI) *per se* (or other metabolic parameters) was detected, the relationship between BMI and plasma leptin levels differed slightly between subjects with the C genotype compared with those with the T allele. This suggests that genetic variation in or near the PPARγ locus could modulate leptin levels in response to variable degrees of body adiposity.

A C → G substitution in the PPARγ2 exon (P12A) was initially reported by Yen *et al.* (1997). They found that the P12A allele frequency varied from 0.03 to 0.12 in several populations (highest in Caucasians, lower in Asians and Africans). In a subsequent study (Beamer *et al.*, 1998), the same group reported that the P12A allele was associated with higher BMI in two Caucasian cohorts. In contrast, no association of P12A with increased obesity could be detected in a cohort of Japanese men, although only 12 heterozygotes and one homozygote were present in the study population of 203 subjects

(Mori *et al.*, 1998). An additional larger study of German subjects found no association of the P12A allele with type II diabetes or obesity (Ringel *et al.*, 1999). In contrast to findings noted above, Deeb *et al.* (1998) recently reported that the P12A allele was associated with lower BMI and improved insulin sensitivity in a cohort of Finnish subjects, and that the wild-type allele was associated with increased type II diabetes incidence in a Japanese American cohort. As discussed below, these same investigators observed functional consequences of the Pro → Ala substitution when comparing the activities of the two recombinant proteins.

A novel PPARγ mutation, P115Q resulting from a G → T substitution, was also recently reported by Ristow *et al.* (1998). In this case, the P115Q allele was present in four of 121 obese German subjects (mean BMI 33.9) but was absent in each of 237 controls of normal weight (mean BMI 25). Of further interest was the fact that three of the four affected obese subjects had type II diabetes and all four were markedly obese (BMI 38–47). Importantly, a functional consequence of the P115Q mutation was also reported (see below). Potential association of the P115Q mutation with metabolic derangement in other patient populations or its potential to underlie familial variation in body adiposity have not yet been examined (C.R. Kahn, personal communication).

Potential for genetic variation in PPARα

PPARα is best known for its key role in regulation of lipid oxidation and metabolism in the liver. Until recently, it was not clear whether PPARα was expressed in human liver. This has now been shown at both the messenger RNA (mRNA) (Palmer *et al.*, 1998) and protein (Su *et al.*, 1998) levels, although hepatic PPARα expression is about 10-fold lower than that in mice or rats.

Of interest is the fact that the two initially reported human PPARα cDNA sequences (Sher *et al.*, 1993; Mukherjee *et al.*, 1994) differ at two amino acid residues; it is not clear whether this discrepancy represents genetic polymorphisms versus cloning artefacts. Tugwood *et al.* (1997) have reported the existence of several additional sequence variants in human PPARα including T71M, K123M, A268V, G296A and V444A. A single cDNA (hPPARα6/29) with multiple mutations, including T71M, K123M and V444A, was shown to be competent for RXR dimerization and DNA binding but was inactive in a cell-based transactivation assay (Tugwood *et al.*, 1997).

An additional potentially important hPPARα sequence variant (hPPARα8/14) has been described which lacks 203 base pairs encoding nucleotides 508–712 (508–712del) at the C-terminal end of the DBD. This variant most likely arises via aberrant splicing of exon 6 (Tugwood *et al.*, 1997). When the PPARα8/14 clone was expressed, the recombinant protein (truncated after nucleotide 512) did not exhibit the ability to form DNA-binding heterodimers with RXR and was inactive in cell-based agonist-induced transactivation (Tugwood *et al.*, 1997). In subsequent studies reported by Palmer *et al.* (1998), this splice variant lacking exon 6 was noted to be present at potentially variable levels (as assessed by RNAase protection) in several samples of human liver. However, expression of a smaller PPARα protein was not noted in another study where human tissue samples were examined by Western blotting with an antibody raised against the receptor N-terminal region (Su *et al.*, 1998).

To date, no published studies have identified possible genetic variation involving human PPARδ.

PATHOPHYSIOLOGY

PPARγ as a potential modulator of adiposity or metabolism

As described above, genetic evidence linking variation in the PPARγ gene to the pathophysiology of altered body adiposity and metabolism is scant at present. However, as discussed elsewhere in this chapter, it is clear that treatment of animals with high doses of PPARγ agonists can promote an increase in body adiposity while providing net benefits in terms of improved insulin sensitivity and reduced triglyceride levels. In addition to consideration of its possible role as a contributor to genetic obesity, and/or insulin-resistant type II diabetes, PPARγ has also been viewed as a plausible candidate gene for congenital forms of rare lipoatrophy and lipodystrophy such as the Berardinelli–Seip syndrome (Moller and O'Rahilly, 1993). However, in several studies reported to date (e.g. Vigouroux *et al.*, 1997) no linkage with, or sequence variation in, PPARγ has been associated with inherited lipodystrophic syndromes.

Attempts to assess the role of PPARγ as a contributor to the pathogenesis of common forms of human metabolic disease have focused on two questions. (1) Are there altered levels of PPARγ expression in relevant tissues from subjects with obesity or insulin resistance? (2) What are the effects of the above noted PPARγ genetic variants on function of the expressed recombinant proteins?

Several studies have employed RNAase protection or quantitative polymerase chain reaction methods to examine PPARγ mRNA expression in adipose tissue or skeletal muscle samples from human subjects with obesity or type II diabetes. Vidal-Puig *et al.* (1997) reported that subcutaneous adipose tissue PPARγ2, but not PPARγ1, mRNA was increased by about 50% in obese versus lean American subjects. In addition to a positive correlation between PPARγ2 mRNA levels and BMI, weight loss in obese subjects was associated with a 25% decrease in adipose tissue PPARγ2 expression. In contrast, no association between altered PPARγ2 (or PPARγ1) mRNA expression in adipose tissue and obesity or type II diabetes was evident in a similar study from France (Auboeuf *et al.*, 1997). Skeletal muscle expression of PPARγ has been examined in two studies. Park *et al.* (1997) found that muscle PPARγ mRNA was increased in subjects with obesity and type II diabetes and that acute *in vivo* insulin stimulation resulted in increased muscle PPARγ expression. Kruszynska *et al.* (1998) noted that muscle PPARγ expression correlated with percentage body fat and with *in vivo* insulin-stimulated glucose disposal in obese subjects. Thus, a potential *in vivo* role for altered (increased) PPARγ expression as a mediator of increased adiposity and increased insulin sensitivity can be envisioned; hyperinsulinemia is a potential factor that might serve to drive higher levels of PPARγ gene expression.

Although the association of the Ala12 PPARγ allele with lower BMI was apparent in only one of several studies cited above (Deeb *et al.*, 1998), the recombinant receptor bearing this single amino acid change was apparently defective with respect to DNA binding and its ability to mediate ligand-stimulated transactivation in transfected cells (Deeb *et al.*, 1998). Given that this variant is relatively prevalent in certain populations, it is possible that it contributes to altered physiology of fat metabolism in humans. Interestingly, the Pro115 → Gln substitution in PPARγ, which was implicated as a potential cause of obesity, is located next to Ser114, which may be an important site of negative regulation via growth factor-mediated phosphorylation (Hu *et al.*, 1996). Indeed, like an artificial Ser114 → Ala mutant (Hu *et al.*, 1996), the naturally occurring Gln115 mutant was apparently less prone to negative regulation since forced overexpression in fibroblasts resulted in a greater degree of adipogenesis than was observed with wild-type PPARγ (Ristow *et al.*, 1998). It is therefore logical to presume that this change in PPARγ function (reduced negative regulation of the Gln115 mutant resulting in a gain of receptor activity) could result in a phenotype of increased adiposity even in patients bearing only one altered allele.

Possible role of PPARγ in other disease processes

In addition to its probable role in the regulation of adipogenesis, lipid metabolism (and insulin sensitivity), PPARγ has also been implicated as a possible regulator of additional physiologic processes such as those involving the immune system, vascular tone, vascular endothelial or smooth muscle cell growth or function, and growth or differentiation of colonic epithelial cells. Among this broad array of additional possible physiologic roles for PPARγ, its presence in the colon has recently been suggested to have possible pathogenic significance.

Given that relatively high levels of PPARγ are present in the colon (Sarraf *et al.*, 1998) and the known effect of cyclooxygenase inhibitors in reducing colonic cancer risk (Williams *et al.*, 1997), it is possible that the production of prostaglandins and other eicosanoids via cyclooxygenase might act to promote tumor formation by providing higher levels of natural PPARγ ligands. In a recently published study, one group of investigators noted that chronic treatment of *Min* mice (a murine model of inherited polyposis) with high doses of two thiazolidinedione PPARγ agonists resulted in a modest increase in colonic tumor number (Lefebvre *et al.*, 1998). In contrast, other studies have reportedly shown that growth of fully transformed human colonic cancer cells can be inhibited by incubation with these same PPARγ-selective agonist compounds (Brockman *et al.*, 1998; Sarraf *et al.*, 1998). Thus, there is no firm evidence to support a pathophysiologic role for PPARγ in the context of colonic tumor formation.

Administration of high doses of thiazolidinedione PPARγ agonists to animals has reportedly resulted in a number of distinct toxic effects such as cardiac enlargement, volume expansion, anemia, and proliferative and/or degenerative changes involving adipose tissue. With the exception of the effects in adipose tissue, none of these effects is a clearly established consequence of PPARγ activation *per se*. Nevertheless, it remains possible that certain disease processes that affect these organ systems may ultimately be shown to be due, at least in part, to pathophysiologic mechanisms involving acquired or genetic changes in PPARγ expression or action.

Potential disease pathophysiology related to PPARα

As noted above, there are several probable sequence variants affecting the PPARα gene. However, to date no specific variants have been reportedly associated with any given disease, nor has the PPARα gene locus been implicated as a disease-associated gene. It is still interesting to speculate about how an acquired or genetic defect leading to loss (or gain) of PPARα function might contribute to selected medical conditions.

As reviewed earlier in this chapter, activation of PPARα is clearly associated with hepatomegaly, hepatic peroxisome proliferation and hepatocarcinogenesis in rodents. Although these effects are not known to occur in humans with therapeutic doses of fibrates (weak PPARα agonists), it is possible that defects involving PPARα might contribute to selected forms of human liver disease. In contrast, loss of PPARα has important pathophysiologic effects which have been documented using PPARα null mice. In particular, these mice have increased body adiposity, altered triglyceride metabolism (trending to higher levels), and reduced levels of hepatic fatty acid oxidation. In a striking example of this effect, PPARα null mice are well compensated until challenged with a fatty acid oxidation inhibitor (etomoxir); this was well tolerated in normal mice but resulted in massive accumulation of lipids in liver and heart, hypoglycemia and death in PPARα null mice (Djouadi *et al.*, 1998). Thus, it is possible that defects that result in reduced PPARα expression or function could ultimately be found to contribute to the pathogenesis of hypertriglyceridemia, dysregulated fatty acid oxidation, abnormal glucose homeostasis, or obesity in humans.

PHARMACOLOGY

Ligands and agonists for PPAR

Demonstrated ligands for PPARs include unsaturated fatty acids, prostanoids, leuko-triene metabolites, hypolipidemic fibrates, thiazolidinedione and other insulin sensiti-zers, some nonsteroidal antiinflammatory drugs (NSAIDs) and a wide variety of substituted carboxylic acids. Owing to their relationship with the peroxisomal fatty acid oxidation pathways, the search for naturally occurring PPAR ligands began with the fatty acids and eicosanoids. Initial studies typically demonstrate PPAR activation in cell transfection models, and several assays demonstrating direct ligand interactions with PPAR have appeared recently.

Natural ligands

Prostaglandins, fatty acids and PPARs

Direct binding of a naturally occurring ligand, 15-deoxy-$\Delta^{12,14}$-prostaglandin J_2 (15d-PGJ2), was demonstrated for PPARγ by Forman *et al.* (1995) and Kliewer *et al.* (1995). The binding affinity of this ligand was reported to be 2.5 μmol L^{-1} (Kliewer *et al.*, 1995). As it is difficult to measure endogenous levels of 15d-PGJ2, it is not clear what the physiologic relevance of this interaction might be. Some lower-affinity binding was seen for PGJ2 and Δ12-PGJ2. Demonstrations of 15d-PGJ2 binding to PPARγ were followed by reports of murine PPARα activation by another prostanoid, carbaprostacyclin (Hertz *et al.*, 1996). Additional cell-based studies showed ligand-mediated activation of PPARγ and/or PPARα by other prostanoids including PGA1, PGA2, PGB2 and PGD2, 15d-PGJ2.

Unsaturated fatty acids and eicosanoid products

Fatty acids composed of more than 18 carbons, polyunsaturated fatty acids and the arachidonic acid metabolite families are all ligands for the PPARs. Long-chain fatty acids were long known to stimulate fatty acid oxidation and were subsequently shown to be PPAR agonists at micromolar concentrations. Natural eicosanoids, synthetic lipoxygenase inhibitors and leukotriene mimics such as ETYA and L-165207 (Berger *et al.*, 1999) have been shown to activate or bind PPARs or to induce classic peroxisome proliferation responses in rodents.

Although a correspondence between ligand binding and agonist activity was demon-strated for some fatty acids (Forman *et al.*, 1997; Kliewer *et al.*, 1997), it has been difficult to demonstrate direct binding because of the very low affinity of these ligands. In keeping with their high physiologic concentrations, fatty acids are tested at 30 and $100\,\mu mol\,L^{-1}$. Several unsaturated fatty acids and C16–18 fatty acids showed strong PPARα activation. Palmitic (C16), oleic (C18Δ^9 cis), petroselenic (C18Δ^6 cis), α-linolenic (C18$\Delta^{9,12,15}$ all cis), linoleic (C18$\Delta^{9,12}$ all cis) and arachidonic (C20$\Delta^{5,8,11,14}$ all cis) acids all activated PPARα at $100\,\mu mol\,L^{-1}$. Linolenic, linoleic and arachidonic acid showed IC$_{50}$ values in the range of $2–20\,\mu mol\,L^{-1}$ on Xenopus PPARα. The contemporary report from Forman et al. (1997) showed similar results for activation of mouse PPARα and PPARγ, with unsaturated fatty acids and eicosanoids, and demonstrated binding through the gel mobility shift assay. The same unsaturated fatty acids linoleic, α-linolenic, γ-linolenic (C18$\Delta^{6,9,12}$ all cis) and arachidonic acids, were shown to be both activators and ligands of mouse PPARα.

Two related spectrophotometric competition assays have been described (Palmer and Wolf, 1998; Lin et al., 1999) using fluorescence shift of all-cis parinaric acid (CPA, C18$\Delta^{9,11,13,15}$ all cis) and all-trans parinaric acid (TPA) for binding to human PPARγ and mouse PPARα respectively. The reported K_d values were dramatically lower than the concentrations required for effective transactivation in cells. A second TPA assay with mouse PPARα centered on the naturally occurring chlorophyll degradation product phytanic acid, which exhibited a K_d of $0.010\,\mu mol\,L^{-1}$, compared with the hypolipidemic fibrate bezafibrate ($0.045\,\mu mol\,L^{-1}$), WY-14643 ($0.004\,\mu mol\,L^{-1}$) and arachidonic acid ($0.083\,\mu mol\,L^{-1}$) (Ellinghaus et al., 1999).

The arachidonic acid metabolite 8(S)-HETE is reported to bind Xenopus PPARα with an IC$_{50}$ of $0.50\,\mu mol\,L^{-1}$ versus [^{3}H]GW-2331 (Kliewer et al., 1997). Studies (±) 8-HEPE and 8(S)-HETE also showed activation and binding to mouse PPARα using gel mobility shift (Forman et al., 1997). With an EC$_{50}$ of $0.20\,\mu mol\,L^{-1}$ for mouse PPARα activation, and an estimated EC$_{50}$ of $0.10\,\mu mol\,L^{-1}$ for binding-induced gel mobility shift, 8(S)-HETE was substantially more potent than WY-14643. Leukotriene B$_4$ is also reported to bind mouse PPARα with K_d values of 0.061 to approximately $0.090\,\mu mol\,L^{-1}$ (Nakanishi et al., 1970; Lin et al., 1999).

Synthetic ligands

Thiazolidinediones and PPARs

Pursuing the evidence of potent adipogenic effects of the glitazone (TZD) insulin sensitizer BRL-49653 (rosiglitazone) (Fig. 10.1), Kliewer et al. (1997) demonstrated that BRL-49653 was a high-affinity ligand for PPARγ (K_d $0.043\,\mu mol\,L^{-1}$). The related TZDs, pioglitazone, englitazone and ciglitazone, were shown in the same study to bind directly to mouse PPARγ with lower affinity. A number of other glitazone structures have since been shown to bind PPARγ and, in some cases, PPARα (Murakami et al., 1998).

The members of this family derive from an effort to improve the hypolipidemic, and possible insulin-sensitizing, activity of the fibrate hypolipidemics (Kawamatsu et al., 1980; Sohola et al., 1982). The original lead, AL-294, evolved into the two related families of thiazolidindione (Hulin et al., 1996a) and α-alkoxyphenylpropionates (Buckle et al., 1996; Hulin et al., 1996), which have provided several potent antihyperglycemic drug candidates. BRL-49653 is considered to be a PPARγ selective agonist while an α-alkoxyphenylpropionate, such as SB 219994, is both a PPARα (EC$_{50}$ $2.5\,\mu mol\,L^{-1}$) and a PPARγ (EC$_{50}$ $0.070\,\mu mol\,L^{-1}$) agonist. The broadest comparison of binding affinities is relative to [^{125}I]SB 236636. This iodinated relative of SB 219994 is used for compara-

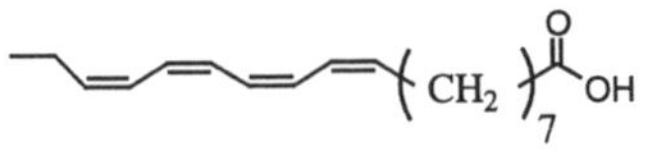

cis Parinaric acid

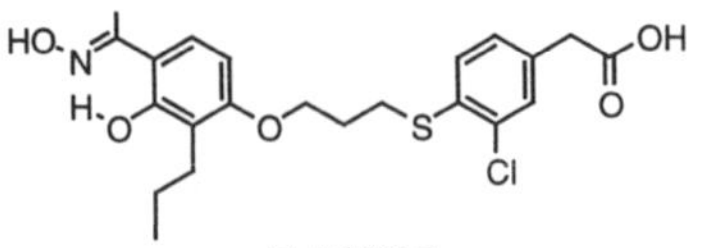

L-165207

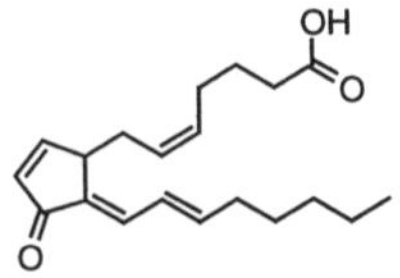

15-deoxy-$\Delta^{12,14}$-PGJ^2

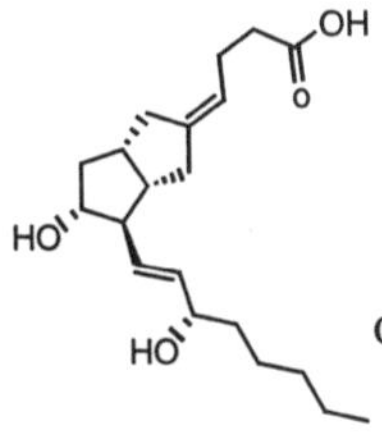

Carbaprostacyclin

PPAR agonist Insulin Sensitizers

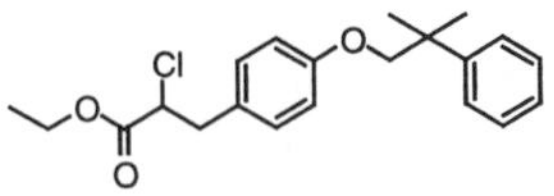

AL- 294

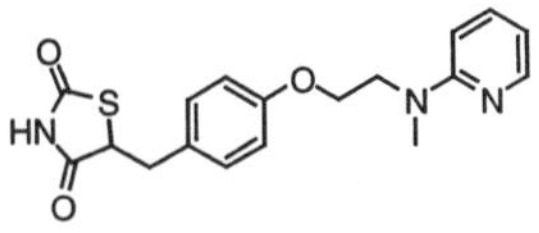

BRL 49653 Rosiglitazone

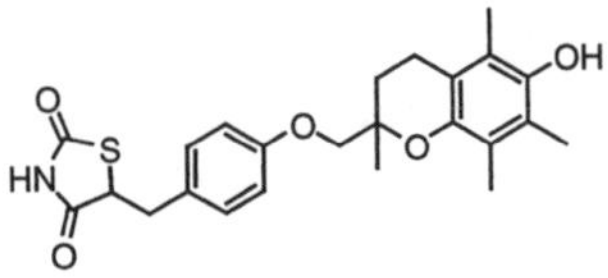

CS-045 Troglitazone

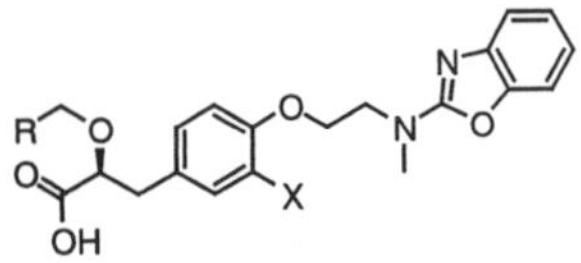

R= Me X= ^{125}I   SB 236636

R= CF$_3$ X= H   SB 219994

Fig. 10.1 Prostanoid PUFA and LT-related families.

tive binding studies on human PPARγ1 (SB 236636, K_d 70 nmol L^{-1}), differentiated 3T3-L1 cells, rat adipocytes and human adipocytes. The binding affinities calculated from the various sources vary by approximately one order of magnitude, but are mostly consistent for rank. The reported order for human PPARγ1 is SB 219994 (*S* enantiomer) at 0.0021 μmol L^{-1}, BRL-49653 (rosiglitazone) at 0.041 μmol L^{-1}, 15d-PGJ2 at 0.045 μmol L^{-1}, SB 219993 (*R* enantiomer of 219994) at 2.77 μmol L^{-1}, WY-14 643 at 4.4 μmol L^{-1}, pioglitazone at 4.83 μmol L^{-1} and troglitazone at 7.97 μmol L^{-1} (Young *et al.*, 1998).

The mechanism of action of these insulin sensitizers is unproven but widely assumed to be principally mediated by PPAR agonism. The class also tends to show a similar animal toxicity profile as noted above. Several thiazolidindione insulin sensitizers have advanced into human studies. Troglitazone (CS-045), rosiglitazone (BRL-49653) and pioglitazone have been marketed.

Fibrates

The study of PPAR pharmacology began some 25 years before the identification and isolation of the PPARs with the study of the anti-hypertriglyceridemic and anti-hypercholesteremic activity of several families of amphipathic carboxylic acids

(Fig. 10.2). Initial studies by Thorp and Waring (1962) with fat-fed rodents led to the development of the archetypal PPAR agonist, clofibrate. Further work on clofibrate gave the peroxisome proliferators their name (Hess *et al.*, 1965). The fibrate nafenopin was used for the identification of a 'peroxisome proliferator-binding protein' and its further characterization as a member of the nuclear receptor superfamily (Lalwani *et al.*, 1987). While this protein became known as mouse PPARα, nafenopin was later also reported to bind to mouse PPARγ with a K_d of 128 μmol L^{-1} (Palmer and Wolf, 1998).

Large numbers of fibrates have been studied in animal models; a subset of these peroxisome proliferators has also been studied for the ability to activate PPARs directly. A further subset has been assayed for binding affinity to the PPARs. The extensive early literature on this family of fibrates is reviewed through 1977 (Witiak *et al.*, 1977) and 1987 (Lalwani *et al.*, 1987). As the prototypical peroxisome proliferators, clofibrate and several congeners have been studied in essentially all of the PPAR activation and binding models. Fibrates are found to be weak agonists and low-affinity ligands. Clofibrate and ciprofibrate showed maximal activation of mouse PPARα at 300 μmol L^{-1} and showed weak agonist activity on mouse PPARγ with no agonist activity on mouse PPARδ (Forman *et al.*, 1997). The same paper demonstrated ligand binding-induced gel mobility shift and agonist activity on mouse PPARα for several, even simpler, alkoxy or (alkylthio) acetic acids.

Several studies have defined the minimal structure requirements for hypolipidemic efficacy of clofibrate relatives (Witiak *et al.*, 1977). The (*S*) monomethyl analog was found to be equipotent with clofibrate in oral efficacy studies (Witiak *et al.*, 1968; Schoonjans *et al.*, 1996b), while an *in vivo* study of FACO induction found that an (*S*) α-Et residue was a minimum requirement (Esbenshade, 1990). A recent SAR study with the clofibrate nucleus demonstrated both receptor activation for mouse PPARα and

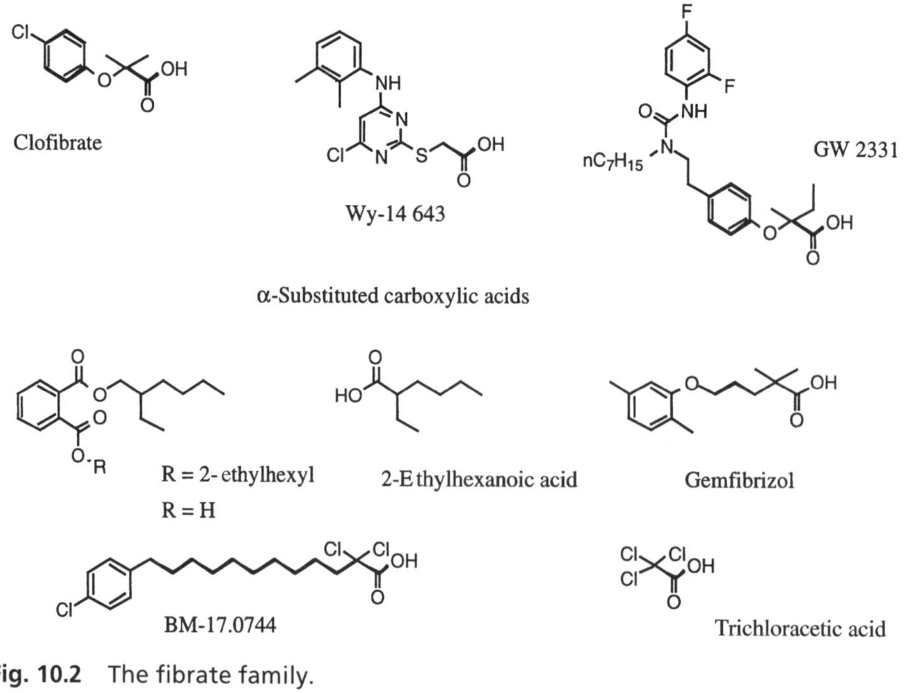

Clofibrate

Wy-14 643

GW 2331

α-Substituted carboxylic acids

R = 2-ethylhexyl
R = H

2-Ethylhexanoic acid

Gemfibrizol

BM-17.0744

Trichloracetic acid

Fig. 10.2 The fibrate family.

induction of peroxisomal β-oxidation in a rat hepatoma cell line for a set of α-alkyl and α-aryl(p-Cl-phenoxy)acetic acids. The active enantiomer was confirmed to be the (S) enantiomer in several cases, with the most potent compound reported to be (S)2-propyl-2-(p-Cl-phenoxy)acetic acid with an EC_{50} of 23 μmol L^{-1} (Rangwala *et al.*, 1997). The structurally simplified clofibrate relative (2,4-dichlorophenoxy)acetic acid is the oft-cited herbicide peroxisome proliferator 2,4-D.

The early hypolipidemic analogs derived from clofibrate tended to elaboration of the α aryl substituent. The more potent examples include the 2-arylthioacetic acid analog Wy-14 643, which has become a benchmark for studies of the activation and ligand binding in the PPARα family. In the same study as clofibrate, WY-14 643 was demonstrated to be a potent agonist and high-affinity ligand for mouse PPARα (estimated K_d 0.6 μmol L^{-1}) as well as a weak agonist on the mouse PPARγ receptor (Forman *et al.*, 1997). A more recently marketed hypolipidemic fibrate, bezafibrate (Monk and Todd, 1987) is reported to show a K_d of 0.045 μmol L^{-1} in the same assay. The more complex structure of GW-2331 is a recently developed member of the fibrate family (Kliewer *et al.*, 1997). This fibrate relative has been shown to bind to both *Xenopus* PPARα and *Xenopus* PPARγ receptors with high affinity (αK_d 0.14 μmol L^{-1}, γK_d 0.30 μmol L^{-1}). Binding was also detected with murine and human PPARα and murine PPARγ. Cell transactivation EC_{50} values are comparable across the species, and range from 0.01 to 0.36 μmol L^{-1} on mouse PPARα (Ellinghaus *et al.*, 1999).

α-Substituted carboxylic acids

The second large family of compounds with broad peroxisome proliferator activity is related to the oft-cited plasticizer, 2-ethylhexylphthalate. This member of the class has been examined in the full range of functional and binding assays for peroxisome proliferation (Lundgren *et al.*, 1987). The compound responsible for activity *in vivo* is believed to be 2-ethylhexanoic acid (EHA). Several esters, or precursors of 2-EHA, were shown to be active peroxisome proliferators *in vivo*. The general structure of an α-substituted carboxylic acid encompasses a considerable number of PPARα and γ activators. Compounds range from EHA, to the hypolipidemic gemfibrozil (Lopid) to the recently reported insulin sensitizer, BM-17.0744. Probably the smallest reported PPAR agonist, trichloroacetic acid belongs to this series (Zanelli *et al.*, 1996).

The hypoglycemic and hypolipidemic compound α,α-dichloroacid BM 17.0744 demonstrates a classic PPARα agonist profile with an increase in carnitine acetyltransferase activity and hepatomegaly in rodents. It is interesting that BM-17.0744 shows potent hypoglycemic and hypolipidemic effects at 0.3–10 mg kg^{-1} day^{-1} in *db/db*, *ob/ob* and yellow KK mice, without the classic toxicity profile of the thiazolidenediones. Body and heart weight gains are not seen after 4 weeks' administration (Pill and Kuhnle, 1999).

Also included in this structural class are the NSAIDs, some of which have been found to activate human PPARγ and PPARα in transfection assays. Indomethacin, fenoprofen, flufenamic acid and ibuprofen showed activation equal to the PPARγ reference compound BRL-49653, or the PPARα reference compound WY-14 643, in the range of 10–100 μmol L^{-1}. The most potent, indomethacin, showed an IC_{50} of 10 μmol L^{-1} against [^{3}H]BRL-49653 (Lehmann *et al.*, 1997).

CONCLUSION

In this chapter we have reviewed the literature on the PPAR subfamily of nuclear receptors. Research with this family began with the discovery and characterization of PPARα and its obvious role in peroxisome proliferation. The identification of PPARγ and PPARδ, and natural and/or synthetic ligands for these receptors, has permitted researchers to explore further the physiologic role of these receptors. Indeed, it appears that all three receptors are capable of binding and being activated by fatty acids. This common feature suggests that the PPARs might act as intracellular sensors monitoring various metabolic pathways. It is clear that the PPARs are important regulators of lipid and glucose metabolism. Thus, they have become drug development targets for the treatment of diabetes, obesity and cardiovascular disease. Further research on PPARs is bound to bring a better understanding of the morbidity associated with these diseases and, hopefully, better therapeutic agents for their treatment.

REFERENCES

Adamo, A.M., Aloise, P.A. and Pasquini, J.M. (1986) A possible relationship between concentration of microperoxisomes and myelination. *Int. J. Dev. Neurosci.* **4**: 513–517.

Aitman, T.J., Glazier, A.M., Wallace, C.A. *et al.* (1999) Identification of Cd36 (Fat) as an insulin-resistance gene causing defective fatty acid and glucose metabolism in hypertensive rats. *Nat. Genet.* **21**: 76–83.

Allan, G.F., Leng, X., Tsai. S.Y. *et al.* (1992) Hormone and antihormone induce distinct conformational changes which are central to steroid receptor activation. *J. Biol. Chem.* **267**: 19 513–19 520.

Allenby, G., Bocquel, M.T., Saunders, M. *et al.* (1993) Retinoic acid receptors and retinoid X receptors: interactions with endogenous retinoic acids. *Proc. Natl. Acad. Sci. USA* **90**: 30–34.

Aperlo, C., Pognonec, P., Saladin, R. *et al.* (1995) cDNA cloning and characterization of the transcriptional activities of the hamster peroxisome proliferator-activated receptor haPPAR γ. *Gene* **162**: 297–302.

Auboeuf, D., Rieusset, J., Fajas, L. *et al.* (1997) Tissue distribution and quantification of the expression of mRNAs of peroxisome proliferator-activated receptors and liver X receptor α in humans. *Diabetes* **46**: 1319–1327.

Beamer, B.A., Negri, C., Yen, C.J. *et al.* (1997) Chromosomal localization and partial genomic structure of the human peroxisome proliferator activated receptor-γ (hPPAR γ) gene. *Biochem. Biophys. Res. Commun.* **233**: 756–759.

Beamer, B. A., Yen, C.-J., Anderson, R.E. *et al.* (1998) Association of the Pro12Ala variant in the peroxisome proliferator-activated receptor γ2 gene with obesity in two Caucasian populations. *Diabetes* **47**: 1806–1808.

Beato, M., Herrlich, P. and Schutz, G. (1995) Steroid hormone receptors: many actors in search of a plot. *Cell* **83**: 851–857.

Berger, J., Leibowitz, M.D., Doebber, T.W. *et al.* (1999) Novel peroxisome proliferator-activated receptor (PPAR) γ and PPARδ ligands produce distinct biological effects. *J. Biol. Chem.* **274**: 6718–6725.

Biardi, L. and Krisans, S.K. (1996) Compartmentalization of cholesterol biosynthesis. Conversion of mevalonate to farnesyl diphosphate occurs in the peroxisomes. *J. Biol. Chem.* **271**: 1784–1788.

Bollen, M., Keppens, S. and Stalmans, W. (1998) Specific features of glycogen metabolism in the liver. *Biochem. J.* **336**: 19–31.

Bosch, H.v.d., Schutgens, R.B.H., Wanders, R.J.A. *et al.* (1992) Biochemistry of peroxisomes. *Annu. Rev. Biochem.* **61**: 157–197.

Braissant, O., Foufelle, F., Scotto, C. *et al.* (1996) Differential expression of peroxisome proliferator-activated receptors (PPARs): tissue distribution of PPAR-α, -β, and -γ in the adult rat. *Endocrinology* **137**: 354–366.

Brandt, J.M., Djouadi, F. and Kelly, D.P. (1998) Fatty acids activate transcription of the muscle carnitine palmitoyltransferase I gene in cardiac myocytes via the peroxisome proliferator-activated receptor α. *J. Biol. Chem.* **273**: 23786–23792.

Brockman, J.A., Gupta, R.A. and DuBois, R.N. (1998) Activation of PPARγ leads to inhibition of anchorage-independent growth of human colorectal cells. *Gastroenterology* **115**: 1049–1055.

Brown, F.R.D., McAdams, A.J., Cummins, J.W. *et al.* (1982) Cerebro-hepato-renal (Zellweger) syndrome and neonatal adrenoleukodystrophy: similarities in phenotype and accumulation of very long chain fatty acids. *Johns Hopkins Med. J.* **151**: 344–351.

Buckle, D.R., Cantello, B.C.C., Cawthorne, M.A. *et al.* (1996) Non thiazolidinedione antihyperglycemic agents. 1: α-Heteroatom substituted β-phenylpropionic acids. *Biorg. Med. Chem. Letters* **6**(17): 2121–2126.

Chen, C. (1998) Troglitazone: an antidiabetic agent. *Am. J. Health Syst. Pharm.* **55**: 905–925.

Collins, F.S., Patrinos, A., Jordan, E. *et al.* (1998) New goals for the US Human Genome Project: 1998–2003. *Science* **282**: 682–689.

Combettes-Souverain, M. and Issad, T. (1998) Molecular basis of insulin action. *Diabetes Metab.* **24**: 477–489.

De Duve, C. and Baudhuin, P. (1966) Peroxisomes (microbodies and related particles). *Physiol. Rev.* **46**: 323–357.

Deeb, S.S., Fajas, L., Nemoto, M. *et al.* (1998) A Pro12Ala substitution in PPARγ2 associated with decreased receptor activity, lower body mass index and improved insulin sensitivity. *Nat. Genet.* **20**: 284–287.

Demoz, A., Vaagenes, H., Aarsaether, N. *et al.* (1994) Coordinate induction of hepatic fatty acyl-CoA oxidase and P4504A1 in rat after activation of the peroxisome proliferator-activated receptor (PPAR) by sulphur-substituted fatty acid analogues. *Xenobiotica* **24**: 943–956.

Devchand, P.R., Keller, H., Peters, J.M. *et al.* (1996) The PPARα-leukotriene B4 pathway to inflammation control. *Nature* **384**: 39–43.

Djouadi, F., Weinheimer, C.J., Saffitz, J.E. *et al.* (1998) A gender-related defect in lipid metabolism and glucose homeostasis in peroxisome proliferator-activated receptor α-deficient mice. *J. Clin. Invest.* **102**: 1083–1091.

Dowell, P., Peterson, V.J., Zabriskie, T.M. *et al.* (1997) Ligand-induced peroxisome proliferator-activated receptor α conformational change. *J. Biol. Chem.* **272**: 2013–2020.

Dreyer, C., Krey, G., Keller, H. *et al.* (1992) Control of the peroxisomal β-oxidation pathway by a novel family of nuclear hormone receptors. *Cell* **68**: 879–887.

Elbrecht, A., Chen, Y., Adams, A. *et al.* (1996) Molecular cloning, expression and characterization of human peroxisome proliferator activated receptors γ1 and γ2. *Biochem. Biophys. Res. Commun.* **224**: 431–437.

Elbrecht, A., Chen, Y., Cullinan, C.A. *et al.* (1999) L-764406 is a partial agonist of human peroxisome proliferator-activated receptor γ. The role of Cys313 in ligand binding. *J. Biol. Chem.* **274**: 7913–7922.

Ellinghaus, P., Wolfrum, C., Assmann, G. *et al.* (1999) Phytanic acid activates the peroxisome proliferator-activated receptor α (PPARα) in sterol carrier protein 2-/sterol carrier protein x-deficient mice. *J. Biol. Chem.* **274**: 2766–2772.

Forman, B.M., Chen, J. and Evans, R.M. (1997) Hypolipidemic drugs, polyunsaturated fatty acids, and eicosanoids are ligands for peroxisome proliferator-activated receptors α and δ. *Proc. Natl. Acad. Sci. USA* **94**: 4312–4317.

Forman, B.M., Tontonoz, P., Chen, J. *et al.* (1995) 15-Deoxy-Δ12, 14-prostaglandin J$_2$ is a ligand for the adipocyte determination factor PPAR γ. *Cell* **83**: 803–812.

Foster, J.D., Pederson, B.A. and Nordlie, R.C. (1997) Glucose-6-phosphatase structure, regulation, and function: an update. *Proc. Soc. Exp. Biol. Med.* **215**: 314–332.

Frohnert, B.I., Hui, T.Y. and Bernlohr, D.A. (1999) Identification of a functional peroxisome proliferator-responsive element in the murine fatty acid transport protein gene. *J. Biol. Chem.* **274**: 3970–3977.

Glorian, M., Franckhauser-Vogel, S., Robin, D. *et al.* (1998) Glucocorticoids repress induction by thiazolidinediones, fibrates, and fatty acids of phosphoenolpyruvate carboxykinase gene expression in adipocytes. *J. Cell. Biochem.* **68**: 298–308.

Gorla-Bajszczak, A., Juge-Aubry, C., Pernin, A. *et al.* (1999) Conserved amino acids in the ligand-binding and tau(i) domains of the peroxisome proliferator-activated receptor α are necessary for heterodimerization with RXR. *Mol. Cell. Endocrinol.* **147**: 37–47.

Gottlicher, M., Widmark, E., Li, Q. *et al.* (1992) Fatty acids activate a chimera of the clofibric acid-activated receptor and the glucocorticoid receptor. *Proc. Natl. Acad. Sci. USA* **89**: 4653–4657.

Green, S. (1995) PPAR: a mediator of peroxisome proliferator action. *Mutat. Res.* **333**: 101–109.

Green, S. and Wahli, W. (1994) Peroxisome proliferator-activated receptors: finding the orphan a home. *Mol. Cell. Endocrinol.* **100**: 149–153.

Greene, M.E., Blumberg, B., McBride, O.W. *et al.* (1995) Isolation of the human peroxisome proliferator activated receptor γ cDNA: expression in hematopoietic cells and chromosomal mapping. *Gene Expr.* **4**: 281–299.

Guan, Y., Zhang, Y., Davis, L. *et al.* (1997) Expression of peroxisome proliferator-activated receptors in urinary tract of rabbits and humans. *Am. J. Physiol.* **273**: F1013–1022.

Heery, D.M., Kalkhoven, E., Hoare, S. *et al.* (1997) A signature motif in transcriptional co-activators mediates binding to nuclear receptors. *Nature* **387**: 733–736.

Hertz, R., Berman, I., Keppler, D. *et al.* (1996) Activation of gene transcription by prostacyclin analogues is mediated by the peroxisome-proliferator-activated receptor (PPAR). *Eur. J. Biochem.* **235**: 242–247.

Hess, R., Staubli, W. and Riess, W. (1965) Nature of the hepatomegalic effect produced by ethyl-chlorophenoxy-isobutyrate in the rat. *Nature* **208**: 856–858.

Hu, E., Kim, J.B., Sarraf, P. *et al.* (1996) Inhibition of adipogenesis through MAP kinase-mediated phosphorylation of PPARγ. *Science* **274**: 2100–2103.

Hulin, B., Newton, L.S., Lewis, D.M. *et al.* (1996) Hypoglycemic activity of a series of α-alkylthio and α-alkoxy carboxylic acids related to ciglitazone. *J. Med. Chem.* **39**: 3897–3907.

IJpeuberg, A., Jeannin, E. *et al.* (1997) Polarity and specific sequence requirements of peroxisome proliferator-activated receptor (PPAR)/retinoid X receptor heterodimer binding to DNA. A functional analysis of the malic enzyme gene PPAR response element. *J. Biol. Chem.* **272**: 20 108–20 117.

Issemann, I. and Green, S. (1990) Activation of a member of the steroid hormone receptor superfamily by peroxisome proliferators. *Nature* **347**: 645–650.

Jiang, C., Ting, A.T. and Seed, B. (1998) PPAR-γ agonists inhibit production of monocyte inflammatory cytokines. *Nature* 391: 82–86.

Jones, B.H., Standridge, M.K., Claycombe, K.J. *et al.* (1998) Glucose induces expression of stearoyl-CoA desaturase in 3T3-L1 adipocytes. *Biochem. J.* 335: 405–408.

Jones, P.S., Savory, R., Barratt, P. *et al.* (1995) Chromosomal localisation, inducibility, tissue-specific expression and strain differences in three murine peroxisome-proliferator-activated-receptor genes. *Eur. J. Biochem.* 233: 219–226.

Juge-Aubry, C., Pernin, A., Favez, T. *et al.* (1997) DNA binding properties of peroxisome proliferator-activated receptor subtypes on various natural peroxisome proliferator response elements. Importance of the 5'-flanking region. *J. Biol. Chem.* 272: 25 252–25 259.

Juge-Aubry, C.E., Hammar, E., Siegrist-Kaiser, C. *et al.* (1999) Regulation of the transcriptional activity of the peroxisome proliferator-activated receptor α by phosphorylation of a ligand-independent *trans*-activating domain. *J. Biol. Chem.* 274: 10 505–10 510.

Kahn, A. (1997) Transcriptional regulation by glucose in the liver. *Biochimie* 79: 113–118.

Kaikaus, R.M., Chan, W.K., Lysenko, N. *et al.* (1993) Induction of peroxisomal fatty acid β-oxidation and liver fatty acid-binding protein by peroxisome proliferators. Mediation via the cytochrome P-450IVA1 omega-hydroxylase pathway. *J. Biol. Chem.* 268: 9593–9603.

Kawamatsu, Y., Saraie, T., Imamiya, E. *et al.* (1980) Studies on antihyperlipidemic agents I. Synthesis and hypolipidemic activities of phenoxyphenyl alkanoic acid derivatives. *Arzneim.-Forsch/Drug Res.* 30: 454–459.

Kelly, L.J., Vicario, P.P., Thompson, G.M. *et al.* (1998) Peroxisome proliferator-activated receptors γ and α mediate *in vivo* regulation of uncoupling protein (UCP-1, UCP-2, UCP-3) gene expression. *Endocrinology* 139: 4920–4927.

Kliewer, S.A., Lenhard, J.M., Willson, T.M. *et al.* (1995) A prostaglandin J_2 metabolite binds peroxisome proliferator-activated receptor γ and promotes adipocyte differentiation. *Cell* 83: 813–819.

Kliewer, S.A., Sundseth, S.S., Jones, S.A. *et al.* (1997) Fatty acids and eicosanoids regulate gene expression through direct interactions with peroxisome proliferator-activated receptors α and γ. *Proc. Natl. Acad. Sci. USA* 94: 4318–4323.

Krey, G., Braissant, O., L'Horset, F. *et al.* (1997) Fatty acids, eicosanoids, and hypolipidemic agents identified as ligands of peroxisome proliferator-activated receptors by coactivator-dependent receptor ligand assay. *Mol. Endocrinol.* 11: 779–791.

Krisans, S.K. (1996) Cell compartmentalization of cholesterol biosynthesis. *Ann. N.Y. Acad. Sci.* 804: 142–164.

Kruszynska, Y.T., Mukherjee, R., Jow, L. *et al.* (1998) Skeletal muscle peroxisome proliferator-activated receptor-γ expression in obesity and non-insulin-dependent diabetes mellitus. *J. Clin. Invest.* 101: 543–548.

Lalwani, N.D., Alvares, K., Reddy, M.K. *et al.* (1987) Peroxisome proliferator-binding protein: identification and partial characterization of nafenopin-, clofibric acid-, and ciprofibrate-binding proteins from rat liver. *Proc. Natl. Acad. Sci. USA* 84: 5242–5246.

Laudet, V., Hanni, C., Coll, J. *et al.* (1992) Evolution of the nuclear receptor gene superfamily. *EMBO J.* 11: 1003–1013.

Lefebvre, A.-M., Chen, I., Desreumaux, P. *et al.* (1998) Activation of peroxisome proliferator activated receptor-γ promotes the development of colon tumors in C57BL/6J-APCMin/ + mice. *Nat. Med.* 4: 1053–1057.

Lehmann, J.M., Moore, L.B., Smith-Oliver, B.B. *et al.* (1995) An antidiabetic thiazolidinedione is a high affinity ligand for peroxisome proliferator-activated receptor γ (PPAR γ). *J. Biol. Chem.* 270: 12 953–12 956.

Lehmann, J.M., Lenhard, J.M., Oliver, B.B. *et al.* (1997) Peroxisome proliferator-activated receptors α and γ are activated by indomethacin and other non-steroidal anti-inflammatory drugs. *J. Biol. Chem.* **272**: 3406–3410.

Le Marchand-Brustel, Y. (1999) Molecular mechanisms of insulin action in normal and insulin-resistant states. *Exp. Clin. Endocrinol. Diabetes* **107**: 126–132.

Lin, Q., Ruuska, S.E., Shaw, N.S. *et al.* (1999) Ligand selectivity of the peroxisome proliferator-activated receptor α. *Biochemistry* **38**: 185–190.

Luers, G., Beier, K., Hashimoto, T. *et al.* (1990) Biogenesis of peroxisomes: sequential biosynthesis of the membrane and matrix proteins in the course of hepatic regeneration. *Eur. J. Cell. Biol.* **52**: 175–184.

Lundgren, B., Meijer, J. and DePierre, J.W. (1987) Examination of the structural requirements for proliferation of peroxisomes and mitochondria in mouse liver by hypolipidemic agents, with special emphasis on structural analogues of 2-ethylhexanoic acid. *Eur. J. Biochem.* **163**: 423–431.

Mangelsdorf, D.J. and Evans, R.M. (1995) The RXR heterodimers and orphan receptors. *Cell* **83**: 841–850.

Mangelsdorf, D.J., Ong, E.S., Dyck, J.A. *et al.* (1990) Nuclear receptor that identifies a novel retinoic acid response pathway. *Nature* **345**: 224–229.

Mangelsdorf, D.J., Borgmeyer, U., Heyman, R.A. *et al.* (1992) Characterization of three RXR genes that mediate the action of 9-*cis* retinoic acid. *Genes Dev.* **6**: 329–344.

Mangelsdorf, D.J., Thummel, C., Beato, M. *et al.* (1995) The nuclear receptor superfamily: the second decade. *Cell* **83**: 835–839.

Martin, G., Schoonjans, K., Lefebvre, A.M. *et al.* (1997) Coordinate regulation of the expression of the fatty acid transport protein and acyl-CoA synthetase genes by PPARα and PPARγ activators. *J. Biol. Chem.* **272**: 28 210–28 217.

Masters, C.J. (1998) On the role of the peroxisome in the metabolism of drugs and xenobiotics. *Biochem. Pharmacol.* **56**: 667–673.

Masters, C. and Crane, D. (1998) On the role of the peroxisome in cell differentiation and carcinogenesis. *Mol. Cell. Biochem.* **187**: 85–97.

McDonnell, D.P., Clemm, D.L., Hermann, T. *et al.* (1995) Analysis of estrogen receptor function *in vitro* reveals three distinct classes of antiestrogens. *Mol. Endocrinol.* **9**: 659–669.

Meirhaeghe, A., Fajas, L., Helbecque, N. *et al.* (1998) A genetic polymorphism of the peroxisome proliferator-activated receptor γ gene influences plasma leptin levels in obese humans. *Hum. Mol. Genet.* **7**: 435–440.

Miyata, K.S., McCaw, S.E., Marcus, S.L. *et al.* (1994) The peroxisome proliferator-activated receptor interacts with the retinoid X receptor *in vivo*. *Gene* **148**: 327–330.

Moller, D.E. and O'Rahilly, S. (1993) Congenital syndromes of severe insulin resistance. In: D.E. Moller (ed.) *Insulin Resistance*, pp. 49–81. John Wiley, Chichester, UK.

Monk, J.P. and Todd, P.A. (1987) Bezafibrate. A review of its pharmacodynamic and pharmacokinetic properties, and therapeutic use in hyperlipidaemia. *Drugs* **33**: 539–576.

Mori, Y., Kim-Motoyama, H., Katakura, T. *et al.* (1998) Effect of the Pro12Ala variant of the human peroxisome proliferator-activated receptor γ2 gene on adiposity, fat distribution, and insulin sensitivity in Japanese men. *Biochem. Biophys. Res. Commun.* **251**: 195–198.

Mueller, E., Sarraf, P., Tontonoz, P. *et al.* (1998) Terminal differentiation of human breast cancer through PPAR γ. *Mol. Cell* **1**: 465–470.

Mukherjee, R., Jow, L., Noonan, D. *et al.* (1994) Human and rat peroxisome proliferator activated receptors (PPARs) demonstrate similar tissue distribution but different responsiveness to PPAR activators. *J. Steroid Biochem. Molec. Biol.* **51**: 157–166.

Mukherjee, R., Davies, P.J., Crombie, D.L. *et al.* (1997) Sensitization of diabetic and obese mice to insulin by retinoid X receptor agonists. *Nature* **386**: 407–410.

Murakami, K., Tobe, K., Ide, T. *et al.* (1998) A novel insulin sensitizer acts as a coligand for peroxisome proliferator-activated receptor-α (PPAR-α) and PPAR-γ: effect of PPAR-α activation on abnormal lipid metabolism in liver of Zucker fatty rats. *Diabetes* **47**: 1841–1847.

Nakanishi, M., Kobayakawa, T., Okada, T. *et al.* (1970) Studies on anti-atherosclerotic agents. I. Studies on physiological activity of aryloxyisobutyrate. *Yakugaku Zasshi* **90**: 921–925.

Nakshatri, H. and Chambon, P. (1994) The directly repeated RG(G/T)TCA motifs of the rat and mouse cellular retinol-binding protein II genes are promiscuous binding sites for RAR, RXR, HNF-4, and ARP-1 homo- and heterodimers. *J. Biol. Chem.* **269**: 890–902.

Nolte, R.T., Wisely, G.B., Westin, S. *et al.* (1998) Ligand binding and co-activator assembly of the peroxisome proliferator-activated receptor-γ. *Nature* **395**: 137–143.

O'Malley, B.W. and Schrader, W.T. (1976) The receptors of steroid hormones. *Sci. Am.* **234**: 32–43.

Palmer, C.N. and Wolf, C.R. (1998) *cis*-Parinaric acid is a ligand for the human peroxisome proliferator activated receptor γ: development of a novel spectrophotometric assay for the discovery of PPARγ ligands. *FEBS Lett.* **431**: 476–480.

Palmer, C.N., Hsu, M.H., Griffin, K.J. *et al.* (1998) Peroxisome proliferator activated receptor-α expression in human liver. *Mol. Pharmacol.* **53**: 14–22.

Park, K.S., Ciaraldi, T.P., Abrams-Carter, L. *et al.* (1997) PPAR-γ gene expression is elevated in skeletal muscle of obese and type II diabetic subjects. *Diabetes* **46**: 1230–1234.

Perichon, R., Bourre, J.M., Kelly, J.F. *et al.* (1998) The role of peroxisomes in aging. *Cell. Mol. Life Sci.* **54**: 641–652.

Pill, J. and Kuhnle, H.F. (1999) BM 17.0744: a structurally new antidiabetic compound with insulin-sensitizing and lipid-lowering activity. *Metabolism* **48**: 34–40.

Puigserver, P., Wu, Z., Park, C.W. *et al.* (1998) A cold-inducible coactivator of nuclear receptors linked to adaptive thermogenesis. *Cell* **92**: 829–839.

Qi, C., Zhu, Y., Pan, J. *et al.* (1999) Mouse steroid receptor coactivator-1 is not essential for peroxisome proliferator-activated receptor α-regulated gene expression. *Proc. Natl. Acad. Sci. USA* **96**: 1585–1590.

Qiao, L.Y., Goldberg, J.L., Russell, J.C. *et al.* (1999) Identification of enhanced serine kinase activity in insulin resistance. *J. Biol. Chem.* **274**: 10 625–10 632.

Rangwala, S.M., O'Brien, M.L., Tortorella, V. *et al.* (1997) Stereoselective effects of chiral clofibric acid analogs on rat peroxisome proliferator-activated receptor α (rPPARα) activation and peroxisomal fatty acid β-oxidation. *Chirality* **9**: 37–47.

Reddy, J.K. (1990) Carcinogenicity of peroxisome proliferators: evaluation and mechanisms. *Biochem. Soc. Trans.* **18**: 92–94.

Reddy, J.K. and Lalwai, N.D. (1983) Carcinogenesis by hepatic peroxisome proliferators: evaluation of the risk of hypolipidemic drugs and industrial plasticizers to humans. *Crit. Rev. Toxicol.* **12**: 1–58.

Ribon, V., Johnson, J.H., Camp, H.S. *et al.* (1998) Thiazolidinediones and insulin resistance: peroxisome proliferator activated receptor γ activation stimulates expression of the CAP gene. *Proc. Natl. Acad. Sci. USA* **95**: 14 751–14 756.

Ricote, M., Li, A.C., Willson, T.M. *et al.* (1998) The peroxisome proliferator-activated receptor-γ is a negative regulator of macrophage activation. *Nature* **391**: 79–82.

Ringel, J., Engeli, S., Distler, A. *et al.* (1999) Pro12Ala missense mutation of the peroxisome proliferator activated receptor γ and diabetes mellitus. *Biochem. Biophys. Res. Commun.* **254**: 450–453.

Ristow, M., Muller-Wieland, D., Pfeiffer A. *et al.* (1998) Obesity associated with a mutation in a genetic regulator of adipocyte differentiation. *N. Engl. J. Med.* **339**: 953–959.

Robinson, C.E., Wu, X., Nawaz, Z. *et al.* (1999) A corepressor and chicken ovalbumin upstream promoter transcriptional factor proteins modulate peroxisome proliferator-activated receptor-γ2/retinoid X receptor α-activated transcription from the murine lipoprotein lipase promoter. *Endocrinology* **140**: 1586–1593.

Sarraf, P., Mueller, E., Jones, D. *et al.* (1998) Differentiation and reversal of malignant changes in colon cancer through PPARγ. *Nat. Med.* **4**: 1046–1052.

Schmidt, A., Endo, N., Rutledge, S.J. *et al.* (1992) Identification of a new member of the steroid hormone receptor superfamily that is activated by a peroxisome proliferator and fatty acids. *Mol. Endocrinol.* **6**: 1634–1641.

Schoonjans, K., Peinado-Onsurbe, J., Lefebvre, A.M. *et al.* (1996a) PPARα and PPARγ activators direct a distinct tissue-specific transcriptional response via a PPRE in the lipoprotein lipase gene. *EMBO J.* **15**: 5336–5348.

Schoonjans, K., Staels, B. and Auwerx, J. (1996b) Role of the peroxisome proliferator-activated receptor (PPAR) in mediating the effects of fibrates and fatty acids on gene expression. *J. Lipid Res.* **37**: 907–925.

Sher, T., Yi, H.-F., McBridge, O.W. *et al.* (1993) cDNA cloning, chromosomal mapping, and functional characterization of the human peroxisome proliferator activated receptor. *Biochemistry* **32**: 5598–5604.

Sohda, T., Mizuno, K., Tawada, H. *et al.* (1982) Studies on antidiabetic agents. I. Synthesis of 5-[4-(2-methyl-2-phenylpropoxy)-benzyl]thiazolidine-2,4-dione (AL-321) and related compounds. *Chem. Pharm. Bull.* **30**(10): 3563–3573.

Souza, S.C., Yamamoto, M.T., Frcanciosa, M.D. *et al.* (1998) BRL 49653 blocks the lipolytic actions of tumor necrosis factor-α: a potential new insulin-sensitizing mechanism for thiazolidinediones. *Diabetes* **47**: 691–695.

Spiegelman, B.M. (1998) PPAR-γ: adipogenic regulator and thiazolidinedione receptor. *Diabetes* **47**: 507–514.

Staels, B., Dallongeville, J., Auwerx, J. *et al.* (1998) Mechanism of action of fibrates on lipid and lipoprotein metabolism. *Circulation* **98**: 2088–2093.

Su, J.-L., Simmons, C.J., Wisely, B. *et al.* (1998) Monitoring of PPARα protein expression in human tissue by the use of PPARα-specific MAbs. *Hybridoma* **17**: 47–53.

Thorp, J.M. and Waring, W.S. (1962) Modification of metabolism and distribution of lipids by ethylchlorophenoxyisobutyrate. *Nature* **194**: 948–949.

Tontonoz, P., Graves, R.A., Budavari, A.I. *et al.* (1994a) Adipocyte-specific transcription factor ARF6 is a heterodimeric complex of two nuclear hormone receptors, PPAR γ and RXR α. *Nucleic Acids Res.* **22**: 5628–5634.

Tontonoz, P., Hu, E. and Spiegelman, B.M. (1994b) Stimulation of adipogenesis in fibroblasts by PPAR γ2, a lipid-activated transcription factor [published erratum appears in *Cell* 1995; **80**: following 957]. *Cell* **79**: 1147–1156.

Torchia, J., Rose, D.W., Inostroza, J. *et al.* (1997) The transcriptional co-activator p/CIP binds CBP and mediates nuclear-receptor function. *Nature* **387**: 677–684.

Torchia, J., Glass, C. and Rosenfeld, M.G. (1998) Co-activators and co-repressors in the integration of transcriptional responses. *Curr. Opin. Cell Biol.* **10**: 373–383.

Tugwood, J.D., Aldridge, T.C., Lambe, K.G. *et al.* (1997) Peroxisome proliferator activated receptors: structures and function. *Ann. N. Y. Acad. Sci.* **804**: 252–264.

Uppenberg, J., Svensson, C., Jaki, M. *et al.* (1998) Crystal structure of the ligand binding domain of the human nuclear receptor PPARγ. *J. Biol. Chem.* **273**: 31 108–31 112.

Versmold, H.T., Bremer, H.J., Herzog, V. *et al.* (1977) A metabolic disorder similar to Zellweger syndrome with hepatic acatalasia and absence of peroxisomes, altered content and redox state of cytochromes, and infantile cirrhosis with hemosiderosis. *Eur. J. Pediatr.* **124**: 261–275.

Vidal-Puig, A., Jimenez-Linan, M., Lowell, B.B. *et al.* (1996) Regulation of PPAR gamma gene expression by nutrition and obesity in rodents. *J. Clin. Invest.* **97**: 2553–2561.

Vidal-Puig, A.J., Considine, R.V., Jimenez-Linan, M. *et al.* (1997) Peroxisome proliferator-activated receptor gene expression in human tissues: effects of obesity, weight loss, and regulation by insulin and glucocorticoids. *J. Clin. Invest.* **99**: 2416–2422.

Vigouroux, C., Khallouf, E., Bourut, C. *et al.* (1997) Genetic exclusion of 14 candiate genes in lipoatrophic diabetes using linkage analysis and 10 consanguinous families. *J. Clin. Endocrinol. Metab.* **82**: 3438–3444.

Wanders, R.J. and Tager, J.M. (1998) Lipid metabolism in peroxisomes in relation to human disease. *Mol. Aspects Med.* **19**: 69–154.

Williams, C.S., Smalley, W. and DuBois, R.N. (1997) Aspirin use and potential mechanisms for colorectal cancer prevention. *J. Clin. Invest.* **100**: 1325–1329.

Willson, T.M., Cobb, J.E., Cowan, D.J. *et al.* (1996) The structure–activity relationship between peroxisome proliferator-activated receptor γ agonism and the antihyperglycemic activity of thiazolidinediones. *J. Med. Chem.* **39**: 665–668.

Witiak, D.T., Ho, T.C., Hackney, R.E. *et al.* (1968) Hypocholesterolemic agents. Compounds related to ethyl α-(4- chlorophenoxy)-α-methylpropionate. *J. Med. Chem.* **11**: 1086–1089.

Witiak, D.T., H.A.I., N. and Feller, D.R. (1977) *Medicinal Research*. Marcel Dekker, New York.

Wolf, G. (1998) Fatty acids bind directly to and activate peroxisome proliferator-activated receptors α and γ. *Nutr. Rev.* **56**: 61–63.

Xin, X., Yang, S., Kowalski, J. *et al.* (1999) Peroxisome proliferator-activated receptor γ ligands are potent inhibitors of angiogenesis *in vitro* and *in vivo*. *J. Biol. Chem.* **274**: 9116–9121.

Xu, H.E., Lambert, M.H., Montana, V.G. *et al.* (1999) Molecular recognition of fatty acids by peroxisome proliferator-activated receptors. *Mol. Cell* **3**: 397–403.

Xu, J., Qiu, Y., DeMayo, F.J. *et al.* (1998) Partial hormone resistance in mice with disruption of the steroid receptor coactivator-1 (SRC-1) gene. *Science* **279**: 1922–1925.

Yamamoto, K. and Fahimi, H.D. (1987) Biogenesis of peroxisomes in regenerating rat liver. I. Sequential changes of catalase and urate oxidase detected by ultrastructural cytochemistry. *Eur. J. Cell Biol.* **43**: 293–300.

Yen, C.-J., Beamer, B.A., Negri, C. *et al.* (1997) Molecular scanning of the human peroxisome proliferator activated receptor γ gene in diabetic Caucasians: identification of a Pro12Ala PPARγ2 missense mutation. *Biochem. Biophys. Res. Commun.* **241**: 270–274.

Young, P.W., Buckle, D.R., Cantello, B.C. *et al.* (1998) Identification of high-affinity binding sites for the insulin sensitizer rosiglitazone (BRL-49653) in rodent and human adipocytes using a radioiodinated ligand for peroxisomal proliferator-activated receptor γ. *J. Pharmacol. Exp. Ther.* **284**: 751–759.

Zanelli, U., Puccini, P., Acerbi, D. *et al.* (1996) Induction of peroxisomal beta-oxidation and P-450 4A-dependent activities by pivalic and trichloroacetic acid in rat liver and kidney. *Arch. Toxicol.* **70**: 145–149.

Zhang, B., Berger, J., Zhou, G. *et al.* (1996) Insulin- and mitogen-activated protein kinase-mediated phosphorylation and activation of peroxisome proliferator-activated receptor γ. *J. Biol. Chem.* **271**: 31 771–31 774.

Zhou, G., Cummings, R., Li, Y. *et al.* (1998) Nuclear receptors have distinct affinities for coactivators: characterization by fluorescence resonance energy transfer. *Mol. Endocrinol.* **12**: 1594–1604.

Zhu, Y., Alvares, K., Huang, Q. *et al.* (1993) Cloning of a new member of the peroxisome proliferator-activated receptor gene family from mouse liver. *J. Biol. Chem.* **268**: 26 817–26 820.

Zhu, Y., Qi, C., Korenberg, J.R. *et al.* (1995) Structural organization of mouse peroxisome proliferator-activated receptor γ (murine murine PPAR γ) gene: alternative promoter use and different splicing yield two murine murine PPAR γ isoforms. *Proc. Natl. Acad. Sci. USA* **92**: 7921–7925.

Zhu, Y., Qi, C., Calandra, C. *et al.* (1996) Cloning and identification of mouse steroid receptor coactivator-1 (mSRC-1), as a coactivator of peroxisome proliferator-activated receptor γ. *Gene Expr.* **6**: 185–195.

Zhu, Y., Qi, C., Jain, S. *et al.* (1997) Isolation and characterization of PBP, a protein that interacts with peroxisome proliferator-activated receptor. *J. Biol. Chem.* **272**: 25 500–25 506.

Chapter 11

Coactivators and Corepressors

David M. Lonard and Zafar Nawaz

INTRODUCTION

The biologic responses to steroid, thyroid and retinoid hormones are mediated through a large group of ligand-activated transcription factors known collectively as the nuclear hormone receptor superfamily. This superfamily comprises the largest class of transcription factors, which include not only receptors for the well known steroid, thyroid and retinoid ligands, but also include orphan receptors for which a cognate ligand has not been identified or does not exist. Ligand-activated nuclear hormone receptors bind to specific DNA sequences in the enhancer region of hormone-responsive genes, termed hormone response elements, where they serve to stimulate or repress transcription of hormone responsive genes. Nuclear hormone receptor coactivators and corepressors are molecules that form either a functional or physical bridge between the receptor and components of the basal transcription machinery that make up the constitutive core of the RNA II polymerase holoenzyme complex. Because of the intermediary role of coactivators and corepressors in effecting physiologic responses to steroid, thyroid and retinoid hormones, mutations to, or perturbations in, the expression of these molecules possess the potential to cause pathologic genetic conditions. Because of the relatively recent emergence of the coactivator field, the link between coactivators and inherited disease states and cancer remains largely uncharacterized. However, rapid growth in the identification of new coactivators and corepressors, and further elucidation of the mechanisms through which they regulate gene expression, promises to expand upon what is known about how these molecules contribute to genetic disease. In this chapter, we discuss the current state of knowledge of coactivator and corepressor function and their possible contribution to genetic disease states.

HISTORICAL PERSPECTIVE

Direct protein–protein interactions have been reported between receptors and general transcription factors (GTFs). TATA-binding protein (TBP) and several TBP-associated factors (TAFs) interact functionally with specific receptors and thus are technically

Nuclear Receptors and Genetic Disease
ISBN 0-12-146160-2

coactivators. Chatterjee and Struhl (1995) showed that fusion of TBP could restore transcriptional activity to transcription factors that had their native activation domains deleted. Yeast two-hybrid and *in vitro* protein-binding assays have detected an association between a portion of TBP and the activation function 2 (AF-2) domain of the retinoid X receptor (RXR) (Schulman *et al.*, 1995; Leong *et al.*, 1998). Also, the progesterone receptor (PR) has been shown to interact with the $TAF_{II}110$ of transcription factor (TF)IID (Schwerk *et al.*, 1995). These reports of direct interactions between nuclear receptors and GTFs suggested that additional intermediary factors might not be necessary for receptor-mediated transcription.

It has become clear, however, that proteins other than GTFs which are present in limiting quantities are playing an important role in receptor-mediated transcription. When coexpressed, the activation of one receptor interferes with the transcriptional capacity of another receptor, a phenomenon known as squelching (Shemshedini *et al.*, 1992). This suggests that a limiting cellular pool of factors exists that are sequestered by activated receptors and indicates that these unidentified common factors serve an important functional link between the receptor and transcription. Additionally, the observation that receptors do not behave the same in different cellular contexts suggests the existence of intermediary factors that are expressed differentially in different tissues that modulate receptor function and can alter the quantitative and qualitative response to hormones and antihormones.

Using a purified ligand-bound estrogen receptor (ER) ligand-binding domain (LBD) to identify ER-interacting proteins from ^{35}S-radiolabeled MCF-7 cell lysates, Halachmi *et al.* (1994) were able to identify two ER-associated proteins, ERAP-140 and ERAP-160, which interacted with ER in a ligand-dependent fashion. Additionally, these two proteins failed to interact with a transcriptionally defective ER mutant and the antiestrogens 4-hydroxytamoxifen ICI 164384 interfered with the ability of ERAP-140 and ERAP-160 to interact with ER. Eggert *et al.* (1995) later identified a 170-kDa protein, glucocorticoid receptor-interacting protein 170 (GRIP-170) which interacted with the GR in a hormone-dependent manner, and an enriched cellular fraction was able to stimulate GR-mediated transcription. Cavailles *et al.* (1994a) identified receptor-interacting proteins (RIPs) of 140 and 160 kDa, and an additional protein of 80 kDa that possessed the same ligand-dependent interaction with ER. Cloning and further characterization of RIP-140 indicated that, in spite of its ligand-dependent interaction with ER, it was unable to act as a coactivator when overexpressed in transient transfections (Cavailles *et al.*, 1994b).

Using a biochemical approach to identify proteins that interact with the thyroid hormone receptor (TR) in a ligand-dependent manner, Fondell *et al.* (1996) identified a number of thyroid receptor-associated proteins (TRAPs). In their approach, an epitope-tagged TR was transfected into cells grown in the presence of thyroid hormone. Sodium dodecyl sulfate–polyacrylamide gel electrophoresis (SDS-PAGE) analysis of proteins that copurified with the epitope-tagged TR revealed a number of proteins that interact with TR in the presence of thyroid hormone. Interestingly, this biochemical approach identified a novel set of proteins that are not part of the steroid receptor coactivator (SRC) family. TRAP-220 was previously identified as TR-interacting protein-2 (Trip-2) (Lee *et al.*, 1995a) and as PPARγ-binding protein (PBP) (Zhu *et al.*, 1997) through yeast two-hybrid screens. Using a similar methodology with an epitope-tagged vitamin D receptor (VDR), Rachez *et al.* (1998), identified a group of VDR-interacting proteins (DRIPs), which are identical to the TRAP complex and likewise do not contain SRC family members, CREB (cyclicadenosine monophosphate (cAMP) regulatory element binding protein) binding protein (CBP)/p300 or other known coactivators. In a cell-free system, depletion of the DRIP complex from cell extracts reduced VDR-mediated transcription. A DRIP of 100 kDa (DRIP-100) was subsequently cloned; however, when overexpressed in transient transfections, it was unable to enhance VDR transactivation. It is possible that each TRAP–DRIP component protein must exist in a precise stoichiometric ratio in order to enhance receptor function. Another interesting finding is that the TRAP–DRIP complex appears to be specific for TR and VDR. The complex failed to interact with an epitope-tagged ER LBD, suggesting that the TRAP–DRIP complex represents a type II receptor-specific complex (receptors that heterodimerize with the RXR such as TR, retinoic acid receptor (RAR) and VDR). Further analysis of a protein complex which associates with p53, the suppressor of rpb/mediator SRB- and MED-containing complex (SMCC) revealed that it is identical to the TRAP–DRIP complex, indicating that this coactivator complex is also associated with non-nuclear hormone receptor superfamily transcription factors (Ito *et al.*, 1999). At the present time, the contribution of the TRAP–DRIP complex members to receptor function or to any genetic disease state remains unknown.

The prototypical coactivator possesses several common features: (1) coactivators interact with receptors in the presence but not in the absence of their cognate ligand; (2) when overexpressed, coactivators potentiate the ligand-mediated transcription from receptors with which they physically interact; and (3) coactivators exist in limiting amounts in the cell as common cofactors for different receptors. Because of this, overexpression of coactivators can reverse the squelching phenomenon of one activated receptor upon another receptor. Additionally, coactivators are able to stimulate gene expression by acting as a bridge between the activated receptor and GTFs or as catalytic enzymes that disrupt the local chromatin surrounding hormone-responsive genes and include histone acetyltransferase and kinase activities (Fig. 11.1).

THE SRC COACTIVATOR FAMILY

The cloning and characterization of steroid receptor coactivator 1 (SRC-1) by Onate *et al.* (1995) was the first demonstration of a *bona fide* coactivator of nuclear hormone receptors that possesses all the features of a prototypical coactivator mentioned above. Using a yeast two-hybrid screen of a human B-lymphocyte complementary DNA (cDNA) library with a bait encoding the PR LBD, SRC-1 was shown to interact with the PR LBD in a ligand-specific manner. Overexpression of SRC-1 was able to stimulate the transactivation of a wide number of nuclear hormone receptors and stimulated the activity of nonsuperfamily transcription factors such as Sp1 and E2F to a lesser extent, indicating that SRC-1 possesses a partial specificity for members of the nuclear hormone receptor superfamily. Additionally, SRC-1 has been shown modestly to coactivate AP-1 (Lee *et al.*, 1998), serum response factor (Kim *et al.*, 1998) and nuclear factor (NF)-κB (Na *et al.*, 1998) indicating its wider role in transcription. SRC-1 was able to reverse the squelching of PR transactivation by coexpressed ER, indicating that it is a common, limiting cofactor recruited by either the ER or PR for efficient transactivation. Additional supporting evidence that SRC-1 is a true coactivator was demonstrated by the observation that the transcriptional activity of PR was diminished by the over-expression of a C-terminal fragment of SRC-1 that lacks an autonomous activation

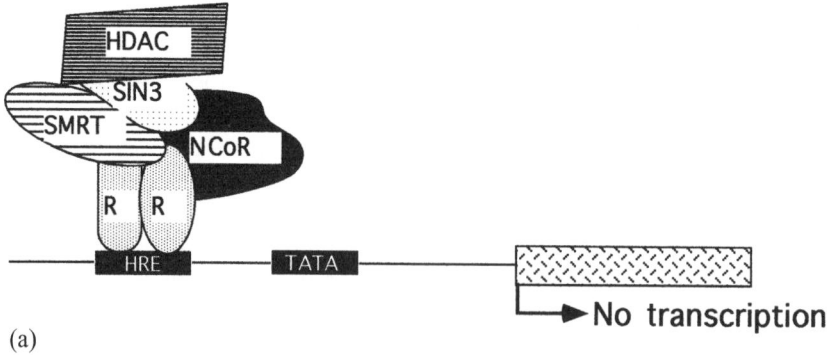

(a)

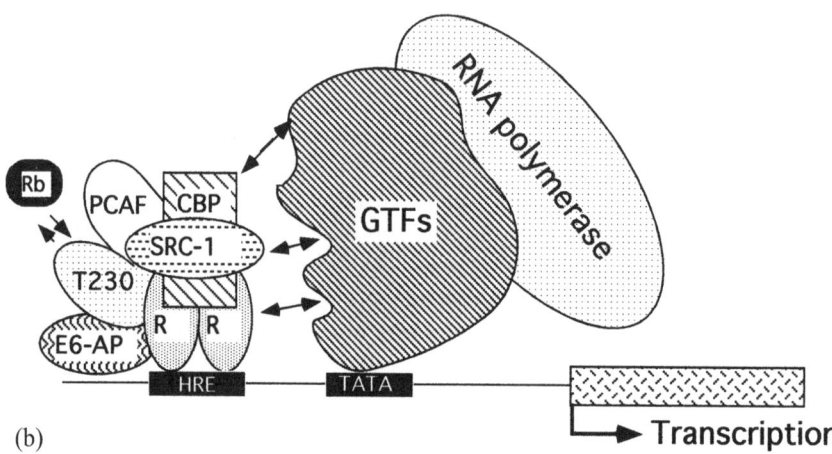

(b)

Fig. 11.1 Model of corepressor and coactivator function in regulating nuclear hormone receptor-mediated transcription. (a) In the absence of hormone or in the presence of antihormones, corepressors such as NCoR and SMRT interact with the receptor (R). These corepressors are part of a complex that also contains SIN3 and histone deacetylase (HDAC), which repress transcription. (b) In the presence of ligand, receptors recruit coactivators which function as a bridge between the receptor and general transcription factors (GTFs) and RNA polymerase, or function as histone acetyltransferases, stimulating transcription.

function but still interacts with receptor, interfering with receptor function by acting as a competitive inhibitor of the full-length SRC-1 with the receptor.

The subsequent cloning of GRIP-1 (Hong *et al.*, 1996, 1997a), transcription intermediary factor 2 (TIF-2) (Voegel *et al.*, 1996) and nuclear coactivator NCoA-2 (Torchia *et al.*, 1997) revealed the existence of other 160-kDa RIPs that are homologous to SRC-1, indicating the existence of a SRC family of coactivator proteins. The highest sequence homology between these proteins and SRC-1 is within in a PAS/bHLH (Per-Arnt-Sim/basic helix-loop-helix) motif residing in the N-terminus of the protein (Hankinson, 1995; Kamei *et al.*, 1996). The contribution of this motif to SRC-1 function remains uncharacterized; however, this domain is thought to play a role in protein–protein interactions. GRIP-1 and TIF-2 associate with the LBD of several receptors in a ligand-dependent manner *in vitro* (Voegel *et al.*, 1996) and *in vivo*, with RARα, ER and PR in the presence of hormone, but not hormonal antagonists (Hong *et al.*, 1996, 1997a; Voegel *et al.*, 1996). Like SRC-1, GRIP-1 and TIF-2 possess autonomous activation domains capable of stimulating transcription when tethered to a heterologous DNA-binding domain (DBD) in yeast and in mammalian cells. Additionally, overexpression of TIF-2 is capable of relieving squelching by ER, and a truncated form of GRIP-1 was shown to inhibit hormone-dependent expression from the mouse mammary tumor virus (MMTV) promoter (Voegel *et al.*, 1996; Hong *et al.*, 1997a; Walfish *et al.*, 1997).

The cloning of a third coactivator, independently identified in a number of laboratories as p300/CBP cointegrator-associated protein (p/CIP) (Torchia *et al.*, 1997), activator of thyroid receptor (ACTR) (Chen *et al.*, 1997), receptor-associated coactivator 3 (RAC-3) (Li *et al.*, 1997), amplified in breast cancer 1 (AIB-1) (Anzick *et al.*, 1997), thyroid receptor activator molecule 1 (TRAM-1) (Takeshita *et al.*, 1997) and SRC-3 (Suen *et al.*, 1998), termed SRC-3 in this review, which bears homology to SRC-1, indicates the existence of another member of the SRC family of coactivators. SRC-3 interacts with and coactivates a wide variety of nuclear receptors in a ligand-dependent manner, including RAR, TR, RXR (Chen *et al.*, 1997), GR, PR (Li *et al.*, 1997) and ER (Anzick *et al.*, 1997). Additionally, SRC-3 has been shown to interact with CREB (Torchia *et al.*, 1997), which was previously shown to be primarily dependent upon the transcriptional cointegrator, CREB-binding protein (CBP).

SRC family members possess an overall sequence similarity of 40%, with the most extensive homology residing in the N-terminus which contains the PAS/bHLH domain. The extensive homology among SRC family members in this region is unique for PAS-containing proteins, which are not normally well conserved (Hankinson, 1995; Kamei *et al.*, 1996). Like other PAS/bHLH-containing proteins, SRC family members are able to form homomultimeric complexes *in vivo* (McKenna *et al.*, 1998). However, the requirement for the PAS/bHLH domain for such interactions remains unclear. Closer inspection of the receptor-interacting domains of RIP-140 and the SRC family members revealed the presence of LXXLL (where L is leucine and X is any amino acid) motifs, termed nuclear receptor (NR) boxes, which are required for coactivator binding to the AF-2 of ligand-bound receptor (Heery *et al.*, 1997). Three NR boxes exist in SRC-1 and an additional NR box is present in the SRC-1e splice variant of SRC-1. Crystal structure determination of the ER LBD–ligand–NR box complex reveals that the hydrophobic leucines of the NR box interact with hydrophobic contacts in helix 3, 4 and 12 of the ligand-bound receptor (Shiau *et al.*, 1998; Mak *et al.*, 1999). The importance of the NR box in mediating ligand-dependent receptor–coactivator interactions is indicated by the fact that mutation of these boxes disrupts the interaction of SRC-1 with receptors and renders the mutant coactivator unable to coactivate receptor-mediated expression (Heery *et al.*, 1997).

SRC FAMILY MISEXPRESSION AND CANCER

SRC-3 is the only member of the SRC family whose altered expression has been well characterized in a genetic disease state thus far (see also Tables 11.1–11.3). During a search for genes residing in chromosome 20 whose expression and copy number were increased in human breast cancer, Anzick *et al.* (1997) identified a protein they termed amplified in breast cancer 1 (AIB-1) and referred to here as SRC-3 (see above), which

Table 11.1 Coactivators involved in inherited genetic disease

Coactivator	Genetic disease	Description of disease	Description of coactivator
E6-AP	Angelman syndrome	Intellectual impairment, ataxia, inappropriate laughter, seizures	General coactivator of nuclear hormone receptors, also characterized as a ubiquitin–protein ligase
CBP	Rubinstein–Taybi syndrome	Intellectual and growth impairment, abnormal face and palate, bone defects	Cointegrator; coactivates nuclear hormone receptors as well as other transcription factors, possesses histone acetyltransferase (HAT) activity

Table 11.2 Somatic misexpression of coactivators and cancer

Coactivator	Genetic disease	Description
SRC-3	Breast, ovarian and pancreatic cancer	Gene amplification and/or overexpression of SRC-3 correlated with ER- and PR-positive breast and ovarian cancers. SRC-3 is a member of the SRC coactivator family, and is also a HAT
ARA-55	Prostate cancer	Overexpression of ARA-55 is correlated with prostate cancer, and is known to enhance the agonist activity of the partial antiandrogen, hydroxyflutamide

Table 11.3 Involvement of corepressors in genetic disease states

Mutation	Genetic disease	Description of corepressor involvement
PML and PLZF gene fusion to RARα	Acute promyelocytic leukemia	Fusion of RARα to PLZF or PML results in recruitment of corepressors to gene promoters that the PML and PLZF genes bind to which are responsible for terminal differentiation of monocytes. The recruited corepressors inhibit expression of these genes, causing the undifferentiated monocytes to proliferate. Treatment with retinoic acid and/or histone deacetylase inhibitors can be very effective in relieving the repression of the genes responsible for monocyte terminal differentiation, allowing for complete remission of the leukemia
Mutations to the hinge or coactivator binding regions of TR	Resistance to thyroid hormone	These mutant TRs either bind corepressor with higher affinity or are defective in binding to coactivators, acting as dominant negative inhibitors of the wild-type TR and constitutively repressing gene transcription

was found to be amplified and overexpressed in four out of five ER-positive breast and ovarian cancer cell lines. Subsequent analysis of 105 primary breast cancer tumors revealed that the SRC-3 gene is amplified 10% of the time and overexpressed 64% of the time in these tissues, suggesting that this coactivator plays a role in breast cancer tumorigenesis. Subsequent analysis of a larger set of breast and ovarian tumors by Bautista *et al.* (1998) revealed that SRC-3 is overexpressed in 4.8% of breast cancers and 7.4% of ovarian cancers. They also went on to show that SRC-3 overexpression is correlated with expression of ER and PR in these cancers, suggesting that SRC-3 contributes to tumor growth by stimulating ER- or PR-mediated transactivation. Ghadimi *et al.* (1999) identified SRC-3 gene amplification and overexpression in six of nine pancreatic cancer cell lines they examined, suggesting that SRC-3 can contribute to tumor growth in other tissues as well. Berns *et al.* (1998) examined the expression level of SRC-1 in 21 primary breast cancer tumors, seven mammary tumor cell lines, 12 fibroblast cultures and six normal breast tissues, and concluded that the highest levels of SRC-1 were found in normal breast tissue with the lowest level existing in breast cancer tumor cell lines –the opposite of what is seen for SRC-3. It is possible that SRC-1 expression may be downregulated in tumors and tumor cell lines where SRC-3 expression is very high. However, Anzick *et al.* (1997) did not see any significant differences in SRC-1 or TIF-2 levels in breast cancer primary tumors or cell lines that overexpressed SRC-3.

Although SRC-1 has not been linked to a genetic disease at this time, targeted disruption of the SRC-1 gene in mice (Xu *et al.*, 1998) is potentially predictive in the phenotype that might be expected when SRC family members are genetically disrupted. Consistent with its characterization as a coactivator of nuclear hormone receptors, SRC-1 disruption resulted in a diminished response to steroid hormones. SRC-1$^{-/-}$ males also possessed smaller testis and SRC-1$^{-/-}$ females had less developed mammary glands and an impaired uterine response to simulated ovarian implantation. However, SRC-1$^{-/-}$ mice are still fertile and possess only a partial insensitivity toward steroid hormones. It is likely that a functional redundancy exists between SRC family members, which can compensate for the loss of SRC-1. Xu *et al.* (1998) demonstrated that TIF-2 expression was increased in SRC-1$^{-/-}$ mice, lending support to the idea that SRC family members can compensate for one another. Targeted disruption of SRC-2 and SRC-3 is predicted to produce a similar phenotype and should provide additional support for dependence of nuclear hormone receptors on coactivator molecules. It is also likely that the simultaneous disruption of multiple members of SRC family will be necessary to observe a more severe phenotype due to the functional redundancy probably displayed by these coactivators.

COACTIVATORS INVOLVED IN UBIQUITIN–PROTEASOME-MEDIATED PROTEIN DEGRADATION

The identification of RIPs, which play a role in ubiquitin–proteasome-mediated protein degradation, suggests that protein degradation represents yet another enzymatic function imparted by coactivators on the regulation of transcription, in addition to the histone acetyltransferase and kinase activities associated with other coactivators. Nawaz *et al.* (1999) identified E6-associated protein (E6-AP), a protein that was previously identified as a ubiquitin ligase enzyme involved in the ubiquitination and degradation of p53 (Scheffner *et al.*, 1993), as a nuclear receptor coactivator. E6-AP possesses an intrinsic activation function that resides in the N-terminus and can reverse squelching between coexpressed PR and ER. E6-AP contains two NR boxes in its C-terminus, which comprises its major receptor interaction domain. The contribution of the ubiquitin ligase function of E6-AP to coactivation remains unclear; E6-AP was still able to function as a coactivator when it was mutated so that its ubiquitin ligase activity was destroyed. E6-AP is closely related to another ubiquitin ligase coactivator, receptor potentiating factor 1 (RPF-1), which is the human homolog of the yeast RSP5 protein that has been shown to interact with RNA polymerase II (Imhof and McDonnell, 1996). E6-AP and RPF-1 also coelute in gel filtration fractions and function together in a synergistic fashion, suggesting that they function in a common coactivator complex (McKenna *et al.*, 1998). Additional RIPs that are involved in ubiquitin–proteasome-mediated protein degradation have been identified, but their contribution to receptor function has not been well established. Lee *et al.* (1995b) have identified a protein, Trip-1, a human homolog of the yeast *sug1* gene that is a component of the 26S proteasome (Rubin *et al.*, 1996), which interacts with TR and RXR baits in a yeast two-hybrid assay in a ligand-dependent manner. Additionally, Sug1 has also been shown to possess helicase activity (Fraser *et al.*, 1997). Masuyama and MacDonald (1998) have shown that Sug1 can promote the proteasome-mediated degradation of the VDR, suggesting that Sug1 is involved in hormone-mediated receptor downregulation. In spite of these two identified roles for Trip-1/Sug1, how either the proteasome or helicase activity of Sug1 plays a role in receptor function remains unclear and the protein has not been shown to stimulate receptor-mediated transcription.

MUTATIONS TO THE E6-AP GENE AND ANGELMAN SYNDROME

Mutations to the maternal allele of the E6-AP gene (*UBE3A*) on chromosome 15 in the 15q11-q13 locus result in Angelman syndrome (AS), a genetic disease characterized by severe intellectual impairment, absence of speech, ataxia, inappropriate laughter, seizures and other behavioral defects (for reviews see Nicholls, 1993; Ledbetter and Ballabio, 1995). AS involves an allelic imprinting mechanism; mutation or deletion of the maternal allele of 15q11-q13 leads to AS. Interestingly, deletion of the paternal allele of 15q11-q13 results in Prader–Willi syndrome, which is phenotypically distinct from AS. Patients with Prader–Willi syndrome also have intellectual impairment, but it is much less severe than that seen for AS. Finer mapping of disruptions of the 15q11-13 region of patients with AS revealed that truncating mutations of *UBE3A* were responsible for the AS phenotype (Kishino *et al.*, 1997; Matsuura *et al.*, 1997; Sutcliffe *et al.*, 1997). Patients with AS appear to possess normal secondary sexual characteristics and reach puberty at a normal age. AS is estimated to occur in one in 15 000 births. In most cases, large-scale *de novo* deletions of the maternal chromosome 15q11-13 lead to AS. However, a number of other abnormalities in the inheritance of the maternal chromosome 15 also contribute to the number of AS cases. A small group of patients with AS possess point mutations within the E6-AP gene itself. Most of these mutations are frameshift or stop mutations that result in premature termination of the protein and abolish the ubiquitin-ligase activity of E6-AP (Nawaz *et al.*, 1999). Paternal uniparental disomy (UPD), whereby two copies of the paternal allele are inherited, can also result in AS, as can alterations in the pattern of methylation of the locus. Expression of E6-AP is biparental in most tissues; however, in specific regions of the brain such as the hippocampus, Purkinje cells and olfactory mitral cells, E6-AP is expressed only from the maternal allele (Rougeulle *et al.*, 1997; Vu and Hoffman, 1997), suggesting that the AS phenotype results from the loss of expression of E6-AP specifically within these tissues of the brain.

Originally, E6-AP was identified as a protein that interacts with the human papillomavirus type 16 and 18 E6 oncoprotein and goes on to interact with and subsequently degrade p53 through the ubiquitin–proteasome protein degradation pathway (Scheffner *et al.*, 1993). E6-AP belongs to a class of proteins known as ubiquitin ligases or E3s, which are part of a chain of enzymes involved in the coupling of ubiquitin to proteins targeted for degradation by the 26S proteasome. For a review of ubiquitin–proteasome-mediated protein degradation, see Pickart (1997).

As discussed above, E6-AP has also been identified as a nuclear hormone receptor coactivator by our laboratory. The connection between coactivation, ubiquitin ligase activity and the AS phenotype remains unclear at this time, however. Jiang *et al.* (1998) disrupted E6-AP in the mouse in order to understand better the contribution that E6-AP makes toward the AS phenotype. E6-AP$^{-/+}$ AS mice in which the targeted allele was inherited from the mother possess motor dysfunction, inducible seizures and a defect in contextual learning and hippocampal long-term potentiation (LTP) consistent with the AS phenotype seen in humans. Interestingly, AS mice possess an abundance of cytoplasmic p53 in their Purkinje neurons, suggesting that the lack of ubiquitination of p53 and its increased abundance in these cells may contribute to abnormal cerebellar function and motor control in these mice. The contribution made by the coactivator function of E6-AP to the AS phenotype is not known. Analysis of point mutants of E6-AP found in human patients revealed that mutant proteins that are defective as ubiquitin ligases are still able to function as coactivators, suggesting that the loss of ubiquitin ligase activity of E6-AP plays a more important role in the manifestation of the AS phenotype (Nawaz *et al.*, 1999). Like SRC-3, E6-AP over-expression has been observed in mouse mammary gland tumors (personal observation), suggesting that its overexpression may also play a role in ER- and PR-mediated breast cancers.

COINTEGRATORS: CBP AND p300

An important coactivator that bears no sequence homology to SRC family members is CREB-binding protein (CBP), which was initially characterized as a coactivator required for efficient activation of cAMP-regulated promoters by the transcription factor CREB (Kwok *et al.*, 1994). A number of studies have implicated CBP as a coactivator for a wide variety of transcription factors, such as p53 (Avantaggiati *et al.*, 1997), NF-κB (Perkins *et al.*, 1997) and nuclear hormone receptors (Chakravarti *et al.*, 1996; Kamei *et al.*, 1996; Fronsdal *et al.*, 1998), providing the impetus to characterize CBP as a cointegrator, which is thought to serve as a limiting cofactor that influences the crosstalk between different groups of transcription factors such as between GR and NF-κB (Sheppard *et al.*, 1998). In addition to its interactions with nuclear receptors, CBP interacts with members of the SRC-1 family, including SRC-1 (Kamei *et al.*, 1996), TIF-2 (Voegel *et al.*, 1998) and SRC-3 (Torchia *et al.*, 1997), indicating that it may form a ternary complex with SRC family members and nuclear receptors. Functional evidence suggests that such a complex may exist; coexpression of CBP and SRC-1 leads to a synergistic increase in ER- and PR-mediated transcription (Smith *et al.*, 1996).

Biochemical evidence indicates that the CBP does not form a stable complex with SRC-1 and that CBP interacts only weakly with the liganded ER (Zhou *et al.*, 1998) and PR (McKenna *et al.*, 1998), suggesting that an initial receptor–SRC-1 complex forms which then subsequently recruits CBP. CBP was also ineffective in restoring activity to a retinoic acid response element (RARE)-linked reporter gene after immunodepletion of SRC-1, further suggesting that CBP might require a preformed receptor–SRC-1 complex that functions as a platform to recruit other intermediary factors (Torchia *et al.*, 1997).

p300 represents another cointegrator molecule that bears a high degree of sequence and functional homology to CBP. Originally, p300 was identified as a major cellular target of the adenovirus E1A gene which functions as a transcriptional intermediary factor (Whyte *et al.*, 1989; Eckner *et al.*, 1994). Like SRC family members, both CBP and p300 possess histone acetyltranferase (HAT) activity (Bannister and Kouzarides, 1996; Ogryzko *et al.*, 1996; Spencer *et al.*, 1997), which likely contributes to its ability to stimulate receptor-mediated transcription.

MUTATIONS OF CBP AND RUBINSTEIN–TAYBI SYNDROME

Mutation of CBP at the 16p13.3 locus has been implicated in Rubinstein–Taybi syndrome (RTS) (Petrij *et al.*, 1995). The condition is characterized by broad thumbs and toes, growth and intellectual impairment, and craniofacial abnormalities (for a review see Cantani and Gagliesi, 1998). Mutation of a single allele of CBP leads to RTS, suggesting that the dosage of CBP is critical for normal cellular function. Tanaka *et al.* (1997) reported that targeted disruption of the CBP gene in mice results in a phenotype similar to that observed for RTS. Additionally, Yao *et al.* (1998) disrupted the p300 gene and showed that, in spite of the high homology exhibited between CBP and p300, the two molecules were only partially redundant. CBP and p300 heterozygotes were less viable and CBP/p300 double heterozygotes invariably died, suggesting that the dosage of CBP and p300 was critical for complete viability or that the functions of p300 and CBP did not completely overlap. CBP and p300 appeared to subserve some different functions, as fibroblasts derived from p300$^{-/-}$ mice exhibited an impaired response to retinoids but not to CREB function (Kawasaki *et al.*, 1998; Yao *et al.*, 1998). Oike *et al.* (1999) reported that disruption of CBP, which resulted in the expression of a truncated form of CBP expressing only the first 1084 amino acids, also exhibited a classic RTS phenotype. These mice exhibited slower growth, defects in bone formation and impaired long-term memory. In spite of the clear connection between genetic lesions to the CBP gene and RTS, it remains unclear what role the nuclear hormone receptor function of CBP plays in the phenotypic manifestation of RTS. Ihara *et al.* (1999) reported the case of a girl with RTS undergoing premature thelarche, suggesting a role for CBP in reproductive development; however, this observation is counterintuitive given the coactivator role that CBP imparts on steroid receptors. The RTS phenotype of short stature and intellectual impairment is also consistent with thyroid resistance; however, Olson and Koenig (1997) showed that thyroid function appears to be normal in patients with RTS, failing to support the possibility that CBP disruption contributes to the RTS phenotype by impairing thyroid hormone action. Further investigation is required to determine how impairment of nuclear hormone receptor function by the loss of CBP contributes to the RTS phenotype.

ANDROGEN RECEPTOR-SPECIFIC COACTIVATORS: POSSIBLE INVOLVEMENT IN PROSTATE CANCER

Cloned as an androgen receptor (AR)-specific coactivator, Yeh and Chang (1996) identified androgen-associated protein-70 (ARA-70) which interacts with AR in a ligand-dependent manner and enhances AR transactivation when overexpressed in transient transfections. Additionally, ARA-70 overexpression was able to alter the ligand specificity of AR such that estradiol could function as an AR agonist, and altered the qualitative response to the antiandrogen, hydroxyflutamide, so that it functions as an agonist. Yeh *et al.* (1999) went on to show that ARA-70 interaction with AR is stimulated by activation of the heregulin-2/neuregulin (HER2/Neu) growth factor pathway. Stimulation of the mitogen-activated protein (MAP) kinase pathway through the HER2/Neu pathway has been implicated in breast and ovarian cancers and in the conversion of tamoxifen from an antagonist of ER into an agonist. MAP kinase pathway stimulation has also been correlated with prostate cancer progression, with very high MAP kinase activity being present in prostate cancers that have progressed to an androgen-independent state (Gioeli *et al.*, 1999). This report suggests that stimulation of MAP kinase through this pathway may allow for ligand-independent association of ARA70 with AR, as well as the conversion of hydroxyflutamide into an AR agonist.

Using a yeast two-hybrid screen, Fujimoto *et al.* (1999) identified ARA-55, which also potentially plays a role in prostate cancer progression. ARA-55 was shown to contain three Lin-II, Isl-I and Mec-3 (LIM) domains in its C-terminus, which may be involved in the interaction of ARA-55 with AR. Overexpression of ARA-55 was able to stimulate AR transactivation in the presence of dihydrotesterone, and to stimulate the partial agonist activity of estradiol and hydroxyflutamide like ARA-70. Interestingly, ARA-55 was shown to be overexpressed in prostate tumor cancer cell lines, suggesting that ARA-55 overexpression may be responsible for the progression of prostate cancer and play a role in agonist–antagonist switching of the AR antagonist, hydroxyflutamide, into an agonist, eliminating its effectiveness as a chemotheraputic agent for the treatment of prostate cancer.

TRIP-230, A COACTIVATOR ASSOCIATED WITH THE RETINOBLASTOMA GENE PRODUCT

A TR-specific coactivator, Trip-230, was originally identified by Chang *et al.* (1997) as a protein that interacts with the retinoblastoma (Rb) gene product. TRIP-230 was localized to the 14q31 locus, which is common to several other genes involved in thyroid function, including Grave's disease and hyperthyroidism, implicating Trip-230 in playing a role in TR function. Trip-230 was shown to interact with TR in a ligand-dependent manner and to stimulate TR-mediated transactivation when overexpressed in transient transfections. Coexpression of Rb abolishes the ability of Trip-230 to coactivate TR, indicating that Trip-230 may play a role in integrating crosstalk between Rb- and TR-mediated transcription.

COREPRESSORS

Horlein *et al.* (1995) identified nuclear receptor corepressor (NCoR), a 270-kDa protein that associates with the unliganded TR and RAR, and potentiates the repressive effect that unliganded TR and RAR have on gene expression. Mutational analysis of TR revealed that NCoR interacts within the hinge region of the receptor, a region they defined as the CoR box. Deletion of the CoR box of TR abolishes its ability to repress transcription in its unliganded state, suggesting that NCoR mediates TR-mediated transcriptional repression. When NCoR was fused to a heterologous DNA-binding motif, it was able to confer repression to a promoter containing Gal4DBD-binding sites, indicating that it contains an autonomous repressive function. Deletion mapping of NCoR revealed that it contains two receptor-interacting domains (RIDs) in the C-terminal portion of the protein. The N-terminus of the protein contains three repression domains (RI, RII and RIII). Seol *et al.* (1996) isolated the same protein, which they named RIP-13, in a screen for proteins that interact with RAR.

Another corepressor of 168 kDa was identified by Chen and Evans (1995) through a yeast two-hybrid screen using RXR as a bait. They termed this corepressor silencing mediator for retinoid and thyroid hormone receptor (SMRT). SMRT was also identified as T_3 receptor-associating cofactor 2 (TRAC-2) on the basis of its interaction with TR, RAR and RXR (Sande and Privalsky, 1996). Considerable homology has been observed between SMRT and the C-terminal half of NCoR, suggesting that SMRT was not a full-length clone. Ordentlich *et al.* (1999) and Park *et al.* (1999) have subsequently cloned larger variants of SMRT which are thought to be full-length forms of the protein. The N-terminus of the full-length forms of SMRT bears strong similarity to the N-terminus of NCoR. Both proteins contain a domain that interacts with mouse seven in abstenia homolog-2 (mSIAH2), identified by Zhang *et al.* (1998a) in a screen for proteins that interact with NCoR. mSIAH2 was identified as a mouse homolog of the *Drosophila SINA* gene, which has been shown to play a role in the ubiquitin–proteasome-mediated degradation of the tramtrack transcription factor. It was demonstrated that RevErb repression of transcription is cell-line specific and depends upon the relative abundance of NCoR in different cell types. While NCoR messenger RNA was equally abundant in both 293 monkey kidney cells and N18 neuroblastoma cells, the NCoR protein was present only in 293 cells. The absence of NCoR in the N18 cells was attributed to mSIAH2 expressed in the N18 cell line (but not in 293 cells), which is responsible for the downregulation of the NCoR protein through a ubiquitin–proteasome-mediated process. Presently, it is not known whether mSIAH2 interacts with the full-length SMRT and promotes its degradation.

COREPRESSORS MODULATE RESPONSES TO RECEPTOR ANTAGONISTS

Smith *et al.* (1997) demonstrated that SMRT blocks the agonist activity of the mixed antagonist, 4HT, to stimulate ER-mediated transcription when overexpressed in HepG2 cells. It was also shown that SMRT interacts with ER *in vitro*, both in the presence of 4-hydroxytamoxifen and estradiol. These data suggest that corepressors may play a role in modulating the antagonist–agonist response of ER to mixed antagonists. NCoR has also been shown to interact with ER in the presence of the partial antagonist, 4-hydroxy-tamoxifen (Lavinsky *et al.*, 1998) and with PR in the presence of RU486 (Wagner *et al.*, 1998). The interaction of NCoR with the ER is thought to depend upon the presence of Sin3-associated protein-30 (SAP-30) (Zhang *et al.*, 1998b), which is a component of a large corepressor complex that contains NCoR, Sin3 and histone deacetylase (Kasten *et al.*, 1997), and is variably expressed in different tissues, suggesting that this molecule may also be involved in the differential response seen to 4-hydroxytamoxifen in different tissues. Also, as mentioned above, Zhang *et al.* (1998a) have shown that NCoR expression varies in different tissues. Further support for the possibility that corepressors modulate the receptor responses to mixed antagonists is the observation that NCoR expression declines in tamoxifen-resistant MCF-7 cells implanted into nude mice (Lavinsky *et al.*, 1998), raising the interesting possibility that diminished corepressor expression can convert tamoxifen from an antagonist into an agonist, eliminating the effectiveness of one of the most common forms of treatment for estrogen-dependent breast cancers.

COREPRESSORS AND RESISTANCE TO THYROID HORMONE

Resistance to thyroid hormone is a syndrome associated with mutations of TRβ, which render the receptor unable to activate T_3-responsive genes. Three mutant forms of TRβ (A234T, R243Q and R243W), which represent mutations to the hinge region of the receptor, have been identified (Collingwood *et al.*, 1997; Yagi *et al.*, 1997). Safer *et al.* (1998) examined these mutants and determined that T_3 was unable to promote release of NCoR, consistent with the thyroid resistance phenotype manifested by these mutants. Furthermore, they showed that, while the binding of T_3 to these mutants was equal in solution, T_3 bound poorly to these mutants when complexed with DNA. The wild-type TRβ was unable simultaneously to bind both NCoR and SRC-1, suggesting that the association of coactivator and corepressor with TRβ is mutually exclusive.

Tagami *et al.* (1998) demonstrated that a patient with severe resistance to thyroid hormone possessed a mutation to TRβ (L454S). This site is within the ligand-dependent transactivation domain which has been shown to interact with coactivators, and this mutant was able to interact only weakly with SRC-1 and GRIP-1, as judged by electromobility shift and mammalian two-hybrid assays. Additionally, this mutant form of TRβ interacted more strongly with corepressors in the absence of T_3 and physiologic concentrations of T_3 were unable to dissociate NCoR from this mutant receptor. These observations suggest that coactivators function, in part, by promoting dissociation of corepressors from TRβ. Together these TRβ mutants may represent two different classes of mutants: those that interact more strongly with corepressors due to mutations in the CoR box of the receptor and another class that retains interaction with corepressors in the presence of ligand due to a disruption in the receptor that renders it unable to recruit coactivators.

COREPRESSORS AND ACUTE PROMYELOCYTIC LEUKEMIA

Although no specific corepressor has yet been identified in playing a role in a genetic disease, the histone deacetylase activity associated with corepressors has been implicated in playing a role in acute promyelocytic leukemia (APL), which is caused by chromosomal rearrangments that result in the fusion of pox virus and zinc finger (POZ) domain-containing proteins such as the promyelocytic leukemia (PML) or promyelocytic leukemia zinc finger (PLZF) genes to the DBD and LBD of RARα (de The *et al.*, 1991; Chen *et al.*, 1993). These fusion proteins then repress transcription of genes required for the terminal differentiation of monocytes, resulting in their uncontrolled proliferation. Promyelocytic leukemias that result from PML–RARα gene fusions are normally very responsive to treatment with all-*trans* retinoic acid (atRA). It is thought that ligand-induced dissociation of corepressors from the PML–RARα fusion protein is responsible for this effect, allowing for terminal differentiation of mononuclear cells to continue. Lin *et al.* (1998) showed that leukemias resulting from PML–RARα fusion were very responsive to atRA treatment. Additional treatment with histone deacetylase inhibitors were able to dramatically potentiate the positive effects of atRA as well, consistent with the notion that deacetylases are involved in this cancer. Leukemias stemming from PLZF–RARα fusions do not respond to atRA treatment, however, unless a histone deacetylase compound is also used, suggesting that atRA treatment by itself is not sufficient to dissociate corepressors from this fusion protein (He *et al.*, 1998). Hong *et al.* (1997b) have shown that SMRT interacts with PML–RARα fusion proteins both through the RARα portion of the protein and with the POZ domain of the PML portion of the protein. Grignani *et al.* (1998) also showed that NCoR associates with PML–RARα and that treatment with atRA releases NCoR from this fusion protein.

CONCLUSIONS

A great deal of information has accumulated since the identification of the first *bona fide* coactivator, SRC-1, in 1995. During this time many other RIPs and coactivators have been identified; mention of them all is beyond the scope of this review. Originally, coactivators were thought to function as intermediary bridging proteins which allowed the receptor to couple with the general transcription machinery. Identification of SRC-1, SRC-3, CBP and p300 as HATs led to an expansion in the way of thinking about receptor function. The identification of coactivators that possess kinase and ubiquitin–ligase activities suggests that other enzymatic functions are involved in receptor-mediated regulation of transcription as well. Corepressors, on the other hand, have been shown to be associated with histone deacetylase activity, which is associated with negative regulation of transcription.

Our state of knowledge about the role played by coactivators in genetic disease states is much more limited at this time, however. Presently, Angelman and Rubinstein–Taybi syndromes represent the only known inheritible genetic syndromes that involve germline disruption of coactivator genes. Misexpression of a number of coactivators such as SRC-3 and ARA-55 in somatic tissue may be responsible for, or involved in, the progression of breast, ovarian and prostate cancers. At this time, the involvement of overexpression of these coactivators in cancer is only correlative; targeted disruption or tissue-specific overexpression of these coactivators in mouse models will be very informative. With further characterization of corepressor proteins and their mechanism of function, it is likely that they will also be recognized as playing a greater role in genetic disease. The identification of gene rearrangements that fuse the LBD of RARα to the PML and PLZF genes led to the realization that these fusion proteins were responsible for APL due to their recruitment of NCoR and SMRT to gene targets that are responsible for terminal differentiation of monocytes, resulting in their uncontrolled proliferation. Resistance to thyroid hormone represents another syndrome that involves mutant TRs which are defective in corepressor release and/or coactivator recruitment. Besides the obvious potential role that coactivator and corepressor dysfunction could play in genetic disease, it is possible that differences in the expression of these proteins among individuals may be responsible for the variation in sensitivity that individuals have for steroid, thyroid, retinoid and vitamin D_3 hormones, and may influence individual differences in reproductive function and metabolism.

REFERENCES

Anzick, S.L., Kononen, J., Walker, R.L. *et al.* (1997) AIB1, a steroid receptor coactivator amplified in breast and ovarian cancer. *Science* **277**: 965–968.

Avantaggiati, M.L., Ogryzko, V., Gardner, K., Giordano, A., Levine, A.S. and Kelly, K. (1997) Recruitment of p300/CBP in p53-dependent signal pathways. *Cell* **89**: 1175–1184.

Bannister, A.J. and Kouzarides, T. (1996) The CBP co-activator is a histone acetyltransferase. *Nature* **384**: 641–643.

Bautista, S., Valles, H., Walker, R.L. *et al.* (1998) In breast cancer, amplification of the steroid receptor coactivator gene *AIB1* is correlated with estrogen and progesterone receptor positivity. *Clin. Cancer Res.* **4**: 2925–2929.

Berns, E.M., van Stavern, I.L., Klijn, J.G. and Foekens, J.A. (1998) Predictive value of SRC-1 for tamoxifen response of recurrent breast cancer. *Breast Cancer Res. Treat.* **48**: 87–92.

Cantani, A. and Gagliesi, D. (1998) Rubinstein–Taybi syndrome. Review of 732 cases and analysis of the typical traits. *Eur. Rev. Med. Pharmacol. Sci.* **2**: 81–87.

Cavailles, V., Dauvois, S., Danielian, P.S. and Parker, M.G. (1994a) Interaction of proteins with transcriptionally active estrogen receptors. *Proc. Natl. Acad. Sci. USA* **91**: 10 009–10 013.

Cavailles, V., Dauvois, S., L'Horset, F. *et al.* (1994b) Nuclear factor RIP140 modulates transcriptional activation by the estrogen receptor. *EMBO J.* **14**: 3741–3751.

Chakravarti, D., LaMorte, V.J., Nelson, M.C. *et al.* (1996) Role of CBP/p300 in nuclear receptor signalling. *Nature* **383**: 99–103.

Chang, K.H., Chen, Y.M., Chen, T.T. *et al.* (1997) A thyroid hormone receptor coactivator negatively regulated by the retinoblastoma protein. *Proc. Natl. Acad. Sci. USA* **94**: 9040–9045.

Chaterjee, S. and Struhl, K. (1995) Connecting a promoter-bound protein to TBP bypasses the need for a transcriptional activation domain. *Nature* **374**: 820–822.

Chen, H., Lin, R.J., Schiltz, R.L. *et al.* (1997) Nuclear receptor coactivator ACTR is a novel histone acetyltransferase and forms a multimeric activation complex with P/CAF and CBP/p300. *Cell* **90**: 569–580.

Chen, J.D. and Evans, R.M. (1995) A transcriptional co-repressor that interacts with nuclear hormone receptors. *Nature* **377**: 454–457.

Chen, S.J., Zelent, A., Tong, J.H. *et al.* (1993) Rearrangements of the retinoic acid receptor alpha and promyelocytic leukemia zinc finger genes resulting from t(11;17)(q23;q21) in a patient with acute promyelocytic leukemia. *J. Clin. Invest.* **91**: 2260–2267.

Collingwood, T.N., Rajanayagam, O., Adams, M. *et al.* (1997) A natural transactivation mutation in the thyroid hormone β receptor: impaired interaction with putative transcriptional mediators. *Proc. Natl. Acad. Sci.* **94**: 248–253.

de The, H., Lavau, C., Marchio, A., Chomienne, C., Degos, L. and Dejean, A. (1991) The PML–RARα fusion mRNA generated by the t(15;17) translocation in acute promyelocytic leukemia encodes a functionally altered RAR. *Cell* **66**: 675–684.

Eckner, R., Ewen, M.E., Newsome, D. *et al.* (1994) Molecular cloning and functional analysis of the adenovirus E1A-associated 300kD protein (p300) reveals a protein with properties of a transcriptional adaptor. *Genes Dev.* **8**: 869–884.

Eggert, M., Mows, C.C., Tripier, D. *et al.* (1995) A fraction enriched in a novel glucocorticoid receptor-interacting protein stimulates receptor-dependent transcription *in vitro. J. Biol. Chem.* **270**: 30 755–30 759.

Fondell, J.D., Ge, H. and Roeder, R.G. (1996) Ligand induction of a transcriptionally activate thyroid hormone receptor coactivator complex. *Proc. Natl. Acad. Sci. USA* **93**: 8329–8333.

Fraser, R.A., Rossignol, M., Heard, D.J., Egly, J.M. and Chambon, P. (1997) SUG1, a putative transcriptional mediator and subunit of the PA700 proteasome regulatory complex, is a DNA helicase. *J. Biol. Chem.* **272**: 7122–7126.

Fronsdal, K., Engedal, N., Slagsvold, T. and Saatcioglu, F. (1998) CREB binding protein is a coactivator for the androgen receptor and mediates cross-talk with AP-1. *J. Biol. Chem.* **273**: 31 853–31 859.

Fujimoto, N., Yeh, S., Kang, H. Y. *et al.* (1999) Cloning and characterization of androgen receptor coactivator, ARA55, in human prostate. *J. Biol. Chem.* **274**: 8316–8321.

Ghadimi, B.M., Schrock, E., Walker, R.L. *et al.* (1999) Specific chromosomal aberrations and amplification of the *AIB1* nuclear receptor coactivator gene in pancreatic carcinomas. *Am. J. Pathol.* **154**: 525–536.

Gioeli, D., Mandell, J.W., Petroni, G.R., Frierson, H.F. Jr. and Weber, M.J. (1999) Activation of mitogen-activated protein kinase associated with prostate cancer progression. *Cancer Res.* **59**: 279–284.

Grignani, F., De Matteis, S., Nervi, C. *et al.* (1998) Fusion proteins of the retinoic acid receptor-α recruit histone deacetylase in promyelocytic leukaemia. *Nature* **391**: 815–818.

Halachmi, S., Marden, E., Martin, G., MacKay, H., Abbondanza, C. and Brown, M. (1994) Estrogen receptor-associated proteins: possible mediators of hormone-induced transcription. *Science* **264**: 1455–1458.

Hankinson, O. (1995) The aryl hydrocarbon receptor complex. *Ann. Rev. Pharmacol. Toxicol.* **35**: 307–340.

He, L.Z., Guidez, F., Tribioli, C. *et al.* (1998) Distinct interactions of PML–RARα and PLZF–RARα with co-repressors determine differential responses to RA in APL. *Nat. Genet.* **18**: 126–135.

Heery, D.M., Kalkhoven, E., Hoare, S. and Parker, M.G. (1997) A signature motif in transcriptional co-activators mediates binding to nuclear receptors. *Nature* **387**: 733–736.

Hong, H., Kohli, K., Trivedi, A., Johnson, D.L. and Stallcup, M.R. (1996) GRIP1, a novel mouse protein that serves as a transcriptional coactivator in yeast for the hormone binding domains of steroid receptors. *Proc. Natl. Acad. Sci. USA* **93**: 4948–4952.

Hong, H., Kohli, K., Garabedian, M.J. and Stallcup, M.R. (1997a) GRIP1, a transcriptional coactivator for the AF-2 transactivation domain of steroid, thyroid, retinoid, and vitamin D receptors. *Mol. Cell. Biol.* **17**: 2735–2744.

Hong, S.H., David, G., Wong, C.W., Dejean, A. and Privalsky, M.L. (1997b) SMRT corepressor interacts with PLZF and with the PML–retinoic acid receptor α (RARα) and PLZF–RARα oncoproteins associated with acute promyelocytic leukemia. *Proc. Natl. Acad. Sci.* **94**: 9028–9033.

Horlein, A.J., Naar, A.M., Heinzel, T. *et al.* (1995) Ligand-independent repression by the thyroid hormone receptor mediated by a nuclear receptor corepressor. *Nature* **377**: 397–404.

Ihara, K., Kuromaru, R., Takemoto, M. and Hara, T. (1999) Rubinstein–Taybi syndrome: a girl with a history of neuroblastoma and premature thelarche. *Am. J. Med. Gen.* **83**: 365–366.

Imhof, M.O. and McDonnell, D.P. (1996) Yeast RSP5 and its human homolog hRPF1 potentiate hormone-dependent activation of transcription by human progesterone and glucocorticoid receptors. *Mol. Cell. Biol.* **16**: 2594–2605.

Ito, M., Yuan, C.X., Malik, S. *et al.* (1999) Identity between TRAP and SMCC complexes indicates novel pathways for the function of nuclear receptors and diverse mammalian activators. *Mol. Cell* **3**: 361–370.

Jiang, Y.H., Armstrong, D., Albrecht, U. *et al.* (1998) Mutation of the Angelman ubiquitin ligase in mice causes increased cytoplasmic p53 and deficits of contextual learning and long-term potentiation. *Neuron* **21**: 799–811.

Kamei, Y., Xu, L., Heinzel, T. *et al.* (1996) A CBP integrator complex mediates transcriptional activation and AP-1 inhibition by nuclear receptors. *Cell* **85**: 403–414.

Kasten, M.M., Dorland, S. and Stillman, D.J. (1997) A large protein complex containing the yeast Sin3p and Rpd3p transcriptional regulators. *Mol. Cell. Biol.* **17**: 4852–4858.

Kawasaki, H., Eckner, R., Yao, T.P. *et al.* (1998) Distinct roles of the co-activators p300 and CBP in retinoic-acid-induced F9-cell differentiation. *Nature* **393**: 284–289.

Kim, H.J., Kim, J.H. and Lee, J.W. (1998) Steroid receptor coactivator-1 interacts with serum response factor and coactivates serum response element-mediated transactivations. *J. Biol. Chem.* **273**: 28 564–28 567.

Kishino, T., Lalande, M. and Wagstaff, J. (1997) UBE3A/E6-AP mutations cause Angelman syndrome. *Nat. Genet.* **15**: 70–73.

Kwok, R.P., Lundblad, J.R., Chrivia, J.C. *et al.* (1994) Nuclear protein CBP is a coactivator for the transcription factor CREB. *Nature* **370**: 223–226.

Lavinsky, R.M., Jepsen, K., Heinzel, T. *et al.* (1998) Diverse signaling pathways modulate nuclear receptor recruitment of N-CoR and SMRT complexes. *Proc. Natl. Acad. Sci. USA* **95**: 2920–2925.

Ledbetter, D. H. and Ballabio, A. (1995) Molecular cytogenetics of contiguous gene syndromes: mechanisms and consequences of gene dosage imbalance. In: C.R. Scriver, A.L. Beaudet, W.S. Sly and D. Valle (eds) *The Metabolic and Molecular Bases of Inherited Disease*, pp. 811–839. McGraw Hill, New York.

Lee, J.W., Choi, H.S., Gyuris, J., Brent, R. and Moore, D.D. (1995a) Two classes of proteins dependent on either the presence or absence of thyroid hormone for interaction with the thyroid hormone receptor. *Mol. Endocrinol.* **9**: 243–254.

Lee, J.W., Ryan, F., Swaffield, J.C., Johnston, S.A. and Moore, D.D. (1995b) Interaction of thyroid-hormone receptor with a conserved transcriptional mediator. *Nature* **374**: 91–94.

Lee, S.K., Kim, H.J., Na, S.Y. *et al.* (1998) Steroid receptor coactivator-1 coactivates activating protein-1-mediated transactivation through interaction with the c-Jun and c-Fos subunits. *J. Biol. Chem.* **273**: 16 651–16 654.

Leong, G.M., Wang, K.S., Marton, M.J. *et al.* (1998) Interaction between the retinoid X receptor and transcription factor IIB is ligand-dependent *in vivo*. *J. Biol. Chem.* **273**: 2296–2305.

Li, H., Gomes, P.J. and Chen, J.D. (1997) RAC3, a steroid/nuclear receptor-associated coactivator that is related to SRC-1 and TIF2. *Proc. Natl. Acad. Sci. USA* **94**: 8479–8484.

Lin, R.J., Nagy, L., Inoue, S., Shao, W., Miller, W.H. Jr. and Evans, R.M. (1998) Role of the histone deacetylase complex in acute promyelocytic leukaemia. *Nature* **391**: 811–814.

Mak, H.Y., Hoare, S., Henttu, P.M. and Parker, M.G. (1999) Molecular determinants of the estrogen receptor-coactivator interface. *Mol. Cell. Biol.* **19**: 3895–3903.

Masuyama, H. and MacDonald, P.N. (1998) Proteasome-mediated degradation of the vitamin D receptor (VDR) and a putative role for SUG1 interaction with the AF-2 domain of VDR. *J. Cell. Biochem.* **71**: 429–440.

Matsuura, T., Sutcliffe, J.S., Fang, P. *et al.* (1997) *De novo* truncating mutations in E6-AP ubiquitin–protein ligase gene (*UBE3A*) in Angelman syndrome. *Nat. Genet.* **15**: 74–77.

McKenna, N.J., Nawaz, Z., Tsai, S.Y., Tsai, M.J. and O'Malley, B.W. (1998) Distinct steady state nuclear hormone receptor coregulator complexes exist *in vivo*. *Proc. Natl. Acad. Sci. USA* **95**: 11 697–11 702.

Na, S.Y., Lee, S.K., Han, S.J., Choi, H.S., Im, S.Y. and Lee, J.W. (1998) Steroid receptor coactivator-1 interacts with the p50 subunit and coactivates nuclear factor κB-mediated transactivations. *J. Biol. Chem.* **273**: 10 831–10 834.

Nawaz, Z., Lonard, D.M., Smith, C.L. *et al.* (1999) The Angelman syndrome-associated protein, E6-AP, is a coactivator for the nuclear hormone receptor superfamily. *Mol. Cell. Biol.* **19**: 1182–1189.

Nicholls, R.D. (1993) Genomic imprinting and candidate genes in the Prader–Willi and Angelman syndromes. *Curr. Opin. Genet. Dev.* **3**: 445–456.

Ogryzko, V.V., Schiltz, R.L., Russanova, V., Howard, B.H. and Nakatani, Y. (1996) The transcriptional coactivators p300 and CBP are histone acetyltransferases. *Cell* **87**: 953–959.

Oike, Y., Hata, A., Mamiya, T. *et al.* (1999) Truncated CBP protein leads to classical Rubinstein–Taybi syndrome phenotypes in mice: implications for a dominant-negative mechanism. *Hum. Mol. Genet.* **8**: 387–396.

Olson, D.P. and Koenig, R.J. (1997) Thyroid function in Rubinstein–Taybi syndrome. *J. Clin. Endocrinol. Metab.* **82**: 3264–3266.

Onate, S.A., Tsai, S.Y., Tsai, M.J. and O'Malley, B.W. (1995) Sequence and characterization of a coactivator for the steroid hormone receptor superfamily. *Science* **270**: 1354–1357.

Ordentlich, P., Downes, M., Xie, W., Genin, A., Spinner, N. B. and Evans, R.M. (1999) Unique forms of human and mouse nuclear receptor corepressor SMRT. *Proc. Natl. Acad. Sci. USA* **96**: 2639–2644.

Park, E.J., Schroen, D.J., Yang, M., Li, H., Li, L. and Chen, J.D. (1999) SMRTe, a silencing mediator for retinoid and thyroid hormone receptors-extended isoform that is more related to the nuclear receptor corepressor. *Proc. Natl. Acad. Sci. USA* **96**: 3519–3524.

Perkins, N.D., Felzien, L.K., Betts, J.C., Leung, K., Beach, D.H. and Nabel, G.J. (1997) Regulation of NF-κB by cyclin-dependent kinases associated with the p300 coactivator. *Science* **275**: 523–527.

Petrij, F., Giles, R.H., Dauwerse, H.G. *et al.* (1995) Rubinstein–Taybi syndrome caused by mutations in the transcriptional co-activator CBP. *Nature* **376**: 348–351.

Pickart, C.M. (1997) Targeting of substrates to the 26S proteasome. *FASEB J.* **11**: 1055–1066.

Rachez, C., Suldan, Z., Ward, J. *et al.* (1998) A novel protein complex that interacts with vitamin D_3 receptor in a ligand-dependent manner and enhances VDR transactivation in a cell-free system. *Genes Dev.* **12**: 1787–1800.

Rubin, D.M., Coux, O., Wefes, I. *et al.* (1996) Identification of the gal4 suppressor Sug1 as a subunit of the yeast 26S proteasome. *Nature* **379**: 655–657.

Rougeulle, C., Glatt H. and Lalande, M. (1997) The Angelman syndrome candidate gene, *UBE3A/E6-AP*, is imprinted in brain. *Nat. Genet.* **17**: 14–15.

Safer, J.D., Cohen, R.N., Hollenberg, A.N. and Wondisford, F.E. (1998) Defective release of corepressor by hinge mutants of the thyroid hormone receptor found in patients with resistance to thyroid hormone. *J. Biol. Chem.* **273**: 30 175–30 182.

Sande, S. and Privalsky, M.L. (1996) Identification of TRACs (T_3 receptor-associating cofactors), a family of cofactors that associate with, and modulate the activity of, nuclear hormone receptors. *Mol. Endocrinol.* **10**: 813–825.

Scheffner, M., Huibregtse, J.M., Vierstra, R.D. and Howley, P.M. (1993) The HPV-16 E6 and E6-AP complex functions as a ubiquitin–protein ligase in the ubiquitination of p53. *Cell* **75**: 495–505.

Schulman, I.G., Chakravarti, D., Juguilon, H., Romo, A. and Evans, R.M. (1995) Interactions between the retinoid X receptor and a conserved region of the TATA-binding protein mediate hormone-dependent transactivation. *Proc. Natl. Acad. Sci. USA* **92**: 8288–8292.

Schwerk, C., Klotzbucher, M., Sachs, M., Ulber, V. and Klein-Hitpass, L. (1995) Identification of a transactivation function in the progesterone receptor that interacts with the $TAF_{II}110$ subunit of the TFIID complex. *J. Biol. Chem.* **270**: 21 331–21 338.

Seol, W., Mahon, M.J., Lee, Y.K. and Moore, D.D. (1996) Two receptor interacting domains in the nuclear hormone receptor corepressor RIP13/N-CoR. *Mol. Endocrinol.* **10**: 1646–1655.

Shemshedini, L., Ji, J.W., Brou, C., Chambon, P. and Gronemeyer, H. (1992) *In vitro* activity of the transcription activation functions of the progesterone receptor. Evidence for intermediary factors. *J. Biol. Chem.* **267**: 1834–1839.

Sheppard, K.A., Phelps, K.M., Williams, A.J. *et al.* (1998) Nuclear integration of glucocorticoid receptor and nuclear factor-κB signaling by CREB-binding protein and steroid receptor coactivator-1. *J. Biol. Chem.* **273**: 29 291–29 294.

Shiau, A.K., Barstad, D., Loria, P.M. *et al.* (1998) The structural basis of estrogen receptor/coactivator recognition and the antagonism of this interaction by tamoxifen. *Cell* **23**: 927–937.

Smith, C.L., Onate, S.A., Tsai, M.J. and O'Malley, B.W. (1996) CREB binding protein acts synergistically with steroid receptor coactivator-1 to enhance steroid receptor-dependent transcription. *Proc. Natl. Acad. Sci. USA* **93**: 8884–8888.

Smith, C.L., Nawaz, Z. and O'Malley, B.W. (1997) Coactivator and corepressor regulation of the agonist/antagonist activity of the mixed antiestrogen, 4-hydroxytamoxifen. *Mol. Endocrinol.* **11**: 657–666.

Spencer, T.E., Jenster, G., Burcin, M.M. *et al.* (1997) Steroid receptor coactivator-1 is a histone acetyltransferase. *Nature* **389**: 194–198.

Suen, C.S., Berrodin, T.J., Mastroeni, R., Cheskis, B.J., Lyttle, C.R. and Frail, D.E. (1998) A transcriptional coactivator, steroid receptor coactivator-3, selectively augments steroid receptor transcriptional activity. *J. Biol. Chem.* **273**: 27 645–27 653.

Sutcliffe, J.S., Jiang, Y.H., Galjaard, R.J. *et al.* (1997) The E6-AP ubiquitin–protein ligase (*UBE3A*) gene is localized within a narrowed Angelman syndrome critical region. *Genome Res.* **7**: 368–377.

Tagami, T., Gu, W.X., Peairs, P.T., West, B.L. and Jameson, J.L. (1998) A novel natural mutation in the thyroid hormone receptor defines a dual function domain that exchanges nuclear receptor corepressors and coactivators. *Mol. Endocrinol.* **12**: 1888–1902.

Takeshita, A., Cardona, G.R., Koibuchi, N., Suen, C.S. and Chin, W.W. (1997) TRAM-1, a novel 160-kDa thyroid hormone receptor activator molecule exhibits distinct properties from steroid receptor coactivator-1. *J. Biol. Chem.* **272**: 27 629–27 634.

Tanaka, Y., Naruse, I., Maekawa, T., Masuya, H., Shiroishi, T. and Ishii, S. (1997) Abnormal skeletal patterning in embryos lacking a single *Cbp* allele: a partial similarity with Rubinstein–Taybi syndrome. *Proc. Natl. Acad. Sci. USA* **94**: 10 215–10 220.

Torchia, J., Rose, D.W., Inostroza, J. *et al.* (1997) The transcriptional co-activator p/CIP binds CBP and mediates nuclear-receptor function. *Nature* **387**: 677–684.

Voegel, J.J., Heine, M.J., Zechel, C., Chambon, P. and Gronemeyer, H. (1996) TIF2, a 160 kDa transcriptional mediator for the ligand-dependent activation function AF-2 of nuclear receptors. *EMBO J.* **15**: 3667–3675.

Voegel, J.J., Heine, M.J.S., Tini, M., Vivat, V., Chambon, P. and Gronemeyer, H. (1998) The coactivator TIF2 contains three nuclear receptor-binding motifs and mediates transactivation through CBP binding-dependent and -independent pathways. *EMBO J.* **17**: 507–519.

Vu, T.H. and Hoffman, A.R. (1997) Imprinting of the Angelman syndrome gene, *UBE3A*, is restricted to brain. *Nat. Genet.* **17**: 12–13.

Wagner, B.L., Norris, J.D., Knotts, T.A., Weigel, N. L. and McDonnell, D.P. (1998) The nuclear corepressors N-CoR and SMRT are key regulators of both ligand- and 8-bromo-cyclic AMP-dependent transcriptional activity of the human progesterone receptor. *Mol. Cell. Biol.* **18**: 1369–1378.

Walfish, P.G., Yaganathan, T., Yang, Y.F., Hong, H., Butt, T.R. and Stallcup, M.R. (1997) Yeast hormone response element assays detect and characterize GRIP1 coactivator-dependent activation of transcription by thyroid and retinoid nuclear receptors. *Proc. Natl. Acad. Sci. USA* **94**: 3697–3702.

Whyte, P., Williamson, N.M. and Harlow, E. (1989) Cellular targets for transformation by the adenovirus E1A. *Cell* **56**: 67–75.

Xu, J., Qiu, Y., DeMayo, F.J., Tsai, S.Y., Tsai, M.J. and O'Malley, B.W. (1998) Partial hormone resistance in mice with disruption of the steroid receptor coactivator-1 (SRC-1) gene. *Science* **279**: 1922–1925.

Yagi, H., Pohlenz, J., Hayashi, Y., Sakurai, A. and Refetoff, S. (1997) Resistance to thyroid hormone caused by two mutant thyroid hormone receptors β, R243Q and R243W, with marked impairment of function that cannot be explained by altered *in vitro* 3,5,3′-triiodothyronine binding affinity. *J. Clin. Endocrinol. Metab.* **82**: 1608–1614.

Yao, T.P., Oh, S.P., Fuchs, M. *et al.* (1998) Gene dosage-dependent embryonic development and proliferation defects in mice lacking the transcriptional integrator p300. *Cell* **93**: 361–372.

Yeh S. and Chang, C. (1996) Cloning and characterization of a specific coactivator, ARA-70, for the androgen receptor in human prostate cells. *Proc. Natl. Acad. Sci. USA* **93**: 5517–5521.

Yeh, S., Lin, H.K., Kang, H.Y., Thin, T.H., Lin, M.F. and Chang, C. (1999) From HER2/Neu signal cascade to androgen receptor and its coactivators: a novel pathway by induction of androgen target genes through MAP kinase in prostate cancer cells. *Proc. Natl. Acad. Sci. USA* **96**: 5458–5463.

Zhang, J., Guenther, M.G., Carthew, R.W. and Lazar, M.A. (1998a) Proteasomal regulation of nuclear receptor corepressor-mediated repression. *Genes Dev.* **11**: 835–846.

Zhang, Y., Sun, Z.W., Iratni, R. *et al.* (1998b) SAP30, a novel protein conserved between human and yeast, is a component of a histone deacetylase complex. *Mol. Cell* **1**: 1021–1031.

Zhou, G., Cummings, R., Li, Y. *et al.* (1998) Nuclear receptors have distinct affinities for coactivators: characterization by fluorescent resonance energy transfer. *Mol. Endocrinol.* **12**: 1594–1604.

Zhu, Y., Qi, C., Jain, S., Rao, M.S. and Reddy, J.K. (1997) Isolation and characterization of PBP, a protein that interacts with peroxisome proliferator-activated receptor. *J. Biol. Chem.* **272**: 25 500–25 506.

Index

Note: Page references in *italics* refer to figures; those in **bold** refer to Tables